Fundamentals of Molecular Biology

Fundamentals of Molecular Biology

Veer Bala Rastogi

Ane Books India

New Delhi ♦ Chennai ♦ Mumbai

Bengaluru ♦ Kolkata ♦ Thiruvananthapuram ♦ Lucknow

Fundamentals of Molecular Biology

Veer Bala Rastogi

First Edition 2008

Revised and Updated 2010

Published by

Ane Books Pvt. Ltd.
4821, Parwana Bhawan, 1st Floor, 24 Ansari Road,
Darya Ganj, **New Delhi** - 110 002, India
Tel.: +91(011) 23276843-44, Fax: +91(011) 23276863
e-mail: kapoor@anebooks.com, Website: www.anebooks.com

Branches

- Avantika Niwas, 1st Floor, 19 Doraiswamy Road, T. Nagar, **Chennai** - 600 017, Tel.: +91(044) 28141554, 28141209
e-mail: anebooks_tn@airtelmail.in
- Plot No. 59, Sector-1, Shirwane, Nerul, **Navi Mumbai** - 400 706, Tel.: +91(022) 27720842, 27720851
e-mail: anebooksmum@mtnl.net.in
- 38/1, 1st Floor, Model House, First Street, Opp. Shamanna Park, Basavannagudi, **Bengaluru** - 560 004, Tel.: +91(080) 41681432, 26620045
e-mail: anebang@airtelmail.in
- Flat No. 16A, 220 Vivekananda Road, Maniktalla, **Kolkata** - 700 006, Tel.: +91(033) 23547119, 23523639
e-mail: anebooks_vsn@bsnl.in
- # 6, TC 25/2710, Kohinoor Flats, Lukes Lane, Ambujavilasam Road, **Thiruvananthapuram** - 01, Kerala, Tel.: +91(0471) 4068777, 4068333
e-mail: anebookstvm@airtelmail.in

Representative Office

- C-26, Sector-A, Mahanagar, **Lucknow** - 226 006
Mobile - +91 93352 29971

ISBN : 978-81-8052-137-9

Printed at : Thomson Press, Delhi

Preface

The enormous knowledge that has accumulated in the field of biology, biochemistry and microbiology during the last century, has totally revolutionised the approach for the study of life and living beings. Life is now regarded as the expression and interaction of biomolecules and all the biological processes are the result of biochemical reactions occurring at the cellular level. Molecular biology owes its origin for this phenomenal integration of chemistry and biology. Molecular biology is concerned principally with those biomolecules that are related to genes and the gene products. It is also associated with understanding and explaining the mechanisms responsible for the storage, transmission and expression of biological information at molecular level.

The present book on molecular biology is an attempt to provide the basic, concise but indepth information about above processes at molecular level. It encompasses molecular structures of biomolecules associated with transmission of biological information, i.e. the DNA, its replication and repair, the molecular mechanism of its expression and regulation, the role of proteins in all these processes and the modification of nascent proteins into functional proteins for targeting and regulating the various cell activities.

The present book, 'Molecular Biology', is not intended as a reference book for faculty members and research scholars, because it does not provide an exhaustive account of this modern discipline. However, it is aimed to effectively introduce the foundation and application of molecular biology and to provide a basic textbook with latest concepts for undergraduate and postgraduate students from a variety of biological disciplines.

The subject matter of this book is divided into 17 chapters. Chapter 1 provides a brief introduction to the subject, its brief history, the contributions of other sciences in the birth and growth of molecular biology, and its relation with other biological sciences.

Chapters 2 and 3 present a brief account of chemical basis of life and the structure and functions of various biological molecules. Chapter 4 is aimed to introduce various methods used for isolation and study of macromolecules. Chapter 5, 6, 8 and 9 provide a detailed account of hereditary material, the nature of DNA, its structure, repair and replication. Chapter 7 provides an insight about how in eukaryotes DNA is compacted with proteins to form nucleoprotein fibre or chromatin. Chapters 10 and 11 cover the details of ribonucleic acids, their structure, transcription, processing and their role in living forms. Chapter 12 deals with genetic code and Chapter 13 explains various steps of protein synthesis or mechanism of translation in prokaryotes and eukaryotes.

Chapter 14 covers gene expression in viruses, prokaryotes and eukaryotes, Chapter 15 is associated with various mechanisms operating in prokaryotes and eukaryotes to regulate gene expression, Chapter 16 covers regulation of gene expression in Eukaryotes, and Chapter 17 gives a brief account of recombinant DNA technology which has evolved as an offshoot of moleculars biology and has become most fruitful and totally indispensible for human existence.

I preferred to give no references or extra reading at the end of the chapters for two reasons. Firstly, I understand that at undergraduate or graduate level neither the students nor the teachers prefer to use such textbook references. The students have enough to deal with in the core material. Secondly, updated reference material is available through Web search on Google Scholar. Moreover, in India, libraries are not equipped with the latest textbooks, journals and research papers and referring to old editions of the books will prove to be confusing .

With long teaching experience and expertise in writing, I have made an attempt to present text with the help of suitable and neatly drawn illustrations. The treatment of the subject matter has been simple, clear and comprehensive. Important and additional information are given in boxes as extension material.

This book would not have been possible without the cooperation and keen interest of a large number of well wishers who guided me at each and every step in developing the manuscript. I am highly indebted to them and express thanks from the core of my heart.

I invite constructive comments and valued suggestions from the students and learned teachers to further improve the book in subsequent editions.

I express my heartfelt thanks to Mr Sunil Saxena and Mr JR Kapoor of Ane Books Pvt. Ltd., for the interest shown while preparing the book and their staff for their efforts in solving day to day problems coming across.

Veer Bala Rastogi

Contents

Chapter 1
Introduction to Molecular Biology 1–23

Chapter 2
The Chemical Basis of Life: Atoms and Molecules 24–42

Chapter 3
Molecules of Life: Biomolecules 43–85

Chapter 4
Methods Used to Study Macromolecules 86–123

Chapter 5

Identification of Genetic Material 124–138

Chapter 6

Deoxyribonucleic Acid (DNA) 139–166

Chapter 7

Organisation of Genome 167–187

Chapter 8

DNA Replication 188–229

Chapter 9

DNA Repair 230–239

Chapter 10

Ribonucleic Acid 240–259

Chapter 11
Transcription and Processing of RNA 260–283

Chapter 12
Genetic Code 289–299

Chapter 13
Protein Synthesis 301–327

Chapter 14

Gene: Its Expression 328–349

Chapter 15

Regulation of Gene Expression or Regulation of Protein Synthesis 350–366

Chapter 16
Regulation of Gene Expression in Eukaryotes 367–386

Chapter 17

Genetic Engineering or Recombinant DNA Technology 387–432

Introduction to Molecular Biology

1
Chapter

IN THIS CHAPTER

Origin of the Term Molecular Biology; Use of Computers; Birth of Molecular Biology; Development of Molecular Biology: Expansion of Molecular Biology; Classical and New Molecular Biology; Favourite Experimental Origanisms: Bacteria, Bacteriophages, Yeasts, Animal Cells; Molecular Biology in Life Science.

Molecular Biology is a branch of biological science that aims to understand and explain the biological processes in terms of molecular interaction. It is the study of biology at a molecular level. Thus, molecular biology is not only concerned with the fine structure of cell, cell organelles and biological macromolecules, but also with their functions and genesis. Molecular biology can be best defined as '**the biochemistry of genes and their products**'. It means molecular biology is the culmination of three fields: **Genetics**, **biochemistry** and **cell biology**. An organism is a group of several thousands to millions of cells. The structure and function of each cell is largely determined by the totality of protein molecules it contains, and the structure and function of a protein is determined by the linear sequence of its constituent amino acids. Finally, the sequence of amino acid molecules in a protein is determined by the sequence of nucleotides in a concerned gene. Or it can be said that molecular biology is principally concerned with the understanding and explaining the mechanism for transmission and expression of genetic information. As a matter of fact, molecular biology is not a new discipline but a new way of looking at living organisms and their biological activities.

Initially, cell biologists, geneticists and biochemists had their independent and clearly defined objectives and research fields. The **cell biologists** were concerned with the structure and functioning of cell organelles. They concentrated to find out how specific proteins participate in the construction and operation of a specialised cell structure. The **geneticists** were concerned with the identification and characterisation of gene products and also with the inheritance of genes. The **biochemists** were engaged in finding out the three-dimensional structure and action of different proteins and enzymes. During the last few decades, molecular biology has catalysed the integration and unification of all the three disciplines.

ORIGIN OF THE TERM MOLECULAR BIOLOGY

The term 'molecular biology' was coined in 1938 by officials at Rockefeller Foundation to describe a programme for the application of tools of physical sciences to biology, biochemistry, cell biology and genetics. Later in 1945, **William Astbury** used this term to study chemical and physical structure of biological macromolecules.

Following three major areas of science have contributed immensely to the development of molecular biology:

1. Instrumentation and techniques.
2. Radioactive and fluorescent labelling.
3. Nucleic acids and enzymology.

USE OF COMPUTERS

The use of **computers** in processing nucleotide sequences in DNA and amino acid sequences in different types of proteins has led to the fast expansion in the field of molecular biology. **Bioinformatics**, **Genomics** and **Proteomics** are latest expansions of molecular biology and biotechnology. Each of these fields utilises computers to store, process and analyse biological information such as sequence of nucleotides in DNA, sequence of amino acids in proteins and the sequence of genes in a chromosome, etc.

BIRTH OF MOLECULAR BIOLOGY

Molecular biology originated when geneticists focused to find out the nature of genes and their mechanism of action and biochemists tried to understand how proteins and enzymes are synthesised in cells and genes intervene in this process.

1. Contribution from Biochemistry

Molecular biology has its incipient beginning during early twentieth century with the replacement of physiological chemistry by biochemistry and when scientists tried to understand the ways of molecular transformation within the cell.

The first biochemical experiment was performed in 1897 by **Edward Buchner**, the German biochemist, to carry out sugar fermentation *in vitro* using cell-free yeast extract. No significant work was carried out after that for quite some time. The first half of twentieth century was marked by the decoding of major metabolic pathways like glycolysis, Krebs cycle, etc. The term **macromolecule** was introduced by a German chemist, **Hermann Staudinger** (1922), to describe biomolecules of the cell.

Study of *in vitro* enzyme action was made possible with advances in physical chemistry. Isolation of crystallised proteins and enzymes, and X-ray diffraction images of such isolated crystals showed that macromolecules have well defined structure and specific active sites. This contributed to the concept of antigen-antibody interaction and development of the **science of immunology**.

2. Contribution from Genetics

Discoveries in the field of genetics equally contributed to the origin of molecular biology. In 1930s, it was established that chromosomes are composed of nucleic acids and proteins. Initially, genes were supposed to be formed of proteins because of great diversity in their form and structure. The nucleic acids were thought either to provide material support or to act as energy reserves.

The first precise relationship between genes and metabolism was described by **Archibald Garrod** (1902) in his book *'The Inborn Errors of Metabolism'*. **Garrod** (1909) described the exact chemical nature of metabolic disorders in man. Thus, he established a relationship between Genetics and Biochemistry.

However, the development of molecular biology had to wait until the biochemical basis of basic biological processes was understood by studying simple systems of bacteria and bacteriophages. In 1941, **George Beadle** and **Edward Tatum** demonstrated that genes control enzyme synthesis in *Neurospora*. They proposed **'One gene one enzyme theory'**. But, founder of one gene one enzyme hypothesis is considered to be **Franz Moewas**, who while working with microorganisms had anticipated Beadle and Tatum's one gene one enzyme theory much earlier.

Discovery of chemical nature of gene by **Avery, MacLeod** and **McCarty** in 1944 further strengthened the association between genetics and biochemistry. This established that every living being from microbes to man has DNA which stores all genetic informations and transmits them to the next generation through their offsprings.

The scientists who worked on bacteriophages between 1940 to 1960 greatly contributed to the knowledge of DNA structure. **Delbruck, Felix d'Herelle** and **Frederick Hershey** and **Chase** immensely contributed to the molecular structure of DNA. The double helical structure of DNA was described by **Watson** and **Crick** in 1953 based on X–ray diffraction studies made by **M. H. F. Wilkins** and his co-worker, **Rosalind Franklin**.

Following four discoveries during the period 1950s to 1960s added to the further expansion of molecular biology:

- Elucidation of structure and properties of DNA,
- Structure and role of different RNAs in protein synthesis,
- Genetic code, and
- Mechanism of gene regulation.

The above mentioned discoveries were the outcome of work on prokaryotes such as bacteria (mainly *Escherichia coli*) and acellular organisms like viruses and bacteriophages. The knowledge about the various aspects of molecular biology of eukaryotic cell is more recent except for (1) the relationship between a genetic defect and the biochemical abnormality in human beings described by **Garrod,** and (2) biochemical mutants of *Neurospora crassa* and their genetic control.

3. Role of Physics and Physicists in the Birth of Molecular Biology

Physicists and technicians contributed immensely to the field of molecular biology by the developments of instruments and techniques for the separation of macromolecules and study of their structure. Physicist, **Szilard**, developed technique for the analysis of gene regulation in bacteria, and **George Gamow** (1954) deciphered the genetic code to explain relationship between sequences of nitrogenous bases in DNA and amino acids in the polypeptide chain.

4. Contribution of Instrumentation and Biological Techniques in the Field of Molecular Biology

Instruments like ultracentrifuge, analytical centrifuge, electron microscope, electrophoresis, chromatography, X-ray diffraction, DNA synthesiser (or gene machine) and computer have immensely helped in the separation and study of molecular structure of biological macromolecules, cell organelles and cell. Use of techniques of spectrophotometry, autoradiography, fluorescent labelling and isolation and determination of role of various enzymes in DNA replication, RNA transcription and translation of RNA into protein have revolutionised the knowledge in the field of molecular biology.

Ultracentrifuge is used to determine the sedimentation rate of a substance and the structure of macromolecules. The unit of sedimentation is 'S'. It is an expression of density of the substance. Analytical centrifuge was developed in 1920s by **Theodor Svedberg**. It is used to determine the molecular weight of macromolecules and small cell organelles or their parts.

The electron microscope was invented by **Knoll** and **Ruska** in 1930s. The structural details of viruses, macromolecules and various cell organelles could be studied with the help of electron microscope. The various methods of **chromatography** perfected during mid 1940–50s helped in the separation and purification of biological macromolecules according to their size. Simple column chromatography was developed by **Mikhail Tswett** in 1906. **Erwin Chargaff** used paper chromatography to determine base composition of DNAs obtained from different sources.

A correlation between cell structure and biochemical activity was achieved by disrupting or fractionating the extracts of broken cells by centrifugation and studying each fraction under electron microscope and analysing it biochemically. Much of the early identification of cell organelles and their biochemical composition and functions were studied by **Claude**, **Christian deDuve, George Palade** and **Keith Porter**.

DEVELOPMENT OF MOLECULAR BIOLOGY

Development of molecular biology occurred mainly during the twentieth century. It includes major breakthroughs in understanding molecular structure of DNA, mechanism of functioning and replication of DNA, mechanism of protein synthesis and its control, manipulation of genetic material, synthesis of recombinant DNA and its introduction in new host'. Some of the methods that made important landmarks in the development of Molecular Biology are:

1. Discovery of Double Helical Structure of DNA by Watson and Crick (1953)

It is based on the following important contributions:

a. Perfection of X-ray diffraction pattern technique by **Lawrence Bragg** and his group that included **F. H. C. Crick**.

b. Study of X-sray diffraction pattern by **Wilkins** and **Franklin**.

c. **Chargaff's** chemical analysis of DNA showing equimolarity in adenine and thymine and in cytosine and guanine.

d. **Gulland's** conclusion that bases of DNA interact through hydrogen bonds.

The double helical model of DNA clearly explains that:

- How the genetic information is coded in DNA through the sequence of nucleotides.
- How genetic information is passed on to the daughter cells during cell division.
- How a DNA molecule is able to replicate accurately generation after generation.

2. Deciphering of Genetic Code

Watson and **Crick** proposed a relationship between sequence of nitrogenous bases in DNA and the sequence of amino acids in the polypeptide chain synthesised from it. **Gamow** elaborated the idea of triplet codons. In 1957, **Francis Crick** and **Leslie Orgel** proposed that codons are read from left to right. The elucidation of entire genetic code was the most spectacular discovery of 1960s.

3. Mechanism of Protein Synthesis

The discovery of mechanism of RNA synthesis (transcription), the biochemical role of different types of RNAs in the synthesis of polypeptide chain, and the mechanism of synthesis of polypeptide chain (translation) have helped in establishing relationship between gene and characters. **Marshall Nirenberg** and **Johann Heinrich Mathei** (1961) synthesised a protein *in vitro* using poly-U segment of RNA. The protein thus synthesised was a chain of only one amino acid, phenylalanine.

4. Mechanism of Regulation of Protein Synthesis

The regulation of gene activity was revealed by **Beadle** and **Tatum** (1941) while working on *Neurospora*. It was followed by the work of **Jacob**, **Monod** and **Lwoff** (1965) who developed the 'operon model' of gene regulation in prokaryotes and were awarded the Nobel Prize.

EXPANSION OF MOLECULAR BIOLOGY

Little progress was made in the field of molecular biology between 1965 and 1972. After 1972, for about two decades, molecular biology witnessed significant advances. The most striking developments of this period were:

- *In vitro* synthesis of enzyme ***ribonuclease A.***

- Discovery of three-dimensional structure of different proteins.
- Complete interpretation of molecular mechanism of replication, transcription and translation and the role of enzyme ***polymerase I*** and ***polymerase-III***.
- Isolation, structure and role of different types of RNAs (mRNA, tRNA and rRNA).
- Knowledge in the field of pathology.
- *In vitro* synthesis of DNA.
- Isolation, manipulation and incorporation of foreign DNA to produce recombinant DNA.

During this period molecular biologists discovered the 'central dogma' and 'dogma reverse' in protein synthesis. **Sanger** determined sequence of amino acids in insulin and later, the sequence of nitrogenous bases in its mRNA. Later, **Har Gobind Khorana** synthesised first gene *in vitro*. Molecular biologists developed techniques of isolation of gene and its purification.

Many molecular biologists concentrated on the study of functioning, development and pathological effect of microbes in higher organisms. These studies led to the origin of new branches of molecular biology, like **pharmacology**, **neurobiology**, **endocrinology** and **immunology**. The molecular basis of differentiation during development also attracted the attention of many molecular biologists. The important landmarks that played important role in the expansion of molecular biology are:

1. Recombinant DNA Technology or Genetic Engineering

The development and perfection of techniques of cell culture of prokaryotic and eukaryotic cells, the isolation and fragmentation of their cell organelles, isolation and fragmentation of eukaryotic DNA and then their ligation with vector DNA enabled molecular biologists in obtaining recombinant DNA. Now the techniques have been developed to transfer the recombinant DNA or clonal DNA (cDNA) into the bacterial or the eukaryotic cells or into the eukaryotic eggs. The bacterial cells with cDNA or recombinant DNA can be used as living manufacturing factories for the synthesis of various human hormones (like insulin, somatostatin and human growth hormone) and novel organic compounds, etc. Technique of gene transfer has enabled scientists to replace defective gene in animals and even in man with the normal gene. This could form a novel way of treating hereditary diseases in man and domestic animals.

2. Discovery of Split Genes and Method of Gene Splicing

Richard Roberts and **Philips Sharp** found the leader sequence on mRNA that is essential for the initiation of translation of mRNA into protein. **Gilbert** reported presence of introns and exons in eukaryotic genes. He propounded the concept that the message for any protein in eukaryotic DNA is coded in segments, separated by noncoding segments. The coding segments are called **exons** and noncoding segments as **introns**. Therefore, eukaryotic genes are called **split genes**. Their mRNA-primary transcript also has exons and introns. Its maturation involves removal of introns and splicing of exons.

The concept of split genes helped molecular biologists to understand and work out:

- Mechanism of transformation of normal cells into cancerous cells.
- Mechanism of ageing.
- Mechanism of the occurrence of coordinated changes during cell differentiation.
- Discovery of catalytic role of RNA.
- Mechanism of mutation.

3. Discovery of Pseudogenes

Claude Jacq (1977) discovered pseudogenes. They were considered as evolutionary residues originated by gene duplication. But now they are regarded to be the result of parasitic action of retroviral reverse transcriptase enzymes formed from mRNA molecules.

4. Discovery of Editing Mechanism

It was discovered that RNA molecules produced in eukaryotic cells by faithful transcription of genes are non-functional. They are called **primary transcripts**. They are transformed into functional form by the removal of all intron segments, splicing of all exon segments, addition of a methyl G cap at 5′ end and addition of a poly A or poly U tail at the 3′ end.

5. Discovery of Oncogenes (Oncogene Paradigm)

Around 1975, **Robert Huebner** and **George Todaro** proposed that some retroviruses exist in all human genomes in silent or nonexpressed form and are transmitted generation after generation. Sometimes, they get activated by the action of some other DNA or RNA viruses and are turned into oncogenes.

The new **'paradigm of cellular oncogenes'** was established between 1981–84, when oncologists and molecular biologists observed that the same genes were associated with the spontaneous cancers and in cancers induced by a chemical agent or virus. It was also noticed that the same oncogene coded for proteins for normal cell growth. Therefore, it was concluded that the normal genes get transformed into cancerous genes. Oncogene paradigm had overflowed with the additional information like:

- Genomes of normal genes and oncogenes were prepared and compared to trace similarities between them.
- It helped in the development of agnostic and prognostic tests to detect cancer in the initial stages.

6. DNA Finger-Printing

Synthesis of DNA fragments *in vitro* using single DNA strands as template and assembly of nucleotides on RNA primer to form complementary strands and radioactive labelled DNA or RNA probes in the last decade has helped in deciphering the entire human genome and also helped in solving many medico-legal cases, identifying rape victims, killers, etc. Gel electrophoresis has immensely contributed in this field.

Today, genomic maps of many viruses, bacteria and multicellular animals including that of man are available on the internet for further research and manipulation. No doubt, molecular biology is one of the most rapidly developing and also most exciting discipline that provides a bright future for those opting for this field.

A list of contributors to the birth, growth and expansion of molecular biology is summarised in Table 1.1.

Table 1.1 Summary of contributions in the field of molecular biology

Year	Contributor(s)	Contribution
1928	**F. Griffith**	Discovered transformation in bacterium, *Diplococcus pneumoniae* that causes pneumonia in rats.
1934	**M. Schlesinger**	Demonstrated that bacteriophages are composed of DNA and proteins.
1941	**Beadle** and **Tatum**	Published results of classical experiments on the biochemical genetics of *Neurospora* and established 'One gene one enzyme' hypothesis.
1944	**Avery, MacLeod** and **McCarty**	Indentified that the transforming principle of *Diplococcus* bacteria was DNA.
1950	**E. Chargaff**	Demonstrated that in DNA, the number of adenine molecules is equal to thymine groups and the number of cytosine is equal to the guanine groups.
1952	**Hershey** and **Chase**	Demonstrated that only the DNA of T2 bacteriophage enters the host, the protein coat remains out. They were awarded Nobel Prize in 1969.
1952	**Zinder** and **Lederberg**	Discovered phage-mediated transduction in bacteria.
1953	**J.D. Watson** and **F.H.C. Crick**	Proposed a double helical model of DNA. The two helices are joined by hydrogen bonds between purines and pyrimidines and are intertwined; were awarded Nobel Prize in 1962.
1955	**Ochoa** and **Grunberg–Manago**	Discovered polynucleotide phosphorylase enzyme.
1956	**Gierer** and **Schramm**	Demonstrated that RNA is genetic material in tobacco mosaic virus (TMV).
1957	**Fraenkel–Conrat** and **B. Singer**	Produced hybrid RNA viruses and supported the view that RNA is genetic material in tobacco mosaic virus (TMV).
1957	**Meselson** and **Stahl**	Confirmed Watson and Crick's semiconservative model of DNA replication.
1957	**Hogland, Zamecnik** and **Stephenson**	Isolated transfer RNA and postulated its function.

▸ 1958	**Beadle** and **Tatum**	Received Nobel Prize for their contribution in biochemical genetice of *Neurospora*.
▸ 1958	**Lederberg** and **Tatum**	Got Nobel Prize for the discovery of recombination in bacteria by conjugation.
▸ 1958	**Kornberg**	Isolated DNA-polymerase from *E. coli* got Nobel Prize in 1959.
▸ 1958	**Crick**	Enunciated the 'central dogma' of molecular genetics.
▸ 1958–1959	**Weiss, Horowitz** and **others**	Discovered DNA–directed RNA polymerase.
▸ 1959	**Sinsheimer**	Isolated single stranded DNA form a small virus Ψ -X-174.
▸ 1959	**Ochoa** and **Kornberg**	Received Nobel Prize for *in vitro* synthesis of nucleic acids.
▸ 1961	**Nirenberg** and **Matthaei**	Cracked the genetic code present on messenger RNA.
▸ 1961	**F.H.C. Crick**	Suggested the three letter codons for genetic code (triplet codons).
▸ 1961	**Jacob** and **Monod**	Put forward 'Operon concept for regulation of gene expression'. Received Nobel Prize in 1965.
▸ 1962	**Watson, Crick** and **Wilkins**	Got Nobel Prize for the contribution of three-dimensional structure of DNA.
▸ 1964	**Temin**	Prepared DNA provirus from RNA-tumour viruses. Received Nobel Prize in 1975.
▸ 1964	**Marcker** and **Sanger**	Discovered aminoacyl-tRNA in *E. coli* which plays main role in chain initiation.
▸ 1965	**F.H.C. Crick**	Proposed Wobble hypothesis for anticodons of tRNA. It shows that several codons meant for same amino acid are recognised by same tRNA.
▸ 1965	**Jacob, Monod** and **Lwoff**	Contributed to microbial genetics; received Nobel Prize for the discovery of protein synthesis mechanism in virus.
▸ 1966	**Khorana, Nirenberg** and **Holley**	Worked out complete genetic code for one simple enzyme (Nobel Prize in 1968).
▸ 1969	**Joshua Lederberg**	Predicted that viruses could be used for introducing new genes in human chromosomes or plant chromosomes.
▸ 1969	**Hershey, Delbruck** and **Luria**	Got Nobel Prize in Medicine for replication and recombination in viruses (Bacteriophages).
▸ 1969	**Britten** and **Davidson**	Proposed the gene-battery model for regulation of protein synthesis in eukaryotes.
▸ 1970	**Howard M. Temin** and **David Baltimore**	Demonstrated synthesis of DNA on RNA template in tumour viruses; was awarded Nobel Prize in 1975 for the discovery of RNA-directed DNA polymerase enzyme for reverse transcription.

Year	Scientist(s)	Contribution
1970	**Nathans** and **Smith**	Isolated the first restriction endonuclease (Nobel Prize, 1978).
1970	**Knippers, Kornberg, Gefter** and **Richardson**	Isolated DNA-polymerase II enzyme.
1971	**Sutherland**	Described role of cyclic AMP in hormone action.
1972	**Mertz** and **Davis**	Demonstrated that cohesive terminals of cleaved DNA molecules could be covalently sealed with *E. coli* DNA by enzyme DNA ligase.
1972	**Stein, Moore** and **Anfinsen**	Described structure and activity of ribonuclease were awarded Nobel Prize.
1972	**Porter** and **Edelman**	Received Nobel Prize for the discovery of chemical structure of antibodies.
1973	**Kim, S.H.**	Proposed three-dimensional structure of tRNA.
1974	**Claude** and **Baltimore**	Described ultrastructure of reverse transcriptase.
1974	**Kornberg** and **Thomas**	Proposed a model for chormatin structure, the 'nucleosome concept'.
1975	**Dulbecco**	Proposed that viruses are causative agent of cancer.
1975	**Southern, E.M.**	Developed 'Southern blotting technique' for the isolation of DNA restricted fragment.
1975	**Pribnow**	Discovered Pribnow box or minus ten sequence in DNA of *E. coli*.
1976	**Tonegawa** and **Hozumi**	DNA rearrangement in antibody formation, their association with antibody diversity; Nobel Prize in 1989.
1977	**Maxam, Gilbert and Coulson**	Described DNA sequencing techniques.
1977	**Chambon, Leder, Flavell**	Studied split genes of β-globin, ovalbumin and tRNA.
1977	**Hogness, David** and **Davidson**	Studied split genes for 28S rRNA in *Drosophila*.
1977	**Sanger** and **Colleagues**	Worked complete 5387 nucleotide's sequence of phage $\psi \times 174$ and got Nobel Prize in 1980.
1978	**Arber, Nathans** and **Smith**	Elaborated restriction endonuclease technology and shared Nobel Prize.
1978–79	**Gilbert**	Used the terms exon and intron.
1978	**Hinnen,** *et al.*	Described the transformation of yeast (*Saccharomyces cerevisiae*) with the help of plasmid of *E. coli*.
1978	**Michael Smith**	Developed a method of directed mutagenesis using synthetic oligonucleotides.
1979	**Khorana**	Reported *in vitro* synthesis of a functional gene.
1979	**Alwin,** *et al.*	Developed 'Northern blotting technique' for separating RNA bands.

▸	1979	**Towbin,** *et al.*	Developed 'Western blotting technique' to isolate protein molecules.
▸	1980	**Gilbert**	Improved recombinant DNA technology and DNA sequencing technology.
▸	1982	**Klug, A.**	Got Nobel Prize in chemistry for providing three dimensional structure of tRNAs and described structure of nucleic acid-protein complexes.
▸	1982	**Palmiter** and **Brinter**	Produced transgenic mice by genetic engineering.
▸	1983	**Mc Clintock**	Introduced 'jumping gene concept' i.e. described mobile elements in the genome of maize; got Nobel Prize.
▸	1983	**Cech** and **Altman**	Established the existence of enzymatic or catalytic role of RNA and received Nobel Prize in 1989.
▸	1984	**Robert Tijan**	Identified a DNA binding protein called SPI associated with eukaryotic gene regulation.
▸	1984	**Kohler** and **Milstein**	Developed monoclonal antibodies.
▸	1985	**Alec Jeffrey**	Discovered DNA finger-printing.
▸	1985	**Karry Mullis**	Discovered polymerase chain reaction (PCR) exploited in gene cloning and thermostable enzyme, called Taq DNA polymerase. Got Nobel Prize in 1993.
▸	1987	**Stanford** and co-workers	Developed the particle bombardment gun to introduce foreign DNA into plant cells.
▸	1988	**Watson**	Coordinated human genome project.
▸	1988	**Black, Elion** and **Hitchings**	Got Nobel Prize for formulating drugs such as 6-mercaptopurine and thioguanine for the inhibition of DNA synthesis and cell division and for effective cancer chemotherapy.
▸	1989	**Tsui, Collins** and colleagues	Cloned cystic fibrosis gene.
▸	1990	**Saiki**	Discovered polymerase chain reaction (PCR).
▸	1991	**Lalji Singh**	Developed a new technique of DNA finger-printing using β KM-DNA probe at CCMB, Hyderabad.
▸	1992	**Edwin G. Krebs** and **Edmond H. Fischer**	Got Nobel Prize for reversible protein phosphorylation as a biological regulatory mechanism.
▸	1993	**Roberts** and **Sharp**	Described split genes and RNA processing and got Nobel Prize.
▸	1993	**Chamberlain**	Proposed Inchsworm model for the elongation of transcript of *DNA template.*
▸	1994	**Gilman**	Structure and function of 'GTP-binding (G) protein'.
▸	1995	**Lewis, Christiane, Nusslein–Volhard** and **Wieschaus**	Shared Nobel Prize for the genetic control of early embryonic development.

▸	1997	**Stanley B. Prusiner**	Received Nobel Prize for his discovery of Prions–a new biological principle of infection.
▸	1999	**Gunter Blobel**	Received Nobel Prize for the work on transport and localisation of protein molecules within the cell because of having intrinsic signals.
▸	2001	**Leland H. Hartwell, Timothy Hunt** and **Paul M. Nurse**	Shared Nobel Prize for the regulation of cell cycle.
▸	2002	**Sydney Brenner, John E. Sulston,** and **H. Robert Horvitz**	Shared Nobel Prize for discoveries about how genes regulate organ growth and process of programmed cell suicide.
▸	2005	**Barry J. Marshall** and **Robin Warren**	Proved that bacteria are the main cause of painful ulcers of the stomach and intestine; was awarded Nobel Prize for their discovery.
▸	2006	**Roger, D. Kornberg**	Got Nobel Prize for his studies on the molecular basis of eukaryotic transcription.
▸	2006	**Andrew Z. Fire** and **Craig C. Mello**	Awarded Nobel Prize in Medicine/Physiology for the discovery of RNA interference-gene silencing by double-stranded RNA.

CLASSICAL AND NEW MOLECULAR BIOLOGY

Classical Molecular Biology

It was the science of observation. It was centered on proteins. Proteins with their three-dimensional structure were considered to be the specific agents of biological processes. They act as enzymes, structural proteins and regulatory proteins.

New Molecular Biology

It is the science of intervention and action. It adopts an experimental approach and is concerned with determining the sequences of nucleotides in genes and from that deducing the sequence of amino acids in proteins.

It is now associated with the identification of families of proteins having similar structure, determining the function of those proteins whose structure is known and the structure of those genes that are responsible for the synthesis of these proteins.

New molecular biologists have worked on the mechanism of regulation of gene expression in both prokaryotic and eukaryotic cells. They have isolated the combination of proteins that bind upstream to the promoter or regulator of the gene for controlling its expression. Molecular biologists are able to demonstrate the effect of induced molecular modifications by introducing the modified or mutant gene into an egg or into a bacterial cell or mouse cells. In such transgenic organisms the function of introduced DNA and the protein synthesised by the introduced gene can be determined.

FAVOURITE EXPERIMENTAL ORGANISMS IN MOLECULAR BIOLOGY

The rapid growth in the field of molecular biology occurred only after the realisation that there exists a basic uniformity in the life processes in all living beings from bacteria and viruses to man. Therefore, simple organisms like bacteria, viruses, bacteriophages, unicellular green algae , yeast, *Neurospora* and other fungi, nematodes, fruitflies and plant *Arabidopsis* were used for molecular studies. Work on bacteria and viruses yielded information on DNA molecules, their relation with proteins, the action of various proteins, role of different enzymes and development of recombinant DNA technology. The various metabolic pathways and the enzymes required for each step in every pathway were also worked out using simple microorganisms. It was later established that the fundamental biological principles that govern the activities of simple organisms equally apply to cells of higher multicellular organisms.

1. Bacteria: The Most Favourite Organisms

Bacteria are simple, acellular or unicellular organisms having a single chromosome, which lies naked in the cell cytoplasm. Bacteria do not have a nucleus as found in eukaryotes. The bacterial chromosome is a single circular double-stranded DNA molecule, highly folded and attached to a protein scaffold. DNA is without basic proteins as found associated with DNA in eukaryotes forming nucleoprotein fibre or chromatin fibre. The bacterial cells are without cell organelles like ER, Golgi apparatus, mitochondria, lysosomes, etc.

Bacteria are most favourite with molecular biologists because of the following reasons:

1. **Simple Organisation:** Bacteria have a very simple structure. The cell organelles like mitochondria, E.R, Golgi complex, lysosomes, chloroplasts, etc, are absent.
2. **Hereditary Material:** Each bacterial cell has a single naked, double-stranded circular DNA molecule which remains embedded in the cytoplasm. The typical bacterial genome has no introns. The protein coding genes are arranged in the form of operons. Each operon has an operator, a regulator and a repressor gene along with one or more structure genes or cistrons. Therefore, bacterial mRNA transcribed from such an operon is polycistronic.
3. **Culture:** Bacteria are easily grown in culture medium whether liquid or solid. They grow rapidly completing one cycle every 20 minutes. For example, a single bacterial cell of *E. coli* becomes 10^9 bacteria in 20 hours under optimum conditions.
4. **Prototrophs and Auxotrophs:** Bacteria can be cultured in a **minimal medium**. A minimal medium has only inorganic salts and a source of carbon which may be glucose. It is without amino acids, vitamins and proteins, etc. Bacteria can synthesise all necessary

organic substances such as amino acids, proteins, enzymes, vitamins and lipids from the inorganic medium. Such bacteria are called **prototrophs.**

Some mutant strains of bacteria need addition of some specific amino acid to their culture medium. Such bacteria are called **auxotrophs**. For example, if amino acid, leucine, is needed in the culture medium, the bacterium is termed as **leucine auxotroph** and its genetic composition is represented by leu^-. Its prototroph is called **leucine prototroph** and is represented by leu^+.

Extension: Minimal Medium

A **minimal medium** is culture medium for growing bacteria, which is without organic compounds other than a carbon source, such as sugars (mainly glucose, glycerol or lactate). The medium has all inorganic ions, like sulphates, phosphates, chlorides and sodium, potassium, magnesium, calcium and ammonium.

Bacteria are grown on solid surfaces:

- Initially bacteria were grown on **potato slice.**
- Potato slice was replaced by **gelatin**, but it was digestible by certain enzymes from bacterial cells.
- Now-a-days bacteria are grown on **agar**.

Agar plating

Agar is a gelling substance. It is obtained from a variety of sea weeds. Warm solution of agar is poured in a sterile petridish and liquid nutrient medium is added to it. As agar cools it gets solidified. The petridish with solid agar and nutrient medium is called a **plate** and the solid growth medium is called **nutrient agar**. The act of depositing bacteria on the agar surface is called **plating**.

On solid agar plate, the daughter cells produced by the divisions of each bacterial cell remain together and form a colony. As a matter of fact, each bacterial colony is a clone of daughter cells, which are derived from one parent cell and have identical genotype and phenotype.

Plating helps in determining whether a bacterium is an auxotroph or prototroph.

5. **Conjugation:** Bacteria exchange genes through **conjugation, transduction** and **transformation.**

- In **conjugation**, the donor bacterium (F^+ or Hfr strain) transfers a copy of its DNA to a recipient (F^-strain) cell while the bacterial cells of two strains remain in close contact.
  - **Transduction** involves a virus which carries a part of host DNA along with its own DNA, and exchanges it with another host's DNA when enters another host cell.
  - **Transformation** involves imbibition of a segment of free bacterial DNA from surrounding medium. The process can be initiated by making bacterial cell membrane and cell wall permeable to free DNA segments present in the medium. This technique of experimental introduction of foreign DNA is called **transfection**.

6. **Metabolic Regulation in Bacteria:** The metabolic activities in bacteria are highly regulated. They do not synthesise substances that are not needed. For example, if amino acid tryptophan or methionine is present in the growth medium, gene for these amino acids are not allowed to synthesise enzymes required for their synthesis. Similarly, the enzyme system responsible for the utilisation of lactose is not elaborated till lactose is absent in the growth medium. So it was comparatively easy to trace out various metabolic pathways in bacteria and to study the mechanism of control on gene action.

Extension: *Escherichia coli,* The Early Workhorse

Most of the early work in molecular biology was carried out on the colon bacterium, *Escherichia coli*. It is commonly mentioned as *E. coli*. It was developed into a genetic tool by **Beadle, Tatum** and **Lederberg**. It is considered as most ideal model for studying many fundamental aspects of biochemistry, molecular biology and molecular genetics because of following reasons:

- *E. coli* can be handled and grown easily *in vitro*.
- *E. coli* can be grown in minimal culture medium.
- Mutant varieties of *E. coli* were successfully produced.
- Because of the study of mutant varieties of *E.coli,* it was possible to construct a genetic map of bacterial genome.
- *E.coli* regularly exchanges genes through conjugation and transduction.
- *E. coli,* under favourable conditions, divides by binary fission after every 20 minutes.
- The process of cell division is very simple.
- The genome of *E. coli* is simple and short. It does not have any nonfunctional nucleotide sequences (introns) as found in eukaryotes. *E. coli* genome has approximately 4 million base pairs and encodes about 4,000 different proteins.
- A bacterial cell also contains small, independent extracellular genetic elements called **plasmids**. Each plasmid has a few genes for its ability to cause disease or resistance to antibiotics. They confer some novel properties to their host bacterial cells and a feature of identifying such cells.

2. Bacteriophages

Viruses that parasitise bacteria are called **bacteriophages** or **phages.** Their structure and life cycle is much simpler than bacteria, yet they possess the most essential attributes of life.

1. **Structure:** The phages may be spherical or comma-shaped but majority of them are tadpole-like in appearance. The **T_4 bacteriophage**, a parasite of bacterium *E. coli* is tadpole-shaped and is differentiated into head, collar and tail.
   - **Head:** The head is polyhedral or hexagonal. It is a hollow structure. The protein coat of head is called head capsid. It is made up of about 2,000 protein units called **capsomeres**. The interior of head capsule is occupied by a circular double-stranded DNA molecule, which has about 75 or more genes.
   - **Tail:** The tail is in the form of a hollow cylinder. It has a central hollow core surrounded by a spring-like contractile sheath, formed of protein units.

- **End plate and tail fibres:** The core of the tail rests on a hexagonal **end plate**. Six tail fibres and six spikes protrude from the end plate. They are formed of polypeptide chains.

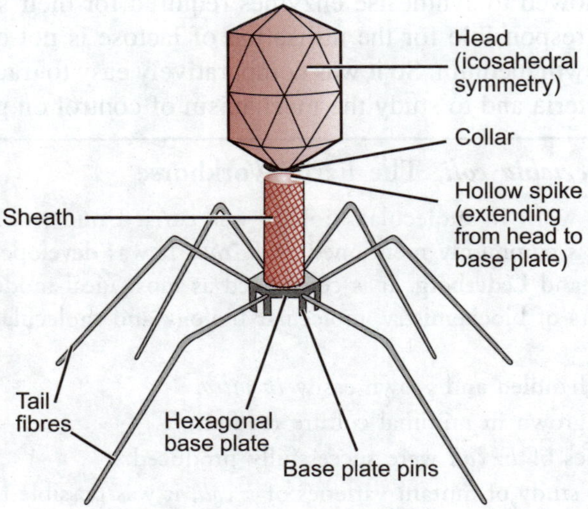

Fig. 1.1 Structure of T_2 -bacteriophage

2. **Genome of Bacteriophage:** The most common phages have double-stranded DNA but phages with single-stranded DNA, single-stranded RNA and double-stranded RNA are also known.
3. **Life Cycle:** Phages can not multiply except in host bacterial cell. Thus, life cycle of a phage begins when it gets attached or adsorbed to the surface of host cell. It does not exhibit asexual and sexual reproduction, but multiplies by lytic life cycle.

   The lytic life cycle of a phage includes following steps:

   - **Adsorption** of a phage particle to the surface of host bacterial cell.
   - **Penetration** of its tail through bacterial cell wall.
   - **Entry** of phage DNA through tail into bacterial cell cytoplasm.
   - **Control** on bacterial synthetic machinery and its conversion to a phage-synthesising factory that synthesises phage's DNA and proteins.
   - **Assembly** of phage particles from DNA and protein molecules.
   - **Lysis** of bacterial cell and release of phage particles.

The bacterial cell on lysis releases several hundred phages. Thus, phages multiply faster than bacteria. A bacterium doubles its number in about half an hour, while a single phage gives rise to more than 100 daughter phages. In two hours, there are four cycles of infection producing about 100^4 phage particles whereas a bacterium in the same period produces 2^4 cells.

The various steps in the lytic cycle of phages are genetically regulated. The study of this regulation by molecular researches contributed immensely to the basic life processes in all the cells.

The metabolic and genetic studies on viruses, bacteriophages and bacteria have been the corner-stones of molecular biology since its inception. The last decade has seen great advances in the understanding of biological functions at molecular level using bacteria and viruses because of their simple structure and short life cycle.

The small genome size and rapid rate of multiplication make viruses and phages an excellent model for molecular studies.

3. Yeasts

Yeasts are also unicellular organisms. They can be regarded as simplest eukaryotes. The genome of *Saccharomyces cerevisai* is the most thoroughly studied one. It consists of 14 million base pairs organised into 16 linear chromosomes. In recent years, the technological advances have enabled molecular biologists to use yeast mutant cells for understanding many regulatory strategies in higher organisms, such as DNA replication, transcription, RNA processing, protein synthesis, protein sorting, protein transport and regulation of cell division, etc.

Other unicellular eukaryotes used by molecular biologists are *Chlamydomonas* (an alga) and *Tetrahymena* protozoan.

Extension: Why Yeast Cells

- Yeast cells can be cultured in the laboratory as colonies from a single cell, in much the same way as bacterial cells except that yeast cells grow in liquid suspension culture medium.
- Yeast cells are unicellular like bacteria. They can be handled easily like bacteria.
- Basic metabolism of yeast and bacteria is the same. However, the overall process of synthesising macromolecules resembles that of eukaryotic cells and is different from bacteria. Thus, study of these processes in yeast provides an insight of synthesis of macromolecules in all eukaryotes.
- Yeast cells have both haploid and diploid phases.
- Haploid yeast cells have two mating types which are equivalent to male and female gametes of higher organisms.

4. Animal Cells

In the last decade, animal cells have been studied widely *in vitro*. This has provided understanding of:

- Process of development of an egg into the adult.
- Hormonal regulation of various processes.
- Hormonal regulation of cell differentiation.
- Differences between normal cells and cancer cells.

Molecular biologists and Geneticists faced some initial problems in culturing animal cells and maintaining their cultures because:

1. Animal cells grow only for a limited number of generations.
2. The number of chromosomes in the progeny cells usually increases beyond the normal diploid number.
3. The cell culture becomes more and more heterogeneous, if grown for a longer period.
4. Animal cells stop dividing once they come in contact with each other. This is called **contact inhibition**. Continuous growth and division of animal cells can be achieved only by removing the confluent layer, separating the cells and placing only a small number of cells in dispersed condition.
5. Only a few types of cells can be cultured, such as reticulocytes, mesenchyme cells and gonadial cells.

MOLECULAR BIOLOGY IN LIFE SCIENCE

The association of molecular biology with different fields of biology can be discussed under the following heads:

1. Genetics and Molecular Biology (Molecular Genetics)

All the phenomena of classical genetics can be explained at molecular level. For example, in terms of molecular biology, a gene is a fragment of DNA. The appearance of a character is the result of functioning of that DNA segment or gene. The DNA segment directs the synthesis of protein that may act as an enzyme, as a carrier protein or as a structural protein.

With the help of techniques developed in the field of molecular genetics and recombinant DNA technology, it is possible to isolate the desired gene or genes responsible for various hereditary diseases in man and replace them with normal genes that may be synthesised *in vitro* or isolated from the normal cell. Current research in human genetics is mainly centered on **'gene replacement therapy'.**

With the help of molecular genetics, the development of transgenic animals and plants is a step to solve food problem and to provide nutritious diet for human population. Synthesis of various enzymes or hormones in the laboratory by introducing concerned human gene or genes in bacterial cells is another exciting field of molecular genetics.

2. Molecular Biology and *in vitro* Genetics

The techniques in molecular biology have made it possible to obtain synthetic DNA of desired nucleotides and the various enzymes required for DNA replication and for protein synthesis, etc. Such functional nucleic acid molecules can be exposed to mutagenesis, selection, isolation and propagation in cell free medium. The phenotypic effect of such mutant DNA molecules can also be studied in the test tube. The study of phenotypic effect of a mutant segment of nucleic acid in test tube is called ***in vitro* genetics**. *In vitro* genetics may have following applications:

- For the analysis and identification of **protein binding sites** on DNA molecules for proteins that are necessary for **transcriptional regulation**.
- In **PCR amplification** of required isolated DNA segment to have its multiple copies. This can generate a new gene pool.
- In transforming the ribozyme in such a way that its substrate specificity is changed.
- In locating protein binding sites on RNA.

3. Genetic Footprinting

Genetic footprinting is an economical and highly efficient method for the functional analysis of DNA sequences on a genomic scale only in microorganisms. Some DNA sequences in the bacterial genome are inserted. Now these bacteria are exposed for mutagenesis and mutant forms are selected for different mutations and are studied for various phenotypes and changes therein produced due to different induced mutations.

4. Molecular Biology and Cell Biology (Molecular Cell Biology)

According to molecular biology, all intracellular structures are the result of assembly of organic macromolecules. Development of very efficient but simple methods of studying the cell, like immunofluorescence, flow-sorting cytometry, immunocytochemistry, autoradiography, gradient centrifugation, fluorescence microscopy and electron microscopy, etc., have helped in understanding cell architecture. Presence of cytoskeleton, microtubules, microfilaments and tonofibrils in the cell cytoplasm and their chemical composition was worked out due to technological advance in the field of molecular biology.

Molecular biology has helped in understanding the mechanism of intracellular and intercellular traffic of molecules and the role of transport proteins like G-proteins, cyclic AMP and ionophores in the intercellular transport.

5. Molecular Biology and Genomics (Molecular Biology's Periodic Table)

The objective of **'new genomics'** is to prepare genetic maps, physical maps and the complete nucleotide sequence of genomes of different animals including that of human beings. The study of human genome is called **'Human Genome Project' (HGP)**. The HGP is viewed as the century's version of the discovery and consolidation of biology's periodic table. From 1869 to 1889, chemists systematically enumerated all atoms and arranged them in a pattern according to their similarities and differences. The HGP in 20th century completed the task of producing biology's periodic table of more than one million human genes. Just as chemists identify atoms by mass and charge alone, biologists will be able to recognise each gene from the sequence of nucleotides and each protein by its amino acids sequence. Now entire data of the genomes of different organisms (prokaryotes and eukaryotes) is available from specific sites on internet for genome project. At present biologists or geneticists are working with genome with following goals:

- Routine re-sequencing of multimegabase regions of human genome.
- Identification of all common variants in human genome.
- Fresh sequencing from other organisms.
- Monitoring of expression of genes at mRNA level.
- Genetic tools for manipulating cell circuitry.
- Monitoring the level and modification state of proteins.
- Systematic catalogues of protein interactions.
- Identification of all basic protein shapes by determining their amino acid sequences.

The transport of protein between cell surface and various cell organelles is carried out by the formation of vesicles. The formation of these vesicles is controlled by signal proteins. Therefore, different transportants are taken up by different transport systems of the cell.

Molecular biologists have identified, isolated, cloned and sequenced the proteins and enzymes that are involved in intracellular traffic or intercellular transport.

Molecular mechanism of transport of signals and excitation of a specific tissue has been understood. The sorting, tagging and targeting of proteins to their appropriate destinations has been worked out.

6. Molecular Biology and Evolution (Molecular Evolution)

Molecular biology has provided evolutionists with some wonderful tools for studying relationship between different groups of living organisms and in understanding the mechanism of evolution. Molecular biology has provided data on proteins and DNA from fossils and has helped in understanding the genetic similarities and variabilities that existed within the members of a species, between species, between genera and so on. Some of the contributions of molecular biology to the understanding of evolutionary mechanisms are as follows:

1. **Origin of First Cell:** Molecular data has enabled evolutionary biologists to contemplate and analyse the events that occurred in past billions of years ago leading to the formation of first cells.
2. **Development of Complex Proteins:** The discovery of **split genes** has helped in visualising the possible mechanism for the development of complex proteins.
3. **Origin of Eukaryotic Cells:** Comparison of **nucleic acids** and **proteins** found in different cell compartments supported the hypothesis of symbiotic origin of first eukaryotic cells about 3.2 billion years ago. According to evolutionists, the eukaryotic

cells evolved due to the fusion of different types of acellular prokaryotes and their symbiotic coexistence in the form of **mitochondria, chloroplasts** or **nucleus.**

4. **Estimation of Genetic Distance:** Collection of vast quantities of molecular data of nucleic acid (DNA) and proteins from vast groups of living forms have made it possible to make a quantitative estimate of genetic distance or degree of similarity between different species.
5. **Molecular Clock:** Knowing that molecular variations or gene mutations are often neutral, it is possible to deduce from genetic distance, a 'divergence time' for various species. This is called **molecular clock**. This has helped in estimating the time of divergence of two closely related species, rate of evolution of a character in a group of species or the rate of evolution of a species or taxon.
6. **Cladism:** Molecular data has helped in tracing new relationships between different animal groups or different plant groups. This has led to the establishment of new branches of phylogenetic tree. This new method of classifying organisms on the basis of similarity in the molecular structure of DNA is called **cladism**.
7. **Relationship between Man, Apes and Monkeys:** Molecular evolution of haemoglobin of man, apes and monkeys, the study of their DNA and their mitochondrial DNA has established relationship between them and also the time of diversification of these groups.
8. Japanese scientist, **Motoo Kimura,** has proposed a 'neutralist model' of molecular evolution. This model minimised the role of selection in molecular evolution. This emphasizes that evolution occurs due to mutation in the molecular organisation of hereditary material of living organisms.

7. Molecular Biology and Developmental Biology

Two essential features of developmental biology are:

1. Cleavage or cell division, and
2. Cell differentiation leading to organ formation.

All organisms develop from a single cell, the fertilised ovum, called **zygote**. In highly organised multicellular organisms, it gives rise to various tissues and organs. It has been a great puzzle for scientists that how an apparently structureless egg is converted into a complex highly organised embryo and how cells arising from an undifferentiated zygote undergo differentiation.

At molecular level, cell differentiation means **variable gene activity** in different cells of the same organism. As a matter of fact, cell specialisation involves preferential synthesis of some specific proteins such as haemoglobin in RBCs, neurotransmitter proteins in nerve cells, enzymes or secretory proteins in gland cells and antibodies in plasma cells, etc. Each eukaryotic cell, although, has whole lot of genes but expresses only a small percentage. Cells of different tissues express different sets of genes and, therefore, possess different proteins. **Jacob** and **Monod** suggested that the structural complexity in higher organisms is achieved during embryonic development due to the existence and interaction of highly complex network of **regulatory genes**. The discovery of **'homeobox genes'** coding for transcription factors and the discovery of operon model in prokaryotes support the view of cellular differentiation.

Thus two types of genes have been identified:

1. **House-keeping Genes:** These genes are active in all types of cells. They are associated with the synthesis of **structural proteins** and are required for the formation of membranes, ribosomes, mitochondria and respiratory enzymes. Studying proteins from chimpanzees and humans, **Mary Claire King** and **Alien Wilson** (1975) concluded that structural proteins and, therefore, associated structural genes were too closely related to find any morphological differences.
2. **Regulatory Genes or Luxury Genes:** The genes that are expressed in different cells or tissues are called **regulatory genes** or **luxury genes**. The differences in the morphology or physiology exist at the level of regulatory genes. For example, genes for globin or ovalbumin and for immunoglobulins express them in different types of cells. The diversification in the individuals occurs due to mutations affecting the regulatory genes.

Molecular biologists have isolated and identified a number of genes involved in the development and differentiation. Modification in the expression of some of these genes has led to the evolution of new morphological structures and new organisations.

A molecular insight into the overall working of the living system with the help of molecular biology was made possible with the landmark discovery of structure and mode of functioning of double-stranded helical molecules of DNA. The phenomenal growth and contribution of molecular biology after the discovery of DNA can be compared with the growth in physics following the acceptance of atomic theory. Today, no field of biology can be understood and can exist and expand without the intervention of molecular biology. It has helped in tracing the fundamental relationship between simple forms like bacteria and viruses with most highly evolved animals and plants, because the fundamental biological principles that govern the life activities of bacteria and viruses do apply to complex cells of highly evolved multicellular animals and plants.

TEXT QUESTIONS

1. What do you understand by molecular biology? Discuss the role of molecular biology in the field of biology.
2. Discuss with suitable examples that molecular biology has helped in understanding mechanism of evolution.
3. Justify the statement 'Molecular biology owes its origin to the marriage of biochemistry and genetics.'
4. Describe important landmarks in the history of molecular biology.
5. Differentiate between house-keeping genes and luxury genes.

6. Discuss in brief the mechanism of cell-differentiation during development.
7. Which discoveries led to the birth of molecular biology?
8. Summarise role of computer in molecular biology.
9. What do you mean by *'in vitro'* synthesis? Which was the first *in vitro* experiment in the field of biochemistry and by whom it was carried out?
10. Give contributions of the following scientists:
    (a) Alec Jeffrey
    (b) Pribnow
    (c) Khorana
    (d) Gilbert
    (e) Beadle and Tatum
    (f) Griffith.
11. Name the scientists for following discoveries:
    (a) Recombination in bacteria
    (b) Provided experimental evidence for semi-conservative replication of DNA
    (c) Discovered polynucleotide phosphorylase enzyme
    (d) Three dimensional structure of DNA
    (e) Ultrastructure of reverse transcriptase
    (f) Northern blotting technique for the separation of RNA bands
    (*g*) Used the term exons and introns.
12. What is the use of Western blotting technique?
13. Who discovered that in some cases RNA acts as a catalyst or who discovered the enzymatic role of RNA?
14. What is Taq DNA polymerase? What is its role?
15. Who developed polymerase chain reaction technique?

2
Chapter

The Chemical Basis of Life : Atoms and Molecules

> **IN THIS CHAPTER**
>
> Atoms, Subatomic Particles, Chemical Bonds: Ionic Bonds, Covalent Bonds; Other types of Bonds : Coordinate Bonds, Hydrogen Bonds, Hydrophobic Interactions; Water is Essential for Life; Acids, Bases and Salts: Acid-Base Balance in Living Organisms; Chemical Reactions.

Everything on our planet earth is made up of atoms and molecules. In living forms they are organised in very particular manner so that the living substance, **protoplasm**, has some specific characteristics which are not seen in non-living objects. Much of the modern biology is being studied at the molecular level. This is the field of molecular biology. The three unifying concepts of biology: the energy transfer and transformation, transmission of information and evolution, all depend upon the interaction and activity of atoms and molecules. In order to understand life processes, we must know the basic principles of biological chemistry. For this reason, before starting the study of molecular biology, knowledge about the interaction of various atoms and molecules found in living beings is essential.

The study of biologically important molecules and metabolic reactions, genetic code and protein synthesis and its control in organisms of varied complexity have led to the emergence of following two generalisations:

1. The same physical and chemical principles govern the living as well as non-living systems.
2. Even though living things are strikingly diverse, their chemical composition and metabolic chemistry is the same (This explains why conclusions drawn from the study of bacteria can be applied to other animals including man).

ATOMS ARE THE BASIC PARTICLES OF ELEMENTS

Elements are composed of units called **atoms**. The word 'atom'comes from Greek word *atomos*, which means 'incapable of division'. The term was used by ancient Greek philosophers who believed that all matter consists of small particles that can not be further sub-divided. There are only one type of atoms in each type of element. Atoms are much smaller than the tiniest particle visible under a light microscope.

Atoms Consist of Subatomic Particles

An atom is formed of three types of subatomic particles: **protons**, **neutrons** and **electrons**. Each proton has one unit of positive electrical charge. The neutrons are neutral and each electron has one unit of negative charge. The number of electrons in an atom is equal to the number of protons, thereby making the atom neutral.

Protons and Neutrons make up the Atomic Nucleus: The number of protons in the nucleus of an atom is called its **atomic number**. The total number of protons and neutrons in the atomic nucleus is the **atomic mass** or **atomic weight** of the atom. For example, an oxygen atom with 8 protons and 8 neutrons has atomic number 8 and atomic mass 16.

Electrons: Electrons occur in orbits around the atomic nucleus. These orbits are called **shells** or **orbitals**. The distribution of electrons in an atom is fixed and is known as **electronic configuration**. This can be represented by a general formula, $2n^2$, where n is the number of shells, *i.e.* 2, 8, 18, etc. from the nucleus, *i.e.* 1, 2, 3.

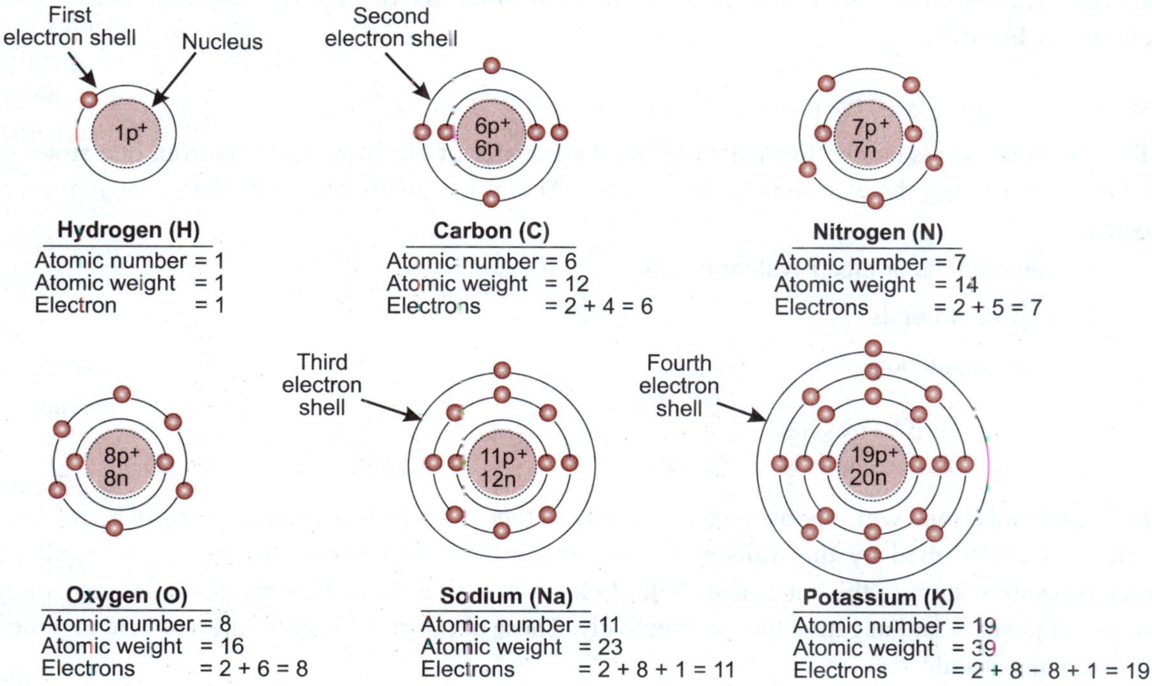

Fig. 2.1 Structure of atoms of hydrogen, carbon, nitrogen, oxygen, sodium and potassium

The first shell has two electrons. The second and also the outermost shell can have maximum 8 electrons. The outermost shell is called **valence shell,** because its electrons take part in chemical bonding. The electrons are held in their orbitals or shell by their attraction to the nucleus by the positive charge of its protons.

Energy is required to move a negatively charged electron farther away from the positively charged nucleus. When energy is added to the system, an electron can jump from one level to the next. During this process, the electron absorbs a discrete packet of energy, known as a **quantum**.

Octet Rule: The presence of 8 electrons in the outermost shell provides stability to the atom. Atoms having less than 8 electrons in the valence shell are with incomplete valence shell. They tend to interact with other such atoms in a way that each partner completes its valence shell. This is achieved by losing, gaining or sharing their electrons. The characteristics of having maximum 8 electrons in the valence shell to attain chemical stability is described as **octet rule**.

CHEMICAL BONDS HOLD ATOMS TOGETHER

Two or more atoms combine chemically to form a **molecule**. When atoms of different elements combine, they form a **compound**. The atoms of a molecule are held together by forces of attraction called **chemical bonds**. Each bond represents a certain amount of potential chemical energy. This is called bond energy and can be defined **as the energy needed to break a chemical bond.**

Chemical Bonds may be Ionic or Covalent

The chemical bonds can be formed either by the transfer of electron/electrons from one atom to other or by sharing the electrons by both atoms. Therefore, bonds can be of the following three types:

1. Ionic bonds or electrovalent bonds
2. Covalent bonds
3. Coordinate bonds.

1. Ionic Bonds or Electrovalent Bonds (Electrons are transferred in Ionic Bonds)

In ionic bonds, ions with opposite charge are held together by electrostatic forces of attraction. The ions are formed by the transfer of one or more electrons from the outermost shell of electropositive atom to the outermost shell of electronegative atom. This results in the formation of stable positive and negative ions, respectively. These oppositely charged ions are held together to form a molecule.

$$\underset{\text{Atoms}}{A + B} \longrightarrow \underset{\text{Ions}}{A^+ + B^{-e}} \text{ or } \underset{\text{Molecule}}{A^+ \bullet B^{-e}}$$

The ions are attracted to the ions of the opposite charge in all directions. So they are omnidirectional. A positively charged ion may be surrounded by negatively charged ions or a negatively charged ion by positively charged ions.

Extension: Characteristics of Ionic Compounds

- In ionic compounds the bonds are formed between metals and non-metals.
- In ionic compounds the metal is always named first and the non-metal follows it, e.g. sodium chloride (NaCl), potassium chloride (KCl), sodium sulphate (Na_2SO_4). Here sodium and potassium are metals and Cl^- and SO_4^- are non-metallic ions.
- Ionic compounds are readily soluble in water and other polar compounds.
- When in solution, ionic compounds easily conduct electricity.
- Ionic compounds separate out in crystalline form from their saturated solutions.
- Ionic compounds have high melting point due to strong electrostatic bonds between their ions.

Example: In sodium chloride, sodium and chlorine ions are held together by ionic bond. A sodium atom has total 11 electrons (2 + 8 + 1). Its outer or the valence shell has only one electron. A chlorine atom has total 17 electrons (2 + 8 + 7) with 7 electrons in its valence shell. When a sodium atom loses its only electron from the valence shell, it is left with 8 electrons in its outer shell. By losing one electron sodium becomes a **cation** or positively charged ion. Thus sodium ion has a charge of 1^+. When this electron is transferred to chlorine, its outermost or valence shell comes to have 8 electrons. Due to an additional electron, chlorine atom changes to an **anion** or negatively charged ion. It has a charge of 1^-. The resulting positive and negative charges pull both the ions together forming an ionic bond.

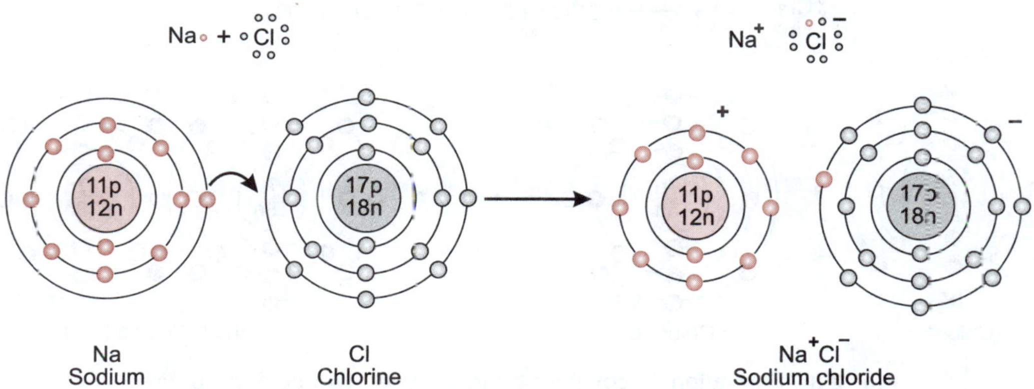

Fig. 2.2 Diagram showing formation of ionic bond in sodium chloride

The compounds formed by electron transfer are called **electrovalent** or **polar compounds**. In a solution, such compounds always exist in ionic form, and are termed as **electrolytes**. Their solution is a good conductor of electricity. Every compound or every molecule with ionic bonds is electrically neutral with zero net charge.

2. Covalent Bonds
(Electrons are shared in Covalent Bonds)

In a covalent bond, two atoms share one, two or three pairs of valence electrons equally. The larger is the number of electron pairs shared between two atoms, the stronger is the covalent bond. The covalent bonds may be formed between atoms of the same or of different elements. They are the most common chemical bonds in the organic compounds but are also found in some inorganic compounds.

A simple example of covalent bond is seen in water where two hydrogen atoms join with one oxygen atom. A simple way of representing the number of electrons in the outer shell of an atom is to use dots placed around the chemical symbol of the elements as shown below:

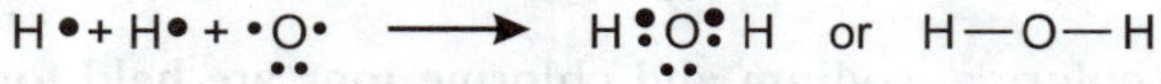

A. Covalent Bonds May be Single, Double or Triple Bonds

Depending upon the number of electron pairs shared, the covalent bonds may be:

1. **Single Covalent Bond:** Two atoms share only one electron pair in single covalent bond. For example, in a hydrogen molecule two hydrogen atoms share their single valence electrons of the first shell or valence shell. Similarly in chlorine molecule two chlorine atoms share two electrons of their outer valence.

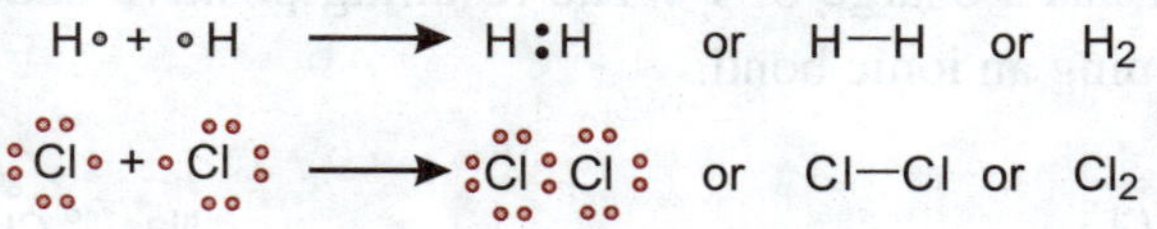

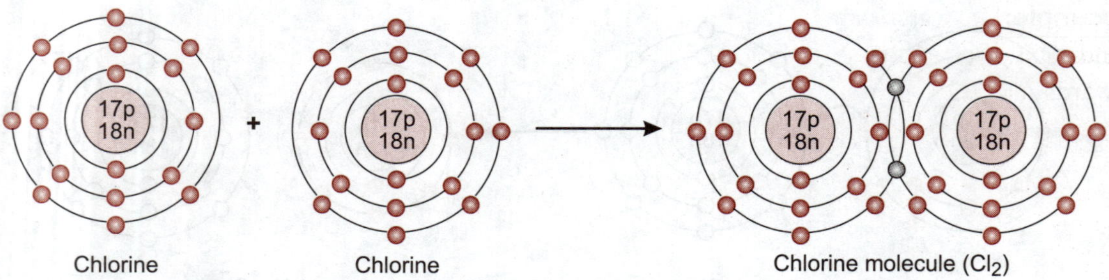

Fig. 2.3 Formation of covalent bond between two chlorine atoms

2. **Double Covalent Bonds:** In double covalent bonds two atoms share two pairs of electrons. In case of oxygen molecule two oxygen atoms share two pairs of electrons.

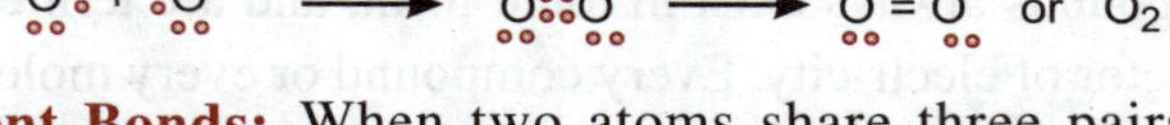

3. **Triple Covalent Bonds:** When two atoms share three pairs of electrons, a triple covalent bond is formed as in a molecule of nitrogen.

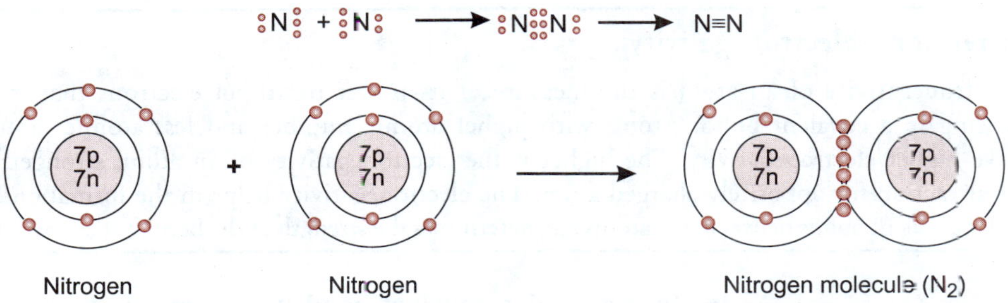

Fig. 2.4 Triple covalent bond between two nitrogen atoms

The number of bonds between the two atoms of a molecule is represented by lines: single bond (–), double bond (=) or triple bond (≡).

The force of bond formation in covalent bonds is the same as in ionic bonds, i.e. electrostatic attraction between two atoms. But the force is generated in a different manner.

B. Covalent Bonds may be Polar or Nonpolar

Covalents bonds are of two types: Polar bonds and non-polar bonds.

1. Polar Covalent Bonds: In a polar covalent bond, the sharing of electrons between two atoms is unequal. The attraction of an atom for electrons in a covalent bond is called **electronegativity**. The atom with higher electronegativity pulls the bonding electrons closer. This atom with more electronegativity acquires a slight negative charge. It is represented as δ^-. The less electronegative atom acquires a small positive charge (δ^+). Therefore, polar covalent bonds represent a transition stage between ionic and covalent bonds.

Example: In water two hydrogen atoms are joined with one oxygen atom by polar covalent bond. Oxygen has the strongest electronegative atoms of all the elements. It attracts electrons towards itself more strongly than hydrogen.

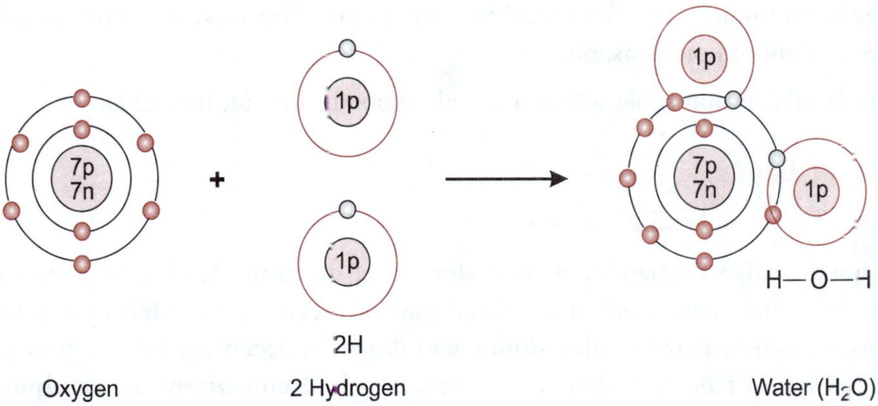

Fig. 2.5 Polar covalent bond in a water molecule

Extension: Electronegativity

Electronegativity of an atom is the measure of its power to attract electrons that it is sharing in a covalent bond. Atoms with higher atomic number and less atomic radius have higher electronegativity. The higher is the electronegativity of an atom, stronger is its attraction for oppositely charged atom. The electronegativity helps in the formation of polar covalent bonds between two atoms and determines the strength of the bond.

2. Nonpolar Covalent Bonds: In a nonpolar covalent bond two atoms share the electrons equally. There is no positive or negative charge. The bonds between two identical atoms are always nonpolar covalent bonds. Also bonds between carbon and hydrogen atoms are nonpolar (as in methane).

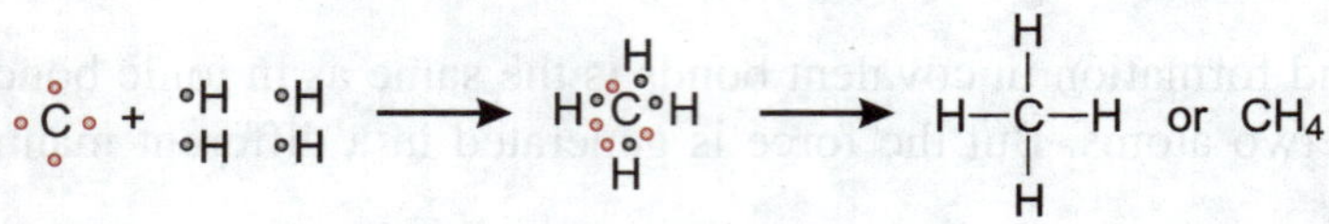

Differences between Ionic and Covalent Bonds

The above two types of bonds differ in following respects:

1. In ionic bonds, the ions with opposite charge (i.e. cation and anion) are held together by electrostatic forces of attraction. In covalent bonds two atoms are held together by sharing of electron pairs.
2. In ionic bond electron or electrons are transferred from one atom to another but in covalent bond electron pairs are shared between two atoms.
3. Ionic bonds are formed mainly in inorganic compounds like NaCl, KNO_3, etc. The covalent bonds are common in organic compounds, although they occur in inorganic compounds also.
4. Ionic compounds are electrovalent and polar. The covalent compounds are non-electrovalent or non-ionisable.
5. Ionic bonds are ionisable whereas covalent bonds are non-ionisable.

Other Types of Bonds

1. Coordinate Bonds or Semipolar Bonds

Coordinate bond is also formed by mutual sharing of electrons but the two electrons that are shared come from the same atom. The shared pair of electrons is called **lone pair**. The atom which provides electron pair is called **donor** and the atom accepting the electron pair is called **acceptor**. This type of bonding between atoms is called **coordinate** or **semipolar**, because slight polarity appears in the molecule due to one-sided contribution.

For example, ammonia molecule has a non-bonded lone pair of electrons on nitrogen which can be donated to an electron-deficient H^+ to form a coordinate bond.

$$NH_3 + H^+ \longrightarrow NH_4^+$$

Coordinate bonds are less common in organic compounds. They are found in **chlorophyll** and **heme**. In them, Mg or Fe are linked coordinately to nitrogen atoms.

2. Hydrogen Bonds
(Weak Attraction between Partially Charged Hydrogen Atoms)

Hydrogen bonds are weak bonds formed between a partially positive hydrogen of one polar molecule with the partially negative atom of another polar molecule. Thus, hydrogen bonds result from the attraction of oppositely charged parts of different molecules. They are just 5 per cent as strong as the covalent bonds. Therefore, they can not bind atoms into molecules but can establish link between molecules or between different parts of a large molecule (such as protein or nucleic acid).

Hydrogen forms the electropositive partner having partial positive charge. In a living system oxygen and nitrogen are partially electronegative partners. Water forms the best example of hydrogen bonds.

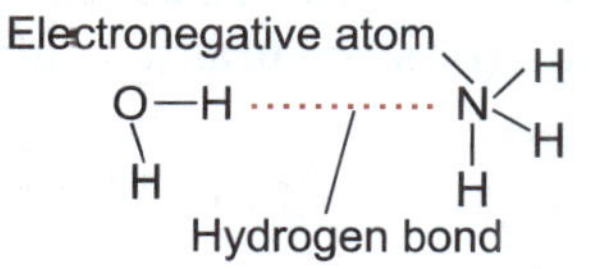

Fig. 2.6 A hydrogen bond between nitrogen atom of ammonium hydroxide molecule (NH_4OH) and hydrogen atom of a water molecule

Neighbouring water molecules are linked through hydrogen bonds. The hydrogen of one water molecule is linked with oxygen atom of another water molecule. This gives cohesion to water molecules creating high surface tension (Fig. 2.7).

Extension: Importance of Hydrogen Bonds

- Hydrogen bonding has an important effect on the properties of water.
- It is important in proteins and nucleic acids and thus in life processes.
- Hydrogen bonding helps in holding the two strands of the double helix of DNA together.
- Unzipping of DNA takes place by breaking of hydrogen bonds.
- Hydrogen bonding is both intramolecular, i.e. within a molecule and intermolecular, i.e. between two molecules as in DNA.
- Hydrogen bonding is highly directional.
- Hydrogen bonds are highly specific because they can form bonds only with those which can donate or accept hydrogen.

3. Hydrophobic Interactions or Nonpolar Interactions

The clustering together of nonpolar molecules or nonpolar groups in water is called hydrophobic interaction. Lipids are hydrophobic molecules. The hydrophobic groups of these biomacromolecules interact to exclude the water molecules and lead to the formation of micelles. The hydrophobic interactions are major driving force in the folding of macromolecules, for the binding of substrate to enzymes and the formation of lipid bilayer in the biological membranes.

Organic Compounds have Carbon Backbones

All compounds are broadly separated into two categories: **inorganic** and **organic.**

1. Inorganic Compounds: Such compounds are relatively small, simple substances with low molecular weight. Except for carbon dioxide (CO_2) and compounds containing carbonates, all other inorganic compounds do not have carbon. Water, mineral salts, simple acids and bases are inorganic compounds. The structure of their molecules is represented by a chemical formula. Inorganic compounds excluding water form about 1 per cent to 1.5 per cent by weight in living cells.

2. Organic Compounds: Organic compounds contain the carbon element. They are generally large and complex and are formed of a chain or ring. Organic compounds are found in living organisms. Carbohydrates, lipids, proteins, nucleic acids and vitamins are examples of organic compounds. The structure of molecules of an organic compound is represented by empirical formula and also by chemical formula (or structural formula).

WATER IS ESSENTIAL FOR LIFE

Water is most important and most abundant inorganic compound in all living systems. About 70 per cent of our total body weight is water and as much as 95 per cent in jelly fishes and certain aquatic plants. Water is most vital for all living organisms because water is the medium for most chemical reactions occurring within living beings. It is either a reactant or a product in many chemical reactions. Water's unique physical and chemical properties have contributed to the appearance, evolution and survival of living things. As a matter of fact life can not exist without water.

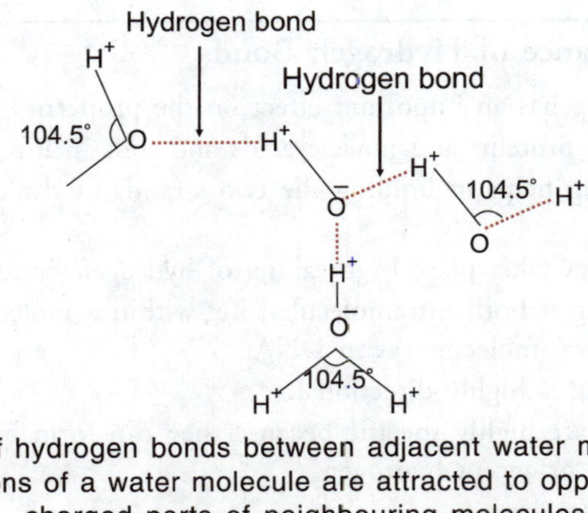

Fig. 2.7 Fomation of hydrogen bonds between adjacent water molecules. The charged regions of a water molecule are attracted to oppositely charged parts of neighbouring molecules

Water Molecules are Polar

In a water molecule two hydrogen atoms are joined with one oxygen atom by covalent bonds. These covalent bonds between hydrogen and oxygen atom are **polar covalent bonds** because the electronegative oxygen nucleus pulls the shared electrons of hydrogen towards itself and away from hydrogen. Therefore, adjacent water molecules are held together by hydrogen bonds (Fig. 2.7) and form a mesh. The hydrogen atom of one water molecule with its partial positive charge is attracted to the oxygen atom of neighbouring water molecule, with partial negative charge and forms a hydrogen bond. Each water molecule thus forms hydrogen bonds with four neighbouring water molecules. The bonding between neighbouring water molecules gives it cohesiveness.

The polar nature of water gives it six characteristic features that make it a useful medium for living cells. These are as follows:

1. Water is an excellent solvent

The polar nature of its molocules makes water an excellent and versatile solvent. It can dissolve both ionic and polar substances. The polar covalent bonds between hydrogen and oxygen induce interaction of each water molecule with several neighbouring ions and molecules of solutes.

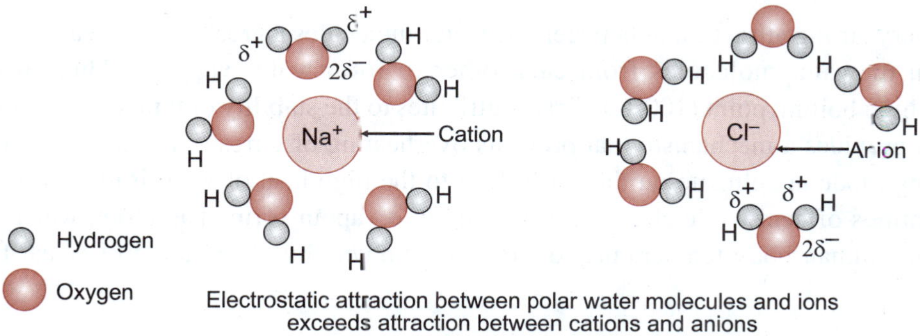

Fig. 2.8 Diagram to show dissolution of ionic solute, NaCl, in water

1. For Ionic Solutes: Ionic solutes are hydrophilic, i.e. they have affinity for water. For example, when sodium chloride is placed in water, its sodium and chloride ions get dissociated. The electropositive hydrogen atoms (H^+) of water molecules attract chloride ions (Cl^-). As a result a number of water molecules surround each chloride ion. Similarly, electronegative oxygen atoms in water molecules attract sodium ions (Na^+) and a number of water molecules surround each Na^+ ion. Thus Na^+ and Cl^- ions being separated and shielded with water molecules are not allowed to reunite.

2. For Polar Compounds: Polar compounds are also water soluble. For example, ammonia is an electrically neutral molecule, but covalent bonds between nitrogen and hydrogen are polar because nitrogen is highly electronegative. The electronegative nitrogen attracts electrons of outer valence from hydrogen. As a result ammonium molecule has opposite charges at its opposite ends. Thus, ammonia like water is polar. Through hydrogen bonding, water has an affinity for ammonia. Therefore, water acts as a solvent for ammonia. Water does not dissolve nonpolar compounds and neutral molecules that have a symmetric distribution of charge.

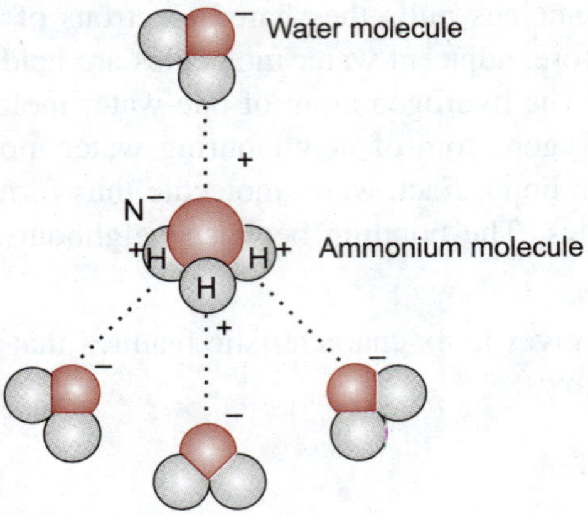

Fig. 2.9 Diagram showing dissolution of polar compound ammonia in water

2. Water has high heat of vaporisation

Due to a very strong attraction in between the water molecules a great deal of heat is required for separation of water molecules from each other to form water vapours. Thus, water has a relatively high boiling point (100°C). This contributes to the stability of temperature in ponds and lakes, and provides a mechanism that prevents overheating of terrestrial organisms and helps in maintaining moderate climate on the earth. Due to the high heat of vaporisation water removes large quantities of heat while changing from liquid to vapour form. Therefore, water also helps maintaining human body temperature low during summers by evaporation of sweat from body surface.

3. Water acts as an excellent temperature buffer because of high specific heat

Due to a relatively strong hydrogen bonding between its molecules, water has high specific heat. This makes water an excellent temperature buffer or a heat bank. It means a slight change in its own temperature causes release or absorption of large amount of heat. It stabilises temperature by absorbing heat from air in warmer conditions and releasing stored heat into air in colder conditions.

4. Density of water is maximum at 4°C

Hydrogen bonding between water molecules affects the density of water. The bonding is maximum when water is at 4°C and so also its density. It starts expanding as temperature decreases. So liquid water expands as it freezes because in ice the hydrogen bonds joining water molecules keep the molecules far apart so that solid ice is about 10 per cent lighter than liquid water at 4°C. For this reason ice floats on water surface. It serves as an insulating layer on the surface of lakes and streams, where life can exist even during cold periods.

5. Water is a good reactant in many chemical reactions

The polarity in water facilitates dissociation or splitting and rejoining of hydrogen (H^+) and hydroxide (OH^-) ions and acts as a **reactant** or **product** in many chemical reactions. Water is the key reactant in all digestive processes, whereby larger molecules are broken down into smaller ones. Water molecules are also involved in synthetic reactions. Water is an important source of hydrogen and oxygen that are used in the synthesis of organic compounds in living cells.

6. Water molecules dissociate slightly

Water molecules ionise slightly into H^+ and OH^- ions. The tendency of water molecules to dissociate is balanced by the tendency of H^+ and OH^- ions to reunite to form water. This characteristic of water can be represented as follows:

$$HOH \rightleftharpoons H^+ + OH^-$$

The concentration of H^+ and OH^- ions is always equal in pure water. Such a solution is called **neutral**, i.e. it is neither acidic nor basic. The concentration of H^+ or OH^- ions in pure water is 0.0000001 (10^{-7}) moles per litre.

ACIDS, BASES AND SALTS

Inorganic salts, when dissolved in water undergo dissociation or ionisation, *i.e.* they break into ions.

1. Acids Yield Hydrogen Ions (Protons)

An **acid** is a substance that dissociates into one or more positive hydrogen ions (H^+) and one or more negative ions (anions).

$$\text{Acid} \longrightarrow H^+ + \text{Anion}$$

It means an acid donates H^+ ions or protons to the medium. Thus, acid can also be defined as **H^+ donor** or **proton donor**. Hydrochloric acid (HCl) and sulphuric acid (H_2SO_4) are inorganic acids, whereas lactic acid ($CH_3CHOHCOOH$) and acetic acid (CH_3COOH) are common organic acids. Their dissociation products are:

$$HCl \longrightarrow \boxed{H^+} + Cl^-$$

$$CH_3COOH \longrightarrow CH_3COO^- + \boxed{H^+}$$

Extension: Summary of Biologically Important Functions of Water

A. In All Organisms

1. **Structure:** Water constitutes a high water content of protoplasm.
2. **Solvent:** Water is an excellent solvent for all organic and inorganic compounds and a medium for diffusion.
3. **Reagent:** Water acts as a reagent during hydrolysis taking place during biological reactions.
4. **Support:** Water provides support and a living environment for all aquatic organisms.
5. **Fertilisation:** Water helps in the transport of gametes to ensure fertilisation.
6. **Dispersal:** Water helps in the dispersal of spores, seeds, gametes and larval stages of aquatic organisms.

B. In Animals

1. **Transport:** Water helps in transport of nutrients, gases (oxygen), hormones, enzymes, etc., from one part of the body to other as part of blood, coelomic fluid and tissue fluid.
2. **Osmoregulation:** Water maintains the composition of internal environment of the body constant.
3. **Maintenance of Body Temperature:** Water cools the body during summer by evaporation, such as sweating and panting and by transporting heat from the region of heat generation to peripheral areas of heat loss.
4. **Lubrication:** Water acts as a lubricant in joints and between the membranes.
5. **Support:** In earthworm and other animals, water acts as a hydrostatic skeleton.
6. **Protection:** In lachrymal fluid and mucus water helps in protection.
7. **Migration:** Oceanic currents help animals swim to far off places.

C. In Plants

1. **Osmosis:** By osmosis, water helps in maintaining the internal environment of plant cells and plant body.
2. **Turgidity:** Water provides turgidity to cells and helps in their elongation, division and growth and opening and closing of stomata.
3. **Reagent in Photosynthesis:** Water on hydrolysis in the presence of sunlight provides H and OH ions for photosynthesis.
4. **Transportation and Translocation:** Water helps in the transport of inorganic ions and in the translocation of food synthesised in the phloem to different regions of plant body.
5. **Germination of Seeds:** The dry seeds imbibe water and swell up to break open the testa leading to sprouting of seeds.

2. Bases are Proton Acceptors

A **base** is a substance that dissociates into one or more positive ions (cations) and one or more negative hydroxy ions (OH^-). The OH^- immediately accepts or combines with H^+ ion or proton. Therefore, base is defined as **'proton acceptor'**.

$$\text{Base} \longrightarrow \text{Cation} + OH^-$$

$$H^+ + OH \rightleftharpoons H_2O$$

For example, sodium hydroxide (NaOH) and ammonium hydroxide (NH_4OH) are common inorganic bases. They dissociate to release OH^- ions.

$$NaOH \longrightarrow Na^+ + OH^-$$

$$NH_4OH \longrightarrow NH_4^+ + OH^-$$

3. Salts Dissociate into an Anion and a Cation

A **salt** is a substance, formed by the reaction between an acid and a base. When dissolved in water it dissociates into cations and anions but neither of them is H^+ or OH^-.

ACID-BASE BALANCE IN LIVING ORGANISMS

Within all the living organisms, in their aqueous environment, acids dissociate into H^+ (cations) and anions. The bases dissociate into hydroxide anions (OH^-) and cations. In case the number of free hydrogen ions is more in a solution, the solution is acidic. The more is the number of free H^+ ions the more acidic the solution is. Conversely, the more is the number of free OH^- ions in a solution, the more basic or alkaline is the solution. But a fairly constant balance of H^+ and OH^- ions is maintained in a living body, because the biochemical reactions are extremely sensitive to slightest change in the acidity or alkalinity of the medium.

Hydrogen Ion Concentration Determines the pH of the Medium

Most of the chemical substances when dissolved in water split into electrically charged particles, called ions. The positively charged ions are called **cations** and negatively charged ions are known as **anions**. Even in pure water there is some dissociation into H^+ and OH^- ions. H^+ ions are cations and OH^- ions are anions. With the addition of some acid (HCl) in water there is excess of H^+ ions while the addition of any alkali (NaOH or KOH) increases the concentration of OH^- ions.

$$H_2O \rightleftharpoons H^+ + OH^-$$

$$HCl \longrightarrow H^+ + Cl^-$$

$$NaOH \longrightarrow Na^+ + OH^-$$

Since ions are charged particles, their presence affects the electrical conductivity of water. The concentration of ions is used to measure the extent of dissociation or ionisation. The concentration of ions is expressed as gram–equivalent per litre of solution. For example, the hydrogen ion concentration of pure water at 22°C is 10^{-7} or $\frac{1}{10^7}$. It means pure water has one gram of H^+ ions in one million litres of water.

Therefore, $$cH \frac{1}{1\,\text{million}} = N = 10^{-7} \text{ gm/litre}$$

Where, *c*H is H ion concentration, N is normality.

As there are equal number of H^+ and OH^- ions in pure water, it is clear that $H^{-7}OH^{-7} = 10^{-14}$ *k* constant. Therefore, by knowing the concentration of H^+ ions it is possible to deduce the concentration of the OH^- ions or vice-versa.

The activity of the solution is expressed by *c*H:

$H^+ = 10^{-7}$ ∴ $OH^- = 10^{-7}$ Solution is neutral

$H^+ = 10^{-6}$ ∴ $OH^- = 10^{-8}$ Solution is acidic

$H^+ = 10^{-9}$ ∴ $OH^- = 10^{-5}$ Solution is alkaline

To avoid the negative representation, **Sorensen** has introduced hydrogen potential or pH. If *c*H = 10^{-7} then *p*H = 7; if *c*H = 10^{-6}, then *p*H = 6.

Significance of Hydrogen Ion Concentration

Hydrogen ion concentration and consequently the pH in living cells and all body fluids is always maintained close to 7 or close to neutral. This is because the structure and functions of molecules in the cells and all the biochemical reactions of metabolism are very sensitive to the concentration of hydrogen (H^+) or hydroxide (OH^-) ions. Even a slight change in pH may prove to be deterimental. Therefore, biological systems resist any change in hydrogen ion concentration. To minimise such changes in H^+ and OH^- ions, living beings have developed various buffer systems.

pH Scale or Scale of Potential of Hydrogen

pH is an abbreviation for 'potential of hydrogen' of a solution. It is a measure of hydrogen ion concentration. pH can be defined as **'the negative log of the hydrogen ion concentration'**. It is represented as – log [H^+].

$$pH = \log\frac{1}{[H^+]} - \log [H^+]$$

Extension: pH is a Logarithmic Function

pH of a solution represents its logarithmic function. It means when pH of a solution decreases by one unit from 4 to 3, the concentration of H^+ ions increases tenfold from 10^{-4} M to 10^{-3} M. Thus one number on pH scale represents tenfold change in the number of H^+ ions. It means a solution of pH 3 is not twice as acidic as of pH 4 but 10 times more acidic. When pH of a solution changes slightly, the actual concentration of H^+ and OH^- ions in the solution changes substantially.

pH Scale

The **acidity and alkalinity of a solution** is expressed on the **pH scale**. It extends from 0 to 14. This scale is based on the concentration of H^+ ions in mols per litre. A pH of 7 means that a solution contains one millionth (*i.e.* 0.0000001) of a mol of hydrogen ions per litre. The number 0.0000001 is written as 1×10^{-7}. To convert this value to pH, the negative exponent (–7) is changed to a positive number (7). It means a solution with H^+ concentration 1×10^{-7} per litre has a pH of 7.

pH 7 is the midpoint of a pH scale. At this point concentration of H^+ and OH^- ions is equal. A solution with pH 7 such as pure water is, therefore, neutral. A solution that has more H^+ ions is acidic. The acidic compounds have a pH range of 0 to 7. A solution that has more OH^- ions than H^+ ions is **basic** or **alkaline**. The pH of basic solutions may range from 7 to 14.

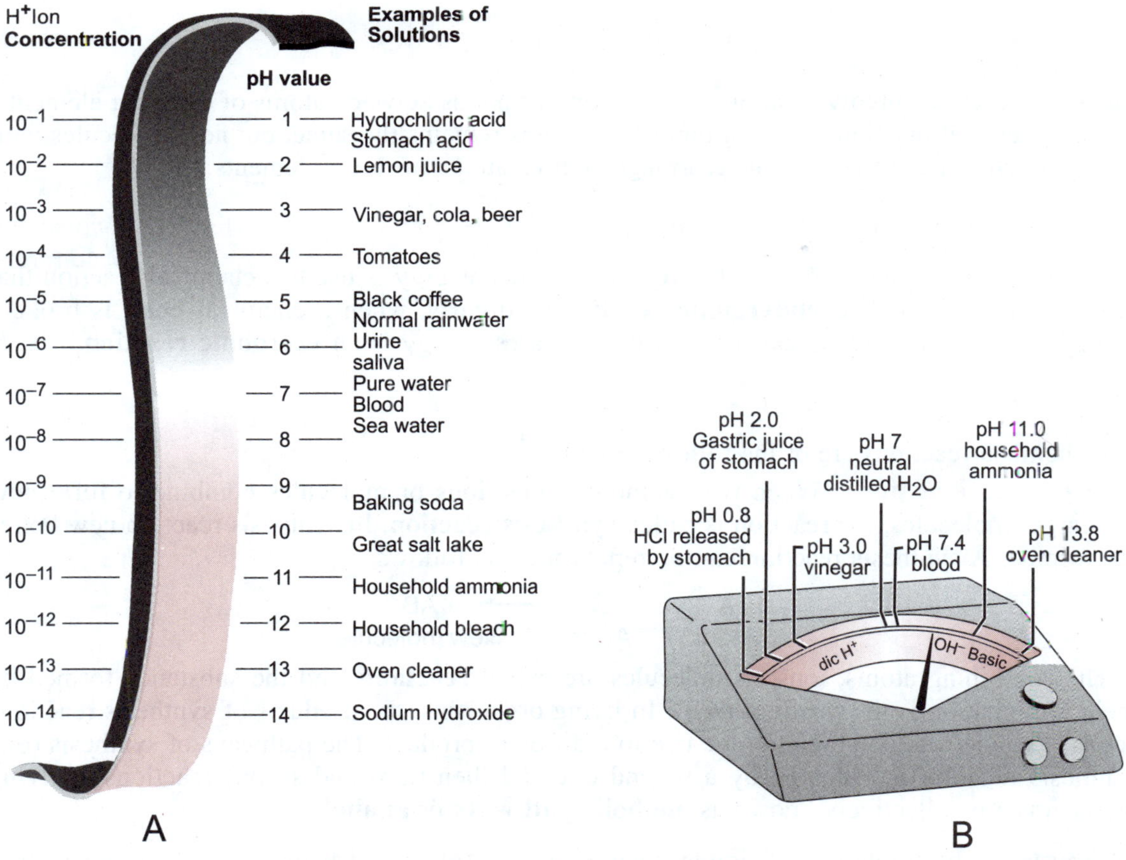

Fig. 2.10 **A.** The pH scale; **B.** pH meter

Table 2.1 pH values of some human body fluids and some common substances

S.No.	Fluid or substance	pH
1.	Saliva	6.35 to 6.85
2.	Gastric juice	1.50 to 2.50
3.	Blood	7.35 to 7.45
4.	Urine	5.00 to 7.80
5.	Milk	6.60 to 6.90
6.	Distilled water	7.00
7.	Lemon juice	2.20 to 2.40
8.	Carbonated drinks	3.00 to 3.50
9.	Tomato juice	4.20
10.	Milk of magnesia	10.00 to 11.00
11.	Lime water	12.30

Most organisms grow best in environments with a pH between 6.5 and 8.5.

CHEMICAL REACTIONS

Chemical reactions involve making and breaking of bonds between atoms of different elements. After a chemical reaction, the total number of atoms remains the same, but new molecules with new properties are formed by the rearrangement of atoms of those elements.

Energy is Required or Released During Chemical Reactions

When a chemical bond is formed between two atoms, energy is used. A chemical reaction that requires energy is called **endergonic reaction**. Similarly, when a chemical bond is broken, energy is released. A chemical reaction that releases energy is an **exergonic reaction.**

Types of Chemical Reactions

The chemical reactions are of the following types:

1. Synthesis Reactions: When two or more atoms, ions or molecules combine to form new and larger molecules, the reaction is called **synthesis reaction**. In synthesis reaction new bonds are formed. A synthesis reaction can be represented as follows:

$$\underset{\text{Atom}}{A} + \underset{\text{Atom}}{B} \longrightarrow \underset{\text{New molecule}}{AB}$$

The combining atoms, ions or molecules are called **reactants** and the substance formed by their combination is the **product** (AB). In living organisms, the products of synthesis reactions act as reactants and combine to give rise to a different product. The pathways of synthesis (one synthesis reaction is followed by a second one and then third and so on) reactions in living organisms are collectively known as **anabolic pathways** or **anabolism**.

Example: The combining of simple sugar molecules (glucose, fructose, etc.) to form starch or of amino acids to form proteins are examples of synthesis reactions. The synthesis of starch and proteins are examples of **anabolism** or anabolic pathway.

2. Decomposition Reactions: The decomposition reactions are reverse of synthesis reactions. In decomposition, bonds are broken and large molecules split into smaller molecules, ions or atoms. It can be represented as follows:

$$\underset{\text{of a compound}}{\underset{\text{Molecule}}{AB}} \longrightarrow \underset{\text{Atom}}{A} + \underset{\text{Atom}}{B}$$

Decomposition reactions occurring in living organisms are collectively known as **catabolic reactions** or **catabolism**. For example, splitting of sucrose into its components glucose and fructose, or of starch into glucose molecules is called catabolism of sugar or catabolism of starch.

3. Exchange Reactions: In exchange reactions both decomposition and synthesis reactions occur simultaneously. An exchange reaction can be represented as follows:

$$AB + CD \longrightarrow AD + BC$$

In this reaction, bonds are broken between A and B and also between C and D. The new bonds are formed between A and D and also between B and C. For example:

$$NaOH + HCl \longrightarrow NaCl + H_2O$$

4. Reversible Reactions: In theory, all chemical reactions should be reversible, i.e. they can occur in both directions. This can be illustrated by the following equation:

$$A + B \underset{\text{Decomposition}}{\overset{\text{Synthesis}}{\rightleftharpoons}} AB$$

Activation Energy

According to collision theory, the atoms, ions and molecules are continuously moving and colliding with one another. The energy transferred by the particles in the collision can disrupt electronic structure and break chemical bonds or cause formation of new bonds. The collision energy required for a chemical reaction is called **activation energy**. Thus, activation energy is the amount of energy needed to disrupt the stable electronic configuration of any specific molecule leading to the rearrangement of electrons.

Enzymes Lower the Activation Energy of a Reaction

Enzymes are large protein molecules that act as biocatalysts inside the living organisms. Similar to a catalyst, an enzyme speeds up a chemical reaction to occur at body temperature without increasing the temperature of reactants or the reaction medium.

Each enzyme molecule has an 'active site' for a specific chemical substance or reactant. The reactant of a particular enzyme is called its substrate. At the time of reaction, an **enzyme-substrate complex** is formed by the temporary binding of enzyme and its reactant. This enables the collision to be more effective and lowers down the activation energy of the reaction. The enzymes, therefore, speed up the biochemical reactions at a temperature compatible with the normal functioning of the cell.

TEXT QUESTIONS

1. What are chemical bonds? Discuss the different types of chemical bonds.
2. Give differences between ionic and covalent bonds and write a note on hydrogen bonds.
3. What are hydrogen bonds? Give any two examples where hydrogen bonds are found.
4. Describe unique features of water acquired because of hydrogen bonds.
5. Explain in brief acids, bases and salts.
6. Explain the significance of hydrogen ion concentration and write a note on pH of a solution.
7. What do you mean by pH of a solution? Why a liquid at pH = 3 is ten times more acidic than a liquid with pH = 4?
8. Describe the structure and extraordinary properties of water.
9. 'Water is a versatile solvent'. Justify the statement.
10. What type of bonding exists between water molecules and within a water molecule?

Molecules of Life: Biomolecules

3
Chapter

IN THIS CHAPTER

Organic Compounds: Functional groups, Role of Functional Groups; Carbohydrates: Chemical Composition, Biological Significance; Classification: Monosaccharides, Oligosaccharides, Polysaccharides; Lipids: Specialised Features of Lipids, True Fats; Fatty Acids; Biological Significance of Lipids ; Classifications of Lipids : Simple Lipids, Derived Lipids; Proteins : Proteins are Macromolecules, Biological Significance of Proteins; Amino Acids: Characteristics of Amino Acids; Essential and Non-essential Amino Acids ; Functions of Amino Acids; Classification of Amino Acids; Protein's Primary, Secondary, Tertiary and Quarternary Structures Classification of Proteins.

All known forms of life are comprised solely of biomolecules. Biomolecules are those molecules that are involved in the maintenance and metabolic processes of living organisms. They in other words can also be called chemical compounds that naturally occur in living organisms. The study of chemicals in living organisms is closely associated with the expansion of molecular biology. Out of the 109 elements found in natures, only 22 are essential for life. They are listed in Table 3.1.

Table 3.1 The elements found in living organisms

S.No.	Chief elements of organic molecules	Symbol
	A. Non-metallic	
1.	Hydrogen	H
2.	Carbon	C
3.	Nitrogen	N
4.	Oxygen	O
5.	Phosphorus	P
6.	Sulphur	S
7.	Chlorine	Cl
	B. Chief Ions in organisms	
1.	Sodium	Na^+
2.	Potassium	K^+
3.	Magnesium	Mg^{2+}
4.	Calcium	Ca^{2+}

S.No.	Trace elements in organisms	Symbol
	C. Trace elements in organisms	
1.	Manganese	Mn
2.	Iron	Fe
3.	Cobalt	Co
4.	Copper	Cu
5.	Zinc	Zn
6.	Boron	B
7.	Aluminium	Al
8.	Silicon	Si
9.	Vanadium	V
10.	Molybdenum	Mo
11.	Iodine	I

*Based on A.L.Lehninger, Worth, N.Y. 1970.

About 98 per cent of the mass of all living beings is composed of six elements. The most common elements are hydrogen, carbon, oxygen and nitrogen, and they account for about 96 per cent of all elements found in our body. The biological importance of hydrogen, oxygen, nitrogen and carbon is due to their having 1, 2, 3 and 4 valencies respectively and their ability to form more stable covalent bonds than all other elements with these valancies.

Extension: Unique Features of Carbon—Fundamental to Life

Carbon has a number of unique features because of which it forms such an enormous variety of organic compounds that the whole branch of chemistry (**organic chemistry**) is devoted for their study. These characteristic features are:

- Carbon is relatively a small atom with a low mass. Its mass number is 12. It means it has six protons and six neutrons in the nucleus.
- The number of electrons is six. They are arranged in two orbits around the nucleus as 2 +4.
- The first orbit of carbon has two electrons. Its second or outer orbit has just four electrons. Therefore, carbon has 4 valency. It can form up to four strong covalent bonds with hydrogen, nitrogen, oxygen, phosphorus and sulphur.
- Carbon can readily form single, double or triple bonds with itself. Therefore, it can build up large carbon skeletons either in the form of ring or chain.
- It can form multiple covalent bonds with itself, oxygen and nitrogen.

ORGANIC COMPOUNDS

Most of the chemical compounds present in living organisms always contain **carbon** and **hydrogen** and their carbon atoms are covalently bonded to form straight chain, branched chain or ring structures. These biological molecules were named **organic compounds**, because it was believed that they are assembled only in living organisms. In 1928, the German biochemist, **Friedrich Wohler** synthesised urea, which made way for the *in vitro* synthesis of almost every organic compound found in living beings. At present more than five million organic compounds are known. This is because carbon can form covalent bonds with many atoms producing a great variety of molecules much more than any other element can do.

Organic compounds are the main structural components of cells and tissues. They are functionally also very important. Organic compounds participate and control all metabolic reactions, transmit information and provide energy for all activities of living beings.

Organic Compounds have Different Functional Groups

The chain of carbon atoms in a molecule of organic compound is called its **carbon skeleton**. When these carbon atoms are bonded with hydrogen, a **hydrocarbon chain** is formed. The bonding of other elements with hydrocarbon chain forms its characteristic **functional groups**. The functional groups are specific groups of atoms that are responsible for most of the physical and chemical properties of particular organic molecules. The various functional groups and the class of compounds in which they are found are represented in Table 3.2.

Table 3.2 Functional groups and compounds in which they are found

S. No.	Functional group	Name of the group	Class of compounds
1.	R—OH	Hydroxyl	Alcohols
2.	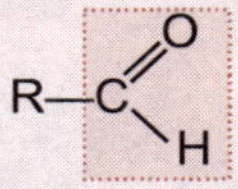	Carbonyl (terminal)	Aldehydes
3.	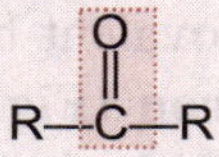	Carbonyl (internal)	Ketones
4.	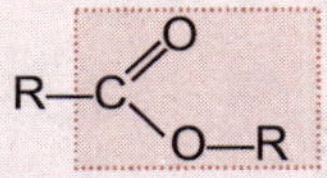	Ester	Esters
5.	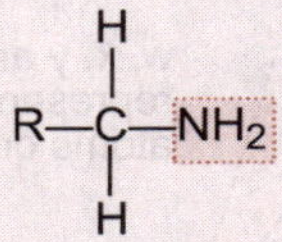	Amino	Amines
6.	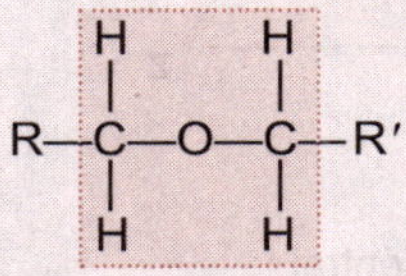	Ether	Ethers
7.	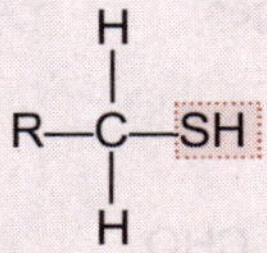	Sulfhydryl	Sulfhydryls
8.	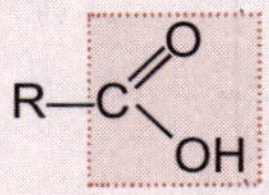	Carboxyl	Organic acids
9.	R—CH_3	Methyl	Many organic compounds
10.	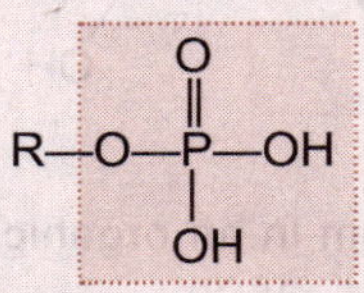	Phosphate	Organic phosphates

Role of Functional Groups

Functional groups form associations with other molecules through hydrogen or ionic bonds and determine the types of reactions in which a particular compound will participate.

Most organic compounds have two or more functional groups. For example, every amino acid has two functional groups: an **amino group** ($-NH_2$) and a **carboxyl group** ($-COOH$). Some amino acids have additional functional groups that determine their specific property associated with their additional group.

Organic Molecules have Three-dimensional Structure

When one carbon atom is attached to four other atoms by single covalent bonds the carbon compound formed has a three-dimensional shape called **tetrahedron (Fig 3.1).** The angle between any two of these bonds is about 109.5 degrees. This bond angle is constant and remains the same for the diverse organic compounds.

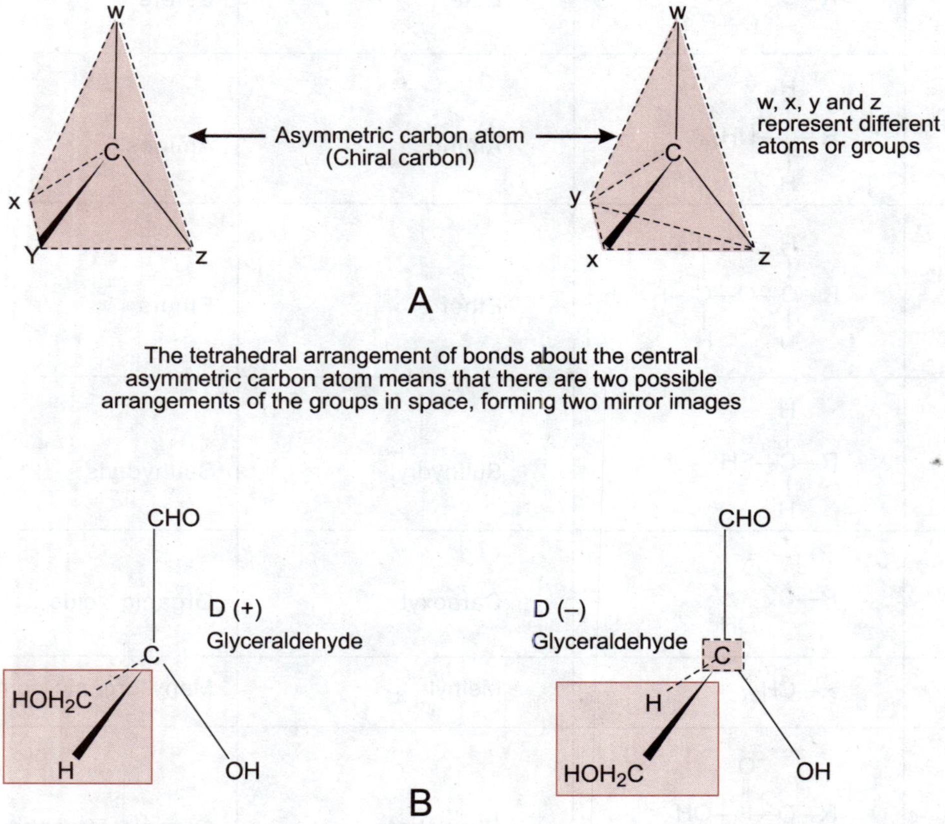

Fig. 3.1 **A.** A tetrahedral carbon; **B.** Carbon atom in an organic molecule with four single covalent bonds, as in glyceraldehyde, has a tetrahedral shape

2. **Asymmetry of Carbon Atom:** A carbon atom which has four different atoms or groups of atoms attached to its four valencies is said to be **asymmetric carbon atom.** All monosaccharides have asymmetric carbon atoms.

3. **Stereoisomerism:** Monosaccharides have one or more asymmetric carbon atoms. These asymmetric carbons are known as **chiral carbon atoms**. Because of asymmetric carbon atoms, monosaccharide molecules in solution occur in two isomeric forms. They have same structural formula but differ in spatial configuration and are called **stereoisomers** or **geometric isomers**.

4. **D- and L-Isomers:** In monosaccharide molecules, when OH group on the second carbon, i.e., adjacent to the carbon of terminal primary alcohol, is on the right, the molecule is represented as **D-isomer**. When OH is on its left, the monosaccharide molecule is **L-isomer**. The naturally occurring monosaccharides are mainly D-isomers.

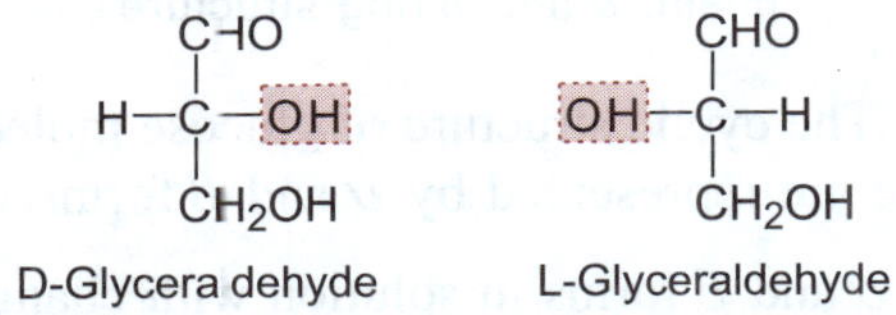

Fig. 3.7 Two isomers of glyceraldehyde: D-isomer and L-isomer

5. **Optical Isomerism:** When a beam of polarised light is passed through a solution exhibiting optical activity, it is rotated to the right or to the left. A compound which causes rotation of polarised light beam to the right is said to be **dextrorotatory** and D or +ve sign is used. Rotation of polarised light to the left is **levorotatory** and is designated by L or –ve sign.

   When in a solution equal number of dextrorotatory and levorotatory isomers are present, their mixture shows no optical activity. Such a mixture is **racemic** or **DL mixture**. Stereoisomerism and optical isomerism are two independent properties.

6. **Epimers:** Isomers formed as a result of interchange of OH^- and H^+ ions on carbon atoms 2, 3, and 4 of glucose are known as **epimers**. The most important epimers of glucose are mannose and galactose formed by epimerisation at carbon 2 and 4.

7. **Cyclic Structure:** The molecular structure of monosaccharides is represented in two different forms: **Open chain structure** and **cyclic** or **ring structure**.

   The straight chain structural formula accounts for its aldohexose character. This was proposed by **Fischer** and is called **Fischer formula** (Fig 3.8).

   Monosaccharide sugars of 5 or more carbon atoms have **cyclic structure**. In case of cyclic structure, the carbonyl group (−C=O) forms a covalent bond with the oxygen of −OH group. D-glucose in dissolved form remains in **hemi-acetal form**. In this form, −OH group present on C–5 carbon atom combines or reacts with the aldehye group on C–1.

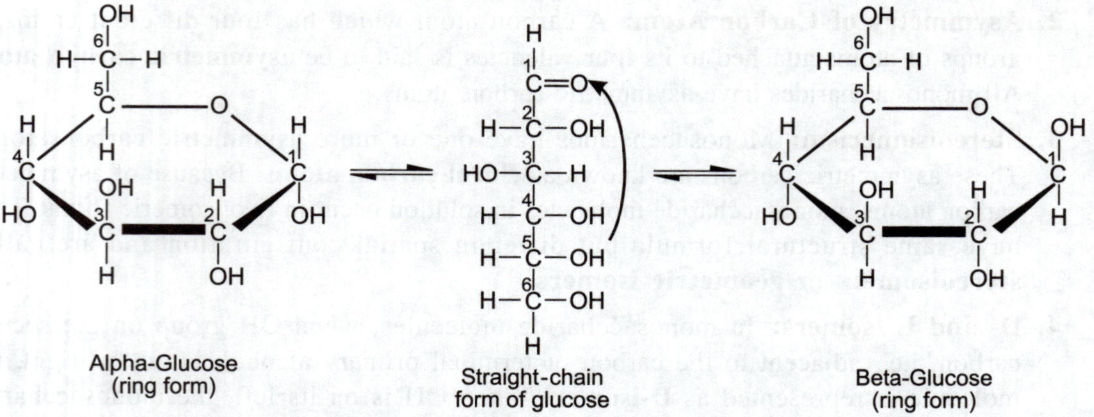

Fig 3.8 Structural formulae of glucose showing straight chain, α and β ring structure

8. α- and β- Anomers: The cyclic structure of glucose molecule also presents two stereo-isomeric forms. These are represented by α and β forms.

Interconversion of α and β forms in solution with change in optical activity is called **mutarotation**. In mutarotation hemi-acetal ring opens and is reformed with change in the position of –H and –OH groups.

9. Pyranose and Furanose Ring Structure: The cyclic or ring like structure of 6 carbon sugars is described as **pyranose ring**. The pyranose ring is formed of 5 carbons and one oxygen with a linkage between C–1 and C–5 carbons and 6th carbon remains outside the ring. The furanose ring is formed of four carbons and one oxygen and the linkage is between C–1 and C–4 carbon atoms. The 5th and 6th carbon atoms remain outside the ring. The pyranose ring is hexagonal while furanose ring is pentagonal.

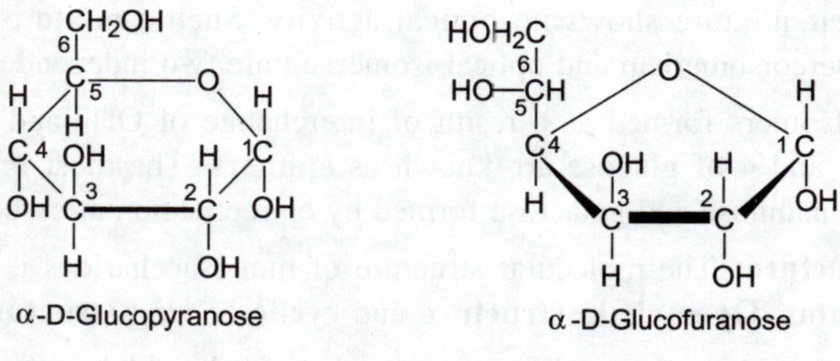

Fig. 3.9 Pyranose and furanose structure of glucose

2. Oligosaccharides (Gk. Oligos =few)

Oligosaccharides have 2–10 Monosaccharide Molecules

Oligosaccharides include carbohydrates formed of 2–10 monosaccharide units joined together by **glycosidic linkage** or **glycosidic bond** (Fig. 3.10). It is a covalent bond. It is formed between two monosaccharide molecules at positions 1 and 4 with the release of one water molecule. This forms a 1–4 linked disaccharide. The monosaccharide units in an oligosaccharide may be similar or dissimilar.

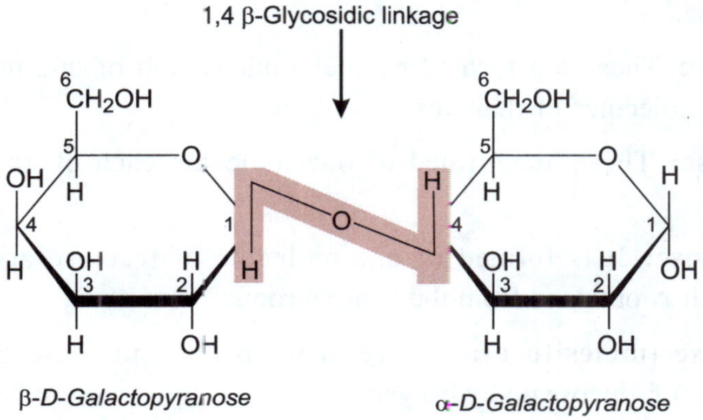

Fig. 3.10 Glycosidic bond between two molecules of monosaccharides in lactose

Classification of Oligosaccharides is based on the Number of Monosaccharide Units

1. **Disaccharides:** These are formed by the condensation of two molecules of same or different monosaccharides with the loss of one molecule of water. Their empirical formula is $C_{12}H_{22}O_{11}$. **Sucrose** and **maltose** in plants and **lactose** in animals are main disaccharides. These sugars are derived by the condensation of hexoses.

$$\underset{\text{Glucose}}{C_6H_{12}O_6} + \underset{\text{Fructose or Glucose}}{C_6H_{12}O_6} \longrightarrow \underset{\text{Disaccharide}}{C_{12}H_{22}O_{11}} + H_2O$$

   (a) **Sucrose:** Sucrose or cane sugar occurs in abundance in plants, such as sugar-cane, beets, carrots and fruits. It is very sweet, crystalline and freely soluble in water. On hydrolysis it forms glucose and fructose in equal proportions.

   (b) **Lactose:** Lactose or milk sugar occurs in milk and is the only carbohydrate for the sucklings. It forms white gritty crystals of the milk whey. It is not very sweet and is not found in plants. On hydrolysis, lactose produces an equimolar solution of glucose and galactose.

   (c) **Maltose (Malt sugar):** It does not occur in nature. It is a product of hydrolysis of starch and glycogen by *amylase* and *ptyalin* enzymes. In the presence of enzyme *maltase* it hydrolyses into two molecules of glucose.

   (d) **Cellobiose:** It is obtained from cellulose and on hydrolysis breaks up into two molecules of glucose.

2. **Trisaccharides:** This type of oligosaccharides produce three molecules of monosaccharides. These include:

   (a) **Mannotriose:** These are formed of two molecules of galactose and one molecule of glucose.

   (b) **Robinose:** These are formed by the condensation of one molecule of galactose and two molecules of rhamnose.

   (c) **Raffinose:** These are formed of one molecule each of glucose, galactose and fructose.

   (d) **Gentianose:** It is formed of one molecule of fructose and two molecules of glucose. It is obtained from the gentian roots.

   (e) **Melibiose (melesitose):** It gives rise to one molecule of fructose and two molecules of glucose on hydrolysis.

3. **Tetrasaccharides:** These are formed of four molecules of monosaccharides. Only two tetrasaccharides are known:

   (a) **Stachyose:** It is composed of D-glucose, fructose and two molecules of galactose. It is found in haricot beans.

   (b) **Scorodose:** It is found in the bulbs of garlic and onion.

3. Polysaccharides are Macromolecules

These are complex sugars formed by condensation of large number of molecules of monosaccharides (glucose) with the loss of equal number of water molecules. The number of monosaccharide units varies from above 10 to many thousands, joined together by glycosidic bonds. They do not have reducing power and never have a sweet taste. The examples are **cellulose**, **starch** and **glycogen**. These on hydrolysis produce molecules of monosaccharides. These can be represented by general molecular formula $(C_5H_{10}O_5)_n$.

Characteristics of Polysaccharides

- Polysaccharides are insoluble in water or form a colloidal solution.
- They are colourless.
- They are tasteless or they are not as sweet as sugars are.
- They are of very high molecular weight.
- Polysaccharides are macromolecules formed by thousands of monosaccharide molecules. They are represented by general formula $(C_6H_{10}O_5)_n$.
- On hydrolysis they produce monosaccharide molecules.

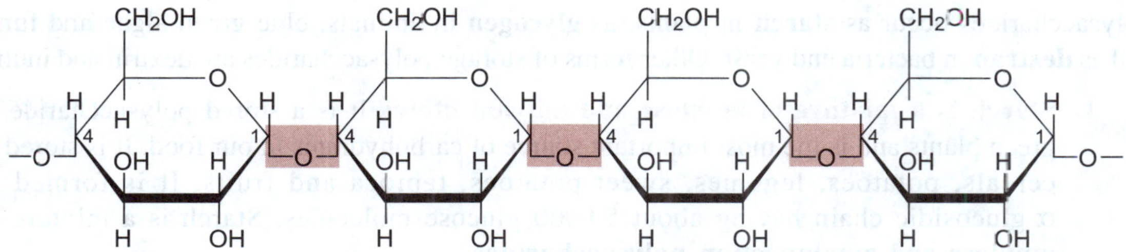

Fig. 3.11 Structure of starch, a polymer of glucose

Polysaccharides are classified in Four Different Ways

A. Based on the Nature of Constituents Polysaccharides are of Two Types

1. **Pentosans:** Formed of pentose units.
2. **Hexosans:** Formed of hexose units.

B. Based on the Name of Constituents Polysaccharides are of Three Types

1. **Glucosans:** Formed of glucose units.
2. **Fructosans:** Formed of fructose units.
3. **Galactans:** Formed of galactose units.

C. Based on Chemical Composition Polysaccharides are of Two Types

1. Homopolysaccharides: These are formed by the polymerisation of only one type of monosaccharide units. For example, glycogen, starch and cellulose are formed by the polymerisation of glucose. **Starch** and **glycogen** serve for the storage of glucose in plants and animals respectively, whereas **cellulose** and **chitin** are structural polysaccharides. Cellulose takes part in the formation of cell wall in plant cells and chitin forms part of exoskeleton in arthropods.

The nomenclature of homopolysaccharides is based on their components. For example, homopolysaccharides formed of glucose units are called **glucans**, of fructose units are **fructans,** of galactose are **galactans** and of xylose molecules are called **xylans**. Some homopolysaccharides are formed of amino sugars like **glucosamine** and **galactosamine**, **monosaccharides** are of simple **glucuronic acid** and **iduronic acid** molecules.

2. Heteropolysaccharides: These are formed of two or more types of monosaccharide molecules, e.g., **mucopolysaccharides, glycoproteins, chitin, peptidoglycans, agar, pectin, hemicellulose,** etc. Hyaluronic acid is a polymer of D-glucuronic acid and N-acetyl D-glucosamine arranged alternately.

D. Based on Functions Polysaccharides are of Three Types

1. Storage Polysaccharides: These are homopolysaccharides. They are polymers of glucose. They are reserved energy sources because glucose is the main source of energy for cell metabolism. They are hydrolysed into glucose monomers as and when required. Storage polysaccharides occur as **starch** in plants, as **glycogen** in animals, blue green algae and fungi and as **dextran** in bacteria and yeast. Other forms of storage polysaccharides are dextrin and inulin.

1. **Starch is a mixture of amylose and amylopectine:** It is a stored polysaccharide in green plants and is the most important source of carbohydrates in our food. It is stored in cereals, potatoes, legumes, sweet potatoes, tapioca and fruits. It is formed of α-glucosidic chain having about 50,000 glucose molecules. Starch is a mixture of **amylose** and **amylopection polysaccharides.**

   (a) **Amylose** forms about 15 – 20 per cent of starch. It is in the form of an unbranched chain coiled like a watch spring. Each chain is formed of about 300 – 400 glucose units. In it glucose molecules are joined by 1 – 4 α-linkages. Its molecular weight is 60,000. It gives blue colour with iodine.

   (b) **Amylopectin:** It forms 80 – 85 per cent of starch. Its molecules are branched and are coiled concentrically. Because of its more complex structure amylopectin is less soluble in water than amylose. It gives red-purple colour with iodine solution.

   Amylopectin molecule has a molecular weight of about 5,00,000 and its one chain has at least 80 branches.

   Starch is stored in plants in the form of starch granules of different shapes and sizes. Raw starch is insoluble in cold water and on heating it forms a colloidal solution in dilute concentration. Concentrated solution on heating forms gelatin- like substance and is used as adhesive paste.

2. **Glycogen is animal starch and a polymer of glucose:** It is stored in liver and muscle cells of all higher animals and human beings. It is a branched polymer of glucose having 5,000 – 10,000 glucose units. Its molecular weight is about 4,00,00,000. In each glycogen molecule, α–D glucose units are joined by α-1-4 glucosidic linkage. The branches are connected with the main chain by α-1-6 linkage.

   Glycogen is soluble in water. It gives red colour with iodine. In adult man about 400 g of glycogen is stored in the muscles. The synthesis of glycogen from glucose is called **glycogenesis** and hydrolysis of stored glycogen into glucose is **glycogenolysis**. Glucose from liver is distributed to all body cells for daily metabolism. Glucose produced in the muscle cells is utilised by muscles only.

3. **Dextrins:** Dextrins are formed from partial hydrolysis of starch by enzyme *salivary amylase*. They have faint sweet taste and form sticky solution in water. On hydrolysis they produce maltose which is further hydrolysed into glucose by the enzyme *maltase*.

4. **Dextrans:** These are stored polysaccharides of yeast and bacteria. Their molecules are in the form of branched chains. In dextrans D-glucose molecules are linked by α 1, 6-glucosidic bond.

5. **Inulin:** It is found as a stored carbohydrate in the plants of family Compositae, like artichock, *Dahelia, Dendeleon*, etc. It is formed by the polymerisation of fructose molecules.

2. Structural Polysaccharides: Structural polysaccharides are also **homopolysaccharides**, formed by the polymerisation of D-glucose molecules. They occur in the form of long and strong fibres in the cell wall of plant cells and in the exoskeleton of arthropods. **Cellulose** and **chitin** are important structural polysaccharides.

1. **Cellulose:** It is the most abundant carbohydrate present on earth. It forms cell wall of plant cells and provides mechanical strength to the cells. It is formed of strong but elastic fibres. Each cellulose fibre is made up of 10 to 15 thousand glucose molecules joined by β, D-glucose molecules by β-1, 4 linkage. A number of cellulose molecules together form a microfibril and a bundle of microfibrils collectively form a **macrofibril**. They have great tensile strength. Cellulose is an important component of human diet. It adds bulk to intestinal contents and stimulates peristalsis in intestinal wall. Cellulose is neither water soluble nor is easily digested by *amylase* present in the digestive juices.
2. **Chitin:** Chitin is found in the exoskeleton of insects and other arthropods and in the cell wall of fungi. It is a polymer of **N-acetyl glucosamine** which is a derivative of D-glucose. Chitin is soft and skin-like but because of the deposition of some proteins and calcium carbonate, it becomes hard. It is insoluble in water and gives definite form to the animals.

3. Complex Polysaccharides: Mucopolysaccharides are complex mucilaginous polysaccharides. These are **heteropolysaccharides** formed of proteins and polysaccharides. These are gel-like and are present in the extracellular matrix. These are secreted by the cells. The sticky or slimy substance formed when isabgol seeds are put in water or when lady-fingers are cut is **mucus** only. It is heteropolysaccharide of galactose and mannose. **Agar** from marine algae and sea weeds, **carageenin** and **aligenic acid** are also mucoid substances. The main muco-polysaccharides found in animals are **hyaluronic acid, chondroitin sulphate, heparin** and **keratin sulphate.**

(a) **Hyaluronic acid:** It is present in the matrix of all tissues, in synovial fluid of bone joints, in the cerebrospinal fluid of brain and spinal cord and in the vitreous humour of eye. It acts as a lubricant. It is formed of alternate units of **glucuronic acid** and **N-acetyl glucos-amine** linked to form thread-like structure.

(b) **Chondroitin sulphate:** It is found in the matrix of connective tissue, cartilage, tendon and skin. It provides strength and elasticity to these structures.

LIPIDS

Lipids are Fats or Fat-like Substances

Lipids are oily, greasy and fat-like substances which are insoluble in water but soluble in ether, chloroform, hot alcohol or carbon disulphide. Lipids constitute major part of stored food material in animal body and a major structural component. The term **'lipid'** was used by German biochemist, **Wilhem Bloor**, in the year 1943 to a group of fats and fat-like substances.

Lipids have Some Special Features

- Lipids are compounds of carbon, hydrogen and oxygen. In them the number of oxygen atoms is much less in comparison to carbon and hydrogen.
- They are esters of fatty acids and polyhydric alcohols.
- Their molecules are polar and hydrophobic. They are insoluble in water but soluble in nonpolar carbonic solvents, like ether, benzene, acetone, petroleum, etc.

True Fats are Triesters of Glycerol and Fatty Acids

True fats, like carbohydrates, are compounds of carbon, hydrogen and oxygen, but the number of oxygen atoms in fat molecules is less than in carbohydrate molecules. True fats are esters of fatty acids and trihydric alcohols and are also known as **triglycerides**.

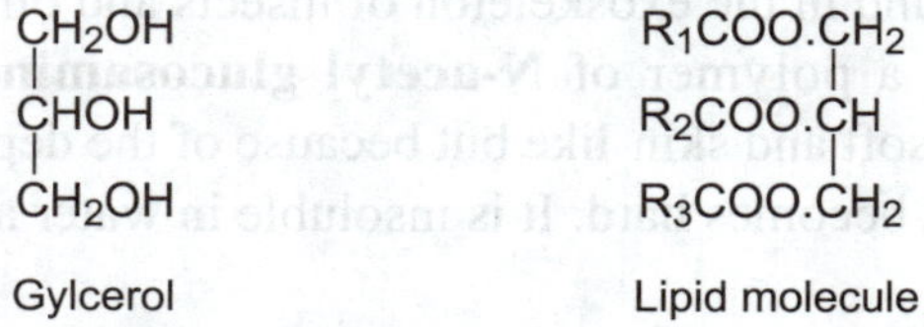

Fig. 3.12 Structure of the molecules of glycerol and lipid

It means in one lipid molecule three molecules of fatty acids are joined to one glycerol molecule. The carboxyl (–COOH) groups of fatty acids form **ester bond** with alcoholic (–OH) group of glycerol molecule. The characteristic of lipids depends on the nature of fatty acids.

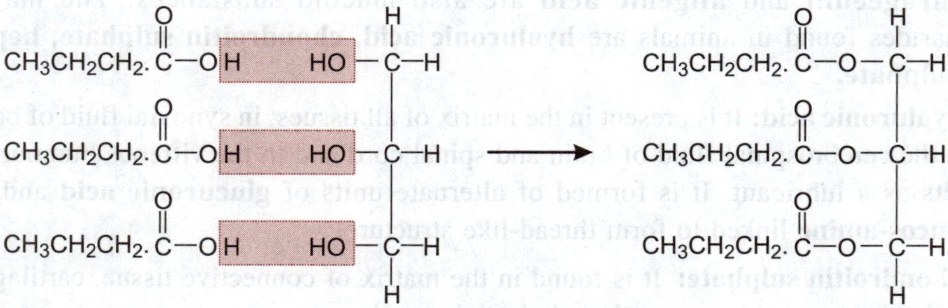

Fig. 3.13 Synthesis of a lipid molecule from one molecule of glycerol and three molecules of fatty acid

Most of the triglycerides which occur in nature are mixed triglycerides, i.e. they contain two or three different fatty acids in one molecule. Since all true fats contain glycerol, their chemical and physical properties are determined by the nature of fatty acid components. Fatty acids may be saturated or unsaturated. Butyric acid, capric acid, caproic acid, stearic acid, cerotic acid and palmitic acid are saturated fatty acids of the lipids. The oleic acid, linoleic acid, linolenic acid and palmitoleic acid, etc., are the unsaturated fatty acids of fats.

Lipids on Hydrolysis or Saponification Produce Glycerol and Three Molecules of Fatty Acids

Lipids can be hydrolysed by heating with alkalies. This process is called **saponification**. On hydrolysis triglycerides split up into glycerol and fatty acids. The hydrolysis occurs in the presence of enzyme *lipase*. Triglycerides serve to store fuel in the form of fat droplets in cells.

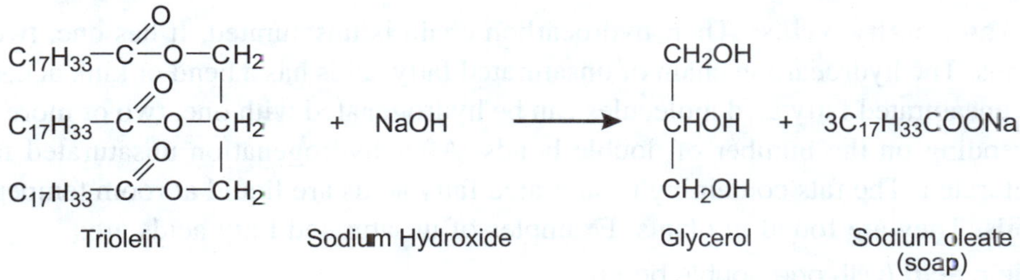

Fig. 3.14 Saponification of a neutral lipid

Fatty Acids

Fatty acids are structural units of lipids. These are obtained by the hydrolysis of fats. They are carboxylic acids of long chain hydrocarbons. They are represented by simple formula $CH_3(CH_2)_nCOOH$. Their carbon chain is long and unbranched. The number of carbon atoms in their chain may vary from 4 – 30 but normally ranges from 16 to 18. The carbon atoms in their chain are always in even number.

One end of each fatty acid has a **carboxyl group** (–COOH) and other end has a **methyl group** ($-CH_3$). The carbon atoms of fatty acids are numbered from the carboxyl carbon. This is numbered 1. The next or number 2 carbon is also called **α-carbon**, number 3 carbon as **β-carbon** and at the end of the chain the methyl carbon is known as **γ-carbon**.

Due to the presence of negative charge on −COOH group, fatty acids are **polar**. Therefore, every fatty acid molecule has **hydrophilic head** or **polar end**. The remaining part of fatty acid is in the form of a long tail and is formed of hydrocarbons. The tails are hydrophobic. Because of hydrophobic tail part fatty acids are insoluble in water and form either surface layer on water or double-layered circles.

Fatty Acids may be Saturated or Unsaturated

The true fatty acids are straight chain compounds and are of two types: **Saturated** and **unsaturated** (Fig. 3.15 A and B).

1. Saturated Fatty Acids: In saturated fatty acids, the hydrocarbon chain is without double bonds. Therefore, their chains are straight. All the carbon atoms in them are fully saturated. Further addition of hydrogen is not possible. They have a high melting point. Therefore, fats containing saturated fatty acids are solid or semisolid at room temperature. Animal proteins such as ghee and fat are formed of saturated fatty acids. They are responsible for increasing cholesterol level in our blood. Some important saturated fatty acids are:

(1) **Palmitic acid** (C = 16) : $CH_3(CH_2)_{14}COOH$

(2) **Stearic acid** (C = 18) : $CH_3(CH_2)_{16}COOH$

(3) **Arachidic acid** (C = 20) : $CH_3(CH_2)_{18}COOH$

2. Unsaturated Fatty Acids: Their hydrocarbon chain is unsaturated. It has one, two or more double bonds. The hydrocarbon chain of unsaturated fatty acids has a bend or kink at each double bond. The unsaturated fatty acid molecules can be hydrogenated with one, two or more hydrogen atoms depending on the number of double bonds. After hydrogenation unsaturated fatty acids become saturated. The fats containing unsaturated fatty acids are liquid at room temperature and occur as oils. They are found in plants. Examples of unsaturated fatty acids are:

(1) **Oleic acid** (with one double bond):

$$CH_3(CH_2)_7CH{=}CH(CH_2)_7COOH.$$

(2) **Linoleic acid** (with two double bonds):

$$CH_3(CH_2)_4CH{=}CH{-}CH_2{-}CH{=}CH(CH_2)_4COOH$$

(3) **Linolenic acid** (with three double bonds):

$$CH_3{-}CH_2{-}CH{=}CH{-}CH_2{-}CH{=}CH{-}CH_2{-}CH{=}CH(CH_2)_7COOH$$

Biological Significance of Lipids

1. **Rich Source of Energy:** Fats provide food of high calorific value (1g fat produces about 9.3 kilo calories of heat).
2. **Food Reserve:** Fats are stored in the body as reserve food material, because these could be readily stored in the body on account of insoluble characteristics. Triglycerides stored in adipocytes (fat cells) of adipose tissue are the principle fat reserve.
3. **Heat Insulators:** Fats deposited in the subcutaneous tissue act as insulators conserving body heat.
4. **Solvents:** Lipids act as solvent for fat soluble vitamins like vitamin A, D and E.
5. **Structural Constituents:** Phospholipids, glycolipids and sterols are structural components of all the membrane systems of the cell (*i.e.*, cell membrane, nuclear membrane, membranes of the endoplasmic reticulum, etc.).
6. **Fat Transport:** Phospholipids play an important role in the absorption and transportation of fatty acids.
7. **Hormone Synthesis:** Adrenocorticoids, sex hormones, vitamin D and cholic acids are synthesised from cholesterol.
8. **Shock Absorbers:** The fat deposited around the visceral organs and underneath the skin acts as cushion and absorbs mechanical shocks.

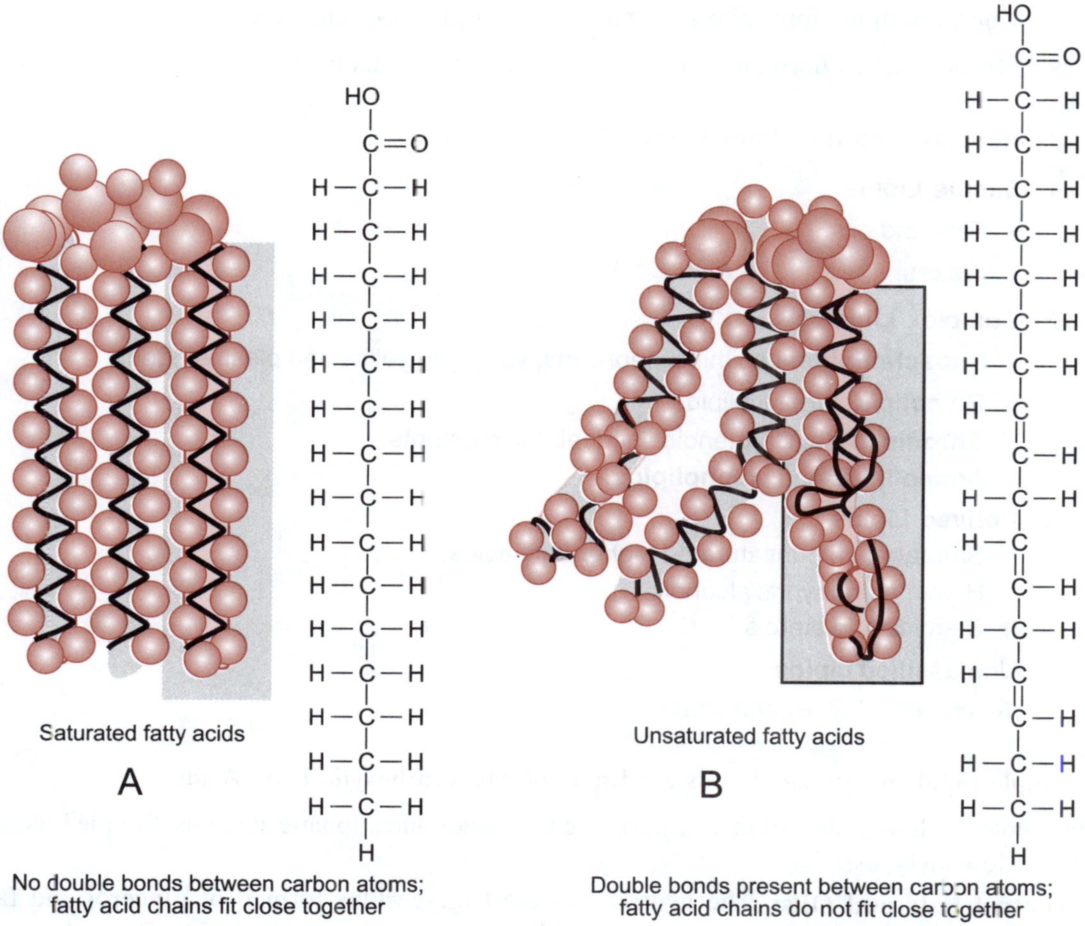

Fig. 3.15 Structure of **A.** saturated and **B.** unsaturated fatty acids

9. **Electric Insulators:** Myelin sheath around medullated nerve fibres forms electric insulation.
10. **Prostaglandins:** They control local activities in the body.
11. **Protective Layer:** Lipids form a protective waxy covering on the aerial parts of plants to check loss of water by evaporation.
12. **Thrombokinase:** Helps blood clotting.
13. **Leukotrienes,** a group of eicosanoids helps in respiration.
14. Some isoprenoids form insect hormones.
15. Some isoprenoids form volatile oils and pigments. Natural rubber is also an isoprenoid.
16. **Glycolipids** help in cell recognition.

17. **Complex lipids** form-phospholipid bilayer of plasma membrane.

18. **Steroids** act as hormones and neurotransmitters in mammals.

Lipids are Classified into Four Groups

1. **Simple Lipids**
   1. Fats and oils.
   2. Waxes.
2. **Complex Lipids**
   1. **Phospholipids:** Lecithins, cephalins, sphingomyelins and plasmalogens.
   2. **Glycolipids** or glucolipids.
   3. **Chromolipids:** Carotenoids and related pigments.
   4. **Aminolipids** and **sulpholipids**.
3. **Derived Lipids**
   1. Saturated and unsaturated higher fatty acids.
   2. Higher monohydric alcohols.
   3. Steroids and sterols.
4. **Unclassified Lipids**

   Tocopherols (Vit. E) and Vitamin K.

1. Simple Lipids or Storage Lipids are Esters of Monocarboxylic Fatty Acids

The simple lipids are esters of monocarboxylic fatty acids and aliphatic alcohols. Simple lipids are of the following types:

1. Neutral Fats and Oils: The neutral fats are triglycerides formed of glycerol and fatty acids. One molecule of glycerol forms ester with three molecules of fatty acids. Depending upon the number of fatty acids in the ester, the glycerides may be **monoglycerides, diglycerides** or **triglycerides**.

Triglycerides are neutral fats. These are lighter than water with a specific gravity of about 0.86. They occur both as solids and liquids.

- **Fats or solid lipids:** Fats are solid at room temperature. Their fatty acids are saturated fatty acids.
- **Oils or fluid lipids:** The lipids which are fluid at room temperature are called **oils**. They have large number of unsaturated fatty acids. The unsaturated fatty acids have double bonds and low melting point.

2. Waxes: Waxes are simple lipids containing one molecule of fatty acid and one molecule of a high molecular weight monohydroxy alcohol. For example, beewax is composed mainly of palmitic acid esterified with either hexacosanol or triacontanol. Waxes are chemically inert and resistant to atmospheric oxidation. These form protective covering at numerous locations in animals, plants and microorganisms.

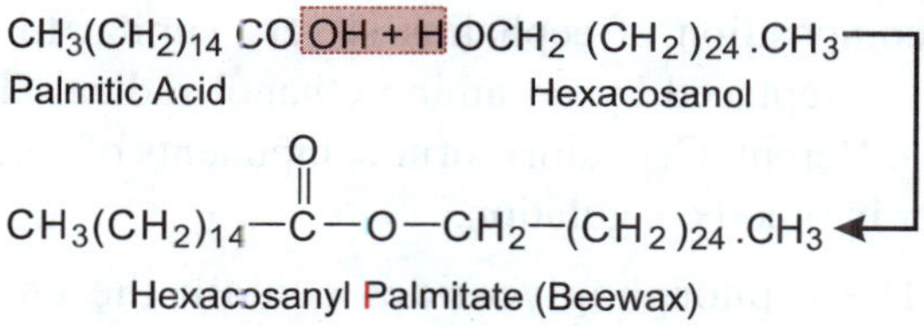

Fig. 3.16 Structural formula of wax

Waxes are insoluble in water and resistant to atmospheric oxidation. These are chemically inert. Therefore, they find several uses. These are used for polishing furniture and automobiles, for the preparation of wax-coated paper which is used as wrapping paper.

2. Complex Lipids or Conjugated Lipids are Esters of Fatty Acids and Alcohols but have Some Additional Group

Complex lipids or conjugated lipids are esters of fatty acids and alcohol but also have a phosphate, carbohydrate or protein group. They are of the following types:

1. Phospholipids or Phosphatides or Phosphoglycerides: Phospholipids are glycerides of fatty acids with phosphorus. Their molecules are polar having hydrophilic and hydrophobic ends. From biological point of view, they are most important lipids. They form plasma membrane and other endomembranes of the cell and are essential for their proper functioning. They regulate cell permeability, control transport and metabolism of synthesised and dietary fat and participate in blood coagulation. These are:

- **Lecithins:** The lecithins are monoamino-mono-phospholipids. They usually occur along with cholesterol in nervous tissue and in egg yolk. The fatty acids most commonly found in lecithins are palmitic acid, stearic acid, oleic acid, linoleic acid, linolenic acid and arachidonic acid. But only two out of so many acids are present at a time in one molecule of lecithin. On account of the presence of unsaturated fatty acids, the lecithins are very unstable and readily undergo oxidation or hydrogenation. They form colloidal dispersion in water.

  Lecithins play an important role in permeability, osmotic tension and surface condition of cells. They also participate in the process of desaturation of fats and act as intermediaries in fat metabolism. **Choline,** a component of lecithins, is closely related with acetylcholine.

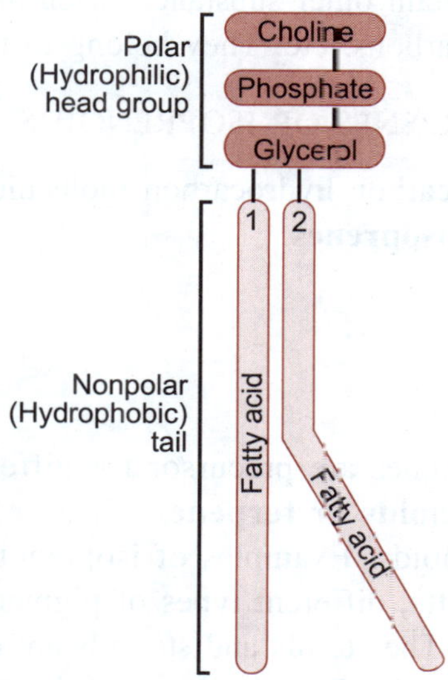

Fig. 3.17 Structure of a phospholipid molecule

- **Cephalins:** The composition of cephalins is very similar to lecithins, but the choline radical of lecithin is replaced by an amino-ethanol radical. The fatty acids present in cephalins are also different. Cephalins form components of certain lipoproteins and form an important factor in blood coagulation.
- **Plasmalogens:** These phosphoglycerides contain the enol form of a long chain aldehyde connected by ether-linkage and replacing one of the fatty acids found in lecithins and cephalins.
- **Phosphosphingoside or sphingomyelin:** Sphingomyelin is devoid of glycerol and on hydrolysis it produces two molecules of base and one molecule of fatty acid. It is a common constituent of brain, nerves and myelin sheath. Therefore, it is absent in plants and microorganisms.

2. Glycolipids: These contain carbohydrate radicals and nitrogen along with fatty acids. These are important constituents of white matter of nervous tissue, but are also found in spleen and egg yolk. **Phrenosin, kerasin, nervone** and **oxynervone** are glycolipids.

3. Chromolipids: This group of fats includes carotenoids and related pigments. The **kerotene** ($C_{40}H_{56}$) is a provitamin A found in deposited fats of animals.

3. Derived Lipids are Fat-like Substances

The derived lipids are fat-like substances produced by the hydrolysis of simple and complex lipids and certain other substances such as sterids, fatty aldehydes, ketones, alcohols, essential oils, hydrocarbons, etc. They belong to two categories: **Terpenes** and **eicosanoids**.

1. TERPENES OR ISOPRENOIDS

The 5-carbon hydrocarbon molecules formed by the condensation of acetyl coenzyme-A are called **isoprenes**.

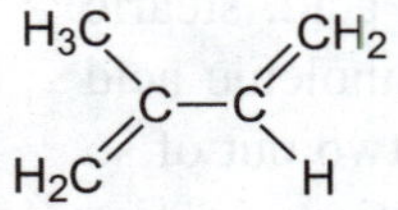

Isoprenes are precursor for different lipids. The lipids derived from isoprenes are called **isoprenoids** or **terpenes**. Two or more isoprene molecules take part in the synthesis of isoprenoids. Examples of isoprenoids are **sterols, steroid hormones, fat soluble vitamins, bile salts,** different types of pigments, like **β-carotene** and the sweet smelling substances in plants. The sterols and steroids are collectively called **sterids**. These are widely distributed in plants, animals and microorganisms. These wax-like lipids are found in free state as well as forming fatty acid esters. In contrast to lipids the sterids contain a ring structure.

1. Sterols: The example of sterols are **ergosterol** from yeast, **stigmasterol, spinasterol** and **sitosterol** from plants and **cholesterol** from animals. These are wax-like solid alcohols, found in most living cells.

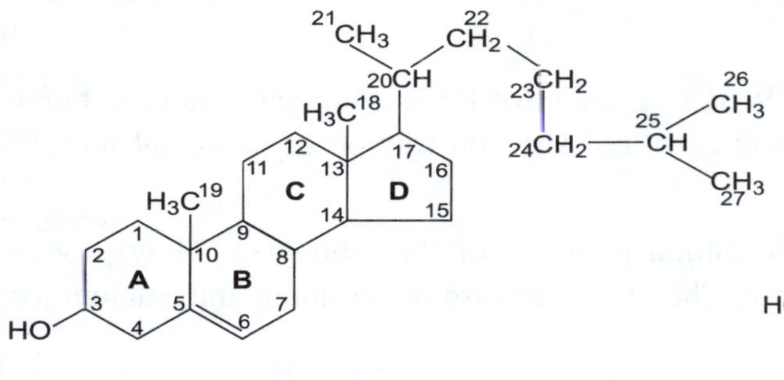

Fig. 3.18 Cholesterol

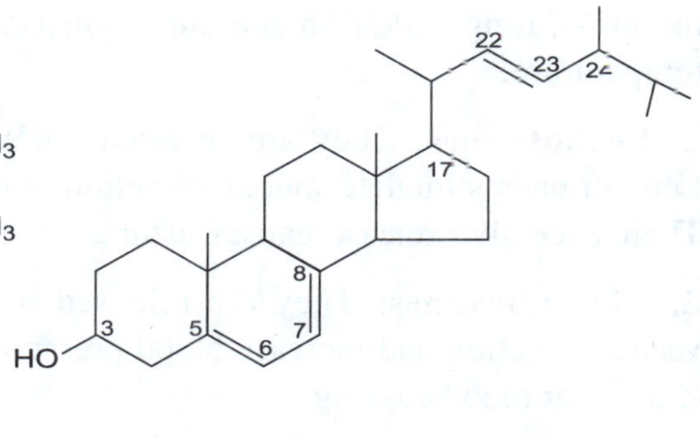

Fig. 3.19 Ergosterol

1. **Cholesterol (The animal sterol):** It is colourless and odourless sterol, found in all the cells. It occurs only in animal fats but not in plant fats. Its formula is $C_{27}H_{45}OH$. Cholesterol has three six-membered rings (A, B, C) and one five membered ring (D).The side chain attached to the fourth ring is formed of 8 carbons.

Biological Functions of Cholesterol

Cholesterol is important in living beings due to the following reasons:

- Cholesterol is an important constituent of all biological membranes.
- All steroid hormones are derived from cholesterol.
- Bile salts are also derived from cholesterol.

2. **Ergosterol (Plant sterol):** It is a **phytosterol**. It occurs in ergot and yeast. It is the precursor of vitamin D. In presence of UV rays ergosterol is converted to vitamin D. Other major phytosterols are **stigmasterol**, **β-sitosterol** and **compesterol**. They are abundant in corn oil, rice bran, wheat germ, soyabean and other vegetables. They bring down the level of LDL cholesterol in the blood.

2. Steroids: Steroids are hormones secreted by the adrenal cortex and by male and female gonads. The hormones secreted by adrenal cortex are **corticosterone, deoxycorticosterone, adrenosterone** and other hormones. Male sex hormones are **testosterone, androsterone** and **dehydro-androsterone**. The female steroid hormones are **esterogenic hormones** and the **progesterone**. Steroids also include vitamin A and D and bile salts, **taurocholate** and **glycocholate**.

2. EICOSANOIDS

Eicosanoids are localised, short-lived paracrine hormones which influence the nearby cells and are not circulated by blood to distant body parts. They are produced virtually in all the mammalian tissues and organs. They are derived from 20-carbon polyunsaturated arachidonic acid and are of the following three types:

1. Prostaglandins: They are derivatives of **arachidonic** fatty acid. They are present in semen of man, menstrual fluid of woman and in the tissue fluid of many tissues. They carry out a number of physiologic and pharmacologic activities. They are responsible for contraction of uterine muscles during childbirth and during menstruation. They generate local pain and increase body temperature.

2. Leukotrienes: They are secreted by WBCs. They have three conjugated double bonds. Leukotrienes stimulate mucus secretion, and carry out vasoconstriction of bronchial muscles. Their excessive release causes asthma.

3. Thromboxanes: They are released by blood platelets or thrombocytes. They produce vasoconstriction and increase blood pressure. They cause release of serotonin and calcium ions (Ca^{++}) for blood clotting.

PROTEINS

Proteins are Macromolecules

Proteins are the most complex organic compounds formed of carbon, hydrogen, oxygen and nitrogen. Some also contain sulphur. They are polymers of amino acids and have high molecular weight. Proteins are the most important, most varied and most abundant organic compounds in the body of all living organisms. Hundreds of different proteins can be found in any single cell. They form 50% or more of the organic compounds in the living beings. Dutch chemist, **Gerardus Mulder** (1838), had used the term protein (Gr, *proteios*, meaning 'primary' or first rank) for these compounds, because they perform most varied functions. The large group of proteins is called **proteome** and the broad study of all cellular proteins as **proteomics**.

Biological Significance of Proteins

Proteins are the most significant compounds in living beings because of the following physiological roles performed by them:

1. Proteins as Enzymes: Proteins act as enzymes or biocatalysts and regulate almost all the life processes. Nearly 2,000 different kinds of enzymes are known, each one catalysing a different chemical reaction. These bring about digestion of food, hydrolysis of ATP, biosynthesis of cell components (DNA, RNA), etc.

2. Proteins as Hormones: A few proteins function as hormones, *e.g.* **somatotropic** or **growth hormone** of anterior pituitary and **insulin**, a hormone regulating glucose metabolism, secreted by pancreas.

3. Proteins associated with Hereditary Characters: Nucleoproteins form chromatin material. During cell division, chromatin condenses to form chromosomes. These are carriers of hereditary characters.

4. Proteins and Interferons: Interferons are low molecular weight, regulatory glycoproteins produced in response to viral infection, endotoxins, antigenic stimuli, rickettesiae and protozoan parasites. Interferons are important therapeutic agents for treating viral diseases such as hepatitis, encephalitis, cancer and common cold.

5. Proteins and Protection: Protective proteins are **antibodies** or **immunoglobulins**. These combine and neutralise foreign proteins (antigens). The blood proteins, **thrombin** and **fibrinogen,** are responsible for blood clotting and thus prevent loss of blood at the time of injury.

6. Proteins and Biochemical Individuality: Proteins confer specific individuality to the living organisms and species. The protein individuality is well illustrated by **antigen-antibody reactions**. The blood of one organism or of individuals of one species can not be transfused into the circulation of the other animals. Similarly, cells and organs from one animal can not be transplanted successfully into the body of other animal. Individual specific proteins, called antigens are present on the surface of cells in all living beings which help in cell recognition.

7. Transport: Some proteins bind and transport specific types of molecules via blood. For example:

- **Serum albumin** binds free fatty acid molecules and transport them between adipose tissue and other cells of the body.
- **Lipoproteins** of blood plasma transport lipids between the intestine, liver and adipose.
- **Haemoglobin** in vertebrates and **haemocyanin** in some invertebrates transport oxygen from lungs to tissue cells (oxygen carrier).
- **Myoglobin** stores oxygen in the muscles.
- **Permeases** are transport proteins. They transport molecules through cell membrane.
- Proteins help to transport organic food through phloem in plants.

8. Toxins: Some proteins such as **snake venom, gossypin** (the toxic protein of cottonseed), **ricin**, (the protein of castor bean) and proteins of pathogenic bacteria (diptheria toxin), etc. are highly poisonous to vertebrates.

9. Contraction: The contractibility of all living tissues and the contractile system of skeletal muscles is due to **actin** and **myosin** proteins.

10. Movement: Locomotory organs of cells, like cilia and flagella are formed of microfilaments, composed of protein, **tubulin**.

11. Receptor Proteins: Some proteins are present in the cell membranes. They receive sensory stimuli from the surroundings.

12. Structural Proteins:

- Proteins form an important part of all membranes and membrane bound organelles of the cell (*i.e.*, plasma membrane, ER, nuclear membrane, lysosome, mitochondria, etc.).
- **Collagen fibres** and **elastin fibres** found in connective tissue and bone are formed of fibrous proteins.
- Proteins **chondrin**, and **ossein** form matrix of cartilage and bone respectively.
- Scales, feathers, hair, fur, wool, hoof, nail, horn, etc., are formed of keratin sulphate, a protein derivative.

13. Cell secretions like mucus are **glycoproteins**. These provide slippery texture and help in protection.

14. Special Secretions: Spiders and silkworms secrete a thick solution of protein **fibroin** which solidifies into thread of exceptional tensile strength and is used for weaving web or cocoon.

15. Storage Proteins: Proteins are stored as reserve food such as egg albumin or in various seeds (**glutellin** in rice, **zeins** in maize and **phaseolin** in pea).

16. Exotic Proteins: These are proteins secreted by different animals or plants for specific purpose. For example:

- Some marine or aquatic animals secrete sticky proteineous substances for adhesion to the rock or substratum.
- Fishes living in freezing cold water of Arctic and Antarctic oceans synthesise **anti-freeze protein** to protect their blood from freezing.
- Some African plants synthesise protein **monellin** which is very sweet and is used as artificial sweetner for diabetic patients.

17. Regulatory Proteins: Proteins regulate cellular and physiological activities of the cell. In addition to enzymes and hormones, proteins control DNA replication, protein synthesis, chain initiation and chain termination, etc. **G-proteins** transmit hormonal signals inside the cells to nucleus.

Proteins are Polymers of Amino Acids

Proteins are linear polymers of amino acids. They are formed of chains of 20 different types of amino acids. Although over 170 amino acids are currently known from cells and tissues, only 26 are found in proteins and only 20 of them are more common. Therefore, amino acids are described as monomers or building blocks of proteins.

AMINO ACIDS

Amino Acids Contain a Carboxyl and an Amino Group

Amino acids are monomers or building blocks of proteins. They are organic acids having tetrahedral structure. Each amino acid has a central carbon atom called **α-carbon**. Following groups are covalently attached to this α-carbon:

(a) a **hydrogen atom** (H)

(b) a positively charged, basic **amino group** ($-NH_2$)

(c) a negatively charged acidic **carboxyl group** ($-COOH$)

(d) a side chain of carbon compound, called **R group**.

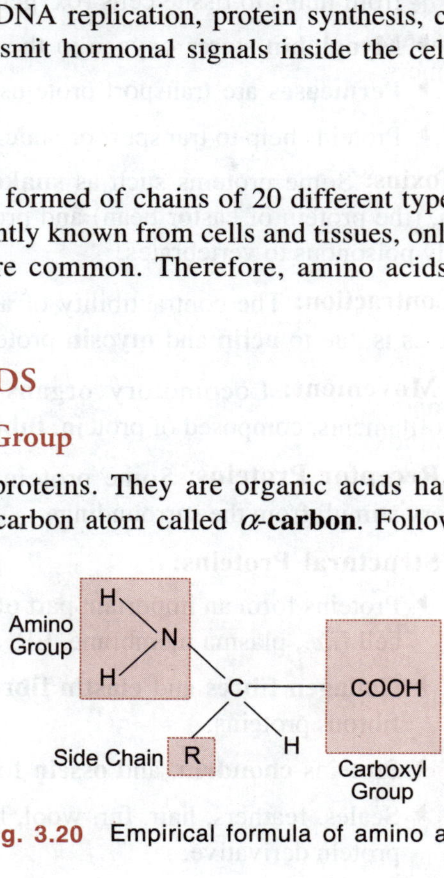

Fig. 3.20 Empirical formula of amino acid

Amino acids differ only in the nature of their R group, which may be hydrogen, aliphatic, aromatic or heterocyclic hydrocarbon group. All amino acids can be represented by the empirical formula given in Fig. 3.20.

The simplest amino acid is **glycine** (NH_2CH_2COOH) in which R group is a hydrogen atom. Amino acids, in which $-COOH$ and $-NH_2$ groups are attached to the first carbon or α-carbon atom, are known as **α-amino acids**. All proteins are formed of α-amino acids only. In nature, β-, γ- or δ- type of amino acids are also found. They are named so based on the position of NH_2 group. The first carbon atom of amino acid with $-COOH$ group is **α-carbon**, second carbon atom is **β-carbon**, third is **γ-carbon** and fourth carbon is **δ-carbon.**

Amino Acids have following Characteristics

1. Stereoisomerism

Except glycine, all amino acids have at least one **asymmetric carbon atom**. Therefore, they occur in two isomeric forms: D-form and L-form. These are mirror images. In **D-form,** $-NH_2$ group is present on the right side and in **L-form,** $-NH_2$ group is present on the left side. These rotate the plane of polarised light. D-amino acids rotate the plane of polarised light towards right side (dextrorotatory) and L–amino acids rotate the plane towards left (levorotatory) and are represented as + and – respectively. The amino acids of animal and plant proteins are all L-isomers, whereas D-amino acids occur only in bacterial cell wall.

D-Amino acid L-Amino acid

Fig. 3.21 Two stereoisomeric forms of an amino acid

2. Amphoteric and Ampholytic Nature

A compound is called **amphoteric** if it can react with both acids and bases. Both amino ($-NH_2$) and carboxyl ($-COOH$) groups of amino acids are ionisable and readily form salts either with acids or alkalies. Depending upon H^+ ion concentration of the medium, amino acids behave either negatively or positively charged.

1. In acidic medium, amino acids carry positive charge and act as base, yielding cations. In electric field, they migrate to cathode or positive pole.
2. In alkaline medium, amino acids carry negative charge and act as acid, yielding anions. In electric field, they migrate to anode or negative pole.

Thus, amino acids may be **amphoteric** or **ampholytic**.

3. Isoelectric pH or Isoelectric Point

The isoelectric point of an amino acid is that pH at which it has no net charge and it does not move to any pole in an electric field. Different amino acids have their own isoelectric pH. This constant forms the basis of isolation and identification of amino acids. For example, isoelectric pH of aspartic acid is 2.9, of alanine 6, histidine 7.59, lysine 9.74 and arginine 10.8.

Isoelectric point can be denoted as follows:

$$p' = \frac{pK_1 + pK_2}{2} \qquad [\because pK_1 = RCOO^-, pK_2 = RNH_3]$$

4. Zwitter Ion

At its isoelectric point an amino acid exists as a **dipolar ion,** having both + and – charges in equal number, but it has no net charge because these charges are internally neutralised. Such a dipolar ion is called **Zwitter ion**. Like a Zwitter ion an amino acid has following characteristics:

- At its isoelectric pH, each amino acid has an equal number of + and – charges.
- In acidic solution an amino acid acts as a base and combines with H^+ ions of the solution.
- In alkaline solution amino acid acts as a base and combines with OH^- ions of the solution.
- Zwitter ion is ampholyte because it acts both as proton donor and proton acceptor.
- In a pure aqueous solution Zwitter ions are not neutral because their acidic and basic strengths are different.

Extension: Essential and Non-essential Amino Acids

Several amino acids are required for the biosynthesis of various tissue proteins, enzymes, hormones and plasma proteins. Some of them are required in the preformed state from diet because they can not be synthesised in the body. These are known as **essential amino acids.** These include **lysine, threonine, tryptophan, leucine, isoleucine, methionine, phenylalanine**. There are seven essential amino acids. These are taken from plants by herbivores and carnivores obtain them indirectly from herbivores. Other amino acids can ordinarily be synthesised in the body and hence their presence in the diet is not essential. Such amino acids are known as **non-essential amino acids.** Plants and microbes can synthesise all amino acids.

Functions of Amino Acids

1. **Formation of Proteins:** Amino acids are linked into long chains through peptide linkage forming polypeptides and proteins. Proteins form structural as well as functional components of the cell.
2. **Formation of Other Compounds:** Some amino acids give rise to other compounds, such as:
   - **Tyrosine** is part of hormone thyroxine and adrenaline.
   - **Glycine** participates in the synthesis of heme.
   - **Tryptophan** produces vitamin nicotinamide and plant hormone indole-3-acetic acid (IAA).

- Amino acid β-alanine is needed in the synthesis of coenzyme-A and vitamin pantothenic acid.

3. **Formation of Antibiotics:** Nonprotein amino acids are components of antibiotics.
4. **Formation of Enkephalins:** These are small peptides (pentapeptides) formed in certain parts of brain. These are associated with the perception of pain and pleasure.
5. **Biological Buffers:** Because of amphoteric nature amino acids act as buffers in the body fluids. They maintain pH either by donating H^+ ions when pH increases or by accepting H^+ ions when pH falls.
6. **Synthesis of Glucose:** When glucose level in blood comes down, glucose is synthesised from amino acids in the hepatic cells. This is called **gluconeogenesis**.
7. **Formation of Histamine:** Amines in the body are derived from some amino acids by the loss of carboxyl group. For example, histamine is derived from amino acid histidine. Histamine is an important vasodilator. It also causes constriction of bronchiole smooth muscles and stimulates gastric secretion.
8. **Prokaryotic Cell Wall:** Some peptides are components of **peptidoglycans** which form bacterial cell wall.

Classification of Amino Acids

The chemical classification of α-amino acids is based upon the composition of the side-chain or R-group, which varies from one amino acid to another. These are classified on the following basis:

I. On the Basis of Reaction in the Solution Amino Acids are of Three TYPES

1. **Acidic Amino Acids (Monoamino-dicarboxylic amino acids):** These have one amino but two carboxyl groups per molecule. The additional carboxyl group is present in the side chain.

   Examples: **Aspartic acid** (ASP) and **Glutamic acid** (Glu).

2. **Basic Amino Acids (Diamino-monocarboxylic amino acids):** These amino acids have two amino groups but only one carboxylic group per molecule. The additional amino group is present in the side chain (R group).

   Examples: **Arginine** (Arg) and **Lysine** (Lys).

3. **Neutral Amino Acids (Monoamino-monocarboxylic amino acids):** These amino acids have only one amino and one carboxyl group per molecule. Their side chain is without any functional group.

   Examples: **Alanine** (Ala), **Glycine** (Gly), **Serine** (Ser), **Threonine** (Thr), **Valine** (Val), **Leucine** (Leu) and **Isoleucine** (Ile).

II. On the Basis of Charge on the Side Chain, Amino Acids are of Three Types

1. **Nonpolar Amino Acids:** These do not have any charge. These are insoluble in water.

   Examples: **Glycine** (Gly), **Serine** (Ser), **Threonine** (Thr), **Cysteine** (Cys), **Tyrosine** (Tyr), **Asparagine** (Asn) and **Glutamine** (Gln).

2. **Polar Uncharged Amino Acids:** These are polar amino acids in which +ve and –ve charges are balanced and amino acids as a whole are uncharged.

   Examples: **Valine** (Val), **Alanine** (Ala), **Leucine** (Leu), **Isoleucine** (Ile), **Phenylalanine** (Phe), **Methionine** (Met) and **Proline** (Pro).

3. **Polar Charged Amino Acids:** These are polar amino acids which have either + ve or –ve charge. These behave either as acidic or basic. These are soluble in water.

   Examples:

   (a) Negatively charged polar amino acids (acidic) are **Aspartic acid** (Asp) and **Glutamic acid** (Glu), and

   (b) Positively charged amino acids are (basic) **Lysine** (Lys), **Arginine** (Arg) and **Histidine** (His).

III. Based on the Characteristics of R Group Amino Acids are of Six Types

1. **Aromatic Amino Acids:** The side chain of these amino acids contains an aromatic ring, as in **Tyrosine** (Tyr), **Phenylalanine** (Phe) and **Tryptophan** (Trp).
2. **Aliphatic Amino Acids:** The side chain of these amino acids is without aromatic ring and is formed of hydrocarbons. Majority of amino acids have aliphatic side chain.
3. **Heterocyclic Amino Acids:** These have nitrogen in the ring of side chain. **Histidine** (His) and **Proline** (Pro) are heterocyclic amino acids.
4. **Sulphur containing Amino Acids:** The side chain of these amino acids has sulphur as in **Cysteine** (Cys) and **Methionine** (Met).
5. **Alcoholic Amino Acids:** The side chain of these amino acids has an alcoholic or hydroxyl group.

   Examples: **Serine** (Ser), **Threonine** (Thr).

6. **Amino Acid Amides:** These are derivatives of acidic amino acids in which one carboxyl group has changed to amide.

   Examples: **Asparagine** (Asn).

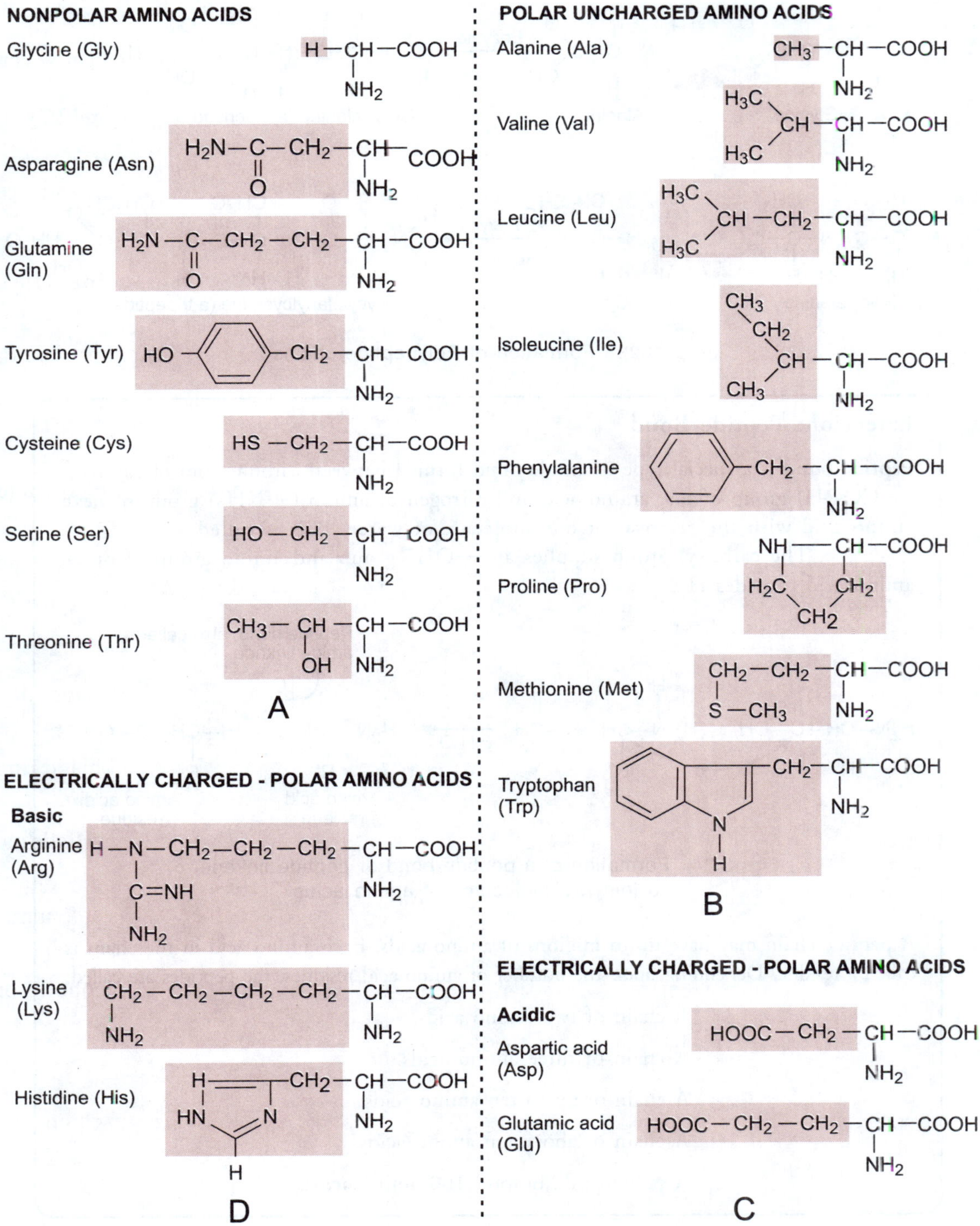

Fig. 3.22 Twenty amino acids, their structural formulae and characteristics of R group

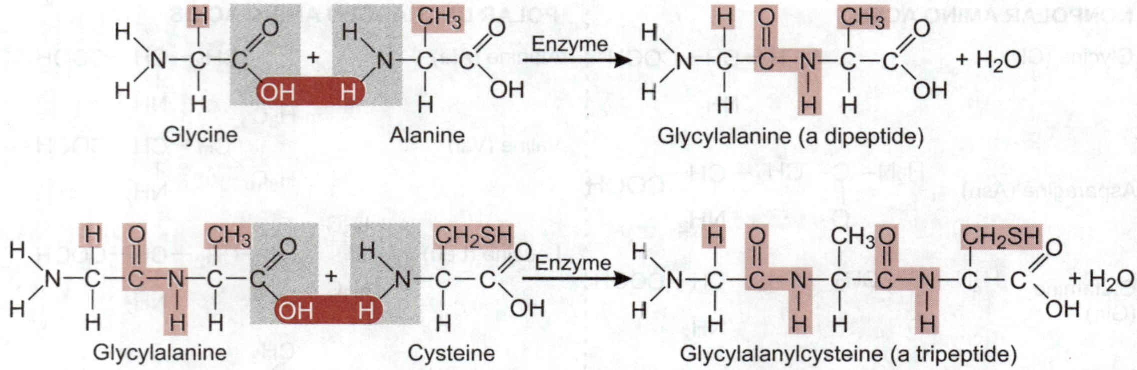

Fig. 3.23 Formation of Polypeptide chains

Extension: Peptide Bond

Peptide bond is a special type of amide bond formed between carbon atom of carboxyl (—COOH) group of one amino acid and nitrogen of amino ($—NH_2$) group of next amino acid with the removal of one molecule of water. This is called
. The carboxyl group supplies an $—OH^-$ group and amino group of other amino acid provides H^+.

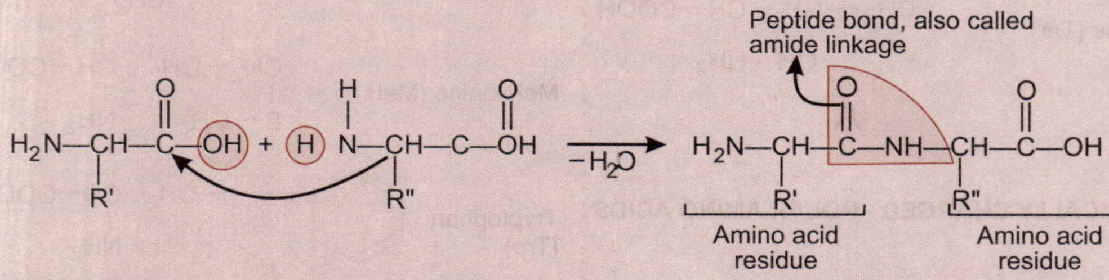

Fig. 3.24 Formation of a peptide bond or peptide linkage to join two molecules of amino acids

A peptide chain may have up to millions of amino acids. Each amino acid in the chain is called a residue. Depending upon the number of amino acid residues, the peptides are called:

A chain of two amino acids.

A chain of three amino acids.

A chain of up to ten amino acids.

A chain of above ten amino acids.

A polypeptide beyond 100 amino acids.

Structure of a Peptide Chain

The sequence of amino acids in a polypeptide chain represents its **primary structure**. The two ends of the peptide chain are marked, the **N-terminal** on the left and **C-terminal** at the right end of the polypeptide. The N-terminal has amino ($-NH_2$) group of left terminal amino acid and is basic (Fig. 3.23). C-terminal has carboxyl ($-COOH$) group of the last amino acid and is acidic. These two terminal groups (one basic and other acidic are ionisable).

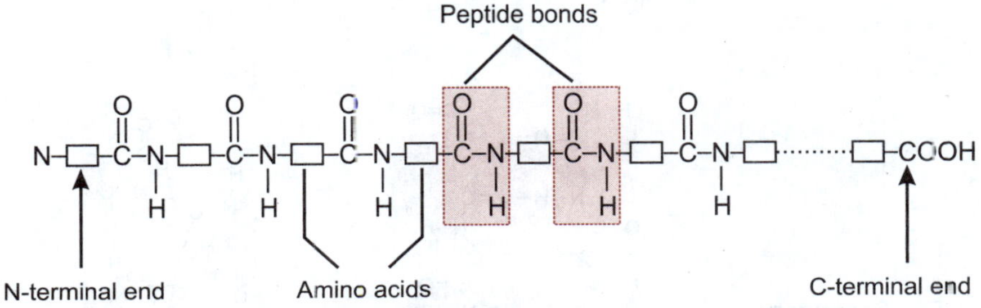

Fig. 3.25 A peptide chain showing peptide bonds between amino acid residues

Proteins Present Primary, Secondary, Tertiary and Quarternary Structure

Proteins vary tremendously in structure. They present a three-dimensional configuration. When a cell makes a protein, it is in the form of a polypeptide chain. It is non-functional and is called nascent. The nascent polypeptide chain gets coiled and folded to assume a functional three-dimensional structure. The purpose of folding is:

- To keep nonpolar amino acids inside and polar side chains on the surface.
- As a result of folding the distant amino acid chains come closer to form active site or sites of the protein and a functional protein molecule is formed.
- The substrate binding site is formed on the surface of protein.

As a result of folding and secondary bonding proteins can assume following four levels of organisation:

1. The linear sequence of amino acids in a polypeptide chain with its peptide linkage presents the **primary structure**.
2. The folding or coiling of a linear polypeptide chain into α-helix or β-pleates represents its **secondary structure**.
3. Three-dimensional configuration showing the bending and entwining of these α-helix or β-pleated sheets with each other represents the **tertiary structure**.
4. The number and nature of such entwined units in the entire protein molecule represents the **quarternary structure**.

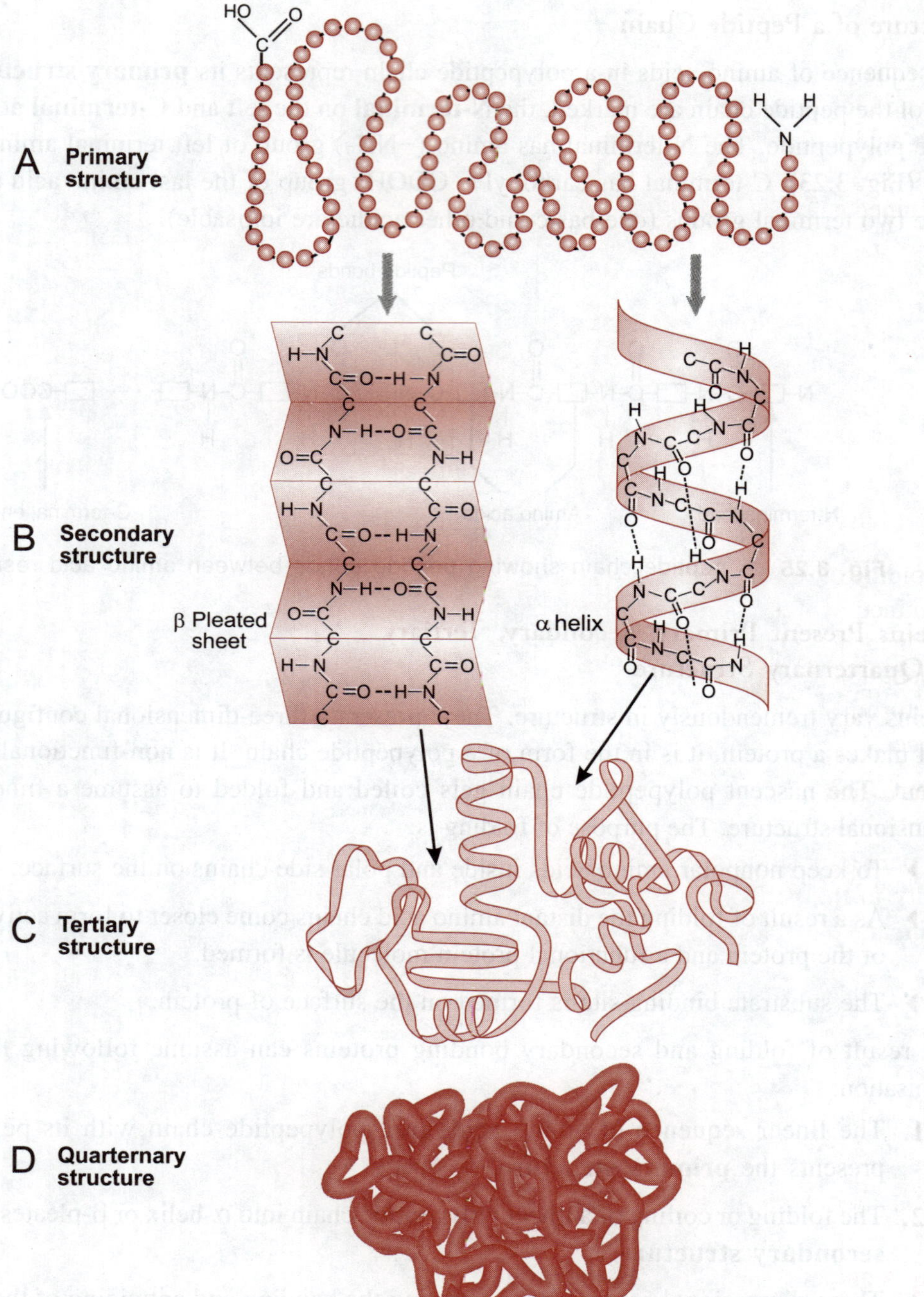

Fig. 3.26 Diagrammatic representation of the orders of magnitude of primary, secondary, tertiary and quarternary structure of protein. **A.** Primary structure; **B.** Secondary structure; **C.** Tertiary structure; **D.** Quarternary structure

The primary structure of a protein refers to the linear sequence of amino acids in its polypeptide chain and location of disulphide bridges if any. It represents the number, nature and sequence of amino acid molecules in the polypeptide chain. Each polypeptide molecule may be formed of a few to several thousands of amino acid molecules linked by peptide bonds. This sequence is determined genetically.

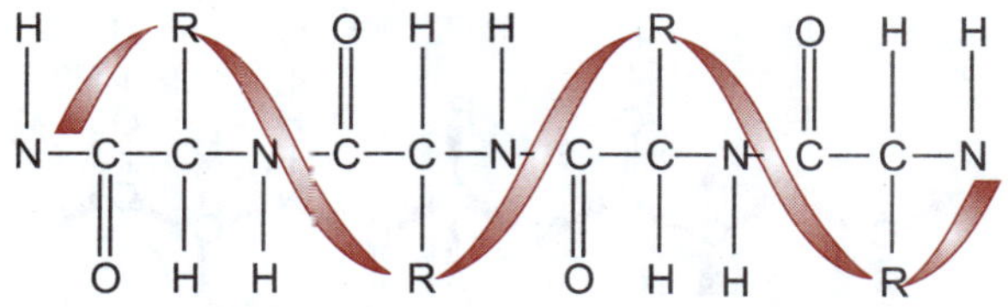

Fig. 3.27 A portion of polypeptide chain showing primary structure

The folding of linear polypeptide chain into a specific coil represents the secondary structure. It is determined by hydrogen bonding between the carboxyl oxygen and amide hydrogen atoms of the component amino acids of the peptide chain. These bonds can occur either between different polypeptide chains of protein to form a pleated sheet or within the molecules of one polypeptide chain forming α-helix. By folding amino acid residues which may lie far apart in their primary sequence are brought in close proximity. This produces a regular helical structure.

1. The α-helix: In α-helix, the polypeptide chain assumes the form of a spiral staircase with 3½ molecules of amino acids per turn. The coils of the helix are held together by hydrogen bonds lying parallel to the main axis of the fibre. Such a structure is flexible and elastic but it gives stability to the peptide chain. The α-helices are found in proteins such as myosin, the muscle protein, and α-keratin in hair and nails.

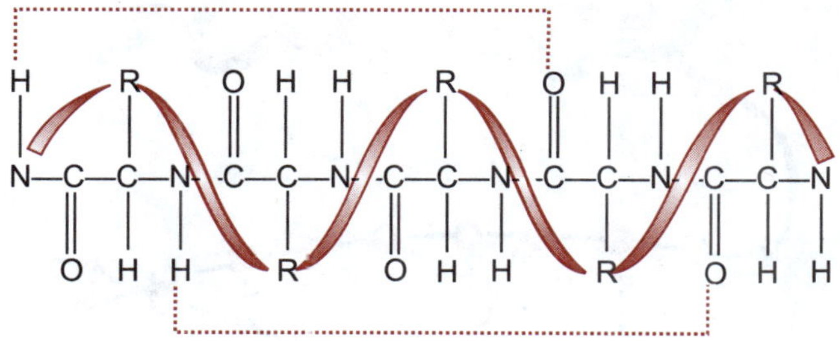

Fig. 3.28 Secondary structure of protein, wherein *α*-helices are held together by hydrogen bonds between H of —NH of one amino acid and O of —C=O of other amino acid

2. The β-pleated Sheet: In pleated sheet structure hydrogen bonding occurs between two or more peptide chains and the bonds are laid down at right angle to the main chains. Fibres containing this type of structure are not elastic because the polypeptide chains are already in extended form. The bonds are very strong and flexible. Silk fibres are best examples of β-configuration. β-keratin found in feathers and claws is another example of this type of protein (Fig. 3.29).

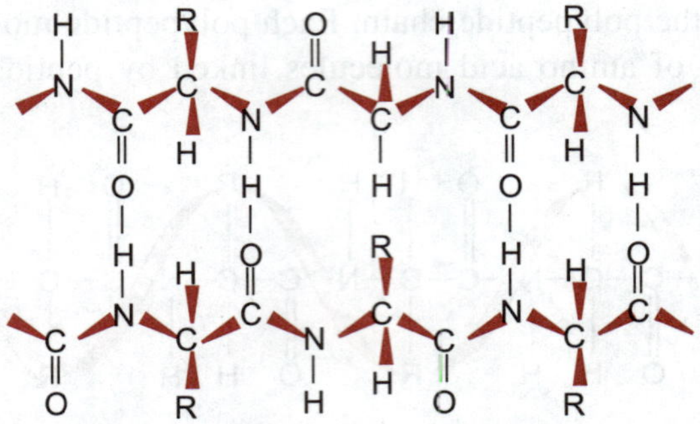

Fig. 3.29 A part of protein molecule showing β-pleated sheet configuration

3. The Collagen Helix: The collagen pattern is formed by the twisting of three helices of polypeptide chains to form a three-stranded fibre. The three helices are intertwined with each other to form a right-handed super helix. This leads to a very strong and rigid structure which can not be bent or extended. Each polypeptide chain is formed of 1,000 amino acids, every fourth amino acid is glycine and is followed either by hydroxy-proline or proline.

3. Tertiary Structure

The tertiary structure is the final three-dimensional structure that a protein molecule assumes under normal conditions by coiling of the long poly peptide chain with or without helical region. **Hydrogen bonds, ionic bonds** and the **hydrophobic bonds** are responsible for tertiary structure (Fig. 3.30). The tertiary structure may be further stabilised by disulphide bonds.

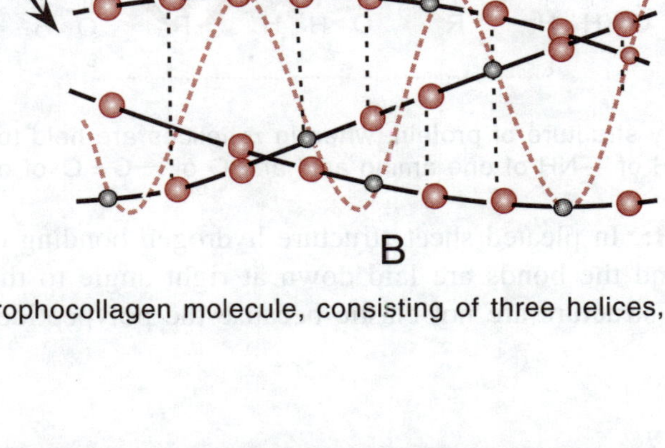

Fig. 3.30 A part of trophocollagen molecule, consisting of three helices, intertwined each other

The proteins with three-dimensional structure are either fibrous or globular. Proteins such as keratin, fibrinogen and myosin are fibrous proteins and albumin and plasma globulins are globular proteins.

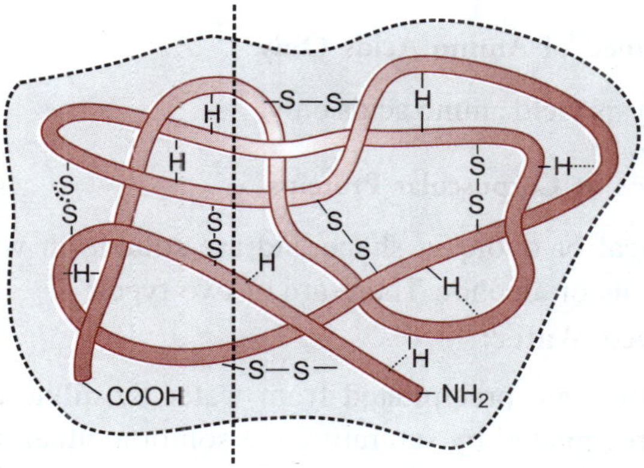

Fig. 3.31 Three-dimensional model of myoglobin molecule showing tertiary structure

4. Quarternary Structure

The quarternary structure involves the noncovalent association of one or more peptide chains. For example, a molecule of haemoglobin has two α-chains and two β-chains. The quarternary structure can be of two types:

(a) In which the protein molecule is formed of several dissimilar peptide chains but only one active site.

(b) In which the protein molecule is formed of similar or identical subunits.

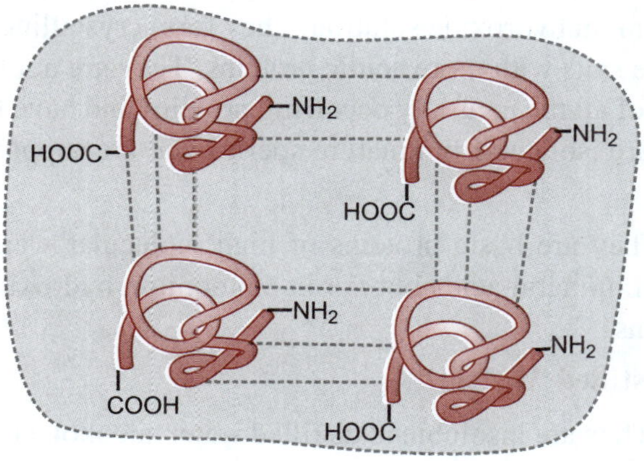

Fig. 3.32 Quarternary structure of a protein molecule

Classification of Proteins

Proteins have been vastly classified on the basis of a number of polypeptide chains, structure, solubility and coagulability.

Proteins are called **monomeric** when formed of single polypeptide chain or **oligomeric** when formed of two or more polypeptide chains. Oligomeric proteins are also described as **multi-subunit proteins**. Based on structure, proteins may be **simple, conjugated** or **derived**.

I. Simple Proteins are formed of Amino Acids Only

Simple proteins on hydrolysis yield amino acids only.

A. Simple Globular Proteins or Corpuscular Proteins

These proteins are spherical or ovoid in shape and are soluble in water or in aqueous media containing acids, bases, salts or alcohol. These are of two types:

1. **Soluble in Distilled Water:**
    1. **Albumins:** They are precipitated from water by dilute acids and alkalies. These can also be precipitated by saturating the solution either with a neutral salt such as Na_2SO_4 in slightly acidic solution or with acidic salts such as $(NH_4)_2SO_4$. In the presence of stronger acids and alkalies they are converted into soluble metaproteins. On heating, albumins get coagulated. Albumins are widely distributed in nature. Examples are egg white or **ovalbumin, blood serum albumin, soyabean albumin, gliadin** (wheat protein), **legumelin** (protein of pulses) and **phaseolin** (protein of kidney bean), **casein** (milk). These are stored as food reserve.
    2. **Pseudoglobulins:** They can be precipitated from water solution with an acid salt $(NH_4)_2SO_4$. They are rare in nature. An example is the pseudoglobulin of milk whey. They are coagulated on heating.
    3. **Protamines:** They are basic proteins, highly soluble in water and in dilute acids and dilute ammonium hydroxide solution. They form crystalline salts with mineral acids, and insoluble salts with more acidic proteins. They are not coagulated by heat. They are simplest of all the naturally occurring proteins and have lowest molecular weight. Protamines are isolated from mature sperms and are supposed to be located in their nuclei.
    4. **Histones:** They are basic proteins of high molecular weight. They are soluble in water and dil. mineral acids but not in ammonium hydroxide. They occur as part of nucleoproteins.
2. **Insoluble in Distilled Water:**
    1. **Glutelins:** They are insoluble in distilled water, alcohol, and neutral salt solution but are soluble in dilute alkalies and acids. They are found exclusively in seeds of cereal grains.
    2. **Prolamines:** They are insoluble in water but are soluble in dilute alkalies and 60–80 per cent alcohol. They are found in plants only. Gliadin (wheat), zein (maize), hordein (barley) etc., are examples of prolamines.

3. **Globulins:** These proteins are insoluble in water but readily soluble in dilute neutral salt solution such as NaCl. These coagulate on heating and heat coagulation is enhanced by the addition of dilute acids.These are found both in animal and plant tissues. **Ovoglobulin** (eggs), **lactoglobulin** (milk), **fibrinogen**, **alpha** and **beta globulins** (in serum), **myosin** and **tropomyosin** (in muscles) are animal globulins. Vegetable globulins include **legumin** (peas), **tuberin** (potatoes) and **edestin** (wheat and hemp seeds).

B. Simple Fibrous Proteins

On account of fibre-like molecular structure the fibrous proteins are insoluble in cold water and any other cold reagent. These are also known as **scleroproteins**. These are found exclusively in animals. Examples of simple fibrous proteins are:

1. **Keratins:** These are found in the outer layer of skin and in hair, feathers, horns, hooves and nails. Keratin is indigestible.
2. **Collagen:** Collagen proteins are found in white fibrous connective tissue which constitutes tendons, aponeurons, duramater and fascia. These also form ground substance of bone and cartilage. These are digested very slowly.
3. **Elastin:** It is found in yellow elastin tissue like ligaments and blood vessels. Elastin is also present in the ground substance of elastic cartilage. It is insoluble and hard to digest.
4. **Fibroin:** This protein is present in silk fibres.

II. Conjugated Proteins are Heteroproteins and have a Nonprotein Prosthetic Group

These proteins are composed of a simple globular protein united with some nonprotein substance. This nonprotein group is known as **prosthetic group.** For example, in haemoglobin, protein globin is combined with an iron-containing porphyrin compound, **heme**. Depending upon the nature of prosthetic group, the conjugated proteins have been divided into several classes:

1. **Chromoproteins:** In chromoproteins, the simple protein combines with a pigment, *e.g.*, **haemoglobin, myoglobin, cytochromes** and **flavoproteins**.
2. **Glycoproteins:** In glycoproteins the simple proteins combine with carbohydrates. For example, **mucin** of saliva, **heparin** of bile juice, **immunoglobulins** of plasma, **mucopolysaccharides** of cartilage and **tendon** are glycoproteins. In them carbohydrate hexosamine is less than 4 per cent.
3. **Mucoproteins:** These are similar to glycoproteins but in them carbohydrate hexosamine is more than 4 per cent. **Haptoglobins** (a-globulins) from blood serum, and **ovomucoid** from egg white are examples of mucoproteins.
4. **Nucleoproteins:** In nucleoproteins the protein molecules combine with nucleic acids. These proteins are protamines or histones. The chromatin material of the nuclei is composed of nucleoproteins.

5. **Lipoproteins:** Lipoproteins are formed by the combination of simple proteins with lipids. These proteins are present in the brain, plasma, eggs, milk, etc.
6. **Phosphoproteins:** Phosphoproteins are the proteins in combination with phosphoric acid or with ortho or pyrophosphate. These are soluble in dilute alkalies and are precipitated by the addition of acids. **Ovovitelline** of eggs and **caseinogen** and **casein** of milk are examples of phosphoproteins.
7. **Flavoproteins:** These are enzymes. Their protein moiety (apoenzyme) is permanently attached to **flavin compound** (**FMN** or **FAD**). Flavin represents the prosthetic group. Flavoproteins are enzymes of Kreb's cycle and participitate in electron transport system of respiratory chain.
8. **Metalloproteins:** These are proteins bound to some metallic ion, like iron, copper or zinc.

III. Derived Proteins are derived from Prexisting Proteins

The derived proteins are actually derived from some previously existing proteins either by hydrolysis or by coagulation. These are:

1. **Metaproteins:** They are derived by the hydrolysis of complex proteins, *i.e.*, by the action of digestive enzymes, acids or alkalies. They are insoluble in water but soluble in dilute acids and alkalies. These may be **acidic metaproteins, basic metaproteins, proteoses** or **albuminoses, peptones** and **peptides**.
2. **Coagulated Proteins:** The coagulated or denatured proteins are formed when ordinary proteins are heated e.g. coagulated egg white.

TEXT QUESTIONS

1. Describe the molecular structure of carbohydrates. Write in brief uses of monosaccharides.
2. Describe structural specialities of monosaccharides.
3. What are polysaccharides? Classify them on the basis of their functions.
4. Which carbohydrates are polymers of organic compounds? Give their general formula. Give some common examples and their uses.
5. What are fatty acids? Describe different types of fatty acids.
6. Describe biological significance of lipids.
7. Give two examples of conjugated lipids. Describe their role in biological membranes.
8. Why amino acids are described as amphoteric molecules?
9. Discuss biological importance of proteins.

10. Describe the structure of a peptide bond. What is the structure of a polypeptide chain? What is the necessity of folding of polypeptide chains?
11. What are amino acids? Which groups of amino acids take part in the formation of peptide bonds.
12. Explain the following terms:
    (a) Homopolysaccharides
    (b) Stereoisomerism
    (c) Optical Isomerism
    (d) Esterification
    (e) Conjugated Lipids
    (f) Flavoproteins
    (g) Protamines.
13. Differentiate between the following:
    (a) Primary and secondary structure of proteins
    (b) Essential and nonessential amino acids
    (c) Nonpolar and polar amino acids
    (d) Saturated and unsaturated fats
    (e) Nucleoside and nucleotide
    (f) Hexosamines and amino acids
    (g) Furanose and pyranose ring structure.
14. Write short notes:
    (a) Zwitter ion
    (b) Peptide bond
    (c) Glucosidic bond
    (d) α-carbon
    (e) Sterols
    (f) Phospholipids
    (g) Saponification
    (h) β-Pleated sheet structure.

4
Chapter

Methods Used to Study Macromolecules

IN THIS CHAPTER

Cell Fractionation: Tissue Homogenisation; Separation of Cell Organelles from the Homogenate; Molecular Separation: Ultracentrifugation; Electrophoresis; Chromatography; Spectrophotometry; Spectrophotometer: Components of Spectrophotometer; Applications of Spectrophotometer; Lambert-Beer Law of Adsorption of Light by Biomolecules; Photometry; Radioisotope Techniques: What are Radioisotopes; Commonly used Radioisotopes; Radioactivity; Role of Radioisotopes; Techniques for the Measurement of Radioactivity: Geiger Mullar Counter; Scintillation Counter; Autoradiography: Procedure of Autoradiography; Preparation of Sections; Application of Photographic Emulsion; Application of Film.

The study of macromolecules was made possible only with the advent of various methods of cell separation, cell fractionation and isolation of their macromolecules. Some of the methods used frequently in molecular biology are:

- Cell Fractionation
- Molecular Separation
- Radioisotope Techniques

CELL FRACTIONATION

Cell fractionation is the separation of different subcellular structures or organelles of cell such as nucleus, mitochondria, plastids, chloroplasts, ribosomes, lysosomes, Golgi complex, etc., by rupturing the cells.

Several methods are used for the disruption of cell. For the study of animal cells, liver is most suitable and is widely used for cell fractionation. Following two steps are involved in cell fractionation:

I. Tissue homogenisation.
II. Separation of cell organelles from the homogenate.

I. Tissue Homogenisation

Homogenisation is the process of destroying or breaking the cell membrane to obtain a homogenous mixture of cell organelles, suspended in a liquid medium. This suspension of ruptured or homogenised cells is called **homogenate**. It contains intact nuclei, mitochondria, plastids, lysosomes, microsomes, Golgi complex and fragments of ER and plasma membrane.

For homogenisation, the liver of an animal is perfused with isotonic saline solution and is cooled with ice to eliminate blood. It is then transferred to 0.25M sucrose solution. The tissue is homogenised in 0.25M sucrose solution either mechanically or chemically. In **mechanical homogenisation (Fig. 4.1)** tissue is homogenised either by grinding in mortar and pestle or by using Waring blender, plunger or Potter-Elvehjem homogeniser. Homogenisation can also be carried out by exposing the tissue suspension to sonic oscillation. In **chemical homogenisation** proteolytic or lipolytic enzymes are used to dissolve plasma membranes of the cells.

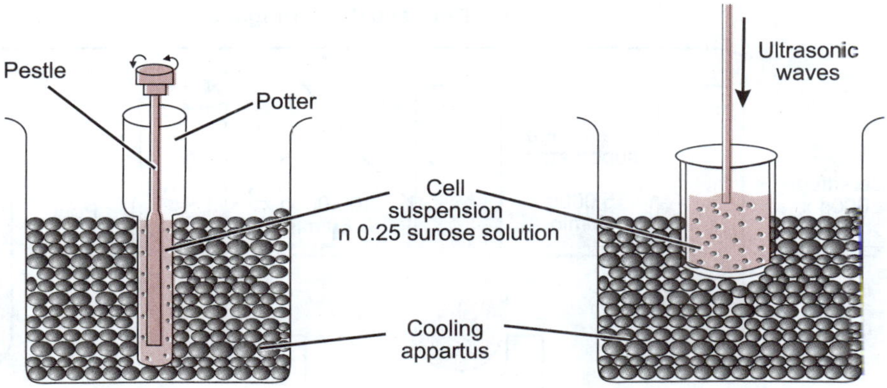

Fig. 4.1 Cell fractionation by mechanical tissue homogenisation

II. Separation of Cell Organelles from the Homogenate

Tissue homogenate contains cell organelles of different sizes and densities. They can be separated either on the basis of their size or on the basis of their density. Following three methods are used for their isolation:

1. Centrifugation
2. Density Gradient Centrifugation
3. Flow Cytometry.

1. Centrifugation or Differential Centrifugation

Centrifugation is the separation or isolation of subcellular particles according to the difference in their size and density. The homogenate in centrifuge tube is rotated in an ultracentrifuge at a definite speed (Fig. 4.2). This exerts a centrifugal force or gravitational pull on the cell organelles present in the homogenate and drives them to settle down at the bottom of the tube according to their size, forming a pellet. Different cell organelles have different densities, which allow them to separate when suspension is centrifuged at increasing speed. Therefore, the homogenate is subjected to different centrifugal forces by rotating the homogenate containing tubes at different speeds.

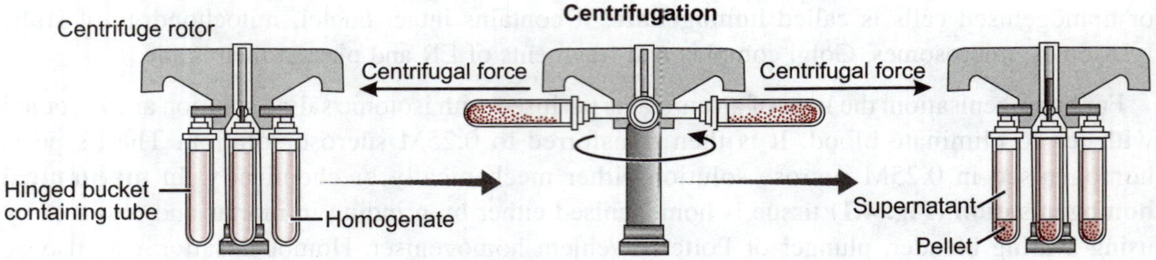

Fig. 4.2 Technique of cell fractionation

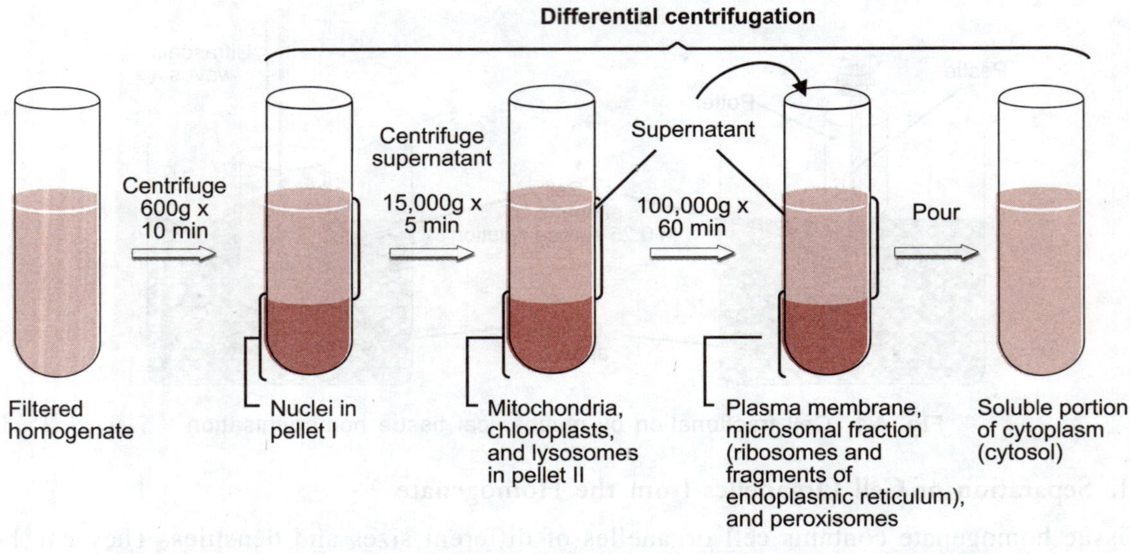

Fig. 4.3 Steps in the process of centrifugation or differential centrifugation

As a result larger particles settle down or sediment first, i.e. at lower speed, and smaller particles sediment at higher speeds. In classical centrifugation (Fig. 4.3), the components sediment into four fractions in the following sequence:

1. **Nuclear fraction** containing unbroken cells and nuclei is seperated after centrifugation at 600g.
2. **Mitochondrial fraction** containing mitochondria settles down after centrifugation at 15,000g.
3. **Microsomal fraction** containing ER, Golgi membranes and ribosomes settles down as sediment after centrifugation at 1,00,000g.
4. **Soluble fraction** containing nucleic acids, proteins, soluble polysaccharides, and lipid droplets, etc.

Extension: Centrifuge

Centrifuge is a device which is used for physical separation of cellular components and macromolecules by using centrifugal force or gravitational pull. The cell organelles or organic macromolecules fall out of the colloidal suspension and sediment according to their size or molecular weight.

Principle of Centrifugation

Centrifuge machine works on the principle of gravitational pull created by centrifugal force. When substances of sufficient mass are spun in a centrifuge, the centrifugal force creates a gravitational field in which heavier materials are pulled to the periphery of the circle. The centrifugal force applied to the sample is equal to the gravitational force, *i.e.* the acceleration. It is measured in gravity (G):

$$G = \omega^2 r$$

where ω represents angular velocity and r is the radius of rotation.

The speed of centrifuge is measured as revolutions per minute (rpm).

Components of a Centrifuge

A centrifuge has two main parts: **head** and **rotor**. The **head** is a revolving disc having four metal cups or four containers for holding the tubes. The sample to be centrifuged is filled in these tubes. The **rotor** is a high speed motor which rotates the head to create centrifugal pull. The speed of rotation is controlled by a speed control switch.

Working of a Centrifuge

A homogenate of particulate matter is prepared. It is poured into the centrifugal tubes. The tubes are then placed in the metal cups of the head. The speed of the rotor is set according to the requirement. The rotor is allowed to run for a required period. The particulate matter or its constituents get settled at the bottom of the tube.

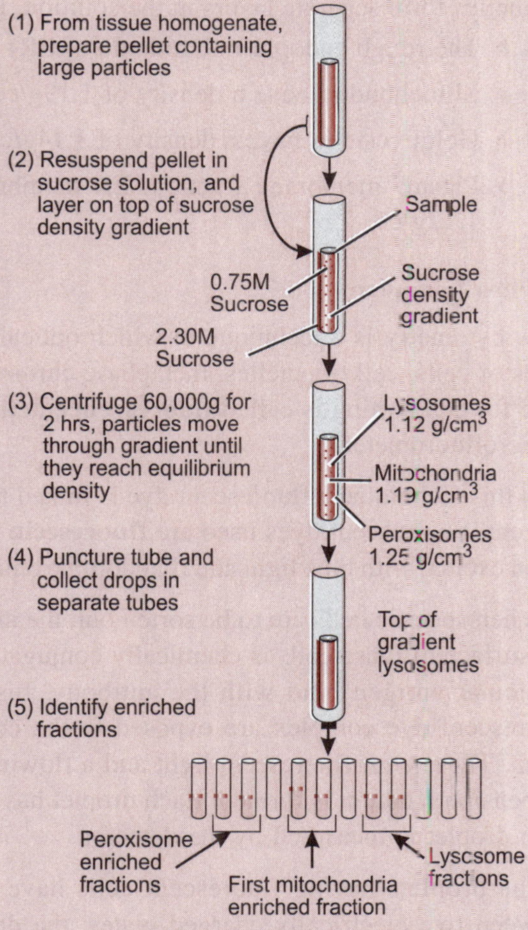

Fig. 4.4 Density gradient centrifugation on sucrose density gradient

2. Density Gradient Centrifugation

The density gradient centrifugation is used to separate cell organelles by their density rather by their size. The centrifuge tube is filled with a gradient of dense substance such as sucrose or glycerol. The gradient solution is very concentrated being most dense at the bottom of the centrifuge tube. Its concentration decreases gradually towards the top. Thus, the solution is least dense at the surface. The tissue homogenate is layered on the top of this solution. The centrifuge tube is centrifuged at high speed (about 40,000 r/min) for several hours. This allows each particle to migrate to position at which its density is equal to the density of the solution. Because of this, this type of separation is also called **isopycnic** (equal density) **centrifugation**. The different cell organelles form separate layers in the solution. In a typical preparation from animal cells:

- The rough endoplasmic reticulum (ER) has a density of 1.20g/cc.
- Mitochondria have a density of 1.19g/cc.
- Golgi vesicles have a density of 1.14g/cc.
- Plasma membrane fraction, ER membranes and lysosomal fraction have a density of 1.12g/cc.

3. Flow Cytometry

Flow cytometry is a technique in which optical system is used to identify and collect different types of cells, cell organelles, metaphase chromosomes or cell macromolecules. The instrument used for cell sorting is called **fluorescence-activated cell sorter** (FACS) or **cytometer**. This is a microfluorometer.

In this technique, a fluorescent dye is linked to an antibody to make it highly specific staining reagent. Fluoroscent dyes used are **fluorescein** and **rhabdomine**. Fluorescein emits green light when excited with blue light and rhabdomine emits red light when excited with green-yellow light.

When specific cells are to be sorted out, the antibody that is specific for the antigen present on the surface of these cells is chemically conjugated with a fluoroscent dye. The cells bearing this particular antigen bind with the antibody. Inside the cell sorter, these cells with antibody-fluorescent dye complex are exposed to the correct wavelength of light, delivered by a laser beam. These emit fluorescent light and a flowing stream of tiny droplets of liquid from the cell suspension medium is formed. Each droplet has at least one fluorescent cell. The fluorescence of each droplet is measured by the detector.

The droplets having fluorescent cells have negative electric charge. When stream passes between two electrically charged plates, the droplets with negative charge get separated. The FACS select these charged cells with specific surface marker in the form of fluorescent antibodies.

Other Uses of Flow Cytometry: In addition to cell sorting, this technique is used in:

- Measuring the size of the cell in the droplet from the amount of scattered light.
- Measuring the amount of DNA and RNA present in the cell from the amount of fluorescence of DNA-binding fluorescent dye.
- Sorting of cells in different phases of the cell cycle.
- Isolation of chromosomes by staining them with DNA-binding fluorescent stain.

MOLECULAR SEPARATION

Macromolecules can be separated or isolated by the following methods:

1. Ultracentrifugation
2. Electrophoresis
3. Chromatography
4. Spectrophotometry.

1. Ultracentrifugation

Modern centrifuges which can generate centrifugal force as great as 7,00,000 times the gravity, are called **ultracentrifuges**. Only ultracentrifuges can be used for the separation of macromolecules and determining their molecular weight. They work on the principle of **velocity sedimentation**.

Velocity Sedimentation: The velocity with which a molecule in a homogenate moves, separates out and settles down is called **velocity sedimentation.** Both the velocity and sedimentation of a molecule depend on two properties of the molecule:

1. **Molecular weight (M):** The velocity of a molecule is directly proportional to its molecular weight. When **M** increases, the velocity of the molecule also increases.
2. **Shape:** The shape of molecules also influences their velocity in the medium. The molecules of different shapes even with the same molecular weight have different velocities because the motion of any particle is impeded by friction.

Theodore Svedberg (1884–1971)

Theodore Svedberg was a Swedish Physical chemist who received **Nobel Prize in 1926** in chemistry for his work on colloids and macromolecular compounds. He contributed to the theoretical and practical development of centrifugation for isolation and determination of molecular weight of large molecules. He worked on nuclear chemistry, radiation biology and photographic processing. He also developed a method of making synthetic rubber during World War II.

Svedberg Unit

The value of sedimentation coefficient is represented in **Svedberg units,** after the name of **Svedberg,** who invented ultracentrifuge. One Svedberg unit (S) is 1×10^{-13}. The value of S for most molecules lies between 1×10^{-13} and 100×10^{-13}. Usually, the macromolecules are represented by their sedimentation coefficient or s value. For example, the prokaryotic ribosomes with sedimentation coefficient as 70s are named 70 S ribosomes.

Sedimentation Coefficient (s): The ratio of molecular velocity to centrifugal force is called **sedimentation coefficient**. It can be represented as:

$$\text{Sedimentation Coefficient(s)} = \frac{\text{Molecular velocity}}{\text{Centrifugal force}}$$

The value of 's' for a particular molecule is the same irrespective of the medium used to determine its 's'. Therefore, any change in 's' for a molecule is indicative of some change in its shape.

The ultracentrifuge can be used for **zonal centrifugation** or for **equilibrium centrifugation**.

Procedure of Zonal Centrifugation

In zonal centrifugation, the centrifuge tube is filled with a sucrose solution or glycerol or some other solute. The concentration of the solute decreases from bottom to the top of the centrifugal tube. The density of the solute increases with the increase in its concentration. Therefore, the centrifugal tube is filled with a solute of decreasing density from bottom to the top. The solution of molecules to be separated by sedimentation is lighter than the top most layer of sucrose solution so that it forms a layer on the surface of sucrose solution in the centrifuge tube. The centrifuge tube with sample on top of sucrose solution is centrifuged in a rotor for a specific period. Once the centrifugation is complete, the tube is left undisturbed for some time for the molecules to settle down. Different cell organelles and macromolecules form seperate zones in sucrose solution (Fig. 4.5), because of density gradient; this procedure is called **zonal centrifugation or gradient centrifugation**. Later on, a tiny hole is made in the bottom of the tube and drops of solutions from different zones or bands are collected in separate tubes. These drops represent successive layers of the homogenate with different molecules (Fig. 4.4) and are assayed for a particular macromolecule.

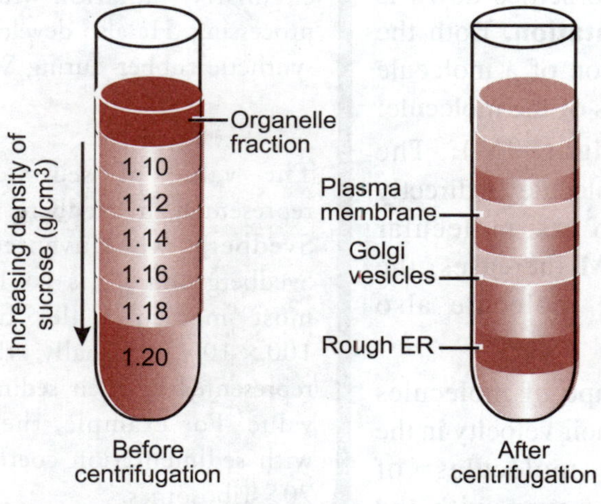

Fig. 4.5 Formation of bands in increasing density of sucrose solution

Procedure for Equilibrium Centrifugation

This is a modified method of **density gradient ultracentrifugation** and is more commonly used for the separation of nucleic acids and viruses. The macromolecules to be separated are suspended in a solution of CsCl. The concentration of CsCl solution taken is such that its density is almost equal to the density of the macromolecules to be separated. The densities of CsCl solution and of macromolecules sample are almost equal so the macromolecules do not sediment initially.

When the above mixture is centrifuged at a high speed for several hours, the ions of Cs^+ and Cl^- achieve an equilibrium concentration distribution and present a nearly linear concentration gradient and hence a nearly linear density gradient in the centrifuge tube. Therefore, the density of ions is maximum at the bottom. The macromolecules, now, begin to sediment according to their density gradient. The macromolecules form a narrow band in the tube. In case the solution has macromolecules of different densities, each type of molecules form a band at the position in the gradient CsCl medium where it matches its own density.

The DNA molecules with a density of 1.708g/cc can be separated from DNA molecules having density 1.722g/cc because of the presence of ^{15}N atoms instead of ^{14}N.

Arne W. K. Tiselius
(1902–1971)

A Swedish physical biochemist, **Tiselius**, developed electrophoresis as a technique for the separation of proteins. He got **Nobel Prize in 1948** in chemistry for the seperation of proteins from blood serum by electrophoresis.

2. Electrophoresis

Electrophoresis is the migration of charged molecules, such as amino acids, proteins, nucleic acids, etc., when placed in an electric field. The process is used for the separation of different proteins from a mixture and was developed by **Arne W. K. Tiselius** in 1937.

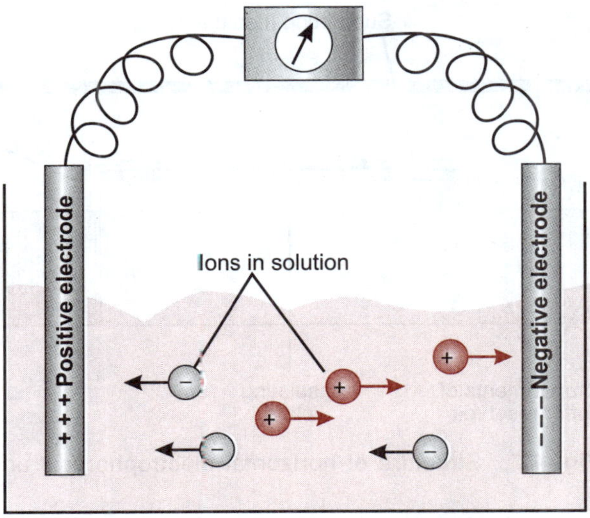

Fig. 4.6 Principle of electrophoresis: In an electric field, charged ions migrate to opposite poles. Since DNA has a negative charge because of phosphate backbone, it moves towards the positive electrode during electrophoresis

Principle of Electrophoresis

Electrophoresis is based on the principle that in a solution, charged molecules under an electric field move towards opposite electrodes, i.e. cations (+ve ions) move towards cathode and anions (–ve ions) move towards anode (Fig. 4.6).

The various biomolecules (amino acids, proteins, enzymes and nucleic acids) have readily ionisable groups. In a solution, these molecules occur as charged ions either as cations or anions. Say, when an electric current is passed through a solution of positively charged protein molecules, protein molecules start moving from positive end of the tube to the negative end. The magnitude of charge on the ions depends upon the pH and composition of the solution and on the nature of molecules. The speed of migration of ions in an electric field depends on the magnitude of charge and shape of the molecules. The mass of molecules plays no role in the rate of migration.

Factors Affecting Migration of Macromolecules

The electrophoretic mobility of biomolecules (like proteins and nucleic acids) depends upon following factors:

1. Electric Field: The separation of ions depends upon electric field which is determined by current, voltage and resistance:

(a) **Current:** The current between electrodes passes through buffer and sample ions. Therefore, the rate of migration is directly proportional to the current. The distance travelled by ions depends upon the time interval for which the current has been kept constant.

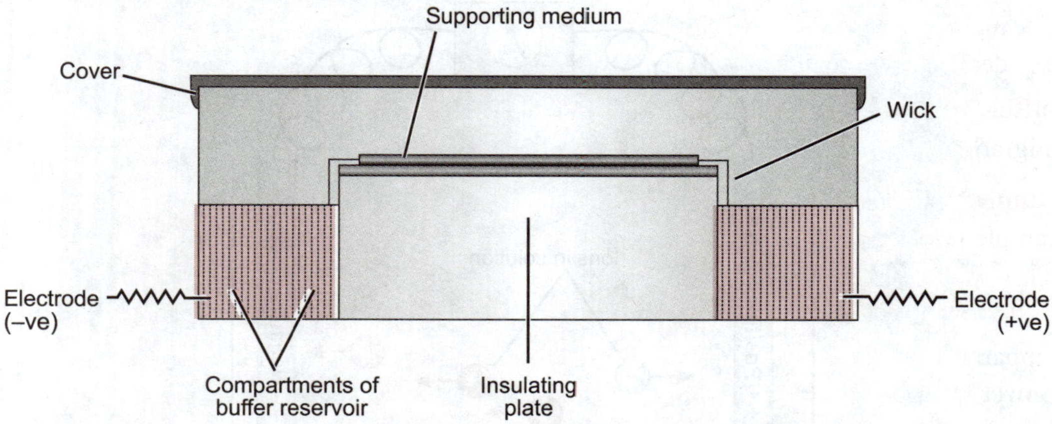

Fig. 4.7 Structure of horizontal electrophoresis unit

(b) **Voltage:** The rate of ionic migration is proportional to the potential difference between the two electrodes. This is called **voltage gradient**. The high voltage is used for the separation of compounds with high molecular weight.

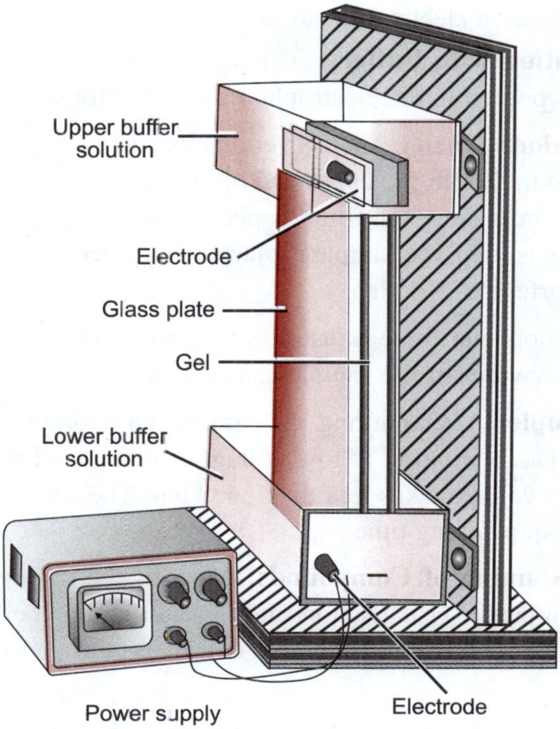

Fig. 4.8 Vertical electrophoresis apparatus

(c) **Resistance:** The rate of migration is inversely proportional to the resistance of the medium. Resistance increases with the increase in the length of the medium, but decreases on increasing the width of medium and concentration of ions.

2. **Buffer:** Buffers are the substances which make the pH of the medium constant and affect the migration rate of ions in many ways.

3. **Sample:** The nature of charge on the molecules and the size and shape of the molecules of the sample affect their migration rate.

Technique of Electrophoresis

The apparatus used for electrophoresis is called **electrophoresis cell**. It has two main units—the **power pack** and **electrophoresis unit**. The electrophoresis unit contains electrodes (one anode and one cathode), two buffer reservoirs and a transparent cover. Each buffer reservoir is divided into two compartments—**electrode compartment** and **wick compartment**.

The sample to be analysed is dissolved in a suitable buffer and the mixture is placed in the electrophoresis cell. The positive and negative electrodes are connected to each limb of the cell. When a current is applied, the migration of protein components begins.

The various steps involved in electrophoresis are:

1. **Medium Saturation with Buffers:** This step is necessary for proper conduction of current. Buffer is poured in the electrophoresis unit before pouring the sample.
2. **Sample Application:** When gel is in a vertical tube as polyacrylamide gel, the sample to be separated is poured from the upper end of cylinder. The sample is placed in the form of a small drop on a piece of filter paper in case of paper electrophoresis. When horizontal gel slab is used, the sample is placed on a piece of filter paper or is poured into a socket at the surface of gel slab.

   Normally, bromophenol blue is used as tracer dye in the sample. By observing the tracer dye, the behaviour of the sample can be noted.
3. **Running of Sample:** After pouring the sample, an appropriate voltage is applied for electrophoresis. The separation at low voltage is completed within 1-2 hours, whereas separation at high voltage takes less than one hour. The separation of proteins by paper electrophoresis requires long time.
4. **Recovery and Staining of Compounds:** Most of the biological compounds are not coloured. For their proper visualisation, and for determining their position, they need to be stained. Staining the medium with a proper dye also stains the compounds which are then clearly identified.

   Enzymes separated in gel are placed in the solution of their substrate. If the products of enzyme substrate reaction become coloured, it indicates that enzyme is present in the coloured part. In case the substance separated is radioactive, autoradiography helps in locating and identifying the separated substances.

Types of Electrophoresis

There are several electrophoretic techniques, based on the ability of charged molecules to migrate when placed in an electric field. These are:

1. Paper Electrophoresis

Paper electrophoresis is the most commonly used electrophoretic method and is used for the resolution and analysis of smaller molecules. This method is not used for the isolation of macromolecules such as DNA, RNA and proteins, etc., because the surface tension and adsorption associated with these molecules can denature them.

Solute samples are used as a small spot after dissolving them in water or in some volatile buffer and kept at the origin line. Compounds of known standards are kept at some other place at the origin line. When the solvent is evaporated the paper is sprayed by an electrophoretic buffer at both ends so that the buffer front of both sides meet at the origin line.

This paper is now kept in electrophoretic unit in such a way that its both ends are in contact with the reservoir of electrophoretic buffer at electrode. The electric field V/Cm or A is applied across the unit. When the sample is separated, the current supply is stopped and paper is removed from the unit. It is now dried and analysed for determining the types and position of various compounds.

2. Gel Electrophoresis

In this method, the molecules are separated in aqueous buffer solution in polymerised gel matrix. The gels used are **starch gel**, **agarose gel, polyacrylamide gel** or charged **sodium dodecyl sulphate (SDS)**.

1. Agarose Gel Electrophoresis: Agarose is a polysaccharide extracted from seaweed Agarose gel is mostly used to separate DNA. When agarose and water are mixed and boiled. agarose forms a homogeneous solution On cooling, it gels to form a meshwork having numerous water-filled pores. The pore size of agarose is suitable for separating nucleic acid polymers (Fig. 4.9).

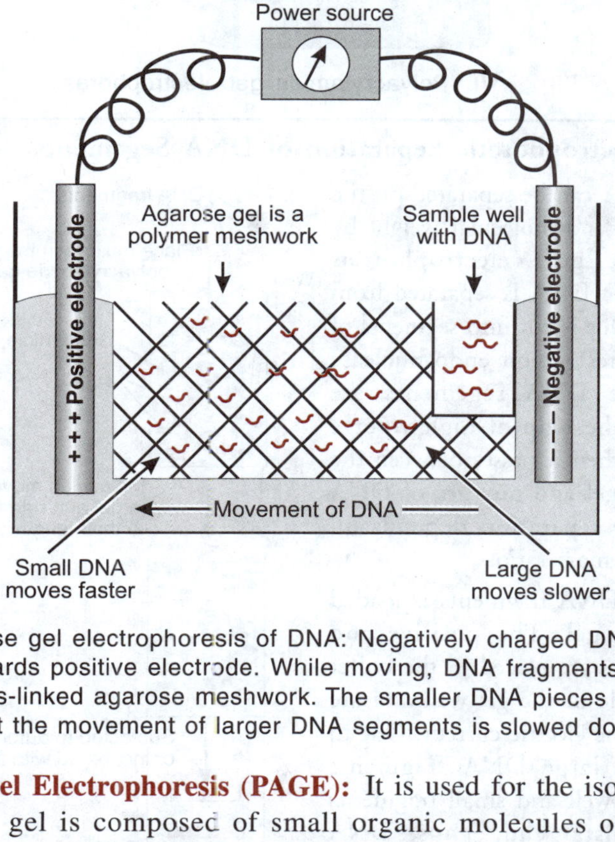

Fig. 4.9 Agarose gel electrophoresis of DNA: Negatively charged DNA segments are attracted towards positive electrode. While moving, DNA fragments are hindered by the cross-linked agarose meshwork. The smaller DNA pieces move faster but the movement of larger DNA segments is slowed down

2. Polyacrylamide Gel Electrophoresis (PAGE): It is used for the isolation of proteins and DNA segments. The gel is composed of small organic molecules of **acrylamide**. These molecules are cross-linked to form a **molecular sieve**. Polyacrylamide gel can be used either as a thin slab between two glass plates or as a cylinder inside a glass tube. Once the gel has polymerised, the slab is suspended between two compartments which are filled with buffer. The electrodes with opposite charge are immersed in the buffer of two compartments. The protein sample is filled in the slots along the top of the gel (Fig. 4.10).

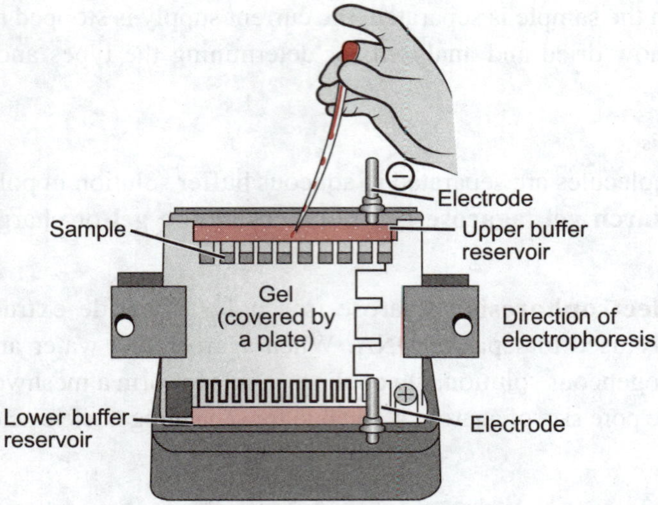

Fig. 4.10 Polyacrylamide gel electrophoresis

Extension: Electrophoretic Separation of DNA Segements

Segments of DNA can be separated on the basis of their size and molecular weight by polyacrylamide gel electrophoresis (PAGE). Genomic DNA is separated from the tissue cells. This is cut into segments of variable size by restriction endo-nuclease enzymes. These DNA segements are separated with the help of high voltage electric current when it passes between the buffers through gel and mixture of DNA segments. Their separation depends on their charge and mass ratio.

The mixture of DNA fragments is loaded in the wells of gel. The positive and negative electrodes are placed in the buffer one on either side of the gel well and are connected with the electric current. Due to electric charge, large DNA fragments remain near the well and small fragments move to the distal end. Thus, DNA fragments of different sizes form different bands which are seen under ultraviolet light (Fig. 4.11).

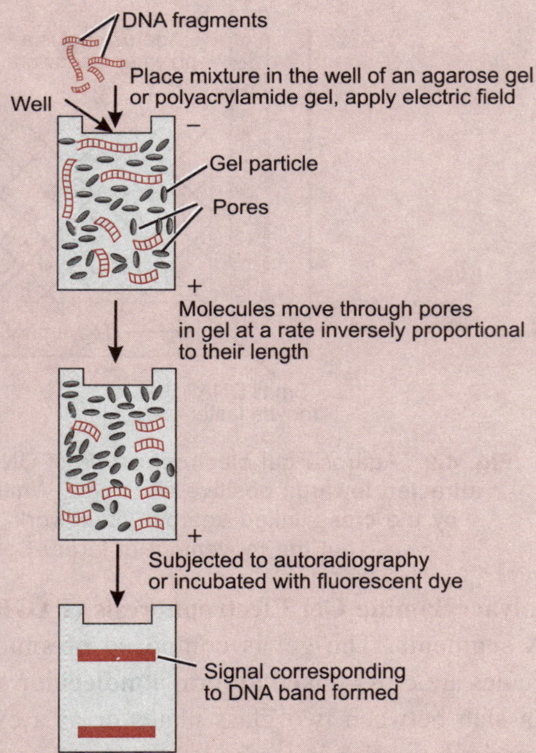

Fig. 4.11 Electrophoretic separation of DNA fragments of different lengths

The protein sample is prepared in a solution of sucrose or glycerol. The density of sucrose or glycerol prevents mixing of the sample with buffer in the upper compartment. When current passes through the buffer compartments it flows across the slab and causes proteins from the slots to move towards the oppositely charged electrodes. Separation of proteins is usually carried out in alkaline buffers, which make proteins negatively charged. The negatively-charged proteins move towards anode at the opposite end of the gel along parallel lanes.

The movement of proteins through gel depends on the size, shape, molecular weight and charge density of protein molecules.

Identification: The gel that emerges after electrophoresis contains isolated proteins. Their location can be revealed by staining or by autoradiography (radioactively-labelled proteins). The isolated proteins can be transferred to nitrocellulose for identification by Western blot method.

3. Pulse Field Gel Electrophoresis (PFGE): It is a modified procedure of simple gel electrophoresis. The principle of pulse field gel electrophoresis (PFGE) is the same as standard agarose gel electrophoresis. In this method instead of one electric field two electric fields situated at 120° with respect to each other are used. The electric field alternates between the two sets of electrodes, A and B. DNA first migrates towards the positive A electrode, then the electric field switches and the DNA migrates toward the positive B electrode, and so on. By constantly switching directions, the larger pieces of DNA do not get caught in the gel meshwork and migrate faster in the gel. As usual, the gel is stained with ethidium bromide and viewed under UV light to see the bands (Fig. 4.12).

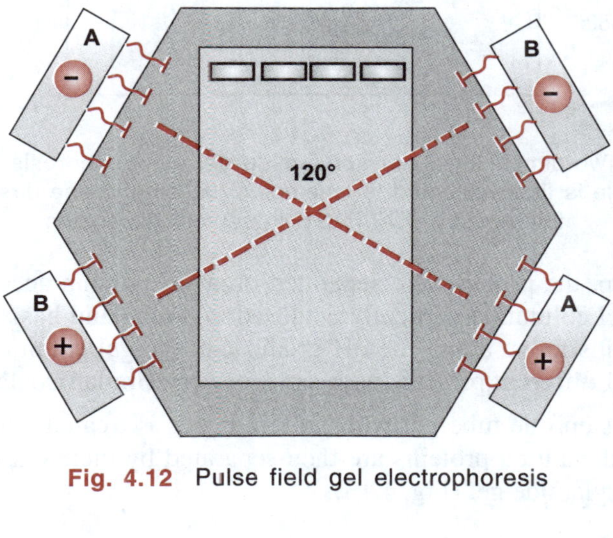

Fig. 4.12 Pulse field gel electrophoresis

3. SDS–PAGE

Polyacrylamide Gel Electrophoresis (PAGE) is usually carried out in the presence of a negatively charged detergent, sodium dodecyl sulphate (SDS). **SDS** (sodium dodecyl sulphate) is a detergent used to unfold proteins and cover them with negative charges for electrophoretic separation of proteins by mass. This gel binds to all types of protein molecules.

The electrostatic repulsion between bounded SDS molecules causes proteins to unfold into rod-like structures. The denatured protein molecules are coated with over all negative charge due to bound sodium dodecyl sulphate (SDS) molecules. The proteins get separated on the basis of their molecular weight. The larger or heavier proteins are held up and lightest proteins reach the lowest end of the column.

Application: SDS–PAGE is used in separating protein from a mixture. SDS–PAGE can also be used to determine the molecular weight of various proteins by comparing the position of bands to the bands of known proteins.

4. Two-dimensional PAGE (2D-PAGE)

Two-dimensional polyacrylamide gel electrophoresis (2D-PAGE) is practised for the separation of large number of proteins (see Fig. 4.13). It involves the following two steps:

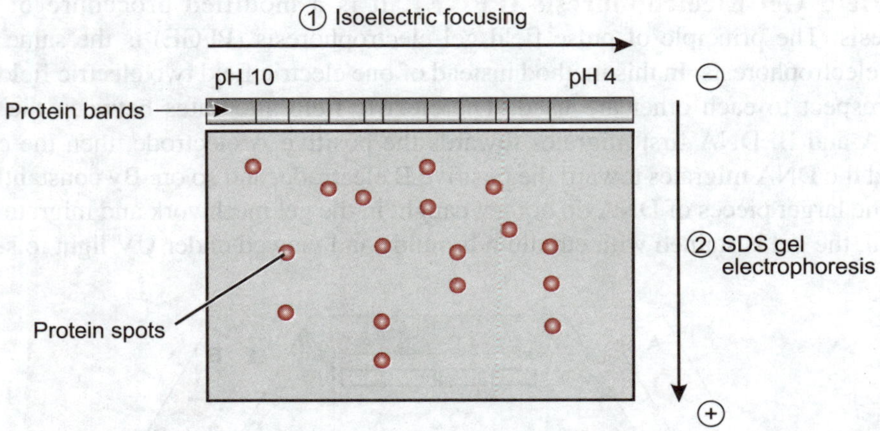

Fig. 4.13 Two-dimensional polyacrylamide gel electrophoresis. The protein sample is first subjected to isoelectric focusing in one direction, followed by SDS-PAGE in second dimension

1. In the first step, the proteins are separated on a gel column according to their inherent charge. The gel column is vertically enclosed in a tube and has a gradient of increasing pH. The proteins move along the pH gradient in the gel column. The bands in the gel contain several different proteins with same or very similar isoelectric point.
2. In the second step, the tube gel with protein bands is treated with SDS to denature the proteins. The denatured proteins are then separated by their size with SDS–PAGE on a slab of polyacrylamide gel (Fig. 4.13).

3. Chromatography

Chromatography is a technique in which molecules of different substances, present in a solution or in a cytosol are separated as they move through some type of porous matrix. The term 'chromatography'(Gr. *Chroma* = colour, *graphein* = written), was coined by **Mikhail Tswett** for which he received Palmer award in 1906.

Chromatography is an analytical process used to separate and purify organic and inorganic susbstances and macromolecules. It is equally important for the fractionation of complex mixtures, separation of closely related compounds, isomers and unstable compounds.

Principle of Chromatography

Chromatography is based on the phenomenon of adsorption and differential migration of components of a mixture or compound. This requires two alternative phases: a **mobile phase** consisting of moving **solvent** and an **immobile** or **stationary phase** called **matrix**. The solvent moves through the matrix. The materials forming the immobile phase contain sites to which molecules from the mobile phase can bind. As individual molecules interact with the material of matrix, their progress through matrix is retarded this is called **impedence**. Thus, greater the affinity of a particular type of molecules for the matrix, the slower is their movement down through the column. Since different components of a mixture have different affinity for the matrix, they are retarded to different degrees. The molecules with the least affinity for the matrix are retarded the least and appear in the first fraction in the matrix. Thus, chromatography works on propelling of molecules through chromatographic column and their selective impedence by matrix.

Applications of Chromatography

Chromatography has the following applications:

- Chromatography is used for separating pigments, proteins, components of proteins (amino acids), DNA, RNA and their nucleotides.
- It is used for the separation of components of colourless compounds or colourless substances from a mixture.
- It is also used for fractioning gases, liquids or dissolved solids.

Table 4.1 Principle types of chromatography

Techniques	Solute property	Solid phase	Solvent
1. Adsorption Chromatography	Size and shape	Hydrated gel	Usually aqueous
2. Ion exchange Chromatography	Ionisation	Matrix with ionised groups	Aqueous buffer
3. Gel Filtration	Adsorption	Adsorbent, usually inorganic material	Nonpolar
4. Paper Chromatography	Adsorption	Whatman's Paper-I	Aqueous
5. Affinity Chromatography	Covalent linkage	Porous agarose gel	Aqueous
6. Partition Chromatography	Solubility	Inert support	Mixture of polar and nonpolar solvents
7. Thin-layer Chromatography (TLC)	Adsorption	Matrix of cellulose, alumina or ion exchange resin	Aqueous
8. Gas Chromatography	Adsorption	Liquid or solid	Gas

Types of Chromatography

There are many types of chromatographic methods. In all the methods, there is interaction between three components: Mixture (to be separated), Solid state or solid phase matrix and Solvent.

1. Adsorption Chromatography or Column Chromatography

Principle:

Adsorption chromatography is also called **column chromatography**. It was developed by **Tswett** in 1903. In adsorption chromatography, the solution or cell extract is passed through a column of solid inert material packed in a glass tube.

The column acts as **impeding medium** or **adsorbent**. The solution of the material to be fractionated is added to the top of adsorbent column. It gradually percolates through the column. The fresh solvent is added from time to time.

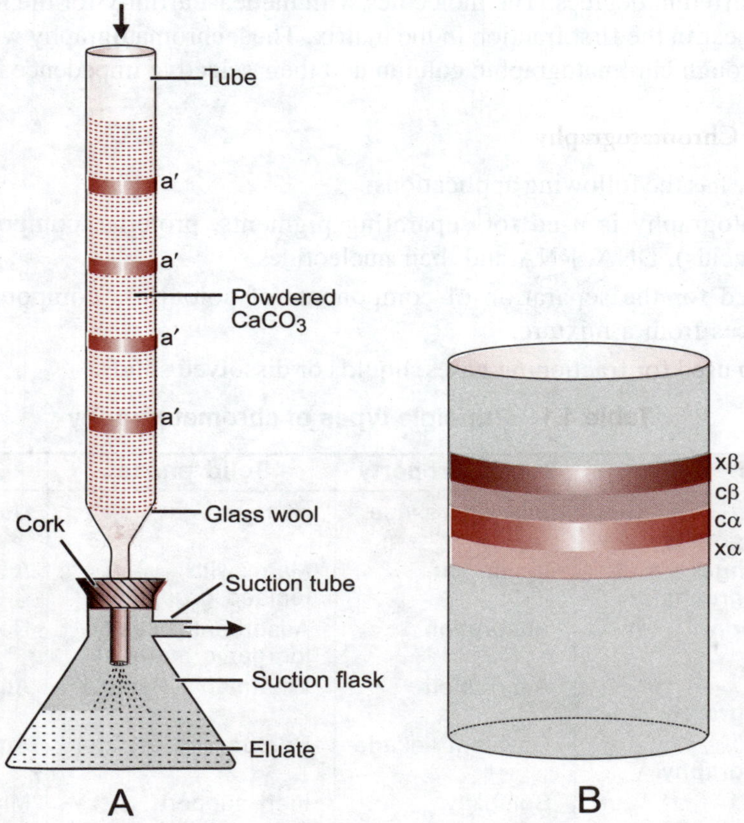

Fig. 4.14 **A.** Chromatographic apparatus for column chromatography; **B.** Chromatogram of chlorophyll

As the molecules of solute of the given sample travel down along the column, they bound at the surface of adsorbent (Fig. 4.14). As soon as the influx of new solvent from above dilutes the dissolved material to less than a certain concentration, the adsorbed molecules are released into the solution. Thus, the substance travels down the chromatographic column as a zone. Slight difference in the chemical structure of the components of the substance leads to differential rate of their adsorption and, therefore, differential rate of travelling down the column.

Adsorbent Medium:

Fuller's earth, charcoal, silica, alumina, cellulose, resin, sucrose, starch, etc., are used as matrix (as adsorbent column). The solvents commonly used are water, alcohols, acetone, hydrocarbons like benzene, ether, chloroform and dilute aqueous solutions of acids, alkalies and salts.

Application:

The main application of adsorbent chromatography is to separate complex fats, carotenoids (plant pigments), steroids, chlorophyll and their derivatives.

2. Ion Exchange Chromatography

Principle:

In ion exchange chromatography, molecules are separated by the difference in their net electric charge. The ion exchange resin used as chromatographic adsorbent consists of porous polystyrene beads (polymers) to which ionisable groups are added chemically. The two most commonly used ion exchange resins are **diethylaminoethyl (DEAE) cellulose, DEAE-sephadex** and **carbomethyl CM-cellulose** or **CM-sephadex**.

DEAE cellulose is positively charged and is called **anion exchanger**. It acts by binding negatively charged molecules. CM cellulose is negatively charged and acts as a **cation exchanger** (Fig. 4.15).

Dissociation of weak acidic resins is very high at pH value less than 5. In such conditions, the cationic exchange becomes negligible. Similarly, the dissociation of weak basic resins is very high at pH value above 9. Under these conditions, the anionic exchange becomes negligible.

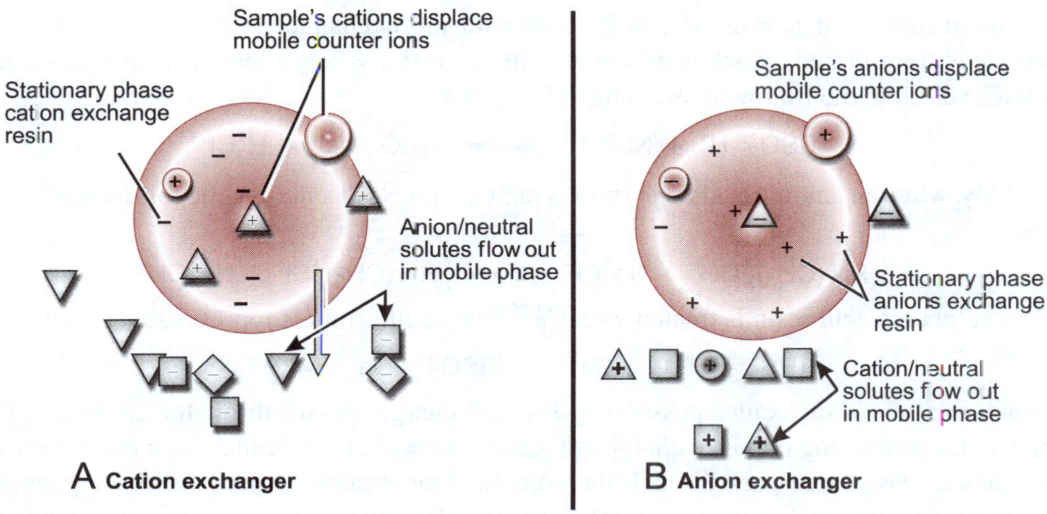

Fig. 4.15 Ion exchange chromatography; **A.** Cation exchanger; **B.** Anion exchanger

The mechanism of a cationic exchange can be shown by the following equilibrium reaction:

$$\underset{\text{Solid}}{x\text{RSO}_3^-\text{H}^+} + \underset{\text{Solution}}{\text{M}^{x+}} \rightleftharpoons \underset{\text{Solid}}{(\text{RSO}_3)^x\,\text{M}^{x+}} + \underset{\text{Solution}}{x\text{H}^+}$$

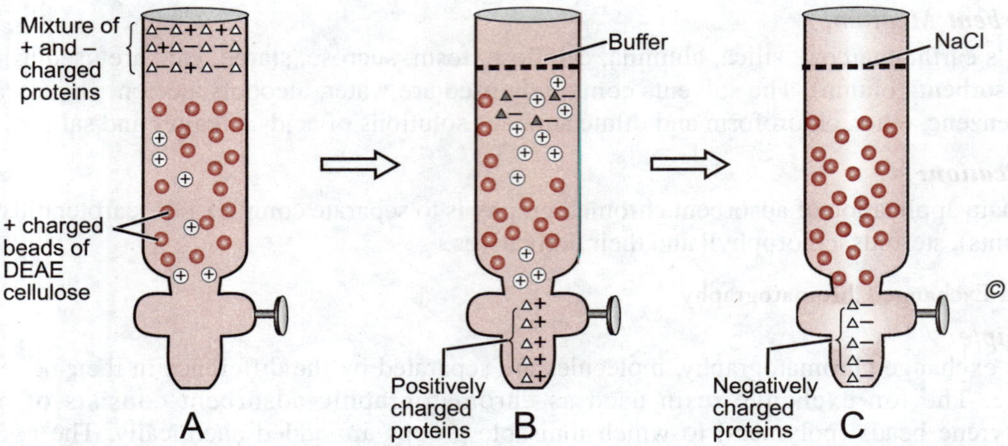

Fig. 4.16 Ion exchange chromatography for the separation of proteins: **A.** Ion exchange resin column with a mixture of positively and negatively charged proteins; **B.** Positively charged proteins percolate down the column; **C.** Negatively charged proteins percolate down the column

where M^{x+} represents cation and R represents a part of resin molecules. Similarly, the mechanism of anionic exchange can be represented by the following equilibrium reaction:

$$\underset{\text{Solid}}{x\text{RN}(\text{CH}_3)_3^+\text{OH}^-} + \underset{\text{Solution}}{\text{Ax}^-} \rightleftharpoons \underset{\text{Solid}}{[\text{RN}(\text{CH}_3)_3^+]_x\,\text{Ax}^-} + \underset{\text{Solution}}{x\text{OH}^-}$$

The identification of mobile ion, associated with ion exchange resin is carried out by the electrolyte of the solution, which is in contact with the resin. When sulphonic acid resin is mixed with NaCl solution, the following exchange takes place.

$$\text{RSO}_3^-\,\text{H}^+ + \text{Na}^+\text{Cl}^- \rightleftharpoons \text{RSO}_3^-\,\text{Na}^+ + \text{H}^+\text{Cl}^-$$

Similarly, when an anionic exchange resin is mixed with NaCl solution, the following exchange occurs:

$$\text{RN}(\text{CH}_3)_3^+\text{OH}^- + \text{Na}^+\text{Cl}^- \rightleftharpoons \text{RN}(\text{CH}_3)_3^+\text{Cl}^- + \text{Na}^+\text{OH}^-$$

When sulphonic acid resin is treated with Ca^{2+}, the equilibrium is represnented as below:

$$(\text{RSO}_3^-\,\text{H}^+)_2 + \text{Ca}^{2+} = (\text{RSO}_3^-)_2\,\text{Ca}^{2+} + 2\text{H}^+$$

When a mixture of molecules possessing different charges passes through a charged column, the molecules possessing opposite charge get associated with the column, while those with same charge or with no charge pass through the column. This process is used for the separation of amino acids, proteins and other macromolecules and the separation is carried out in a tall glass column (Fig. 4.16).

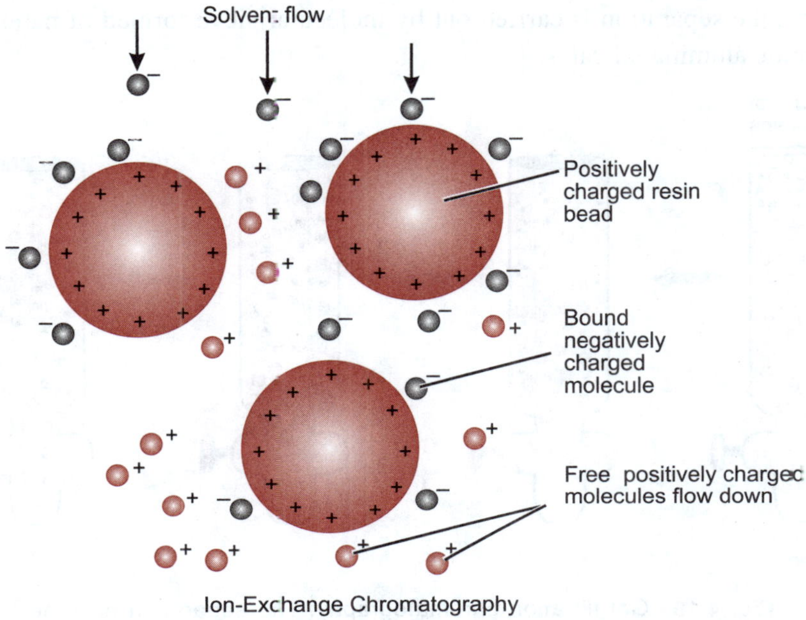

Fig. 4.17 The principle of adsorption ion exchange chromatography: Negatively charged protein molecules are adsorbed to the surface of positively charged resin beads

***Procedure*:**

The ion exchange material (resin) is placed in an appropriate buffer for swelling. After swelling, the resin is poured into a column and an equal amount of proper solvent is added. The sample or protein solution is allowed to percolate through the column in a buffer. Proteins bound to the resin are displaced by increasing the ionic strength of the buffer or by changing its pH or by both. Displacement of bound proteins can be carried out in a stepwise manner by adding a series of different buffers or by adding a solution of continually changing ionic strength or of changing pH. The separated fractions are collected in a series of test tubes and are analysed.

3. Gel Filtration Chromatography

***Principle*:**

In gel filtration chromatography, the separation of proteins and nucleic acids depends upon their size, shape and molecular weight. It basically depends on the ability of molecules to enter the pores within the beads of beaded gel. The adsorbent material consists of tiny beads that are packed into a column in a glass tube and act as molecular sieve. The solution of proteins slowly percolates through the beads. The beads are composed of cross-linked polysaccharides like dextrans or agarose of different porosity and form 3D network of cross-linked chains.

Gel chromatography is also called **gel permeation chromatography** or **exclusion chromatography** or **molecular sieve chromatography**. In gel permeation chromatography, the separation is according to the size of the solute particles. In exclusion chromatography too the separation is in accordance with the size of the solute particles. In molecular sieve chromatography, the separation is carried out by molecular sieve formed of natural or synthetic zeolites or metallic alumina-silicates.

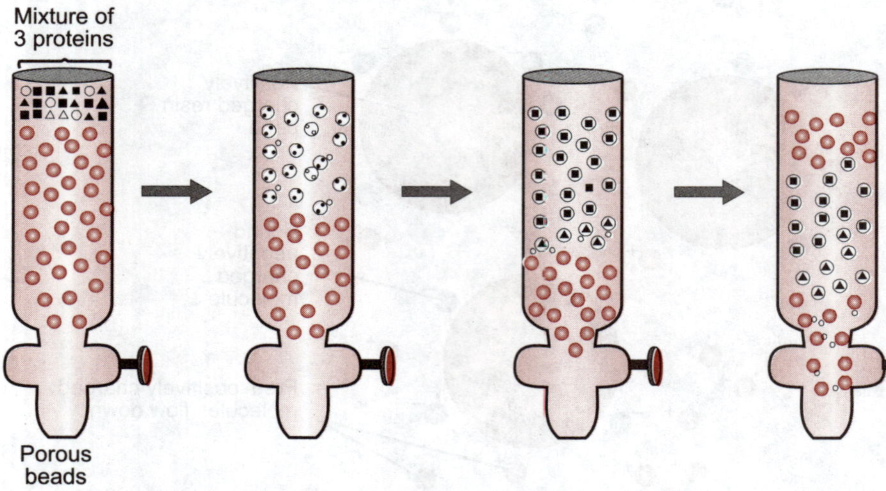

Fig. 4.18 Gel filtration chromatography: The separation of three types of proteins on the basis of their molecular weight

Procedure:

Gel filtration chromatography is different from other chromatographic methods because in this method particles come out of the column in the form of uncharged gel.

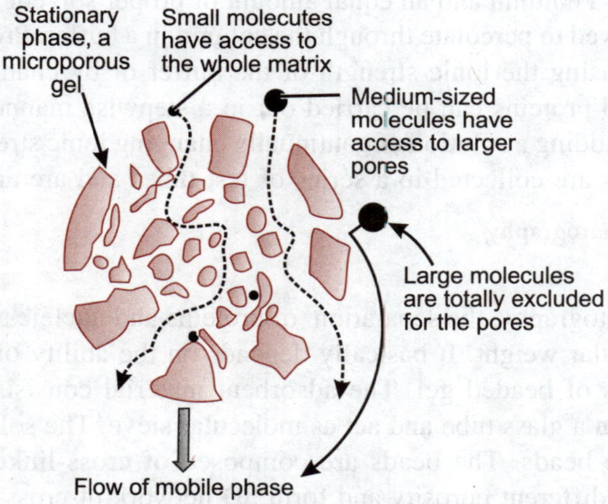

Fig. 4.19 Gel permeation chromatography

The gel material swells up in proper solvent. Due to this, space between the polymer chains increases. Gel molecules have a critical molecular size, i.e. molecules only of a particular size can enter the gel molecules. The larger molecules can not enter the gel and easily pass through the column, whereas movement of the smaller molecules gets retarded. This retardation depends upon the size of the particles and the adsorption property of the gel.

In this method, the mixture of larger and smaller molecules is placed over the column formed of swollen gel. The smaller molecules enter the gel material (larger path). So their flow is retarded. The larger molecules pass down rapidly because they are unable to penetrate the gel molecules. The smaller molecules fall down slowly. So the molecules of different sizes are separated.

Application:

The gel chromatography is mainly used in the purification of macromolecules such as proteins, enzymes, antibodies, hormones and nucleic acids, etc. It is also used for the determination of molecular weight of proteins and enzymes.

4. Paper Chromatography

Principle:

The technique of paper chromatography is based on the principle of fractionation of a compound between two liquid phases. In an atmosphere containing water vapours, the filter paper absorbs moisture. Here water is a solvent and represents stationary state. When a nonaqueous solvent (mobile phase) is passed over this paper by capillary action, it gets distributed between two phases. This distribution occurs in a definite ratio which represents the characteristic distribution coefficient of the solution. The solubility of the solute in mobile phase is proportional to the distance covered by it during distribution or spread on the paper.

The distance travelled by a solute is represented as Rf value. It can be represented as:

$$Rf = \frac{\text{Distance travelled by solute}}{\text{Distance travelled by solvent}}$$

Under standard conditions, for a particular solute, this value is constant and reflects **distribution coefficient of the substance**.

In paper chromatography, separation of substances of a mixture is carried out by flowing the solution over a special type of filter paper (Whatman's No.1 paper). It is the extended version of partition chromatography because substances are separated due to the difference in their partition coefficient leading to their differential migration on the paper.

Paper chromatography was invented by **Sehon Bein** (1961) but was made popular by **Gorden, Martin** and **Synge** (1964).

Types of Paper Chromatography

Three methods are normally followed for paper chromatography:

1. Descending chromatography, and
2. Ascending chromatography,
3. Uni-dimensional and Bi-dimensional chromatography.

In these methods, the solvent is placed at the base of a glass jar or closed tank, so that the chamber is filled up with the fumes of the solvent.

1. **Descending Paper Chromatography:** A small drop of the substance to be fractionated (i.e. the eluate) is applied at a marked point 5 cm from the end of the sheet. The sheet is hung with the drop at top in a closed box saturated with water vapours. The top edge of the paper is bent over to dip into a trough containing an organic solvent mixture in water. The solvent creeps down the paper by capillary action. After the solvent has travelled the length of paper, the components of the drop are separated and distributed as a series of spots along the edge of paper. This is known as **descending paper chromatography**.

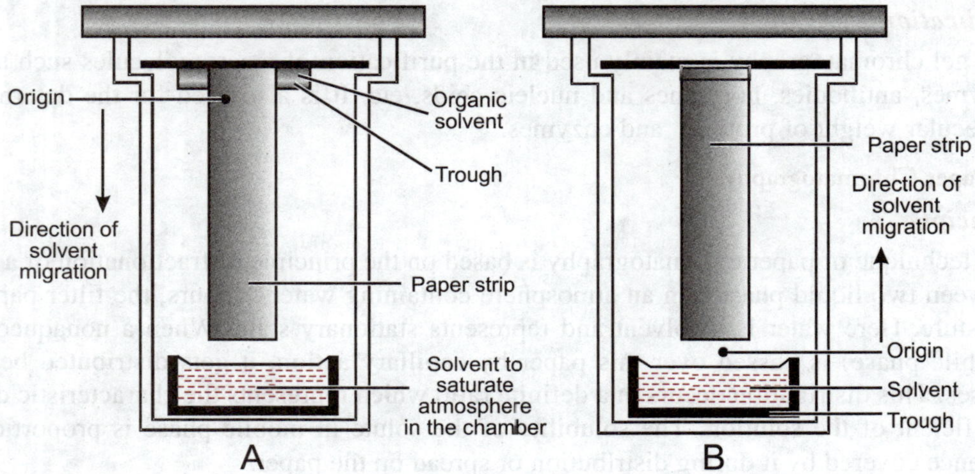

Fig. 4.20 Apparatus for paper chromatography: **A.** Descending paper chromatography; **B.** Ascending paper chromatography

2. **Ascending Paper Chromatography:** In this method, a strip is suspended from the top of the chamber in such a way that lower end containing the substance spot may dip into the trough at the bottom of the chamber. The substance spot, called **origin**, remains above the level of solvent mixture. In this case the solvent flows up along the paper due to absorptive and capillary forces and compounds of the drop are separated. This is known as **ascending paper chromatography**.

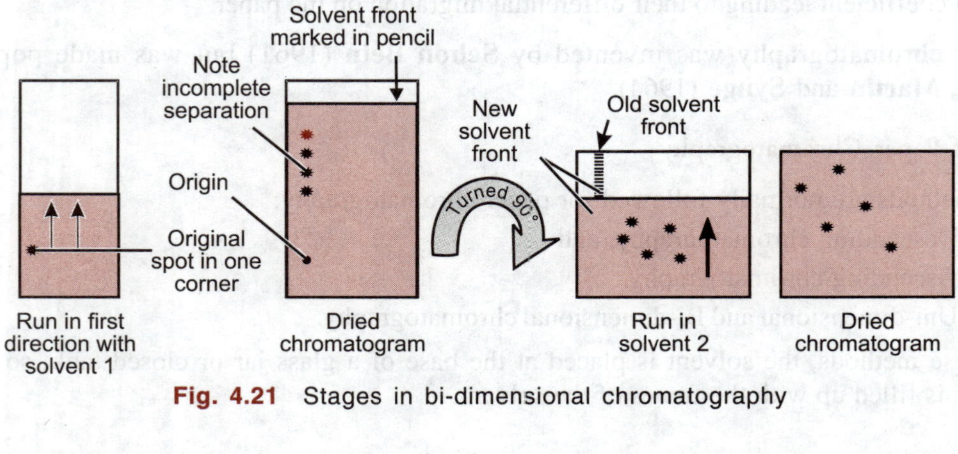

Fig. 4.21 Stages in bi-dimensional chromatography

3. **Uni-dimensional and Bi-dimensional Paper Chromatography:** The ascending and descending paper chromatographic methods are examples of uni-dimensional chromatography. In bi-dimensional paper chromatography, a square sheet of filter paper is used. The drop of the solution is placed on the upper left corner roughly one inch away from the margin. It is then developed by the ascending or descending process for several hours. After this, the sheet with chromatogram is taken out and dried to remove the solvent. Then the sheet is turned through 90° and is chromatographed for the second time in a second solvent mixture. The substance develops in one direction in one solvent mixture and in other direction in the second solvent mixture. In this way all the components of the substance are separated on the surface of paper sheet. The position of coloured substance is itself evident but for the colourless substances either ultraviolet light is used or they are stained with specific dye or stain.

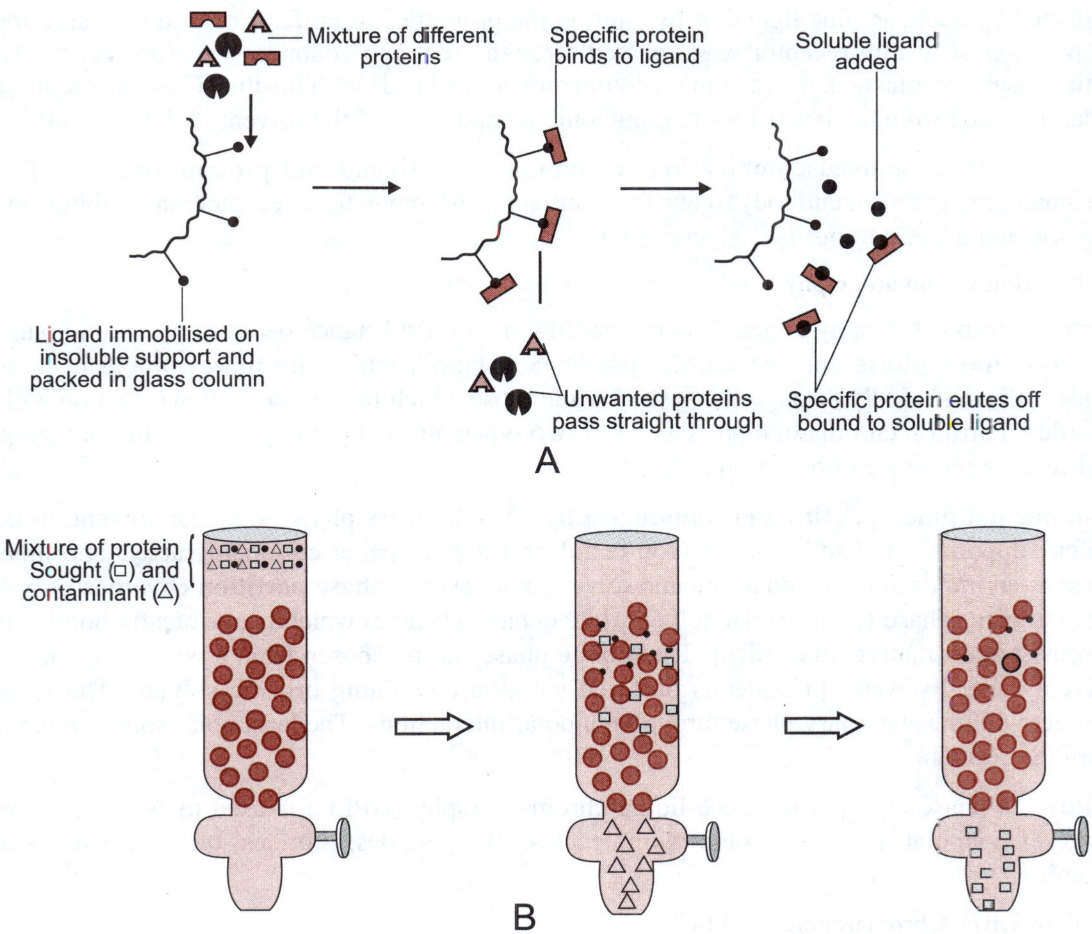

Fig. 4.22 Affinity chromatography: **A.** Agarose gel beads coated with ligand to which only specific protein can combine; **B.** Steps in affinity chromatography

Application of Paper Chromatography:

Paper chromatography is very popular on account of its simplicity, rapidity and high resolving power. It is being widely used in separating the constituents of blood and urine, separation of antibodies, study of new amino acids in bacteria and the study of various steps of photosynthesis.

5. Affinity Chromatography

This technique is used for the purification or fractionation of proteins. It is based on the unique property of protein which allows it to be specifically withdrawn from the solution. Proteins interact with specific compounds, such as enzymes with substrates, receptors with ligands and antibodies with antigens. Each of these types of proteins can be removed from solution when a mixture of these proteins passes through a column containing immobilised ligand. The protein of interest binds to the ligand, whereas all other proteins pass through. All non-specifically bound proteins are removed by washing the column with buffer. The specific proteins are washed and collected by using soluble ligand or by altering the properties of buffer. For example, an impure preparation of insulin receptor when passed through a column containing agarose gel beads to which insulin is attached, the insulin receptors bind to the beads with insulin. These molecules are then displaced from the beads by changing ionic composition of the solvent in the column.

Commonly employed combinations of immobilised ligand and protein used in affinity chromatography are an antibody to purify an antigen, a hormone for a receptor, an inhibitor for an enzyme and a lectin to purify a glycoprotein.

6. Partition Chromatography

Partition chromatography is based on the partitioning of a substance between two liquid phases, the **stationary phase** and the **mobile phase**. Substances which are more soluble in mobile phase will pass rapidly through the system while those which favour the stationary phase will be retarded. Partition chromatography can be of two types: normal phase partition chromatography and reverse phase partition chromatography.

In **normal phase partition chromatography**, the stationary phase is a polar solvent, usually water, supported by a solid matrix (e.g. cellulose fibers in paper chromatography) and mobile phase is an immiscible, nonpolar organic solvent. For **reverse phase partition chromatography**, the stationary phase is a nonpolar solvent (like octadecylsilane) which is chemically bonded to a porous support matrix (like silica). The mobile phase can be chosen from a wide range of polar solvents, usually water or aqueous buffered solution containing organic solvent. The solutes interact with the stationary phase through nonpolar interactions. The least polar solutes elute last from the column.

Reverse phase high performance liquid chromatography (HPLC) is used to separate a broad range of nonpolar, polar and ionic biomolecules, like peptides, protiens, oligosaccharides and vitamins.

7. Thin Layer Chromatography (TLC)

Thin layer chromatography is a modified version of column chromatography. In column chromatography, adsorbent is filled in a glass tube forming a column,whereas in thin layer chromatography adsorbent is spread on a glass sheet to form a thin layer. The adsorbent used in TLC may be cellulose, alumina, silica, calcite, starch-sephadex or cellulose ion exchange resin (see Table 4.2)

Table 4.2 Adsorbents and their application in thin layer chromatography

Adsorbent	Used to separate
Silica	Amino acids, alkaloids, sugars, fatty acids, lipids, essential oils, inorganic anions and cations, steroids, terpenoids.
Alumina	Alkaloids, food dyes, phenols, steroids, vitamins, carotenes, amino acids.
Kieselguhr	Sugars, oligosaccharides, dibasic acids, fatty acids, triglycerides, amino acids and steroids.
Celite	Steriods.
Cellulose powder	Amino acids, food dyes, alkaloids, nucleotides.
Ion exchange cellulose	Nucleotides.
Starch	Amino acids.
Sephadex	Amino acids, proteins.

8. Hydrophobic Interaction Chromatography (HIC)

Biomolecules differ in the amount of hydrophobicity that is present on their surfaces. This difference in the hydrophobicity can be used to separate biomolecules using hydrophobic interaction chromatography (HIC). This was developed mainly to purify proteins by exploiting their surface hydrophobicity and presence of non-polar amino acid residues. These hydrophobic residues are scattered over the surface of different proteins in different patterns. This gives characteristic properties to each protein. In aqueous solution, these hydrophobic regions on protein molecules are covered with an ordered film of water molecules that effectively mask the hydrophobic groups. These can be exposed by the presence of high salt concentration facilitating the protein-protein interaction. The presence of hydrophobic groups attached to a suitable matrix

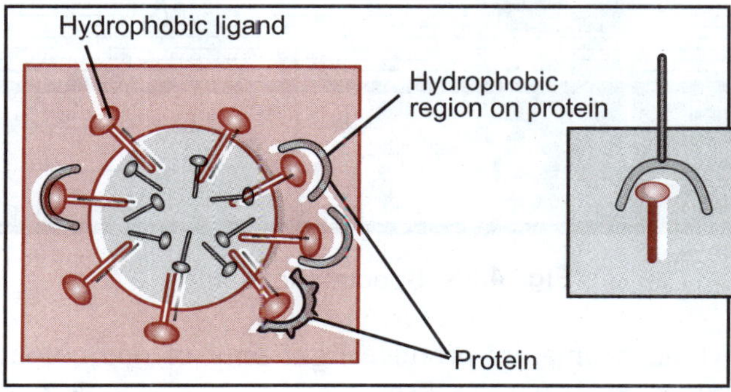

Fig. 4.23 Hydrophobic interactions between ligands and hydrophobic amino acid residues

facilitates protein–matrix interaction. The most commonly used stationary phases are alkyl (hexyl, octyl) or phenyl groups attached to an agarose matrix.

4. Spectrophotometer

Spectrophotometer is the instrument that measures the absorption of radiation in the visible and UV regions of spectrum. It is a sophisticated type of colorimeter. Spectrophotometer is a combination of two words: spectro and photometer. **Spectro** implies the whole range of continuous wavelength that is produced by a light source and **photometer** is a device for measuring the light. It means it can measure the light rays of wavelength ranging over UV and visible regions of the spectra and can obtain absorption spectra.

Components of Spectrophotometer

The basic components of spectrophotometer are:

1. Source of Light: The light source provides electromagnetic radiation in the UV and visible regions of the spectrum. A tungsten lamp is used to obtain light in the visible range. For UV region, a hydrogen lamp or deuterium lamp or xenon lamp is used. For infra-red rays either a Nernst glower or globar is used. Nernst glower is a hollow rod of yttrium and zirconium oxide. On being heated to about 1450°C, it produces infra red rays. **Globar** is a rod of silicon carbide heated to about 1200°C. A small slit allows the passage of a thin light beam to pass through and reach the monochromator.

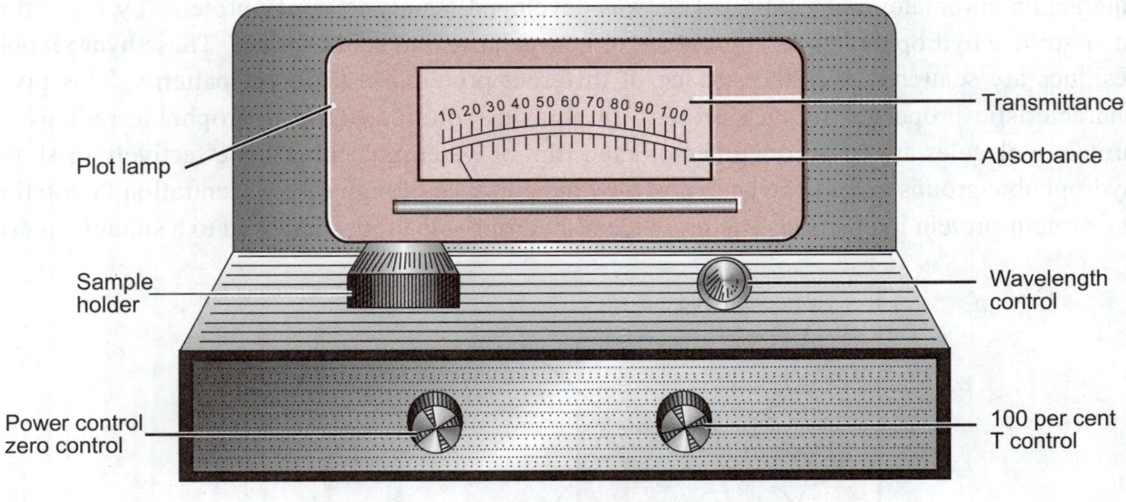

Fig. 4.24 Spectrophotometer

2. Monochromator: Monochromators produce light only of one colour or of one particular wavelength. They are either optic filters or diffraction gratings or both. Glass prisms are used for visible light and quartz prism is used for UV light (glass absorbs radiation below 400 nm). The instrument has a provision for selecting required wavelength by adjusting the knob. The slit S-II allows a thin beam of desired band width to passthrough and reach the sample.

3. Absorption Cell: This is the area where sample is placed in a cuvette. The cuvette used for optical light is made up of glass and is optically transparent, but for UV light it is made up of quartz. It has optical path length of 1 cm and contains 2.5–3.0 ml sample solution.

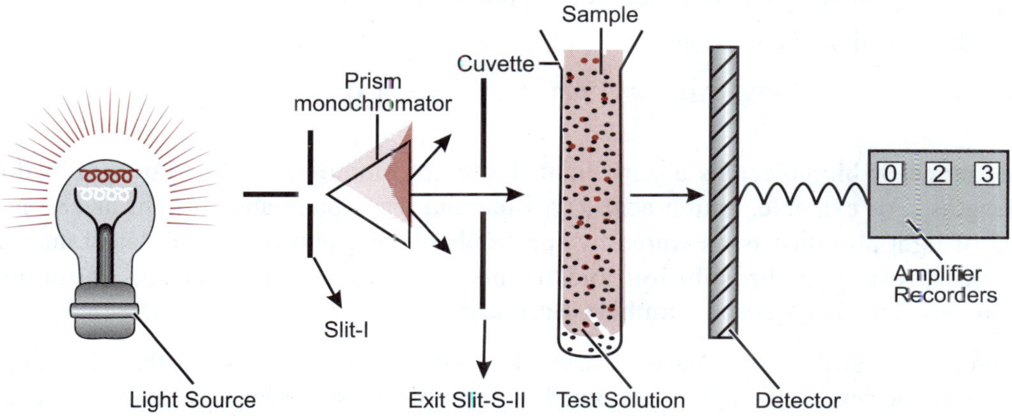

Fig. 4.25 Components of UV-VIS spectrophotometer

4. Detector: It measures absorbance of light rays or UV rays transmitted by the sample solution. It is formed of a **photocell**, which compares the intensity of light transmitted by the pure solvent (I) with that transmitted by the sample solution (I_0). For the visible and UV light the photoelectric detectors or **photocells** are used. For infra-red rays thermal detectors like **bolometers** are used. In the photocell, photons impinging on the metal surface in vacuum cause emission of electrons. They are attracted by a positive electrode. Thus, a current of electrons flows which causes a potential difference across a resistor present in the system.

5. Potentiometer: The potential difference is recorded on the potentiometer. It is calibrated to read absorbance and transmittance. The value of absorbance A/I= log I_0/I and the value of transmittence t = I/I_0.

The transmittance is recorded as a percentage and may vary from 0 to 100. The range of absorbance recorded ranges from 0 to 2.0.

6. Amplifier: In an amplifier photomultiplier tubes are used instead of photocell. Because in photomultiplier tubes the emitted electrons are accelerated by high potential and produce secondary electrons by collision with gas molecules. This results in enhancing the current. Therefore, even very small changes in electron concentration can be measured.

Applications of Spectrophotometer

Spectrophotometer is used in:

- Qualitative and quantitative estimation of biomolecules like proteins, nucleic acids, cytochromes and pigments.

- Measuring the growth kinetics.
- Structural studies of proteins and nucleic acids.
- Enzyme kinetics and assays.
- Measuring effects of pH, ionic strength, etc.
- Testing purity and homogeneity of samples and compounds.
- Identification of compounds.

LAMBERT-BEER LAW OF ABSORPTION OF LIGHT BY BIOMOLECULES

A wide range of biomolecules absorb light. Different biomolecules absorb light at different wavelengths. For example, amino acids, tyrosine and tryptophan absorb light at 280 nm. The amount of light absorbed is measured by spectrophotometer and is used to detect and identify molecules present in the solution and to measure their concentration. Photometry or spectrophotometry is based on **Lambert-Beer law**.

According to **Lambert-Beer law,** the intensity of transmitted light is less than the intensity of incident light because some light is absorbed by the molecules present in the solution. Therefore, the **Lambert-Beer expression** is:

$$\log \frac{I_0}{I} = \epsilon cl, \text{ here}$$

(i) ε (epson) is the molar extinction coefficient. It is expressed in units of litres per mol-centimeter.

(ii) c is the concentration of molecules of absorbing sample. It is expressed in moles per litre.

(iii) l is the path length of the light absorbing sample. It represents thickness of the absorbing sample and is expressed in cm.

The expression $\log \frac{I_0}{I}$ is called **absorbance** and is designated by **A**. The absorbance (**A**) of a solution or sample depends on the following features:

1. **Wavelength of Light:** Biomolecules absorb light at characteristic wavelengths. According to Lambert-Beer Law, the incident light absorbed is parallel and monochromatic. It is formed by light rays of single wavelength.
2. **Nature of Absorbing Molecules:** The amount of incident light absorbed by a solution at a given wavelength depends on the nature of light absorbing biomolecules.
3. **Concentration of Absorbing Solute Molecules:** The absorbance of a given solution is directly proportional to the concentration of absorbing solute molecules in the absorbing layer of fixed path length in the solution.
4. **Thickness of Absorbing Layer:** The amount of incident light absorbed by a given solution at a given wavelength is related to the thickness of the absorbing layer. The thickness of absorbing layer is also called **path length**.

Extension: Photometry

Photometry is the science that deals with the measurement of light intensity. It measures intensity of incident light and intensity of transmitted light and provides transmittance of a biomolecule.

The light that enters the solution of biomolecules is called **incident light** and the light that comes out is called **transmitted light**. The intensity of incident light is represented by $\mathbf{I_0}$ while that of transmitted light by **I**.

RADIOISOTOPE TECHNIQUES

Radioisotope techniques are basically used in analytical biochemistry to detect various macromolecules produced during different metabolic pathways or to trace the future path taken by a molecule or macromolecule during cell metabolism. For this purpose **radioactive isotopes** of different elements like carbon, hydrogen, sulphur, phosphorus or nitrogen are used.

What are Radioisotopes

Radioisotopes are radioactive isotopes of a normal element. They have same atomic number as of the normal elements but differ from them in having more atomic weight. The difference in the atomic weight is due to difference in the number of neutrons in their nucleus. The number of protons and electrons is the same as in their normal atoms. Radioisotopes are also called **radionuclides** and are formed during radioactive decay of a radioactive element. Due to their being unstable they emit α-particles or β-particles.

Extension: Radioactivity

Radioactivity is a phenomenon in which the nuclei of atoms of unstable isotopes decay or disintegrate spontaneously by releasing α-particles, β-particles or γ-rays.

The atomic nucleus of any element is made up of protons (P) and neutrons (N) and the number of protons determines the atomic number and also the number of electrons in the energy orbits. In a normal atom, N:P ratio is 1:1. In a nucleus where N:P ratio exceeds 1:1, a nuclear reaction occurs by releasing β-particles. The β-particles are electrons released from the nucleus by the decay of a neutron into a proton and an electron till the neutron:proton ratio is restored to 1:1.

If a nucleus is heavy and its atomic number is above 82, it releases both neutrons and protons and acquires a more stable arrangement. The release of both neutrons and protons occurs by the emission of α-particles. These particles are relatively large and have double positive charge. By attracting electrons, they cause ionisation effects.

Commonly used Radioisotopes

The commonly used radioisotopes in the study of cell biology are radioactive hydrogen $^{3}\mathbf{H}$, radioactive carbon ($^{14}\mathbf{C}$), radioactive phosphorus ($^{32}\mathbf{p}$), radioactive sulphur ($^{35}\mathbf{S}$), radioactive potassium ($^{42}\mathbf{K}$) and iodine ($^{131}\mathbf{I}$), etc. Hydrogen has three isotopes:

1. **Normal hydrogen (H)** having an atomic weight 1.00814
2. **Deuterium (^{2}H)** with an atomic weight 2.01474
3. **Tritium (^{3}H)** with an atomic weight 3.01701

All three forms of hydrogen have the same atomic number and occupy the same position in the periodic table. Similarly ^{14}C represents the isotope of normal carbon ^{12}C.

Role of Radioisotopes

Radioisotopes are used in analytical biochemistry for localisation, investigation and estimation of various compounds in the biological systems. Some of the uses of radioisotopes are:

1. **Investigating Metabolic Pathways:** Radioisotopes are used for tracing metabolic pathways. This involves:
   - Adding a radioactive substrate (with radioisotopes),
   - Taking samples of experimental material at regular intervals, and
   - Extracting and separating the products chromatographically.

   The steps involved in metabolic pathways of photosynthesis and metabolism of glucose, etc., have been worked out by using radioisotopes. With the help of radioisotopes, it is possible to predict the fate of individual carbon atoms (^{14}C) through TCA cycle.

2. **Study of Absorption, Accumulation and Translocation of Compounds:** Radioisotopes are widely used to study the mechanism and rate of absorption, translocation and accumulation of organic compounds in the cells and also in the organisms. This helps in knowing the location and accumulation of molecules of biological interest.

3. **Enzyme and Ligand Binding Studies:** Virtually any enzyme reaction occurring in the cell can be assayed using radiotracer methods. Radiotracer-based enzyme assays provide high degree of accuracy. Radioisotopes are used in the study of mechanism of enzyme action and also in the ligand binding studies.

4. **Radioimmunoassay:** Radioisotopes are used in quantitative analysis of hormones, steroids or other organic compounds in the cells or in the organism. This is called **radioimmunoassay**.

5. **Isotope Dilution Study:** The compounds or elements which are present in very small amount in the cells and can not be assayed accurately by conventional methods, are assayed or measured by radioisotope dilution. This method has proved to be very useful for assaying trace elements in the cells.

6. **Location of Sites of Synthesis of Various Organic Compounds:** Radioisotopes are used to trace the sites in the cells or cell organelles, where synthesis of particular compounds take place.

7. **Radioisotopes in Molecular Biology:** Radioisotopes have been used for the identification of genetic material, the study of molecular structure of genetic material, its replication, its functioning and mechanism of protein synthesis.

Recent advances in the field of molecular biology that led to the advances in genetic manipulation have developed only by the use of radioisotopes in DNA and RNA sequencing, in DNA replication, transcription, synthesis of cDNA and recombinant DNA technology.

Extension: Radioactive decay

The radioactive isotopes of an element are unstable. They emit neutrons from their nucleus. By gradually loosing their neutrons or neutrons and protons, they finally change into a stable non-radioactive element. This process of loosing neutrons or protons is called **radioactive decay**.

The rate of disintegration of a radioactive isotope of an element is always constant and remains unaffected by any changes in the environment or by the presence of any other radioactive element in the vicinity.

Each radioactive element has a specific **half life**, which can be defined as the time taken by a specific radioactive element to decay to one half of its original amount. For example, ^{3}H has half life of 19 days, ^{14}C has 5579 years and ^{238}U (uranium) has half life of 4.5 billion years (= 4.5×10^9 years).

Techniques for the Measurement of Radioactivity

Two types of counters are used for measuring radioactivity. These are:

1. Geiger Muller counter, and
2. Scintillation counter.

1. Geiger Muller Counter

Geiger Muller counter is used to detect radioactivity by ionisation, produced in the air or gas by radioactive disintegration. It consists of a hollow metal cylinder which acts as a cathode. It is enclosed in a glass cylinder. A wire running along the central axis of the cathode cylinder acts as an anode. The cylinder is filled with a counting gas. It is a mixture of neon and argon under low pressure or helium with a small amount of halogen or the organic gas. For accurate quantitative analysis, Geiger Muller tube is enclosed in a lead block to shield the tube and chamber from outside radiation.

***Working*:**

The radioactive element present in the cylinder produces radiation. On entering the tube these radiations cause ionisation of the detector gas leading to the formation of pairs of ions (positive ions and electrons). As a result, a pulse of current flows between the electrodes together with the emission of UV radiation. If high potential difference is applied across the electrodes these ions are accelerated toward the electrodes of opposite charge. The accelerated ions also react with gas atoms of the tube to produce more ions. This chain reaction continues to produce an avalanche of ions. This amplification of ions produces a flow of electrons in the external circuit and produces a measurable potential of 1–10 volts.

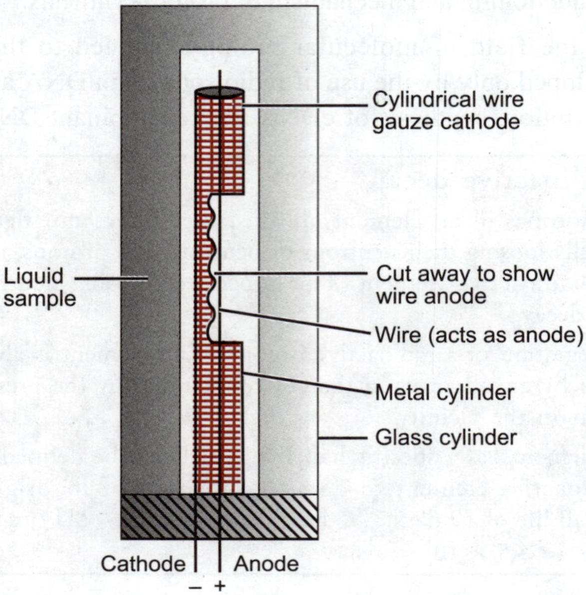

A Geiger Muller tube for measuring the radioactivity of the liquid sample

The electronic circuits associated with the tube are designed either to indicate the average current flow or to measure total number of electrical pulses. By Geiger counter α and β particles are detected. This type of counting is called **proportional counting** because the size of signal is proportional to the amount of ionisation produced by incident radiation. The ability to induce ionisation in gas decreases in the following order:

$$\alpha > \beta > \gamma = (10{,}000 > 100 > 1)$$

Therefore, *α* and *β* particles can be detected by gas ionisation method but γ radiations are poorly detected.

2. Scintillation Counter

Scintillation counting is a technique of detecting radiation via production of light and measuring radioactivity. The energy of radiation is used to produce light pulses which are counted.

Some substances respond to ionising effect of α and β particles by emitting light flashes. Such substances are called **fluors** or **scintillants**. They produce fluorescence.

Scintillation Counting Principle

Liquid scintillation counters have two light shielded **photomultiplier tubes**. The space between the two is filled with the sample. A scintillant mixture contained in a glass vessel is also placed inside. A range of scintillants are commercially available, each showing particular light emission characteristic. Scintillant mixtures contain a primary scintillant such as 1-phenyl-4- phenyloxazole (PPO) and often contains a secondary scintillant such as 1, 4, di–2-5 phenyloxazoyl benzene (POPOP). They are dissolved in a suitable solvent, e.g. toluene or xylene for organic samples and dioxane for aqueous samples.

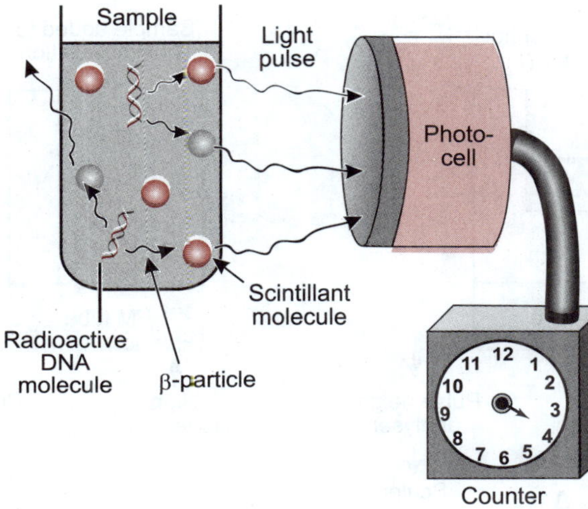

Fig. 4.27 Scintillation counter is used to measure radioactivity

***Working*:**

The photomultiplier tubes transform the emitted photons into electrical signal (Fig. 4.28). The magnitude of this signal is increased by more than a million times. The pre-amplifier further multiples the photomultiplier output. This allows high counting rates to be achieved using scintillation counting.

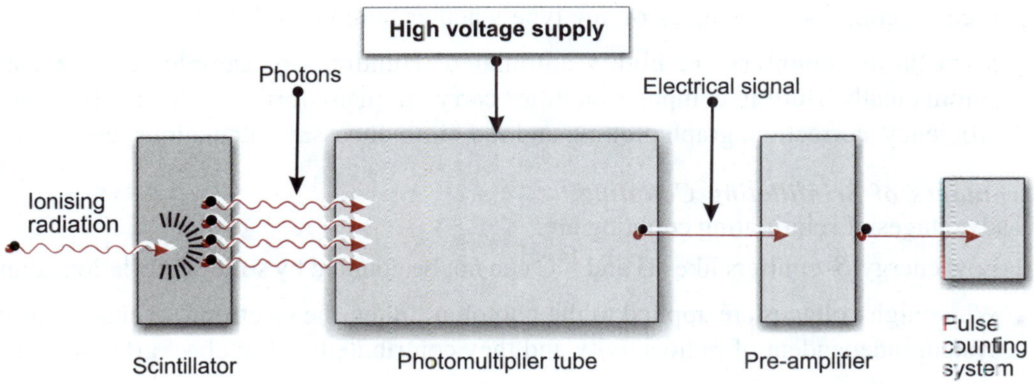

Fig. 4.28 Simplified mechanism of scintillation counting

The **pulse height analyser** differentiates between energies of the incident. It classifies pulses according to their height or amplitude.

Types of Scintillation Systems

Depending on whether the scintillant is solid or liquid there are two systems:

1. Solid scintillation counting system, and
2. Liquid scintillation counting system

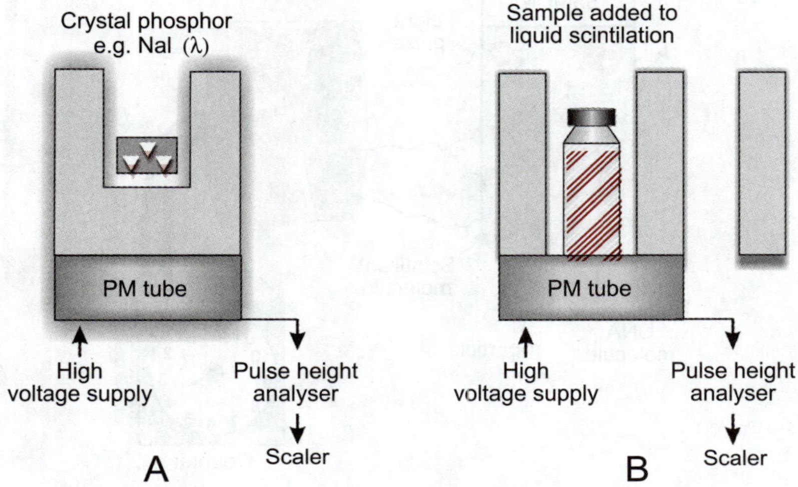

Fig. 4.29 Diagrammatic illustration of solid **(A)** and liquid **(B)** scintillation counting methods. PM tube is photomultiplier tube

Advantages of Scintillation Counting

The scintillation counting is widely used in biological work indicating that it has several advantages over gas ionisation counting. These advantages are:

- It provides very high counting rates.
- It can accommodate samples of any type whether liquid or solid.
- Scintillation counters are highly automated. Hundreds of samples can be counted automatically. Built in computer facilities carry out many forms of data analysis, such as efficiency, correction, graph plotting and radioimmunoassay calculations, etc.

Disadvantages of Scintillating Counting

The disadvantages of scintillating counting are:

- Low energy β- emitters like ^3H and ^{14}C can not be counted by solid scintillation counting.
- When high voltages are applied to the photomultiplier, the electronic events occur in the system independent of radioactivity and they contribute to a high background count.
- When energy transfer process suffers interference, it is called quenching. Correcting the quenching is an expensive solution.
- Chemiluminescence also causes problem. It may react with the samples and produce light emission unrelated to excitation.

AUTORADIOGRAPHY

Autoradiography is a technique of detection and demonstration of radioactive isotopes in tissue sections by the reduction of silver grains in a photographic plate, film or emulsion. Autoradiography is used to trace the route and conversion of molecules in different biochemical reactions. This technique was developed by **Loudon** in 1904.

Radioactive isotopes of various elements such as ^{14}C, ^{3}H, ^{32}P and ^{35}S, etc., are introduced into the culture medium and labelled metabolites are then incorporated in the cells or tissues. These cells or tissues labelled with radioactive isotopes are placed in contact with a photographic emulsion for some time. The ionising radiations emitted by the incorporated isotopes cause blocking of the emulsion and produce the image. This image is known as autoradiograph. The position of radioisotopes can be localised in radioactive material.

Procedure of Autoradiography

The procedure of autoradiography involves following three steps:

Preparation of Sections

1. Small pieces of tissues having radioisotopes are fixed in absolute alcohol for 1–3 hours.
2. These are cleared in benzene for 15 minutes, followed by warm benzene (56°C).
3. These are then transferred to four changes of molten paraffin wax at 56°C for 30 minutes in each case.
4. Each piece is embedded in paraffin for block making. The block is cut into 5μ thick sections. These are allowed to float on warm water.
5. Sections are picked up on to slides which have previously been dipped in an aqueous solution of 1 per cent gelatin and 0–1 per cent chrome alum, and then allowed to dry.

Application of Photographic Emulsion

A variety of emulsions are available having silver halide crystals. X-ray film is generally suitable for macroscopic samples. For microscopic observations, very sensitive films are necessary. Following methods are in use for the application of photographic application.

1. **Contact Method:** The wax from the sections is removed. The tissue sections are then placed in contact with a photographic plate and are stored in dark room for exposure. Contact between sections and the film is maintained for appropriate time. The film is developed afterwards.

2. **Stripping Film Method:** This method is based on the uses of a thin photographic emulsion on a gelating base which must be stripped from the plate or celluloid film. Special stripping films are used for autoradiography. This film consists of two layers, a gelatin layer facing the glass and 4 μ thick photographic emulsion upon it. The emulsion is

very sensitive to scratching. The film is cut into small rectangular pieces. A piece is stripped off and is kept in a trough filled with distilled water. It floats on distilled water with emulsion side directed downward. The deparaffinised slide with sections is placed in water with sections directed upward. The film should cover the sections on the slide. The slide is stored in dark room for exposure. The sections are developed and studied under the microscope.

Application of Film

1. The film is cut into squares. Leave the pieces in dark for 5–10 minutes.
2. Strip one square slowly, starting at one corner. Float it on dust-free, fresh distilled water at 23–25°C. Keep the emulsion side of the film downward for 2–3 minutes.
3. Place the slide with deparaffinised sections in water with sections or squash facing upward under the film. Lift the slide out of water in such a way that the film covers the sections or the squash.
4. Dry the slide at room temperature either under a fan or with a hair-dryer.
5. Store the slide in dark at room temperature.
6. Complete the photographic processing, rinse with distilled water and then in running tap water for one hour.
7. The sections are stained with haem-alum and differentiated and mounted.
8. For phase contrast microscopy, sections are mounted in glycerine jelly or glycerogel without staining.

TEXT QUESTIONS

1. What is electrophoresis? Discuss the principle and types of electrophoresis.
2. Describe the principle on which electrophoresis is carried out. Discuss its applications.
3. Describe the procedure of electrophoresis.What factors influence migration velocity of macromolecules?
4. What are sedimentation coefficient and velocity sedimentation? How is sedimentation coefficient determined and what are the units used to express it?
5. Describe the principle of centrifugation. What is the basic difference between centrifugation and ultracentrifugation? Which of the two is used for separation of macromolecules?
6. Give differences between zonal centrifugation and equilibrium centrifugation.
7. Describe the various techniques used in electrophoresis.
8. Describe the principle, procedure and application of gel electrophoresis.

9. What is the role of CsCl in the separation of macromolecules?
10. Describe the complete procedure for electrophoretic separation of DNA segments.
11. What is chromatography? Describe its applications. Discuss the principle of chromatographic separation of macromolecules.
12. Write a brief account of ion exchange chromatography. What is the principle of this technique?
13. What are radioisotopes? Summarise the role of radioisotopes in molecular biology.
14. Describe the various methods of counting or measuring radioactivity.
15. What is scintillation counting system? Discuss the principle, working and advantages of scintillation counting system.
16. Discuss in brief the advantages and disadvantages of scintillation counting system.
17. Explain the terms—radioisotopes, radioactivity and radioactive decay. What is half life of a radioactive element?
18. What is photometry? What is its use in the study of biological molecules?
19. Discuss Beer–Lambert Law of absorption of light by biomolecules. Discuss the various uses of spectrophotometer.
20. Write a paragraph on each of the following:
    - (a) Gel electrophoresis
    - (b) Spectrophotometry
    - (c) Polyacrylamide gel
    - (d) Quantitative autoradiography
    - (e) Affinity chromatography
    - (f) Ion–exchange resins.

5
Chapter

Identification of Genetic Material

IN THIS CHAPTER

The Genetic Material: Characteristics of the Genetic Material; Nature of Genetic Material; DNA or Nucleic Acid as Hereditary Material; Experimental Evidences that DNA is the Genetic Material: Griffith's Experiment on Bacterial Transformation; Avery's Experiment that DNA is the Transforming Principle; Leaderberg's Experiment of Genetic Recombination in Bacteria; Hershey-Chase Experiment of Bacteriophage Infection to demonstrate that DNA is Hereditary Material; Transduction Experiment to demonstrate that DNA carries Genetic Information; Evidence from the Effect of Ultraviolet Light; Biochemical Evidences for DNA as Genetic Material: Indirect Evidences that DNA is Hereditary Material; Evidences that RNA can be Genetic Material.

THE GENETIC MATERIAL

The genetic material of any organism is the substance that stores information about structure, function and development of various characteristics of a living organism. It is responsible for the transmission of genetic information for all the characteristics of living beings from parents to the progeny.

Following the rediscovery of Mendel's laws, geneticists were able to conclude that organism's characters are controlled by genes, that genes are able to reproduce themselves or replicate without losing their information, that genes are arranged on the chromosomes in a linear fashion, and that the genes are transmitted from parents to offsprings generation after generation almost unaltered.

Characteristics of the Genetic Material

Based on the above conclusions, geneticists were able to assign following characteristics to the genetic material:

1. Storage of Genetic Information: Genetic material should have the ability to store all the genetic information about structure, function, development and reproduction in all the living cells of the organisms.

2. Accurate Replication: It should possess the power of accurate replication so that the same genetic informations as present in parent cell are passed on to the daughter cells.

3. Transcription: It should possess power of accurate transcription so that the stored genetic informations are transmitted to the cell as and when needed.

4. Chemical Stability: It should have physical and chemical stability so that the stored informations are not lost.

5. Variations: It should be capable of undergoing modifications by mutations and recombinations so that it can contribute to the variations and adaptations leading to the evolution of living organisms.

Nature of Genetic Material

After knowing that hereditary (= genetic) material resides in the nucleus and that the chromatin material composing the genes is a complex of proteins and nucleic acids, the most important question to be solved was whether proteins form the genetic material or the nucleic acids.

Protein as the Hereditary Material

Till the beginning of twentieth century, proteins were considered to be the only suitable material for storing and transmitting genetic information, because:

1. Proteins have the required **structural diversity** necessary to perform the hereditary role and to store the hereditary information.
2. Proteins are **most abundant** in the body, constituting about 16 per cent of the total body weight.
3. Proteins form the structural framework of the cell and are associated with the performance of all sorts of cellular and body activities and have a regulatory effect (**functional diversity**).
4. Proteins occur in varied forms (**structural diversity**). The possible combinations of 20 amino acids in proteins can be 20^n, where n is the number of amino acids in a polypeptide chain.

DNA or Nucleic Acid as Hereditary Material

Though, known for long, DNA was not considered to be the possible genetic material. It was found to be a simple polymer of tetranucleotide sequences repeated over and over again. Therefore, DNA was thought to be too simple to form a gene. Rather, it was regarded to be a molecular framework or scaffold for the special proteins (genes) that carried genetic information.

Swiss biochemist, **Johann Friedrich Miescher** (1869), collected white blood cells from the pus from surgical bandages. By treating their nuclei with alkali, he prepared an extract and from that extract separated a nonprotein compound rich in phosphoric acid. Based on the source, this compound was named **nuclein**. It was largely formed of DNA. The quantitative analysis of chromosomes established that chromosomes are formed of 40 per cent DNA and 60 per cent of basic proteins or histones. The histones are small protein molecules rich in basic amino acids. They were named **'protamines'**.

EVIDENCES THAT DNA IS THE GENETIC MATERIAL

Experimental Evidences

The following experiments conducted by the molecular biologists with microorganisms, viruses and bacteriophages provided direct evidence of **DNA** being the genetic or hereditary material:

1. Bacterial transformation
2. Bacterial recombination
3. Bacteriophage infection
4. Transduction.

1. Griffith's Experiment on Bacterial Transformation

Frederick Griffith, a bacteriologist of England (1928), provided the first clear cut evidence that **DNA** is the hereditary material (Fig. 5.1). He found that the pneumonia-causing bacteria, *Diplococcus pneumoniae,* earlier known as *Pneumococcus* exists in two strains:

1. Capsulated Virulent Strain or Smooth Strain (S-III): The pneumococcal cells of virulent strain are surrounded by a thick shiny capsule of polysaccharides. They cause fatal pneumonia when injected in mice. The colony of these cells has glistening appearance and is called **smooth (S)**. The shiny appearance is due to the presence of capsule whose formation is under the control of its genotype.

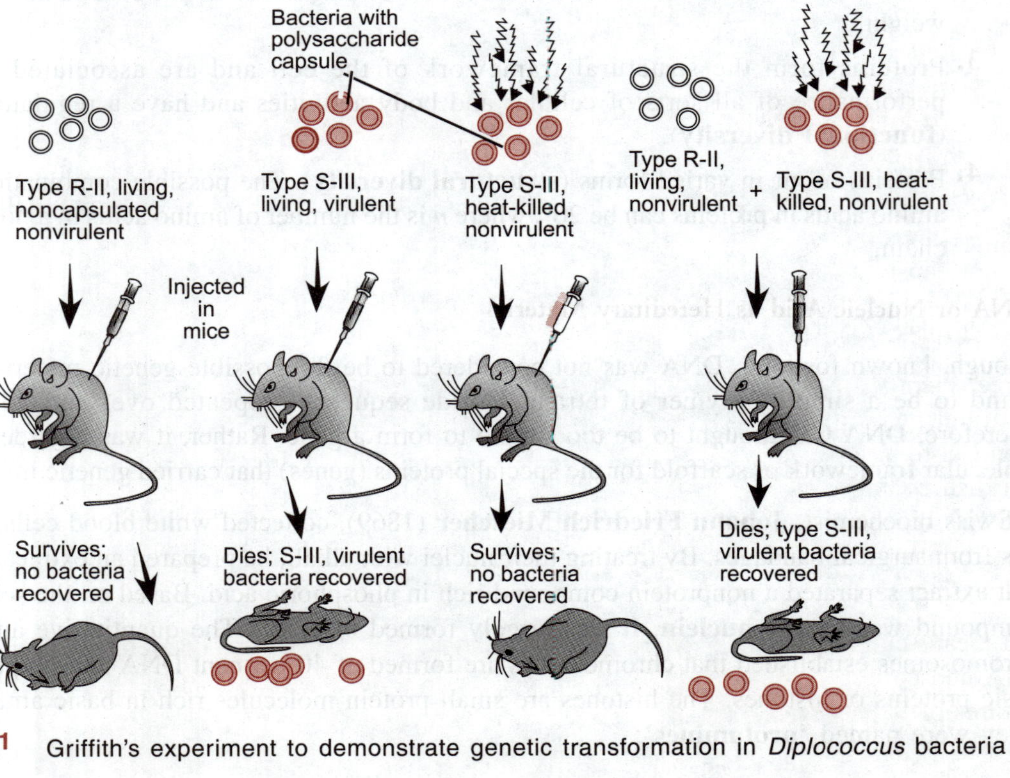

Fig. 5.1 Griffith's experiment to demonstrate genetic transformation in *Diplococcus* bacteria

2. Non-capsulated Avirulent Strain (R-II): The pneumococcus cells of this strain do not produce symptoms of pneumonia. They lack polysaccharide capsule. The colonies of these cells have irregular appearance and are known as **rough (R) colonies**.

Frederick Griffith (1879–1941)

Griffith was a British medical officer. While working to make a vaccine against *Diplococcus* to prevent pneumonia epidemics after World War I, Griffith concluded the presence of some 'Transforming principle' from heat inactivated S-III strain and later on, identified this factor as DNA. Griffith was killed in an air raid in London while at work. He was holding a page with some formulae that seemed to be some breakthrough, but no one succeeded to interpret them.

- When **R-II** (non-virulent) bacteria were injected in mice, mice suffered no illness.
- When **S-III** (virulent) bacteria were injected in mice, mice developed pneumonia and died.
- **S-III** (virulent) bacteria were heated to 60°C and killed. These heat-killed bacteria were injected in mice. No pneumonia symptoms developed and mice remained healthy.
- A mixture of **R-II** (non-virulent) and heat-killed **S-III** bacteria was injected into mice. Some mice developed pneumonia and died.
- Bacteria isolated from these dead mice were found to contain a mixture of capsulated **(S-III)** and noncapsulated **(R-II)** forms. **Griffith** concluded that the presence of heat-killed **S-III** bacteria caused transformation of some **R-II** bacteria into virulent, capsulated **S-III** bacteria. This was described as '**Griffith effect**' or '**bacterial transformation**'. The transformed bacterial cells retained their infectious property and capsule in successive generations.

Avery's Experiment that DNA is the Transforming Principle

Three scientists, **Oswald Avery, Colin MacLeod** and **Maclyin McCarty** (1944) of Rockfeller Institute, performed the crucial experiment of separating and identifying the transforming chemical to be DNA (Fig. 5.2). A culture of nonvirulent type **R-II Diplococci** was grown in the petridish containing agar and the microbiological culture medium. The capsulated bacteria of virulent strain type **S-III** were then ground finely and were divided into three sets. Set **A** was treated with protein degrading enzymes (*proteolytic enzymes*) and set **B** was mixed with RNA-hydrolysing enzyme (*ribonuclease*) and set **C** with DNA-hydrolysing enzyme (*deoxyribonuclease*).

It means culture **A** of S-III bacteria was left with DNA and RNA, culture **B** with proteins and DNA and culture **C** with protein and RNA. These three cultures were added to three separate cultures of R-II bacteria. The new cultures of **R-II** bacteria with extract from **A** and **B** cultures contained a mixture of virulent **S-III** and nonvirulent **R-II** bacteria. It means some **R-II** bacteria were transformed into **S-III** bacteria by DNA of S-III type. The third culture of R-II bacteria with extract from culture **C** (having S-III proteins and RNA) contained only **R-II** type bacteria. This showed that no transformation has occurred because of the absence of DNA. It means DNA is the transforming principle and not the protein or RNA. This phenomenon of transferring characters of one strain to another by using DNA extract is known as **transformation** and DNA as the **transforming principle**. When this extract was treated with (DNAase hydrolysing enzyme), the transforming ability was lost and when treated with *proteases* (the protein hydrolysing enzymes), the transforming ability was retained.

Oswald T. Avery (1877–1955)

Born on 21 Oct.1877 in Canada, **Avery**, studied microbiology at Columbia University. From 1933 he worked at Rockfeller institute. **Avery** was a dedicated bacteriologist. He worked on pneumonia-causing bacteria and identified polysaccharides specific to the capsule of S-bacteria. He also recognised the isolated principle from heat-killed S-bacteria, responsible for the transformation of R-bacteria (non-virulent) into S-bacteria (virulent) to be DNA. He used the following analytical procedures: 1. Electrophoresis, 2. Ultracentri-fugation, 3. Ultraviolet absorption, 4. Spectrophotometry and 5. Chemical analysis.

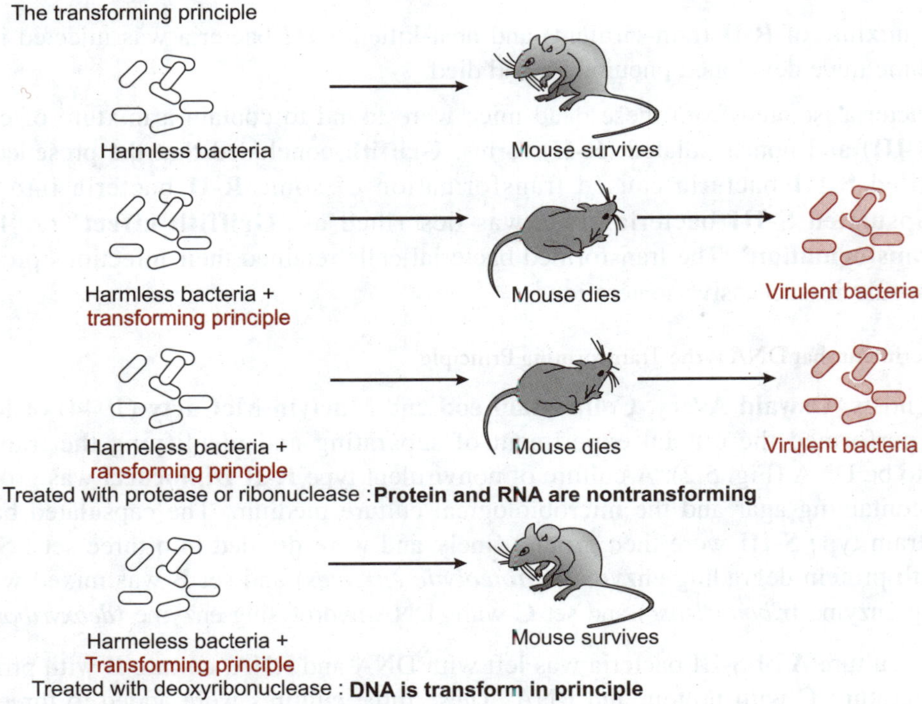

Fig. 5.2 Genetic transformaton in *Streptococcus penumoniae* using extract from S-III heat-killed virulent bacteria, conducted by Avery, MacLeod and McCarty

2. Lederberg's Experiment of Genetic Recombination in Bacteria or Bacterial Conjugation to show that DNA is Hereditary Material

Lederberg and **Tatum** (1946) discovered genetic recombination in bacteria from two mutant strains through the process of conjugation (Fig. 5.3). Bacterium, *Escherichia coli,* can grow in minimal culture medium containing minerals and sugar only. It can synthesise all the necessary vitamins from these raw materials. **Lederberg** found its two mutant strains that lacked the ability to synthesise some of the vitamins necessary for growth. They could not grow in the minimal medium till the particular vitamins were added in the culture medium.

- **Mutant strain A** (used as male strain) had the genetic composition **Met$^-$, Bio$^-$, Thr$^+$, Leu$^+$, Thi$^+$**, $\cdots$. It lacked the ability to manufacture amino acid methionine (Met$^-$) and vitamins biotin (Bio) and could grow only in a culture medium which contained these vitamins in addition to sugar and minerals.
- **Mutant strain B** (used as female or recipient strain) had a genetic composition **Met$^+$, Bio$^+$, Thr$^-$, Leu$^-$, Thi$^-$**$\cdots$. It lacked the ability to manufacture amino acids- threonine (Thr$^-$), leucine (Leu$^-$) and thionine (Thi$^-$) and could grow only when these were added to the growing medium.

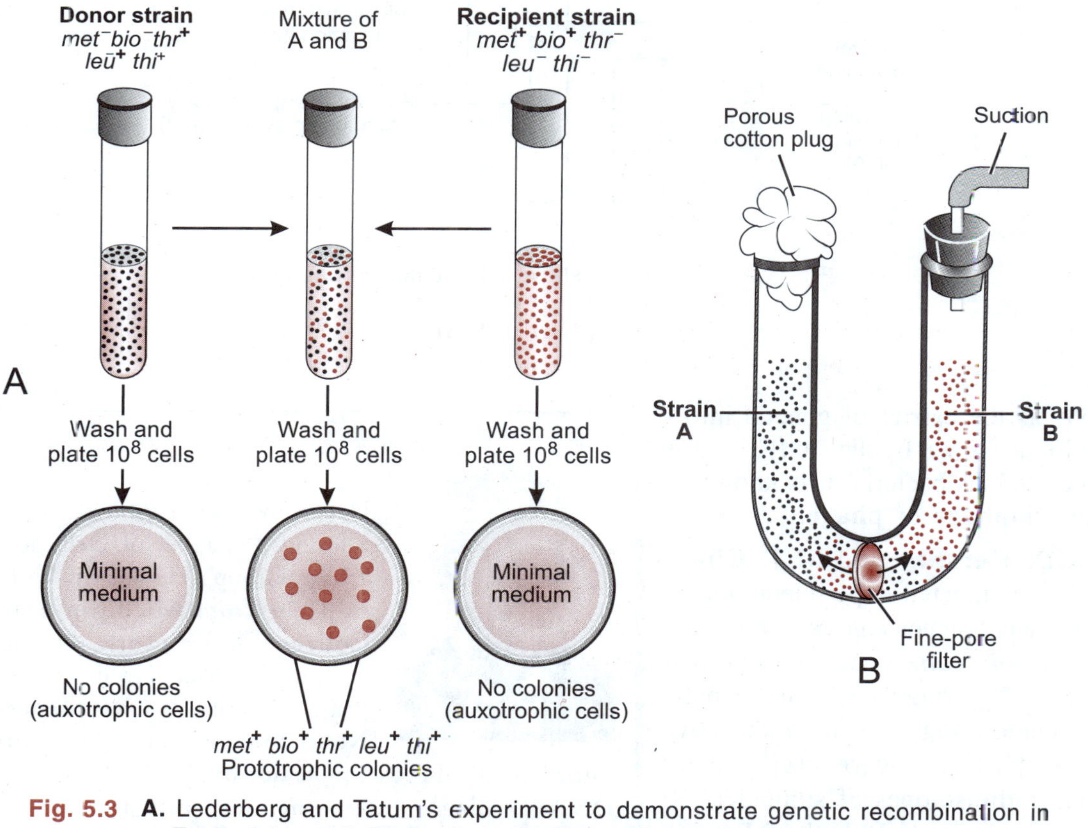

Fig. 5.3 **A.** Lederberg and Tatum's experiment to demonstrate genetic recombination in *Escherichia coli*; **B.** Davis U–tube experiment showing physical contact between the two strains

When grown separately, these two strains of *E. coli* were unable to grow in the minimal culture medium. But when a mixture of these two strains was allowed to grow in the same minimal medium, a number of colonies were formed. This indicates that the portion of donor DNA containing information to manufacture threonine, leucine and thionine has been transferred and incorporated in the recipient's genotype during conjugation.

The experiment conducted by **Lederberg** and **Tatum** (Fig. 5.3) showed that the conjugation results in the transfer of genetic material (DNA) from F^+ bacterium to F^- bacterium. During conjugation, a cytoplasmic bridge is formed between two conjugating bacteria (Fig. 5.4). When these conjugating bacteria were placed in Waring blender at different intervals of time after the formation of cytoplasmic bridge, the amount of DNA transferred was found to vary with the duration of conjugation. Further, there was a direct proportion between the length of DNA and the amount of genetic information transferred.

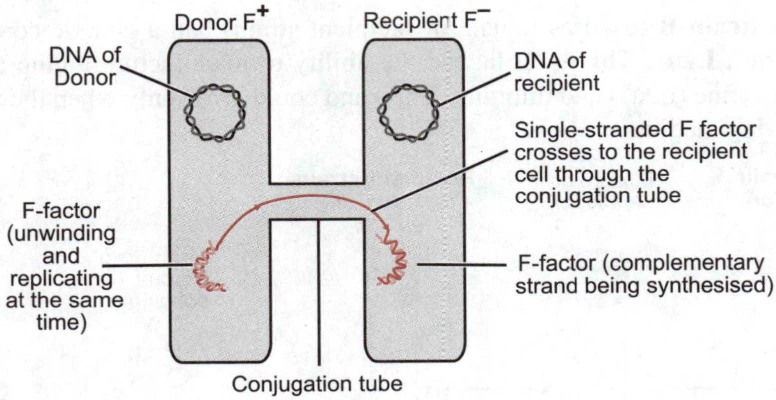

Fig. 5.4 Diagram to show conjugation in bacteria

3. Hershey–Chase Experiment of Bacteriophage Infection to demonstrate that DNA is Hereditary Material

DNA as the carrier of genetic information is shown by the study of virus which infects bacteria. It is known as **bacteriophage** or **phage**.

A.D. Hershey and **M.J. Chase** (1952) conducted experiments on T_2 phage, a parasite on common bacterium, *Escherichia coli,* to demonstrate that DNA of T_2 phage carries the genetic information (Fig. 5.6). The radioactive phage particles were prepared by using radioisotopes of sulphur (^{35}S) and phosphorus (^{32}P) in the following ways:

Alfred Hershey
(4th Dec. 1908)

Hershey was a microbiologist and a founder member of Delbruck's 'Phage Group'. He got his Ph.D. in 1934 from Michigan State College for his work on bacterium, *Brussella*. He discovered phage life cycle and was awarded Nobel Prize in 1969 along with **Luria** and **Delbruck** for demonstrating that only the DNA of phage enters the host cell and acts as hereditary material its protein coat is left outside.

Extension: Life Cycle of Bacteriophage

The phage gets attached to the bacterial cell with the help of its tail. The DNA present inside the head is transferred into the bacterial cell, whereas the protein coat of head and the tail are left outside. Inside the bacterial cell, the phage DNA replicates and produces large number of new phage DNA molecules. Each of these DNA molecules develops its own protein coat forming daughter phage particles which are released by the rupture of host cell.

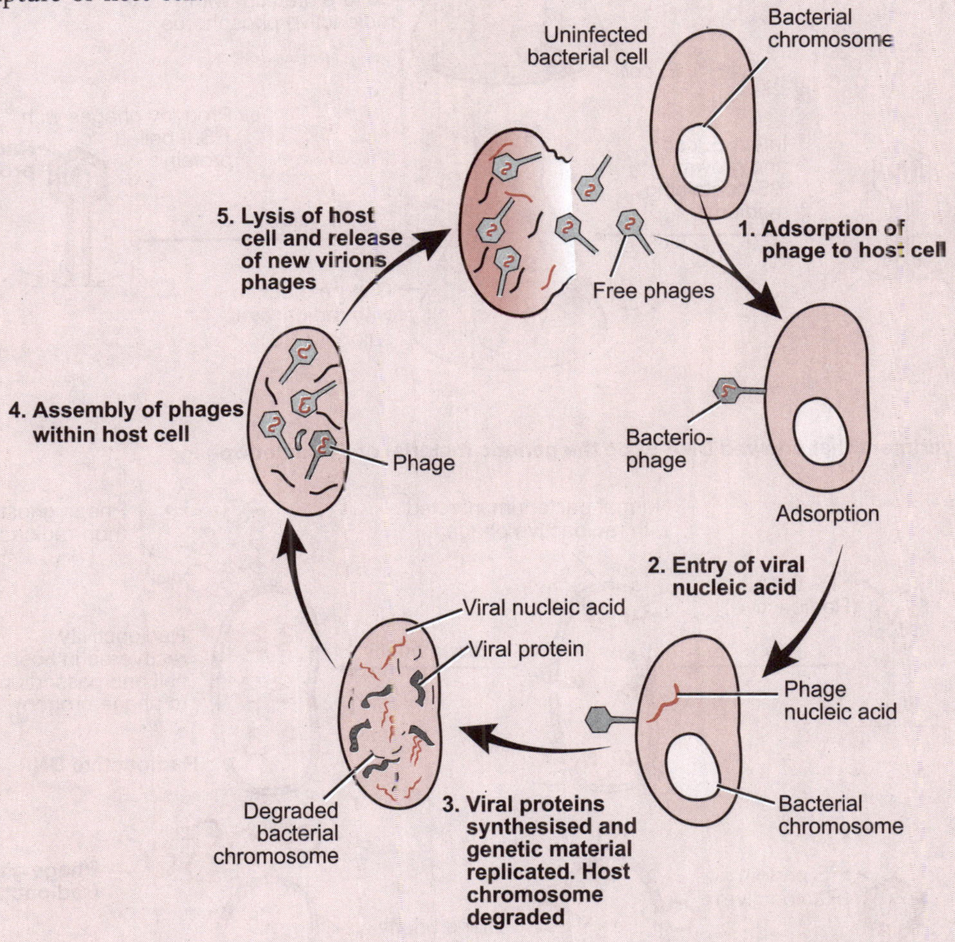

Fig. 5.5 Lytic life cycle of T_2 bacteriophage of *Escherichia coli*

1. Some phages were grown in bacteria containing **radioactive sulphur**. Sulphur is incorporated in the formation of amino acids. These radioactive amino acids formed the protein coat of the phages.
2. Some other phages were grown in bacteria containing **radioactive phosphorus**. Phosphorus is an important constituent of DNA. Therefore, radioactivity in these bacteriophages was restricted to DNA.

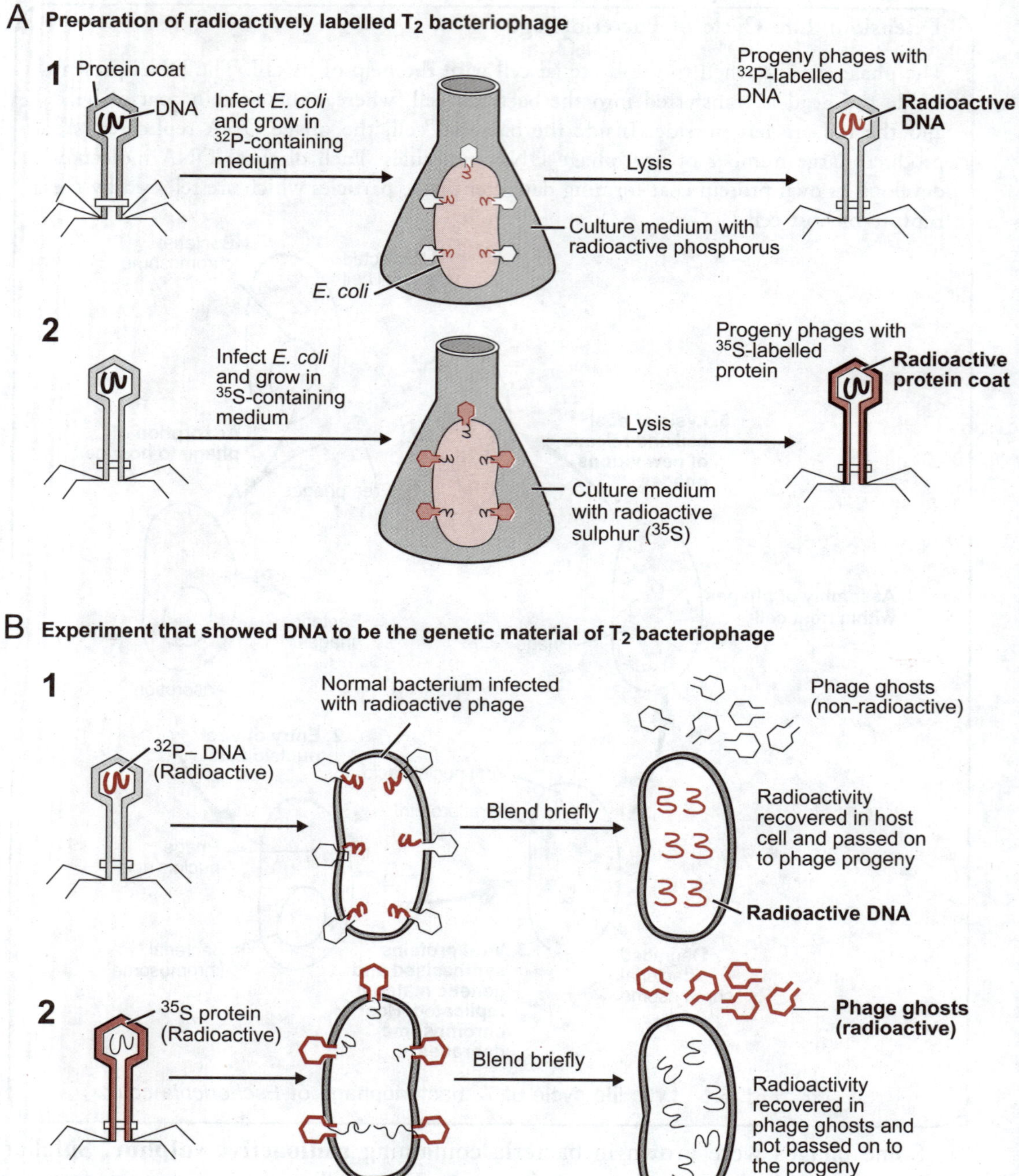

Fig. 5.6 Diagram to represent Hershey's experiment to demonstrate that in T_2 bacteriophage the hereditary material is DNA: **A.** Bacteriophage grown in radioactive phosphorus medium transmits its radioactivity to bacterial cells; **B.** Bacteriophage grown in radioactive sulphur medium does not transmit radioactivity to bacterial cells

These two types of radioactive phages were made to infect normal bacteria in two separate samples. About 20 minutes after the infection, the bacterial cells ruptured liberating daughter phages. When these daughter phages were tested for radioactivity it was found that:

1. The phages grown in radioactive phosphorus passed their radioactivity to the daughter phage particles.
2. The radioactive phages containing **radioactive sulphur** did not transmit their radioactivity to their daughter phages. Their **radioactive sulphur** was found in the remains of bacteriophages (the protein coat left outside the bacterial cells).

These results clearly indicated that **DNA** constitutes the hereditary material.

4. Transduction Experiment to Demonstrate that DNA Carries Genetic Information

Transduction is the transfer of a small fragment of DNA from one bacterial cell to another by means of a phage. The phage, while multiplying within the bacterial cells, occasionally encloses a

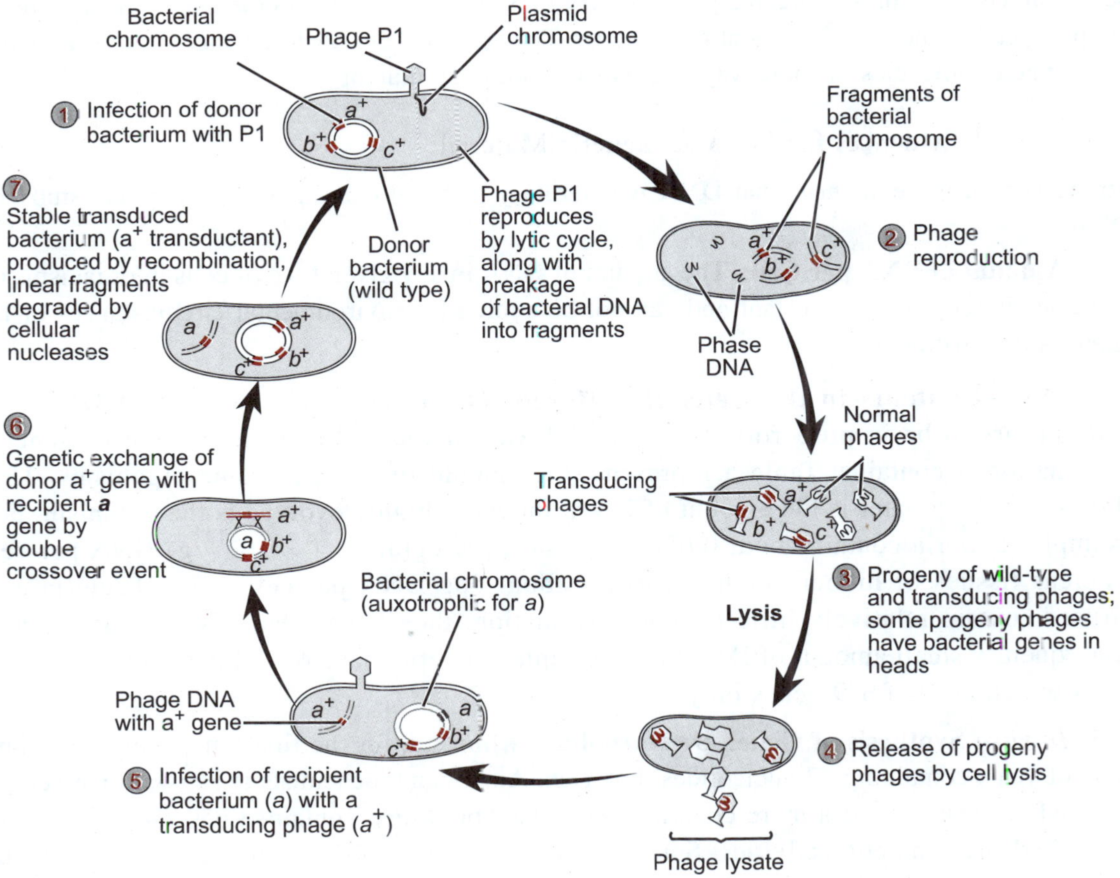

Fig. 5.7 Diagrammatic representation of transduction in bacteria

DNA fragment from the bacterial chromosome. Subsequently, the phage comes out of the host cell and attaches to a new bacterium. While infecting the new bacterium it releases its own DNA and also liberates the fragment of bacterial chromosome from the previous host. This segment may undergo crossing over with the new host's chromosome and thus incorporates genes from the previous host (Fig. 5.7). Since, only DNA is transferred, this provides a clear evidence that genes are made up of DNA.

5. Evidence from the Study of Effect of Ultraviolet Light on DNA

Nucleic acids absorb ultraviolet light very strongly and the maximum absorption occurs at a wavelength of 260 millimicrons. The production of mutations by ultraviolet light in a variety of organisms is maximum at 260 millimicrons. When the number of mutations produced per unit of energy delivered and the wavelength at which that energy is delivered are compared, an action spectrum for mutations is observed. There is a close correlation between the action spectrum for the production of mutations and the absorption spectrum for nucleic acids. This phenomenon can be explained by assuming that the genes are composed of nucleic acid molecules. The mutations are produced as these molecules absorb energy. The absorbed energy is effective in changing the nucleic acid molecules in some way and thus in producing mutant genes.

Biochemical Evidences for DNA as Genetic Material

Strong supporting evidences that DNA is hereditary material come from biochemical study of DNA:

1. Amount of DNA per Cell: The amount of DNA in any given type of cells in organisms of a species is remarkably constant and can not be altered by environmental circumstances or by changes in nutrition, etc.

2. DNA Contents in the Cells of Different Organisms: The amount of DNA per cell appears to be in proportion to the complexity of the cell and the amount of genetic information it contains. Table 5.1 presents the amount of DNA in some organisms. This shows that the greater is the amount of DNA, the more highly evolved is the organism. For example, bacteria contain about 0.02pg (1pg or 1 picrogram = 1×10^{-12} g) DNA per cell whereas tissues of higher animals contain about 6pg DNA per cell. Since bacteria and viruses contain relatively little genetic information, they have relatively few genes and consequently small amount of DNA. For example, bacteriophage $\phi\times174$ has only 1.8 dalton DNA which codes for 9 genes only.

3. *In vitro* Synthesis of Gene: **Dr Hargobind Khorana**, for the first time, synthesised one gene of yeast containing 77 nucleotides. Later on, American biochemists at Harward University successfully synthesised a more complex gene of rabbit which contained 650 nucleotides and controlled the synthetic activities. Scientists now are successfully synthesising genes in the laboratory. These facts leave no doubt that genes are DNA molecules.

Table 5.1 Approximate DNA content per cell in some organisms including viruses

S.No.	Species	DNA amount pg/per cell	Number of nucleotide pairs (in millions)
1.	Bacteriophage T_4	0.0024	0.7
2.	Bacteria	0.002–0.06	2
3.	Fungi	0.02–0.17	20
4.	Sponges	0.1	100
5.	Molluscs	1.2	1 000
6.	Crustaceans	3	2,800
7.	Fishes	2	2,000
8.	Amphibians	7	6,500
9.	Birds	2	2,000
10.	Reptiles	5	4,500
11.	Mammals	6	5,500

4. Base Equivalence in DNA: **E. Chargaff** (1949–53) made a quantitative analysis of four bases in hydrolyzates of DNA specimens isolated from different organisms. From these studies he presented following facts:

- Base composition of DNA varies from one species to another.
- DNA specimens isolated form different tissues of the same species have the same base compostion.
- The base composition of DNA in a given species does not change with age, nutritional state or change in environment.
- The number of adenine residues is always equal to the number of thymine residues (A = T) and number of guanine residues is always equal to the number of cytosine residues (G = C) or purine residues are equal to the sum of pyrimidine residues (i.e., A + G = T + C). This is called **'Chargaff's rule'**.
- The DNAs extracted from closely related species have almost similar base composition whereas those of widely different species are likely to have widely different base composition.

Indirect Evidences that DNA is Hereditary Material

DNA is genetic material is further confirmed by the following facts:

- The total DNA content in all diploid cells of an organism and in cells of all the organisms of a species is the same.
- The gametes contain half the amount of DNA present in their parent cells or in the somatic cells.
- Fertilisation between male and female gametes doubles the amount of DNA in zygote which is the same as found in the somatic cells of the parent.
- Polyploidy brings about a corresponding increase in the amount of DNA in the cell.

5. DNA maintains its integrity during cell metabolism and during the entire cell cycle except when it undergoes mutation.
6. Composition of DNA is the same in all organisms of a species, but different in the organisms of different species.
7. Physical and chemical agents, which bring about mutations in genes change the chemical structure of DNA.

EVIDENCES THAT RNA CAN BE GENETIC MATERIAL

Fraenkel Conrat and Stanley's Experiment with Tobacco Mosaic Virus

Fraenkel Conrat and **Stanley** at the University of California conducted experiments on Tobacco Mosaic Virus (TMV) and demonstrated that RNA in some viruses acts as genetic material (Fig. 5.8). Most of the plant viruses and some animal viruses are formed of RNA and proteins.

The TMV particles were collected from the infected tobacco leaves. Their proteins and nucleic acid (RNA) were then extracted out separately. It was observed that:

- When cell debris, consisting of protein coat of virus was introduced into a healthy tobacco leaf, it remained healthy.
- When cell filtrate containing nucleic acid was injected into tobacco leaf, it got infected with virus and died.

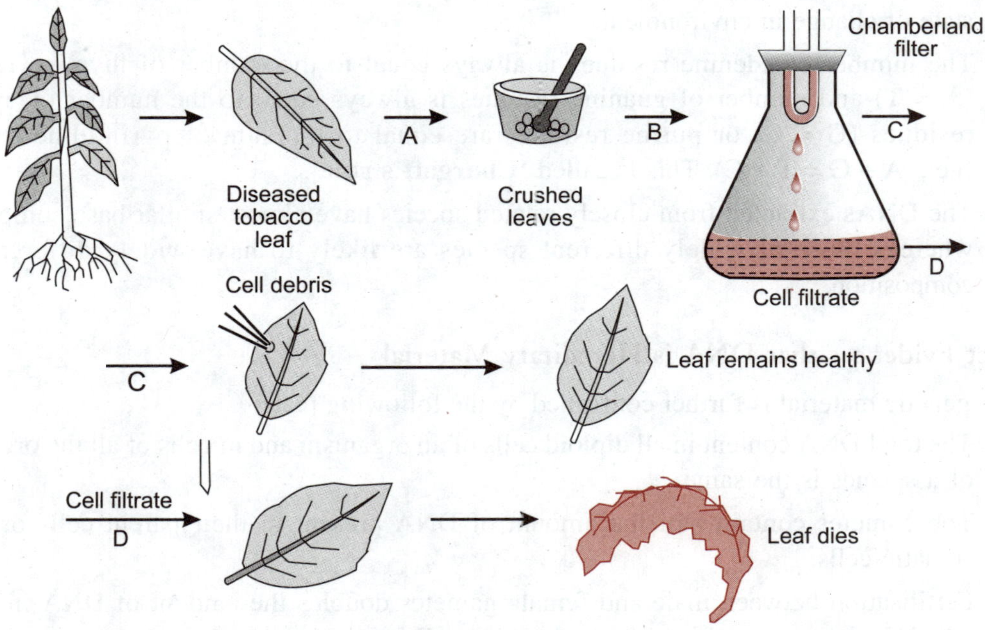

Fig. 5.8 Fraenkel Conrat and Stanley's experiment on TMV to show that RNA can also be the hereditary material

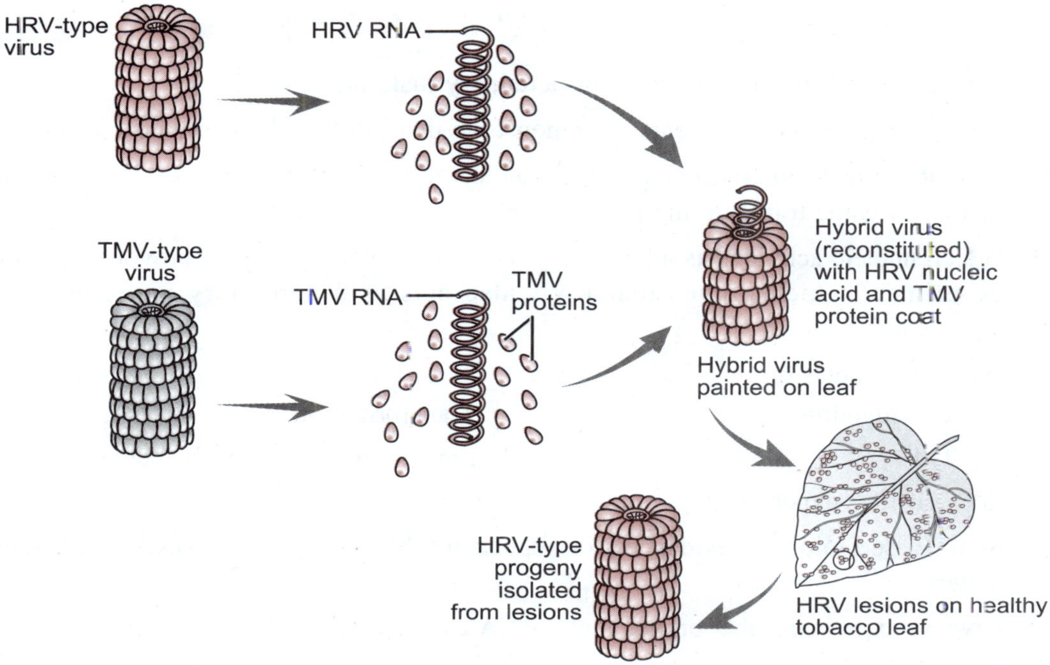

Fig. 5.9 Reconstruction of hybrid virus from two different varieties of viruses and their progeny resembling to the parent from which RNA was received

Above results show that RNA produces the infection but not the proteins. When these two substances were mixed together, they reconstituted TMV particles. These reconstituted TMV particles were capable of causing infection.

It was also possible to synthesise chimaeric virus particles using RNA from one variety and protein from another (Fig. 5.9). The hybrid viruses were prepared from tobacco mosaic virus (TMV) and Holmes ribgrass virus (HRV). These two viruses differ in following features:

1. Their proteins have different frequencies and sequences of amino acids.
2. Their RNA also has different sequence of nitrogenous bases.
3. The TMV particles produce mottling of leaves of tobacco plant, whereas HRV particles produce distinct ring patterns.

The reciprocal hybrid viruses were obtained by using protein from one virus and RNA from other. These chimaeric virus particles produced infection pattern similar to that variety from which they had received RNA. The protein coat in progeny virus particles corresponded to the virus from which RNA was derived.

In later experiments, **Fraenkel Conrat** found that the isolated nucleic acid alone (without any added protein) has the viral activity. From these experiments it is evident that in plant viruses RNA acts as the hereditary material. In some RNA viruses RNA is single-stranded (ssRNA) while in other RNA viruses it is double-stranded (dsRNA).

TEXT QUESTIONS

1. Summarise the essential features of hereditary material.
2. Describe any two experiments to demonstrate that DNA is the hereditary material.
3. Describe transformation experiments conducted by Griffith and Avery to demonstrate that DNA is the transforming principle in pneumococcal bacteria.
4. Which nucleic acid acts as a hereditary material in plant viruses? Can you describe any experiment which proves that RNA also acts as a hereditary material in some organisms?
5. Write short notes on:
   (a) Transduction
   (b) Transformation
   (c) TMV
   (d) Fraenkel-Conrat's experiment
   (e) Griffith's experiment.
6. Summarise biological evidences for supporting the concept that—DNA is a hereditary material.
7. Give an experiment that proves that—RNA can also have a hereditary function.
8. Who discovered nucleic acid and from where?
9. Name the scientist who conducted experiment on bacterial transformation.
10. Which bacterium was used to demonstrate bacterial transformation?
11. Proteins exhibit more structural diversity, but they are not the hereditary material, why?
12. Give reasons why proteins were initially considered as hereditary materials?
13. Define Griffith effect.
14. Name the scientists who proved that transforming principle that changes the nonvirulent strain of *Diplococcus* into a virulent strain was nothing but DNA.
15. Define auxotrophic strain of *Escherichia coli*.
16. What is the difference between auxotrophic and prototrophic/autotrophic strain of bacteria?
17. Who discovered conjugation in bacteria?
18. Differentiate between transduction and transformation.
19. Give one example where RNA acts as a hereditary material.
20. What are chimaeric virus particles?

Deoxyribonucleic Acid (DNA)

6
Chapter

IN THIS CHAPTER

Nucleic Acids: Deoxyribonucleic acid: Chemical Composition; Molar Ratio of Nitrogenous Bases in DNA; Molecular Structure of DNA; Watson and Crick's Double Helical Model of DNA: Important Features of Watson and Crick's Model; Biological Significance of Watson and Crick's Double Helical Model of DNA; Forms of DNA: B-DNA; A-DNA; C-DNA ; D-DNA; Z-DNA; Denaturation–Renaturation; Heteroduplex DNA:Formation of Heteroduplex DNA; Fate of Heteroduplex DNA; Single-stranded DNA; *In vitro* Synthesis of DNA; Supercoiling of DNA in Viruses and Prokaryotes; Repeated DNA Sequence in Eukaryotic Genome: Highly Repeated DNA Sequences; Moderately Repeated DNA Sequences; Non-repeated DNA Sequences; Selfish DNA and Junk DNA; Palindromes and inverted repeats; mtDNA, ctDNA, Catalytic DNA or deoxyribozymes.

NUCLEIC ACIDS

Two types of nucleic acids are found in the cells of all living organisms. These are:

1. Deoxyribonucleic acid – **DNA**
2. Ribonucleic acid – **RNA**

Deoxyribonucleic acid (DNA) contains the genetic instructions for the development and functioning of living organisms. All living things contain DNA genomes. A possible exception is a group of viruses that have RNA genomes, but viruses are not normally considered living organisms. The main role of DNA in the cell is the long-term storage of information. The genome is often compared to a set of blueprints, since it contains the instructions to construct other components of the cell, such as proteins and RNA molecules. The DNA segments that carry this genetic information are called genes, but there are DNA sequences that have structural purposes, or are involved in regulating the expression of genetic information by acting as deoxyribozymes.

Ribonucleic acids perform different functions during protein synthesis. mRNA copies message from DNA and comes to the site of protein synthesis. tRNA molecules bring amino acids to the ribosomes and rRNAs are constituents of ribosomes.

Table 6.1 Contributions to the knowledge of DNA up to 1953

Year	Scientist	Contribution
1869	Miescher	Discovered DNA in an extract prepared from pus cells.
1894	Kossel and Neumann	Identified three of the four bases in DNA.
1900	Hammerstein	Showed that DNA contains a pentose sugar.
1929	Levine	Demonstrated that the sugar is 2′- deoxyribose.
1934	Caspersson	Proved that DNA is a macromolecule and the tetranucleotide theory of DNA structure was overthrown.
1944	Avery, MacLeod and McCarty	Discovered that the transforming principle is DNA, *i.e.* the first indication that DNA is genetic material.
1950	Chargaff	Published his results demonstrating consistency in the base ratios in DNA from different sources.
1950	Maurice Wilkins and Raymond Gosling	Prepared first images of DNA using X-ray diffraction.
1952	Hershey and Chase	Showed that phage genes are made up of DNA.
1952	Franklin and Wilkins	Produced X-ray diffraction pictures indicating that DNA is a helix.
1953	Watson and Crick	Deduced the double helix structure of DNA.

DEOXYRIBONUCLEIC ACID: DNA

DNA is Hereditary Material in all Living Organisms

The deoxyribonucleic acid DNA is found in the cells of all living organisms except plant viruses, where RNA forms the genetic material and DNA is absent. In bacteriophages and viruses there is a single molecule of DNA, which remains coiled and is enclosed in a protein coat. In bacteria, and in mitochondria and plastids of eukaryotic cells DNA is circular and lies naked in the cytoplasm. In the nuclei of eukaryotic cells DNA occurs in the form of long spirally coiled and unbranched threads. The number of DNA molecules is equivalent to the number of chromosomes per cell. In them, DNA is found in combination with proteins forming nucleoproteins or the **chromatin material** and is enclosed in the nucleus.

Chemical Composition

The chemical analysis has indicated that DNA is composed of three different types of compounds:

1. **Sugar molecules:** Sugar is represented by a pentose sugar, the **deoxyribose** or **2′-deoxyribose.**
2. **Phosphoric Acid.**
3. **Nitrogenous Bases**: These are nitrogen containing organic cyclic or ring compounds. These are of following four types:
   (i) **Adenine** represented by – **A**
   (ii) **Thymine** represented by – **T**
   (iii) **Cytosine** represented by – **C**
   (iv) **Guanine** represented by – **G**

The above four nitrogenous bases are separated into two categories: Single-ringed compounds (pyrimidines) and double-ringed compounds (purines).

Extension: Deoxyribose a Pentose Sugar

The name 2′-deoxyribose (a pentose sugar) indicates that it is derivative of ribose sugar by the replacement of hydroxyl group (–OH) at carbon atom 2′ with a hydrogen group (–H). The number of carbon atoms in ribose and deoxyribose are prime (the dash on each number is called prime). The prime is used to distinguish the carbon atoms in the sugar from carbon and nitrogen atoms in the rings of nitrogenous bases.

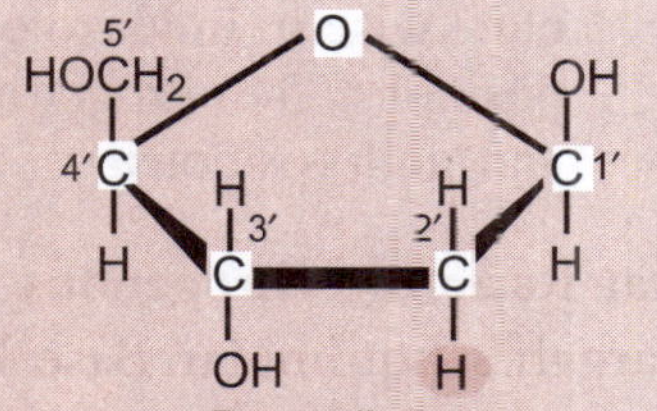

Fig. 6.1 A molecule of deoxyribose

1. **Pyrimidines:** Each pyrimidine molecule is formed of one ring only. The ring is hexagonal and heterocyclic. It is formed of four carbon and two nitrogen atoms. The atoms in the ring are numbered clockwise. Nitrogen atoms are present at the first and third positions and rest of the positions are occupied by carbon atoms.

   Pyrimidine nitrogenous bases found in DNA are **cytosine** and **thymine** (Fig. 6.2). But in RNA they are **cytosine** and **uracil**. In all the three nitrogenous bases, carbon atom present at second position (C-2) has an oxygen atom attached to it by a double bond. The three pyrimidine nitrogenous bases differ in functional groups present on C-5 carbon atom and presence of C=O or C$-NH_2$ group at C-4 carbon. **Cytosine** is 2-oxy, 6-aminopyrimidine, **uracil** is 2,6-dioxypyrimidine and **thymine** is 5-methyle uracil.

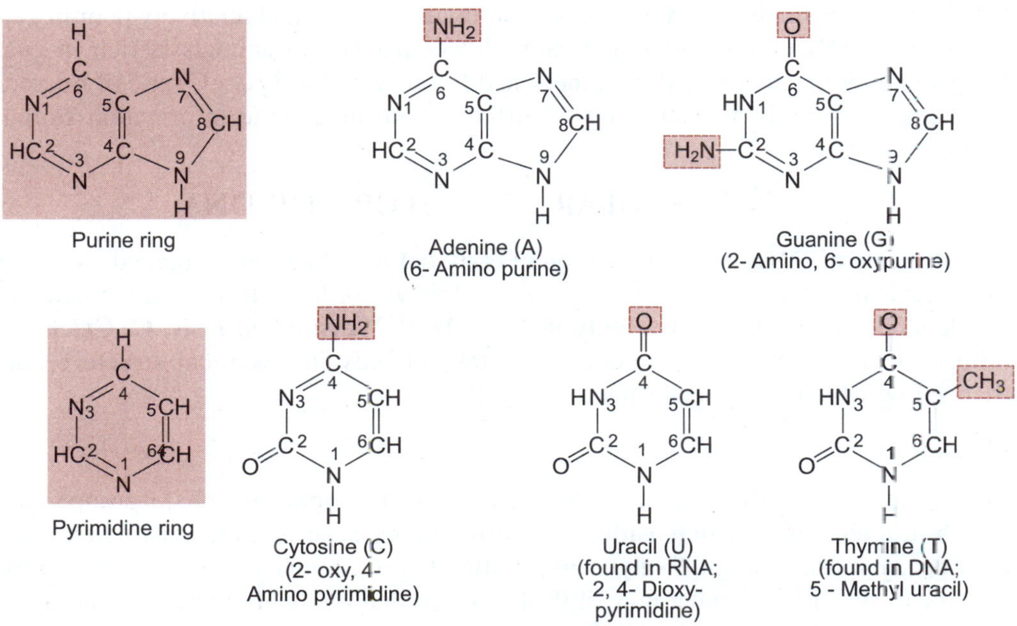

Fig. 6.2 Structural formulae of the nitrogenous bases in DNA and RNA. The parent compounds are purine (top), and pyrimidine (bottom). Differences between the bases are highlighted

2. **Purines:** Purines are two-ringed nitrogen compounds. Each purine molecule is **dicyclic**, formed of nine atoms (Fig. 6.2). There is one **pyrimidine ring** of six atoms and an **imidazole ring** of five atoms. The atoms are numbered anticlockwise in pyrimidine ring but clockwise in imidazole ring. The two rings share carbon atoms present on 4th and 5th positions. The nitrogen atoms in purine ring are four in number present at first, third, seventh and ninth positions.

Molar Ratio of Nitrogenous Bases in DNA (Chargaff's Equimolar Base Ratio)

Chemical analysis of DNA revealed another fundamental feature of DNA. It was described by **Chargaff** in 1950 and is called **Chargaff's equimolar base ratio**. According to this ratio:

1. Regardless of the source, the purine and pyrimidine components occur in equal amounts in a DNA molecule.
2. The amount of adenine (**A**) is equivalent to the amount of thymine (**T**) and of cytosine (**C**) is equivalent to that of guanine (**G**).
3. The base ratio between **A = T pair** and **G ≡ C pair** may vary in the DNA of different groups of animals but is constant for a particular species. Therefore, this base ratio has been used to identify the DNA from a particular source.

Chargaff's equimolar rule is applicable almost universally from viruses and bacteria to all plants and animals. But there is no rule governing the ratio of total concentration of guanine and cytosine to adenine and thymine, i.e. (G) + (C)/(A) + (T). In fact, there is great variation in this ratio ranging from 0.37 to 3.16 for different species of bacteria. Generally, it is found that DNA of higher plants and animals is rich in adenine and thymine (A:T) and relatively poor in guanine and cytosine (G:C). DNA from microorganisms, lower plants and animals is rich in guanine and cytosine and poor in adenine and thymine.The difference in A = T/G ≡ C ratio of microorganisms and higher organisms is an indication of difference in the genetic information stored in these molecules.

MOLECULAR STRUCTURE OF DNA

The constituents of DNA were isolated quite early but how these are arranged so as to carry out their cytological and genetical activities was not known for long. DNA is a long chain **polymer** was understood in late 1930s. But only in 1953, **D. S. Watson** and **F. H. C. Crick** presented a working model of DNA. This model illustrates not only its chemical structure but also the mechanism by which it duplicates itself.

Nucleosides

A nitrogenous base with a molecule of deoxyribose (without phosphate group) is known as **deoxyribonucleoside** or **nucleoside**. The nitrogenous base is attached to first carbon atom C–1′ of deoxyribose through **N–glycosidic bond** (Fig. 6.3) attached to nitrogen number one of pyrimidine (C and T) or number nine of the purine (A and G). In all, there are four nucleosides in a DNA molecule. These are:

1. **Deoxyadenosine** – Adenine + Deoxyribose

2. **Deoxyguanosine** – Guanine + Deoxyribose
3. **Deoxycytidine** – Cytosine + Deoxyribose
4. **Deoxythymidine** – Thymine + Deoxyribose

Extension: Nucleosides and Nucleotides

In nucleic acids, nitrogenous bases are always attached to the C-1′ carbon of the pentose sugar by a covalent **β-N-glycosidic bond.** The pyrimidine bases are bonded at their nitrogen1, while purine bases are bonded at the nitrogen 9. A nitrogenous base with a pentose sugar molecule forms a **nucleoside.**

The phosphate group is attached to the C-5′ carbon atom of pentose sugar. The compound: sugar + nitrogenous base + phosphate together form a **nucleotide.** In DNA, the nucleosides and nucleotides are called **deoxyribonucleosides** and **deoxyribonucleotides**, while in RNA these are called **ribonucleosides** and **ribonucleotides,** respectively.

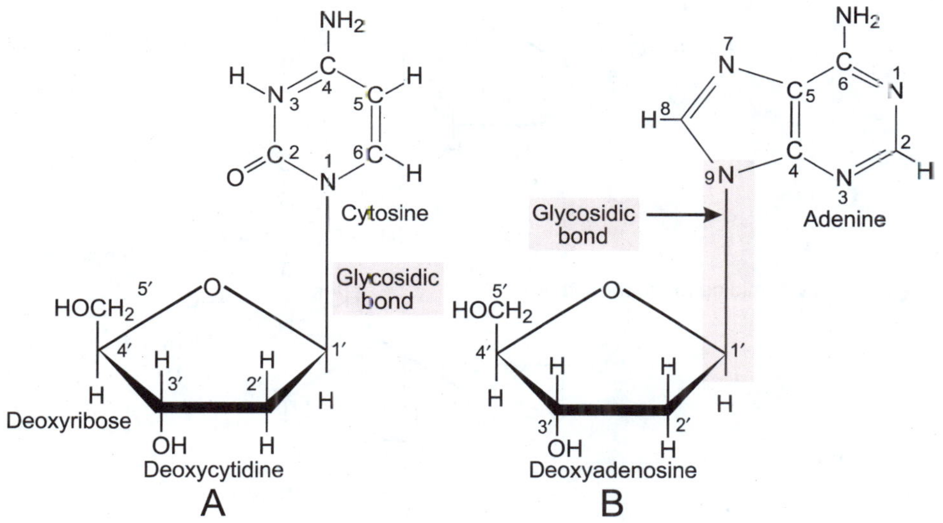

Fig. 6.3 **A.** Linkage of a nitrogenous base cytosine to C–1′ of deoxyribose; **B.** Linking of adenine with C–1′ of deoxyribose through N–glycosidic bond in a nucleoside

Nucleotides are the Monomers of DNA

A nucleotide is formed of one molecule of deoxyribose, one molecule of phosphoric acid and one of the four nitrogenous bases. The phosphate group is attached to 5′-carbon of sugar deoxyribose. Thus a nucleotide is a **nucleoside monophosphate** (dNMP) (Fig. 6.4). Since, there are four nitrogenous bases, there are four types of nucleotides namely:

1. **Deoxyadenylic acid** or **dAMP** = Adenine + Deoxyribose + Phosphoric acid
2. **Deoxyguanylic acid** or **dGMP** = Guanine + Deoxyribose + Phosphoric acid
3. **Deoxycytidylic acid** or **dCMP** = Cytosine + Deoxyribose + Phosphoric acid
4. **Deoxythymidylic acid** or **dTMP** = Thymine + Deoxyribose + Phosphoric acid

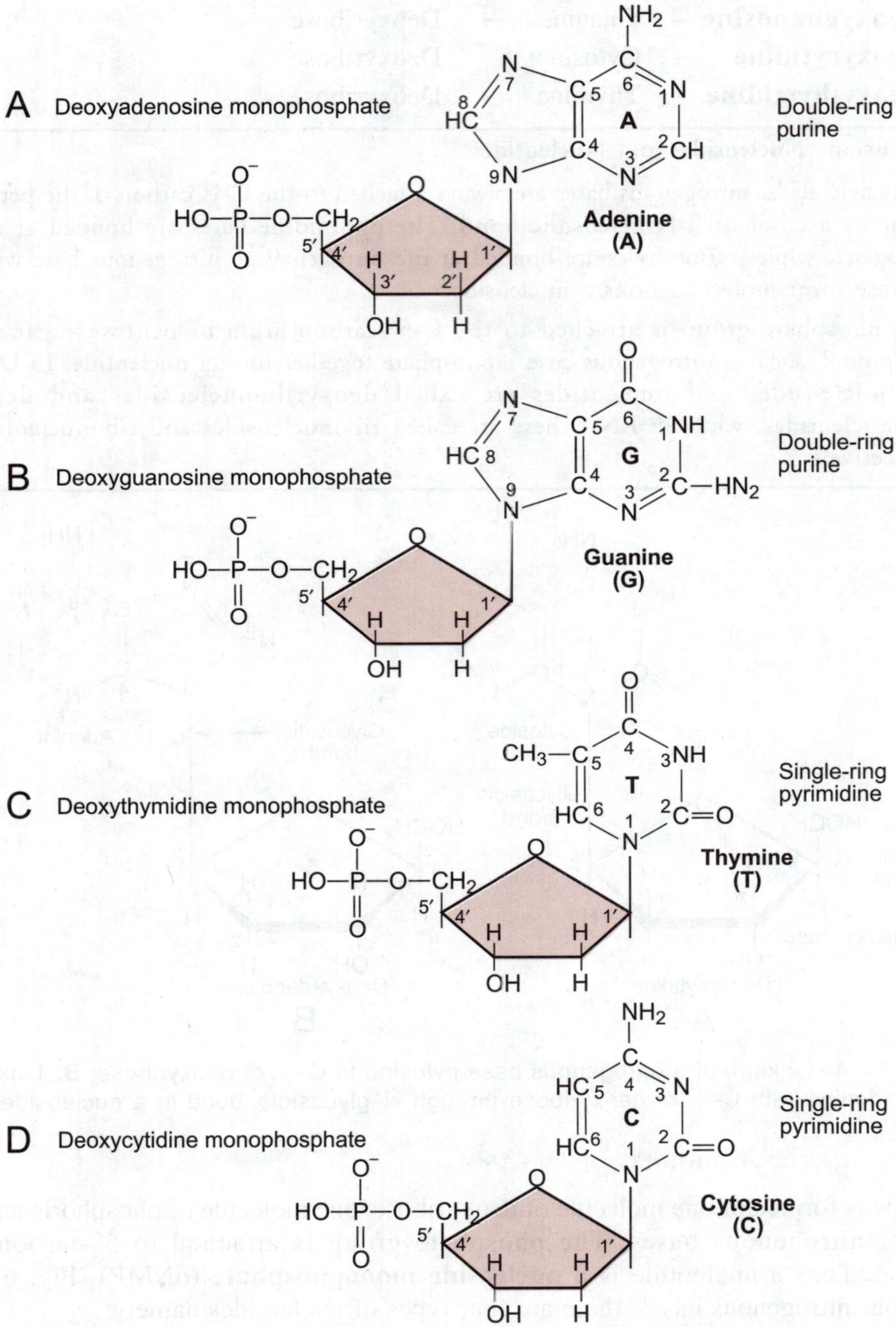

Fig. 6.4 Structural formulae of all the four nucleotides found in DNA molecule: **A.** Deoxyadenosine monophosphate or Adenylic acid; **B.** Guanosine monophosphate or Guanylic acid; **C.** Deoxythymidine monophosphate or Thymidylic acid and **D.** Deoxycytidine monophosphate or Cytidylic acid

Nucleotides are Covalently Linked to form Polynucleotide Chain

DNA is a macromolecule formed by the linking of several thousands of nucleotides. These are called **monomers** or **building blocks** of DNA. In a nucleotide, the phosphate (phosphoric acid) molecule is attached to fifth carbon atom (C–5′) of the deoxyribose molecule. The phosphate molecule of a nucleotide is joined to the third carbon atom of the deoxyribose of the next nucleotide by **covalent bond**. Such 5′–3′ linkages between adjacent nucleotides are called **phosphodiester bonds**. Thus, the adjacent nucleotides are connected together forming the sugar-phosphate-sugar-phosphate chain in which sugar and phosphate molecules are arranged in alternate fashion. The nitrogenous bases are attached to the first carbon atom of the deoxyribose and are directed at right angle to the long axis of the polynucleotide chain. They are stacked one above the other like the steps in a ladder (Fig. 6.6).

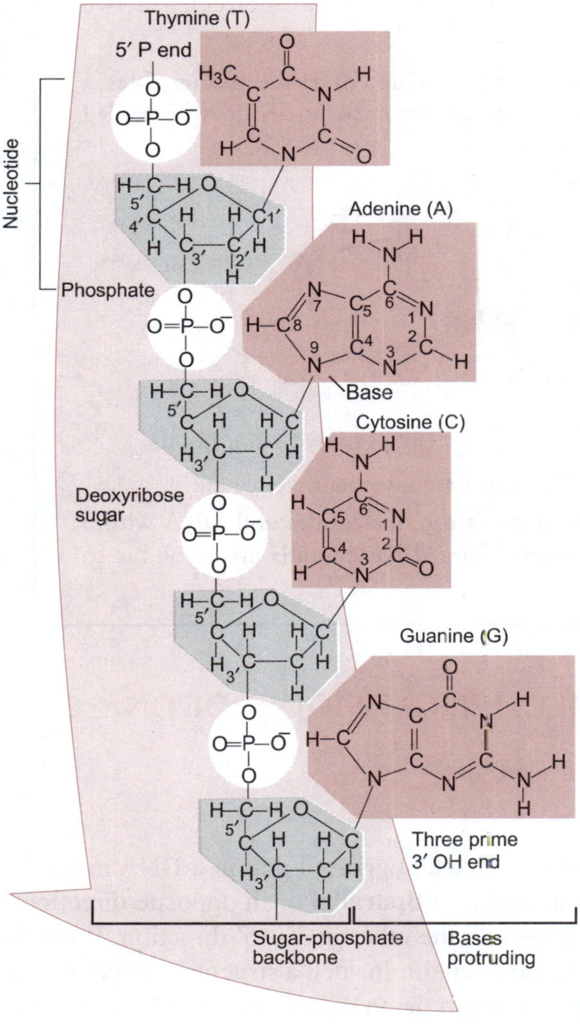

Fig. 6.5 A polynucleotide chain

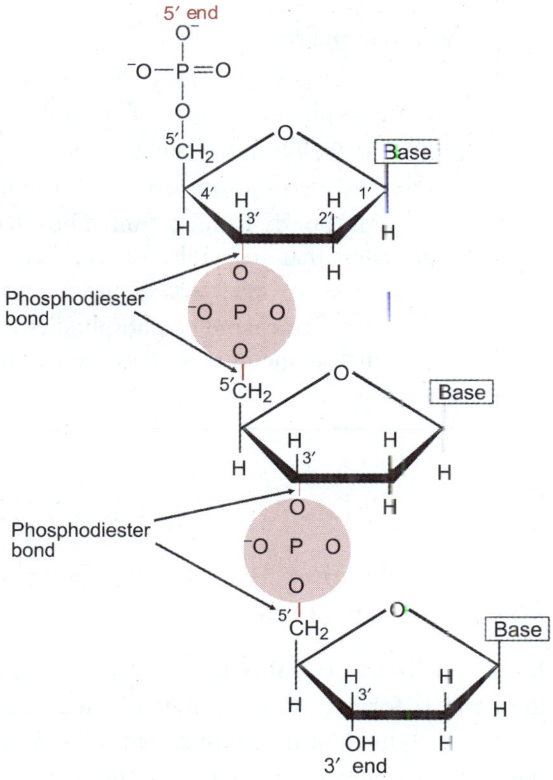

Fig. 6.6 Linking of two nucleotides in a polynucleotide chain through a phosphodiester bond

Each Polynucleotide Chain has marked 3′ and 5′ Ends

Each polynucleotide chain has marked ends. Its top end has a sugar residue with free 5′ carbon atom which is not linked to another nucleotide. The triphosphate group is still attached to it. This end is called the head end or **5′ end** or **5′–P terminus**. The other end of the chain ends in a sugar residue with **C-3′** carbon atom not linked. It bears **3′–OH** group. This end of polynucleotide chain is called tail end or **3′ end** or **3′–OH terminus**. It means polynucleotide chain has direction and is marked as **3′–5′**.

Extension: Phosphodiester Bond

The linkage between two nucleotides is called **phosphodiester bond**, because one phosphoric acid molecule joins with hydroxyl groups of two sugar molecules of two nucleotides forming **ester linkage**.

The joining of two nucleotides forms a **dinucleotide**, of three nucleotides a **trinucleotide** and so forth. Short chains formed of less than 20 nucleotides are called **oligonucleotides** and chains longer than 20 nucleotides are referred as **polynucleotides**.

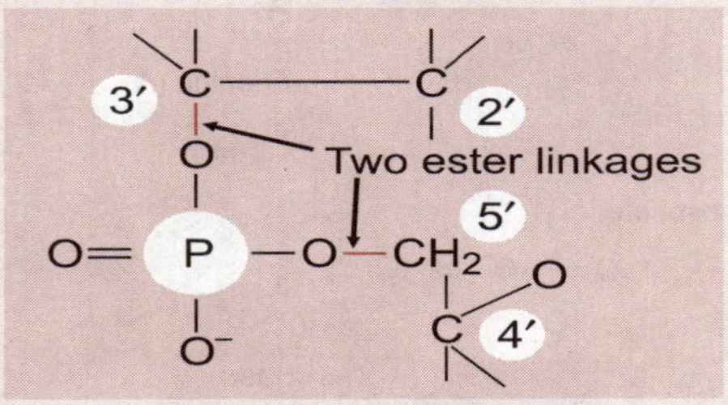

Detail of two ester bonds

Why Phosphates

In DNA, the nucleotides are connected across phosphates because of the following properties of phosphate molecules:

- Phosphate group can form linked bonds.
- Linking bonds formed from phosphates are esters that are stable, but can easily be hydrolysed by enzymes. It means nucleotides can be conveniently isolated.
- The negative charge of phosphodiester bond keeps the nucleotides and DNA within the nuclear membrane, because the negatively charged compounds are insoluble in lipids.

WATSON AND CRICK'S DOUBLE HELICAL MODEL OF DNA

DNA is made up of Two Spirally Twisted Anti-parallel Polynucleotide Strands

Based on X-ray crystallographic studies, **Watson** and **Crick** suggested that in a DNA molecule the helix is formed of two polynucleotide chains arranged **antiparallel** or in opposite directions, i.e. one polynucleotide chain runs in 5′→3′ direction and the other in 3′→5′ direction. It means the 3′ end of one chain lies beside the 5′ end of the other chain. In such a structure the phosphate groups of nucleotides in each polynucleotide chain or strand lie on the outside of the deoxyribose and the nitrogenous bases are directed inward. The nitrogenous bases of the two chains are

Extension: Discovery of Double Helical Structure of DNA

Once it was established that DNA is the hereditary material, the molecular biologists started working to determine DNA structure inconfirmity with its role in inheritance. **Linus Pauling** in California, **Maurice Wilkins** and **Rosalind Franklin** in London and American scientist, **James D. Watson** and British scientist, **Francis Crick** contributed to the discovery of three-dimensional structure of DNA.

Watson studied X-ray diffraction photographs of DNA prepared in the laboratory of **Maurice Wilkins** at King's College, London. Those were produced by Wilkins's colleague, **Rosalind Franklin.** On this basis **Watson** and **Crick** concluded that DNA has a helical structure and the helix had a uniform width of 2 nm with its purine and pyrimidine bases stacked at a distance of 0.34 nm.

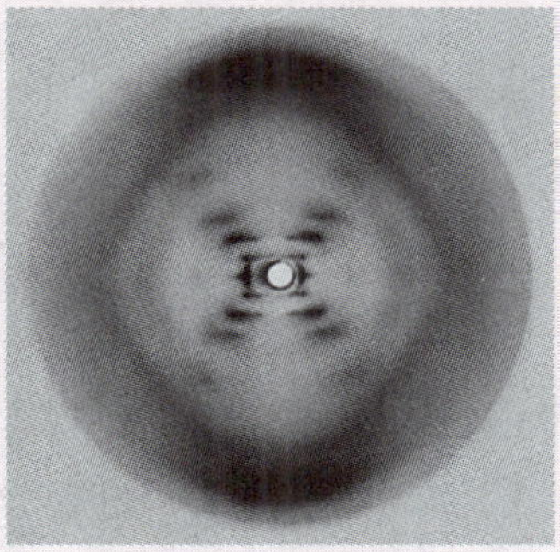

Fig. 6.7 X-ray diffraction photograph of DNA.The shape of X is formed from X-ray diffraction of helical object

Watson and **Crick** tried to make models of a double helix in confirmation with the X-ray measurements, putting sugar-phosphate chains on the inside of the molecule and nitrogenous bases outside. Later, **Watson** tried to put the sugar-phosphate chain on the outside and nitrogenous bases inside the DNA molecule. Franklin's X-ray data suggested that the one full turn of helix is 3.4 nm long and that there are ten nucleotide pairs per turn.

At first, **Watson** presumed that only identical bases form pairs *i.e.,* adenine pairs with adenine, cytosine with cytosine or guanine with guanine. But such pairs could not fit with the X-ray data. Therefore, it was concluded that in a molecule of 2 nm width, a purine on one strand could pair with a pyrimidine on the opposite strand. **Watson** and **Crick** also realised that the base pairing is also dictated by the structure of nitrogenous bases so that adenine can form hydrogen bonds only with thymine and cytosine with guanine only.

linked through hydrogen bonds formed between oxygen and nitrogen atoms of the adjacent bases (Fig. 6.8). The unique feature of pairing between bases is:

1. **Purine** (adenine and guanine) pairs with **pyrimidine** (cytosine and thymine), and
2. **Adenine** pairs with **thymine,** and
3. **Cytosine** pairs with **guanine**.

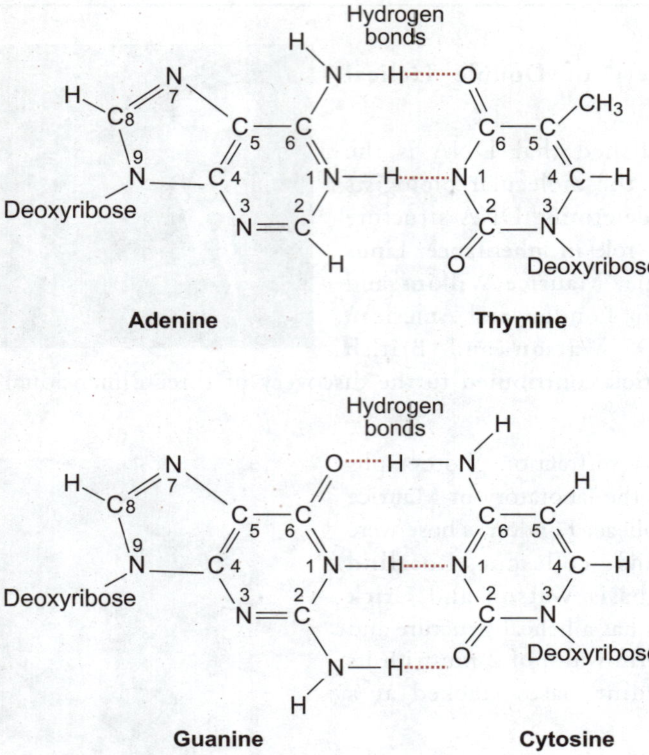

Fig. 6.8 Association of base pairs through hydrogen bonds between opposite nucleotides of two polynucleotide chains of DNA

Extension: Reasons for Purine–Pyrimidine Pairing in DNA

There are definite reasons for such a specific pairing of nitrogenous bases (purine–pyrimidine pairing) in DNA polynucleotide chains:

- Such pairing forms a perfect match between hydrogen donor and hydrogen acceptor sites on the two molecules. Adenine and thymine share two hydrogen bonds, whereas cytosine and guanine are joined by three hydrogen bonds.
- Such a pairing is further supported by the occurrence of constant diameter of DNA. In a limited area, a two ringed (purine) molecule joins a single-ringed pyrimidine molecule maintaining a constant and roughly equal distance. The distance between the different bonds are indicated in Fig. 6.9. They vary 0.29 nm to 0.30 nm in size, and maintain the width of DNA helix constant. A and G pair will be rather too large to fit inside the helix and C and T would appear to be far apart.

- The concept of specific base pairing explains Chargaff's rule. The amount of guanine is equal to the amount of cytosine and of adenine to thymine or amount of pyrimidines being equal to purines.

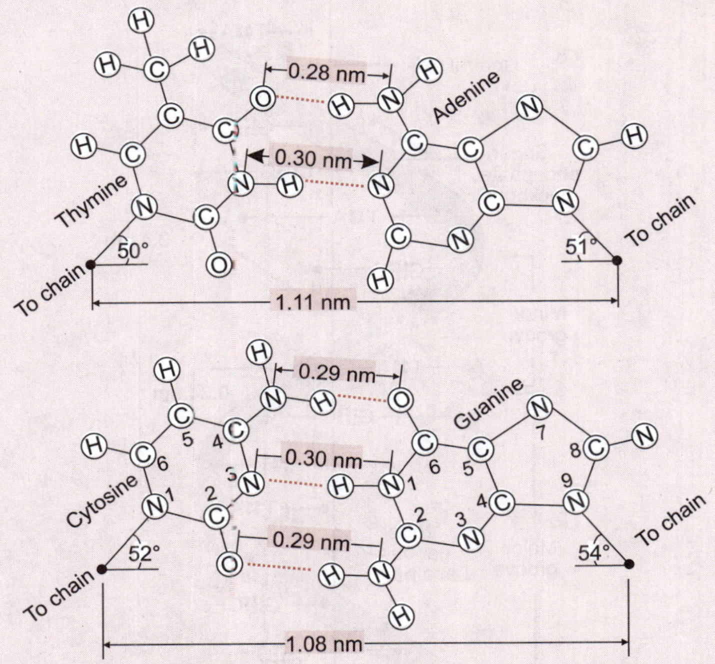

Fig. 6.9 Depiction of the distance between complementary base pairs of two helices of a DNA molecule. *Note:* the width of Adenine–Thymine pair is 1.11nm and of G–C pair is 1.08 and the bond size between any two nitrogenous bases varies from 0.28nm to 0.30nm

Due to this type of base pairing the two strands are **complementary** to each other. It means if a chain has a region with a sequence of nitrogenous bases, **thymine–cytosine–adenine–cytosine–guanine**, then the corresponding region in the complementary chain will have the base sequence **adenine–guanine–thymine–guanine–cytosine.**

Extension: Important Features of Watson and Crick's Double Helical Model of DNA

There are several important features of DNA double helix:

- The double helix comprises of two complementary polynucleotide chains.
- The two polynucleotide chains are wound clockwise around a central axis to form a circular stair-case type structure.
- The two strands (polynucleotide chains) of double helix are antiparallel, i.e. they run in opposite direction. One chain is aligned in $5' \rightarrow 3'$ direction and its antiparallel chain is aligned in $3' \rightarrow 5'$ direction.

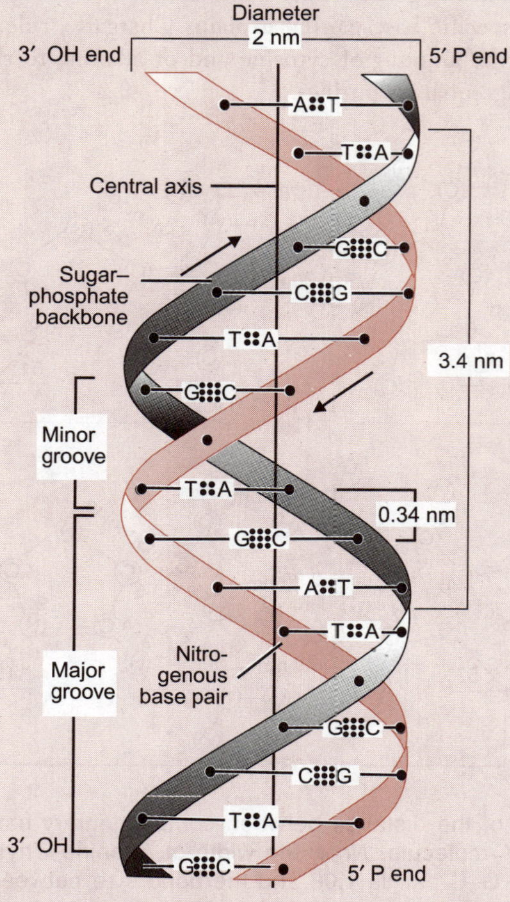

Fig. 6.10 Double helical structure of a part of DNA molecule. The two polynucleotide chains of DNA twisted spirally in a helical fashion

- Each polynucleotide chain has a sugar–phosphate 'back bone' with nitrogenous bases directed inside the helix, i.e. towards central axis. The sugar-phosphate backbone is formed of polar, charged and covalently connected deoxyribose and phosphate groups and, therefore, occur on the outside of the helix. The purine and pyrimidine nitrogenous bases being hydrophobic avoid water and remain towards the inside of helix.
- The phosphate groups give the molecule a negative charge.
- The nitrogenous bases lie almost perpendicular to the long axis of the molecule. Therefore, these are stacked one on top of another like a pile of saucers. The hydrophobic interactions between stacked bases provide stability to DNA.
- The distance from the phosphorus atom of the backbone to the centre of the axis is 10Å. Therefore, the width or diameter of the double helix is 20Å or 2nm.
- The two polynucleotides in a double helix are complementary. The sequence of nitrogenous bases in one determines the sequence of the nitrogenous bases in the other. Complementary base pairing between two polynucleotide strands of DNA is of fundamental importance in molecular genetics.

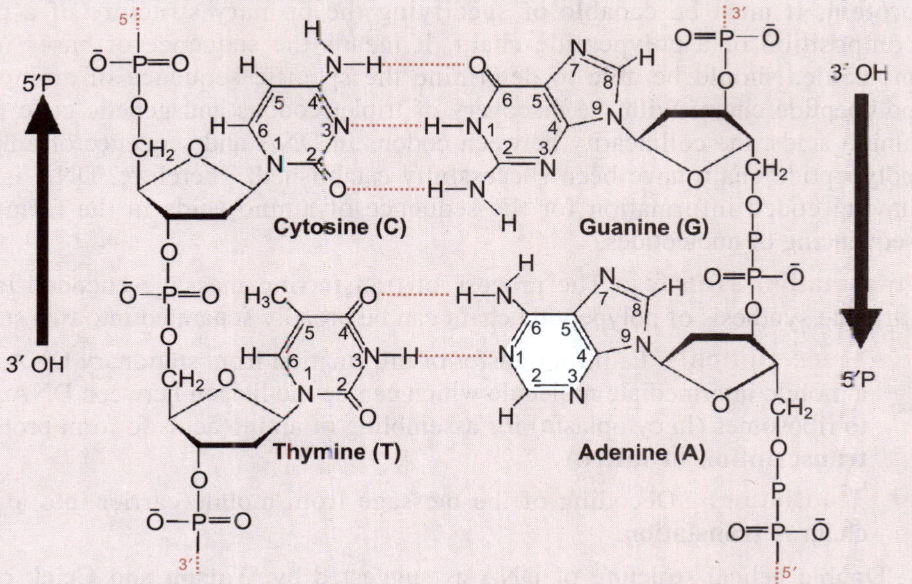

Fig. 6.11 Nucleotides in a strand are linked through deoxyribose and phosphoric acid molecules to form polynucleotide chain that joins the nucleotides of anti-parallel polynucleotide chain by hydrogen bonds

- The nitrogenous bases of two anti-parallel polynucleotide strands are linked through hydrogen bonds. There are two hydrogen bonds between A and T, and three between G and C. These are called **complementary base pairs.** The hydrogen bonds are the only attractive force between the two polynucleotides of double helix. These serve to hold the strands together.
- Ten base pairs occur per turn of helix (abbreviated 10bp). The spacing between adjacent base pairs is 3.4Å or 0.34nm. Therefore, each turn of helix is 34Å or 3.4nm.
- The double helix has two external helical grooves: a major groove which is deep and wide and a minor groove which is shallow and narrow, but large enough to allow protein molecules to come in contact with nitrogenous bases.

DNA consists of two complementary chains twisted around each other forming a **right-handed helix**. One turn of helix measures about 34Å. It conatins 10 pairs of nucleotides placed at regular intervals of 3.4Å. The diameter of the helix is roughly 20Å. A narrow **minor helical groove** and a wide **major helical groove** run along the length of DNA helix. The **narrow groove** is the distance between the paired molecules while the **major groove** is the space between successive turns when the pair is wound into a helix.

Biological Significance of Watson and Crick's Double Helical Model of DNA

The biological significance of Watson and Crick's double helical model of DNA is as follows:

1. **Information Storage:** Genetic material is expected to act as a repository of information. Its structure must be related to the intimate features of gene-product, i.e. the

protein. It must be capable of specifying the primary structure of a protein, i.e. composition of a polypeptide chain. It means the sequence of bases in the DNA molecules should be able to determine the specific sequence of amino acids in a polypeptide chain. With the discovery of triplet codons and genetic code for different amino acids, the collinearity between codons in DNA and sequence of amino acids in polypeptide chain have been successfully established. Therefore, DNA is like a tape having coded information for the sequence of amino acids in the form of specific sequencing of nucleotides.

2. **Information Transfer:** The process of transferring message encoded in DNA and ultimate synthesis of polypeptide chain can be broadly separated into two steps:
   - **Transcription:** Faithful transfer of information from stationary DNA molecule to a mobile intermediate molecule which can be the liaison between DNA (in nucleus) to ribosomes (in cytoplasm) for assembling of amino acids to form protein is called **transcription of mRNA.**
   - **Translation:** Decoding of the message from mobile carrier into a polypeptide chain is **translation.**

   Double helical structure of DNA as suggested by **Watson** and **Crick** conveniently explains the transfer of information. Since the information is coded in the sequence of nitrogenous bases, the same sequence is copied in messenger RNA (mRNA). Thus, a complementary strand of mRNA is synthesised along one of the two strands of DNA. This mobile copy of genetic message moves out of the nucleus to the site of protein synthesis (ribosome) in cytoplasm.

   Transfer RNA (tRNA) molecules pick up specific amino acids and align them on mRNA in a triplet sequence of nitrogenous bases (the codon) recognised by their anticodon region.

3. **Self Replication:** The structure of DNA provides the most convenient and the only possible device, whereby a particular molecule with a particular sequence of base pairs could be duplicated easily. Each strand of double helix of DNA can construct its complementary strand by acting as a template, because the sequence of nitrogenous bases in one strand could easily determine the laying down of bases of the second or complementary strand.
4. **Variations:** Genetic material should be capable of undergoing change in its composition. Its structure should be able to undergo mutations so as to produce diversity in the protein-making information. **Watson** and **Crick's** model easily explains the mechanism of appearance of mutations in DNA either by substitution of nitrogenous base pairs or by their addition or deletion in the polynucleotide chain.
5. **DNA Repair:** DNA molecule is relatively fragile and is easily damaged simply by bending or shearing forces or by local changes in pH. Its base sequence changes by chemical agents from environment and also by X-ray and ultraviolet radiations. Still it preserves its integrity after several millions of replications. Actually, the double helical structure of DNA presents the most stable structure. It is most suited for the long term preservation of the coded information. Because the base sequence in the two

polynucleotide chains is complementary, the coded information in a DNA molecule are preserved in duplicate. If any how, the sequence of bases on one chain is disturbed, the correct sequence of the other chain may help in correcting the errors that occur in the complementary strand and sends the correct message.

FORMS OF DNA

Six different morphological forms of DNA double helix have been described. These are A, B, C, D, E and Z forms. Only B and Z conformations occur as cellular DNA, other forms are found rigidly in controlled, experimental conditions. All DNA molecules of any source, natural or artificial, can take on A or B conformation. Some DNA forms are interconvertible also. The differences in these DNA forms are associated with:

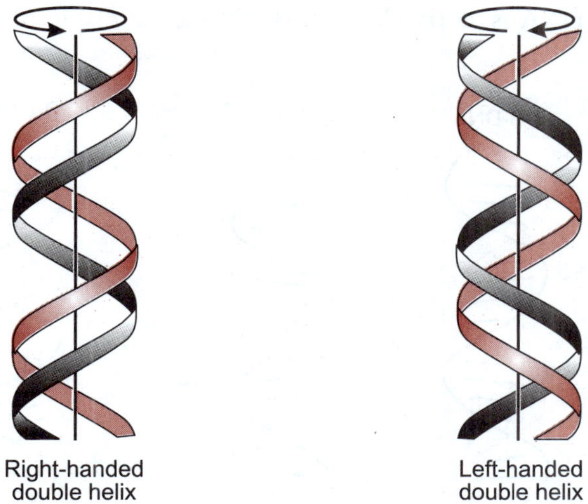

Fig. 6.12 The right- and left-handed helical forms of DNA. *Note* that they are mirror images of one another

- The number of base pairs present in each turn of DNA helix.
- The pitch or angle between each base pair.
- The helical diameter of DNA molecule.
- The handedness of coiling of double helix.

1. B-DNA (Right-handed DNA)

This form of DNA occurs in all living beings under normal condition (*i.e.* naturally occurring form). It is fully hydrated form and is also called **'wet' DNA.** It occurs under low salt concentration and high degree of hydration (at about 92 per cent relative humidity). β-helix is thinner and taller with a wide major groove and narrow minor groove. Its each coil or turn measures 34Å or 3.4 nm. Each turn has 10 base pairs (bp) and each base pair occupies 3.4Å. Its two strands or polynucleotide chains are wounded in right-handed coils and each strand follows a clockwise path. **Rosalind Franklin** described this form of DNA. The double helical structure described by **Watson** and **Crick** was also of B-DNA.

2. A-DNA (Right-handed DNA)

It is the dehydrated form of DNA. It occurs under conditions of decreased hydration and increased concentration of Na^+ ions. Its polynucleotide chains are coiled clockwise and have right-handed configuration. A-DNA is more compact and more bulky. Its helix is short and wide with narrow and deep major groove and a wide, shallow minor groove. It has 10.9 or 11 base pairs per turn of the helix and a diameter of 23Å. The distance between the base pairs is 2.7Å instead of 3.4Å. The orientation of bases is also different. These are tilted and displaced laterally in relation to the axis of the helix.

It is presumed that A-DNA is derived from B-DNA as a result of hydrophobic molecules under dry conditions.

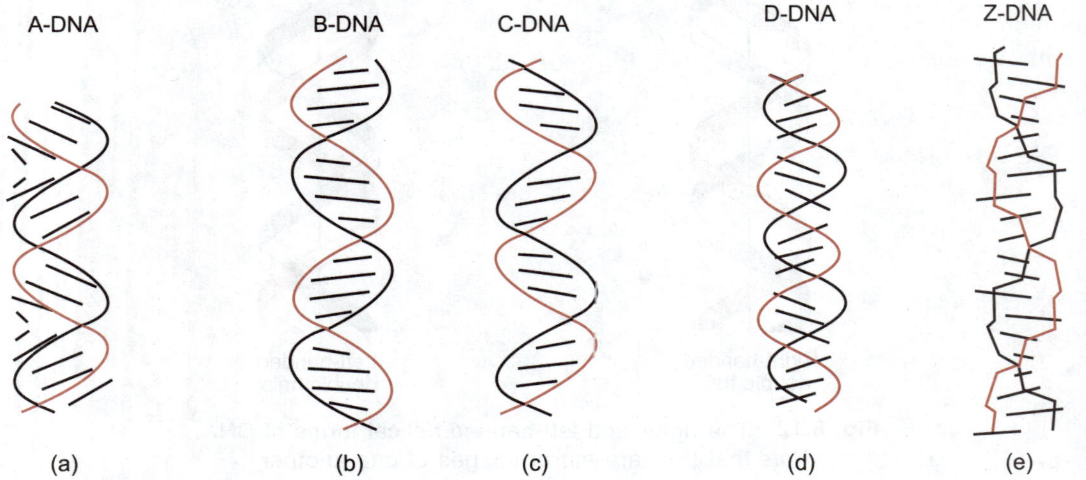

Fig. 6.13 Diagram showing right-handed and left-handed forms of DNA

3. C-DNA (Right-handed DNA)

C-DNA is also right-handed DNA. It occurs under even greater dehydration which is usually not found in cells under normal conditions. It has only 9.3 base pairs per turn of helix instead of 10. The base pairs are tilted relative to the axis of helix as in A-DNA. The helix is only 19Å in diameter. It means C-DNA is narrow and less compact than A-DNA and B-DNA. It occurs when humidity is 66 per cent and medium contains Lithium ions.

4. D-DNA and E-DNA(Right-handed DNA)

These two forms of DNA are also right-handed with slightly different conformation. These occur in helices which lack guanine, i.e. which have alternating A–T base pairs. These have 8 and

7 nucleotides per turn respectively. These are seen under special environment and do not occur *in vivo*. D-DNA has been detected in nature only in bacteriophage T_2.

5. Z-DNA (Left-handed DNA)

It is the only left-handed DNA. It occurs in a small proportion in cellular DNA. It is called Z-DNA because the phosphodiester backbone of its polynucleotide strands passing through phosphate groups follows a zigzag course.

Z-DNA was discovered by **Alexander Rich** (1979). It is the skinniest DNA having 12 base pairs per turn. One complete helix is 45Å as against 34Å in B-DNA. The diameter of Z-DNA is 18Å against 20Å in B-DNA. It has only one groove. The major grooves are absent. Z-DNA is less stable than B-DNA because the negatively charged phosphate groups lie closer and cause repulsion, leading to the conversion of Z-DNA to B-DNA. It is formed only when purines and pyrimidines are present alternately in the chain (GC or AT). Its repeating units are dinucleotides as against mononucleotides in B-DNA. Z-DNA is more compact. In Z-DNA sugar residues have alternating orientation. The Z-conformation is stabilised by high salt concentration or specific cations. When B-DNA is brominated or methylated it changes into Z-DNA.

Table 6.2 Differences between A-DNA, B-DNA and Z-DNA

Character	A-DNA	B-DNA	Z-DNA
1. Direction of helix	Right handed	Right handed	Left handed
2. Appearance	Short and wide	Longer and thinner	Long and thin
3. Width or diameter of helix	23Å	20Å	18Å
4. Base pairs per turn	10.9 or 11	10	12
5. Length of each coil of helix	34Å	34Å	45Å
6. Distance between adjacent nucleotides	2.7Å	3.4Å	3.75Å
7. Major groove	Extremely narrow and very deep.	Wide and of intermediate depth.	Flattened out on helix surface.
8. Minor groove	Very wide and shallow.	Narrow and of intermediate depth.	Extremely narrow and very deep.
9. Location of axis of helix	Through major groove.	Through base pairs.	Through minor groove.
10. Base inclination from helix	13.0	2.0	8.8
11. Occurrence	Occurs in dehydrated conditions.	Occurs in hydrated conditions.	Occurs in dehydrated high salt concentration.

According to **Alexander Rich,** certain control sites in the genes are stabilised in Z-configuration by methylation and provide site for the binding of regulatory protein. This in turn keeps the functional gene turned off. Demethylation might switch the site to B-conformation releasing regulatory protein molecule and in turn switching on the gene. Thus, Z-DNA has a regulatory function.

The existence of Z-DNA in *Drosophila* has been demonstrated using antibodies that recognise and bind specifically to Z-DNA. In human DNA also Z-DNA forming regions are dispersed throughout the genome.

Table 6.3 Differences in the parameters of various conformations of DNA molecules

Conformation	Base pairs per turn	Vertical rise per base pair	Angle turn per base pair	Tilt of base pair	Groove width		Groove depth		Approximate diameter
					Minor	Major	Minor	Major	
A	11	2.56	32.7°	+20°	3.7	10.9	13.5	10.9	23
B	10.1–10.6	3.38	36.0°	–6°	11.7	5.7	8.5	5.7	20
C	9.3	3.52	36.6°	–8°	10.5	4.8	7.5	4.8	19
D	8	3.04	45.0°	–16°	8.9	1.3	5.8	1.3	—
E	7.5	3.25	48.0°	—	—	—	—	—	—
Z	12	3.70	30.0°	–7°	2.0	8.8	13.8	8.8	18

C Value and C Value Paradox

The length of DNA molecules varies greatly in different organisms. The length of DNA molecules is described in terms of number of nucleotides. The DNA of a virus may be as small as a few thousand base-pairs. The DNA of *E. coli* has about four million base pairs. The human genome is estimated to contain about 3 billion base pairs and measures one to two metres. The total amount of DNA present in a haploid set of chromosomes is termed as its **C value.** The C value for eukaryotic cells ranges from 5 to 10 thousand times the C value of a bacterial cell.

Surprisingly, the number of gene in *E. coli* is estimated to be 4,000 whereas, in human beings their number varies between 10,000 to 50,000. It means in human beings the genes are 10 times the number of genes in *E. coli,* but the amount of DNA or base pairs in human cells is about thousand times of DNA in bacterial cell. It means eukaryotic cells contain more DNA than needed for coding genes. This is called **C value paradox.** One main reason for this paradox is the existence of noncoding DNA which are called introns. The message in eukaryotic genes is divided into coding segments or exons interrupted by noncoding segments called introns.

DENATURATION–RENATURATION

The double-stranded structure of DNA can be melted in solution by increasing temperature or decreasing salt concentration or by using certain denaturation reagents. The phenomenon of separation of two strands of DNA molecule by heating is known as **denaturation** or **melting**.

Since the two strands of DNA molecule are held together by weak hydrogen bonds, these get separated easily.

If native DNA molecules (i.e. naturally occurring double-stranded DNA molecules) are heated to nearly 100°C, the hydrogen bonds between nitrogenous bases of two antiparallel polynucleotide strands break and two polynucleotide chains separate. This is called **denaturation** and DNA is called **denatured.** The denatured DNA is single-stranded.

The strands of a DNA molecule separate over a temperature range. The midpoint of this range is called **melting temperature** and is represented by T_m. It is influenced by base composition. DNA rich in G-C pairs melts at a higher temperature.

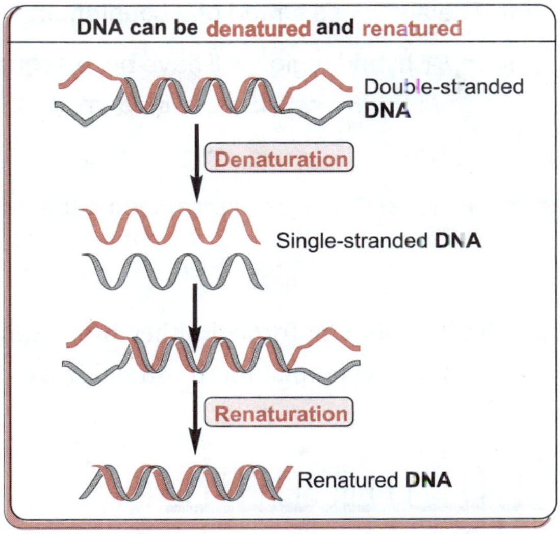

Fig. 6.14 Denatured single strands of DNA can renature to give the duplex form

Denaturation of DNA may be caused by some chemicals which either break hydrogen bonds or weaken hydrophobic interactions. These are called **denaturants.**

When heated DNA solution is slowly cooled, the two strands pair again forming double helix. This is known as **renaturation** or **reannealing.** This property of DNA was discovered by **Julius Marmur** and coworkers at Harvard University in 1960. This has helped in synthesising hybrid DNA. The technique of hybrid DNA synthesis is utilised in establishing genetic similarity between various taxonomic groups. More length of hybrid DNA denotes taxonomic nearness.

HETERODUPLEX DNA

Hybrid DNA formed as a result of recombination is called **heteroduplex DNA.** It contains mismatched base pairs or heterologous base sequences. Heteroduplex DNA is also called **heterozygous DNA.**

Example: In a genetic hybrid $m_1+ / + m_2$, m_1 and m_2 differ from their + counterparts by a single base.

(a) Its m_1 sense strand is

5′ – – – A C A A T – – – 3′.

(b) Its + counterpart will be

5′ – – – A C A G T– – – 3′.

(c) The antisense strand of + counterpart will be 3′ – – – TGTCA – – – 5′.

(d) m_1/ + hybrid region will have base sequence

5′ – – – A C A **A** T – – – 3′

3′ – – – T G T **C** A – – – 5′.

The m_1/ + hybrid DNA has mismatched base pair **A-C.**

Formation of Heteroduplex DNA

Heteroduplex DNA is formed either by reciprocal recombination (by bridge migration) or by non-reciprocal recombination (by strand insertion).

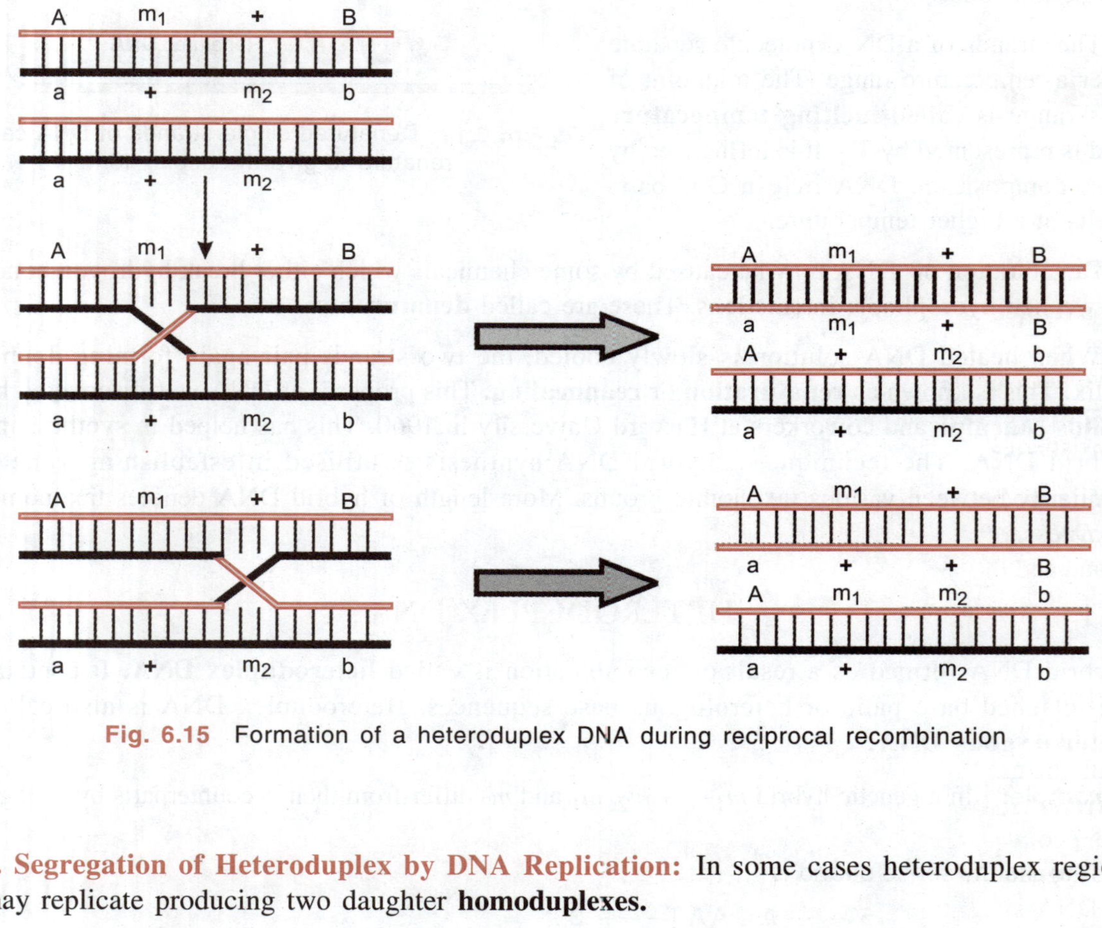

Fig. 6.15 Formation of a heteroduplex DNA during reciprocal recombination

1. Segregation of Heteroduplex by DNA Replication: In some cases heteroduplex region may replicate producing two daughter **homoduplexes.**

2. Mismatch Repair: In other cases one strand of heteroduplex is removed and replaced by a new matching segment producing **homologous DNA.**

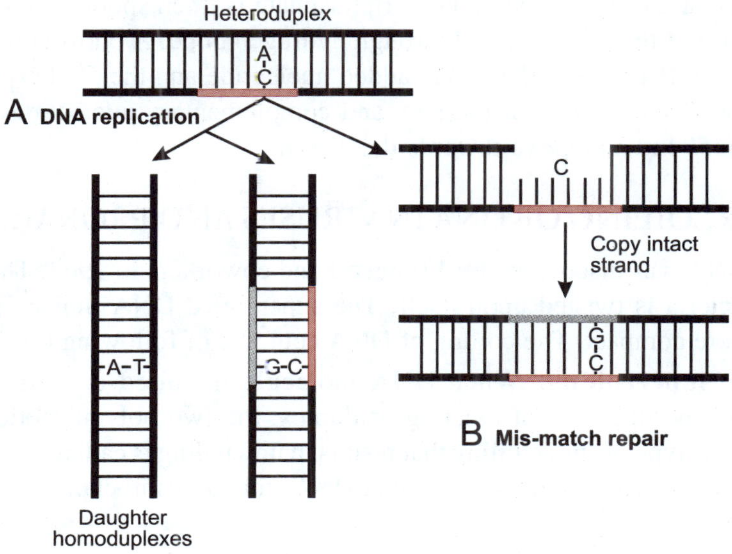

Fig. 6.16 Fate of heteroduplex DNA: **A.** By replication of DNA; **B.** By mis-match repair

SINGLE–STRANDED DNA

Sinsheimer (1959) isolated unusual form of DNA from a small bacteriophage $\phi \times 174$ which attacks *E. coli*. This DNA contains 5,500 nucleotides and is formed of a single strand. It does not have helical configuration. Its molar ratio of bases does not show the equivalence of **A** and **T** or **G** and **C.** Subsequent evidences indicated that the DNA of bacteriophage $\phi \times 174$ is circular and consists of a single strand.

The process of replication of single-stranded DNA is also very interesting. On entering *E. coli*, its *DNA polymerase* synthesises a 'minus strand'. This is complementary to the plus parental strand and the two together or a **'Plus-minus' DNA helix.** This is known as **replicative form (RF).**

IN VITRO SYNTHESIS OF DNA

In 1967, **A. Kornberg** and his coworkers were able to synthesise DNA molecules outside the cell in a test tube. They isolated enzyme ***DNA polymerase*** from *Escherichia coli* bacteria that could be used for *in vitro* synthesis of DNA. The reaction mixture in the test tube contained the ***polymerase enzyme,*** four different types of **deoxyribonucleotides** in the form of triphosphates (i.e. deoxyadenosine triphosphate, deoxythymidine triphosphate, deoxycytidine triphosphate and deoxyguanosine triphosphate), magnesium ions and some previously formed DNA. The amount of DNA synthesised was about 20 or more times greater than the amount of initial DNA. The enzymes hook together the freely available nucleotide units into a DNA strand. The deoxyribonucleotide triphosphates are added to the hydroxyl group at number 3′ carbon atom of deoxyribose sugar at one end of the chain.

The nitrogenous base of the newly added triphosphate is then spontaneously bounded to the complementary base of the primer strand through hydrogen bond. A phosphodiester bond is then formed between 3′–OH group of the newly added nucleotide and the 5′–PO_4 group of the next arriving nucleotide. This shows that the new and complementary nucleotides are added to the 5′→3′ strand from 5′ end and move towards the 3′ end.

SUPERCOILING OF DNA IN VIRUSES AND PROKARYOTES

Supercoiling of DNA was discovered by **Vinograd** and coworkers in 1963. The DNA molecule in viruses and bacteria is twisted upon itself. The supercoiled DNA molecules sediment more rapidly and are more compact. The coiling of DNA could be of following two types:

1. **Negative Supercoiling:** When DNA molecule is coiled or twisted in a direction opposite to the direction of twisting of duplex, the two polynucleotide strands tend to unwind. This type of supercoiling that results in unwinding is called negative supercoiling. Negative supercoiling is found in circular DNAs found in mitochondria, viral and bacterial chromosomes.
2. **Positive Supercoiling:** When DNA molecule is overwound it presents positive supercoiling. This is found in eukaryotic chromosomes. It makes eukaryotic DNA to be compacted into a limited space in the nucleus.

Enzymes ***topoisomerases*** uncoil or unwind supercoiled DNA in both prokaryotes and eukaryotes. Supercoiling plays key role in allowing DNA to be compacted so as to fit inside the limited space.

REPEATED DNA SEQUENCES ARE A FEATURE OF EUKARYOTIC DNA

Roy Britten and **David Kohne** found that mammalian genome has numerous DNA fragments that have repeated nucleotide sequences. These are separated into three classes:

1. Highly repeated fractions
2. Moderately repeated fractions
3. Non-repeated fractions

Extension: Repetitive Sequences

Most genes are present as single copies repeated. In them nucleotide sequences are represented once only. These are called **unique sequences**. Bacterial and viral DNAs have only unique sequences. But in higher organisms unique sequences may constitute as little as 20 per cent of the total DNA. For example, in human genome unique sequences are 65 per cent and in frogs only 22 per cent. The remaining DNA has **repeated sequences** or **repetitive sequences**. These are nucleotide sequences repeated many times throughout the genome. The number of nucleotides per repeated sequence may vary from 5 to 1,000. When these sequences lie next to each other closely, they are called **tandem repeats**. The **interspersed sequences** are scattered throughout the genome spread out separately.

1. Highly Repeated DNA Sequences

In vertebrate genome about 10 per cent of total DNA is formed of **repeated fractions.** These fractions are formed of a few hundred nucleotides in which a given sequence of nucleotides is repeated over again and again without interruption. These show tandem arrangement. The highly repeated sequences fall into three categories: **satellite DNAs, minisatellite DNAs** and **microsatellite DNAs.**

1. **Satellite DNAs:** These are short nucleotide sequences having 5 to 1,000 base pairs. These sequences are clusteres of tandem repeats. They are repeated over again and again forming regions of inert DNA fragments with up to 100 million base pairs. These are called **satellite DNAs** because DNA fragments with these sequences can be separated into distinct satellite bands during density gradient centrifugation. Three different satellite sequences have been identified in the genome of *Drosophila virilis*. These are formed of repeated segments of seven base pairs with a consensus sequence of ACAAACT. The other repeats are formed of 5, 10 and 12bp. In mammals, the satellite sequences are more complex and more varied.

   Satellite DNA is inert and is permanently coiled to form heterochromatin. The satellite DNA around the centromere of chromosomes is presumed to have some structural role providing attachment to the spindle fibres during cell division.

2. **Minisatellite DNAs or VNTRs:** Minisatellites are DNA segments consisting of short tandem repeats and have only fewer copies. They are also called **variable number of tandem repeats (VNTRs)** because the number of repeats in a given minisatellite varies greatly among individuals.

   Typically each VNTR is formed of 5 to 50 tandem repeats and each repititive segment of about 15bp or more. For example, one human VNTR has 64bp. Some hyper-variable VNTRs have as many as 1,000 repeats. Minisatellite sequences are considerably shorter than satellite sequences.

   Minisatellites are found scattered over the genome, but are concentrated close to the telomeres. They are non-coding DNA segments. However, because of highly variable nature VNTRs are used for the identification of individuals by **DNA fingerprinting**.

3. **Microsatellite DNAs:** These are shortest nucleotide sequences formed of 2–5 base pairs and are repeated about 100 times or less. These are scattered throughout DNA length. At least 30,000 different microsatellites are known in human genome. Their position and number of copies are highly variable from person to person and even in the members of the same family. Large number of copies of microsatellite sequences is found to be associated with some abnormalities or genetic disorders. For example, in human beings a gene FMRI under normal conditions has 5 to 60 copies of a triplet CCG. If their repeat number rises above 200,

the individual is mentally retarded. This condition is called **fragile X syndrome** because the microsatellite is present in the fragile site on X-chromosome.

> **Significance of Microsatellites**
>
> The role of micro and minisatellites in DNA structure or function is not known but their presence has helped in the construction of genetic maps, in the isolation of genes responsible for causing human diseases and in the development of technique of DNA finger-printing. This is because the length of short tandem repeats varies from person to person.

2. Moderately Repetitive DNA Sequences

This fraction of DNA includes sequences that are repeated in the genome from a few times to several hundred thousand times. It may form 20 to 80 per cent of the total DNA. The moderately repeated sequences are of coding type and noncoding type.

1. **Moderately Repetitive Noncoding DNA Sequences:** These form bulk of moderately repeated DNA sequences. These do not code for any product. These are found scattered throughout the genome and do not form clusters as in the case of highly repetitive sequences. These are mostly separated into two classes, namely **short interspersed elements (SINEs)** of DNA with less than 500 base pairs and **long interspersed DNA elements (LINEs)** with more than 1,000 base pairs.
2. **Moderately Repetitive Coding DNA Sequences:** These fragments of DNA include multiple copies of those genes that code for transfer RNAs, ribosomal RNAs and chromosomal proteins, the **histones**. This DNA constitutes about 25 per cent of human DNA. Each type of these repeated coding sequences are present in tandem array. Since tRNAs and rRNAs are needed in large quantities, multiple copies of their genes are found in the genome. For example, prokaryotic cells have about half a dozen copies of genes for rRNA and tRNA, whereas in eukaryotic cells, hundreds or thousands of copies of these genes are present.
3. **Slightly Repetitive DNA Sequences:** These DNA sequences exist as few copies. In some cases these are not perfectly repetitive. Their products have slight differences in amino acid sequences. For example, there are five different genes for five human globin chains (α, β, γ, δ and ε). These chains and their genes are all of about the same length. All of them share the same exon-intron structure (3 exons separated by 2 introns). The three-dimensional structures of these chains are also identical. But genes for β, γ and δ chains are closely placed on chromosome 11 (eleven). Thus, they are repetitive genes or repetitive DNA sequences.

3. Non-repeated or Non-repetitive DNA Sequences

> **Extension: Noncoding DNA**
>
> The moderately repetitive noncoding DNA is more abundant than the moderately repetitive coding DNA in eukaryotic cells. It is also called '**junk DNA**'. It is represented in two forms:
>
> - **Short Interspersed Elements** (**SINEs**) are highly repetitive DNA segments, consisting of thousands to millions of copies of noncoding sequences. They form about 10% of human genome. The best known example of SINE is the 300 bp long **Alu element.**
> - **Long Interspersed Elements** (**LINEs**) are larger and more numerous than SINEs. Mammalian genome has about 20,000–50,000 copies of the LINE–I family. They are considererd to be derived from retrovirus-like ancestors. LINEs contain sequence rearrangements that disrupt the coding sequences rendering them nonfunctional.

These are single copy DNA sequences which include functional or coding sequences. It means these sequences represent genes that code for virtually all enzymatic proteins associated with inheritance. These are always localised to a particular site on a particular chromosome. These are also called **unique sequences**. They constitute only a few per cent of the DNA.

SELFISH DNA AND JUNK DNA

Transposable segments of DNA, that are solely concerned with their own survival and replication and are of no use to the host cell in any way, are referred to as **'selfish DNA'**. It accumulates in the genome of slow growing multicellular organisms. On the other hand **junk DNA** is the defective selfish DNA that can no longer move from its location and can not express itself or its gene. It consists of repetitive noncoding sequences, including some introns.

Origin of Selfish DNA and Junk DNA

The eukaryotic cells can have foreign DNA from two different sources:

1. Viral DNA that gets inserted into the chromosome of eukaryotic host cell.
2. Transposable elements of mobile DNA.

The accumulation of selfish DNA depends on replication and reinsertion of selfish DNA. In rapidly dividing, single-celled organisms, selfish DNA tends to be eliminated, whereas in slowly dividing, multicellular organisms it gets accumulated. Though, originally useless, some defunct selfish DNA sequences may take up the role of maintaining chromosome structure.

The origin of repetitive DNA or junk DNA involves following processes:

- The original sequences of either viral DNA or transposable elements get inserted into the chromosome of eukaryotic cell.
- Both types of elements replicate and the duplicate copies get inserted at more locations in the chromosomes of infected cell.

- Many of these sequences mutate producing family of related sequences, most of which become nonfunctional because of mutations.

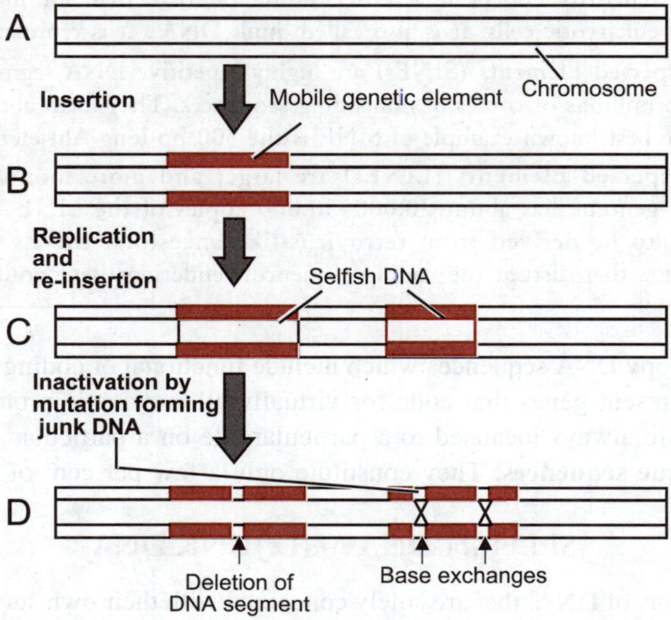

Fig. 6.17 Origin of selfish DNA and junk DNA

Palindromes and Inverted Repeats

Palindromes or palindromic sequences are sequences of nitrogenous bases that read the same backward or forward (*Note*: palindromes are words or phrases that read the same backward and forward). Since DNA is double-stranded, two types of palindromes are theoretically possible.

1. **Mirror-like palindromes** that are similar to palindromic words and that read the same forward and backward on the same strand.
2. **Inverted repeat type of palindromes** are those DNA sequences that are the same when read in 5′→3′ direction in both the strands of a DNA, *i.e.* the sequence reads the same forward on one strand and backward on the complementary strand.

Inverted repeats are more common and of major biological significance. They act as recognition sites on DNA for the binding of a variety of proteins. Regulatory proteins, restriction enzymes and also the modification enzymes recognise inverted repeats.

Mitochondrial DNA (mtDNA)

Mitochondrial DNA (mtDNA) is a double-stranded circular DNA. It occurs in two to six copies per mitochondrion and is attached to the inner mitochondrial membrane and lies in the mitochondrial matrix. The total amount of mtDNA is much less than 1 per cent of the total cellular

DNA of a eukaryotic somatic cell. In size a mtDNA resembles the DNA of a small bacteriophage. It has a molecular weight of only about 10 million.

mtDNA differs in base composition from nuclear DNA of the same cell. mtDNA is responsible for the synthesis of mitochondrial proteins. Therefore, it carries out synthesis of mRNA, tRNA and rRNA. However, the transcription and translation activities of mtDNA are under the control of nuclear DNA.

Mitochondrial DNA is presumed to have originated from some free-living bacteria-like ancestors. Its size gradually reduces from lower to higher organisms. For example, yeast mtDNA is five times longer than human mtDNA. A small fraction of DNA occurs as catenated dimers, having two interlocked DNA circles. These are believed to be formed during replication of mtDNA.

Chloroplast DNA (ctDNA)

Like mitochondria, chloroplasts also possess double-stranded circular DNA. It is represented by several copies that lie in the matrix. The functioning of ctDNA is controlled by nuclear DNA.

Catalytic DNA (Deoxyribozymes)

In all living organisms, except few viruses, DNA is exclusively used to store genetic information and has never been found to serve as an enzyme. Enzymes are mainly proteins but some RNAs also have cleaving function and act as enzymes. However, single-stranded DNA molecules are found to greatly accelerate transformations in a substrate, a variety of chemical reactions occurring in the cells and reactions during DNA cloning. DNA are found to catalyse cleaving of RNA molecules using cofactors such as histidine.

The single-stranded DNA molecules, acting as deoxyribozymes form intricate structures by forming binding pockets for small molecules. This enhances their catalytic activity.

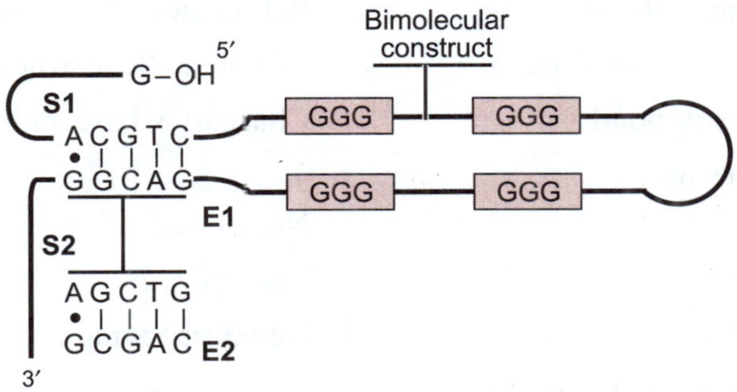

Fig. 6.18 RNA cleaving DNA's amino acid utilising deoxyribozymes phosphorylating DNA with DNA capping

Significance of Catalytic DNA

The study of catalytic DNA can enhance our understanding of enzymatic catalysis and biopolymer function in general and also in the development of useful biotech tools.

TEXT QUESTIONS

1. Write an essay on the structure and functions of DNA.
2. What are the chemical building blocks or monomers of DNA? Draw the structural formulae for each one of them. Name the bonds by which the adjacent monomers are connected in the polynucleotide chain.
3. Why DNA needs to have a double helical structure to perform the functions attributed to it?
4. What is Watson and Crick's model of DNA structure? How does it explain the biological properties of DNA?
5. Give evidences in favour of double helical structure of DNA.
6. Enumerate the important postulates of Watson and Crick's model of DNA.
7. Differentiate between B-DNA and Z-DNA.
8. What do you understand by renaturation and denaturation?
9. Describe Chargaff's rule of base equivalence in DNA.
10. Explain how the two ends of polynucleotide chain are marked.
11. What evidences support the double helical structure of DNA?
12. Outline two major types of experimental analysis that laid the foundation for the deduction of the structure of DNA by Watson and Crick.
13. Describe the biological significance of Watson and Crick's model of DNA.
14. Enumerate the important features of double helical structure of DNA molecule.
15. Explain the following:
    (a) Complementary base pairing
    (b) Polynucleotide chain has direction
    (c) Left and right handed DNA
    (d) Double helical structure of DNA
    (e) Phosphodiester bond
    (f) Chargaff's base ratio.
16. Write short notes on:
    (a) Nucleoside
    (b) Nucleotide
    (c) Pyrimidines
    (d) Purines
    (e) Watson and Crick's model
    (f) Deoxyribozymes.
17. Differentiate between the following:
    (a) Nucleoside and nucleotide
    (b) Polymerases and ligases
    (c) Purine ring and pyrimidine ring
    (d) Ribose and deoxyribose
    (e) Major and minor grooves
    (f) 3′ and 5′ terminus.

Organisation of Genome

7 Chapter

IN THIS CHAPTER

Genome: Viral Genome; Prokaryotic Genome; Bacterial DNA; Mitochondrial Genome; Eukaryotic Genome; Organisation of Chromatin: Supersolenoid Structure of Eukaryotic Chromosome; Structure of Interphase Chromatin; Structure of Anaphase Chromosome; Different Regions of Anaphase Chromosome and their Functions; Special Types of Chromosomes: Lampbrush Chromosomes; Polytene Chromosomes; Physiology of Chromosomes.

GENOME

The genome is the entire collection of genes and all other functional and non-functional DNA sequences in an organism in a haploid set of chromosomes. It includes genes, regulatory genes and nonfunctional nucleotide sequences. Therefore, genome comprises of:

- The **structural genes** are DNA segments that code for some specific RNAs or proteins. They encode for mRNAs, tRNAs, snRNAs and scRNAs.
- The **functional sequences** or **regulatory genes** occur as regulatory elements or as initiation sites, promoter sites, operator sites and enhancer sites, etc.
- The **nonfunctional sequences** include introns and repetitive sequences. They are needed for coding, regulation and replication of DNA. Their number is found to be much more than the functional sequences.

The genome of all living beings represents their hereditary material and is formed of DNA. In simple prokaryotic cells, the genomic DNA forms a single circular chromosome. It is without basic proteins and lies in the cell cytoplasm in the nucleoid region. In eukaryotic cells, DNA is associated with basic proteins **(histones)**. They form long **chromatin fibres**. The chromatin fibres form a network and are enclosed in a double-layered nuclear envelope. This structure is called **nucleus**. Its chromatin condenses into chromosomes during cell division.

I Viral Genome

In viruses and bacteriophages a single molecule of DNA or RNA represents the **viral chromosome**. It may be single-stranded or double-stranded.

Table 7.1 The size and structure of the genomes of some viruses

Organisms	Types of genome	Molecular weight ($\times 10^{-6}$ daltons)	Approximate no. of genes
DNA viruses			
1. ϕ × 174 coliphage	Circular single-stranded DNA	1.7	6 (8)
2. Polyoma virus	Circular double-stranded DNA	3.0	5
3. SV 40 virus	Circular double-stranded DNA	3.2	6
4. Adenovirus type 12	Linear double-stranded DNA	21–12	38
5. Herpes simplex	Linear double-stranded DNA	96	160
RNA Viruses			
1. Potato spindle tuber virus	Linear single-stranded RNA	25–0.11	< 1
2. Polio virus	Linear single-stranded RNA	2.6	4 (3)
3. Influenza virus	Linear single-stranded RNA	3.6	12
4. Mouse leukemia virus	Linear single-stranded RNA (can be dissociated into fragments)	11.2	37
5. Reo virus	Linear double-stranded RNA	15	25

1. DNA-Containing Viral Chromosomes

These are 20Å wide chromosomes corresponding to the width of DNA molecule. These may occur in two forms:

1. Linear DNA: In most cases DNA occurs as linear duplexes, but it may be single-stranded linear DNA and in some viruses like ϕ × 174 bacteriophage, is circular and single-stranded. It has **cohesive ends.** The linear viral DNAs present in phage λ is a linear duplex DNA with a molecular weight of 32 million and a contour length of about 17.2 μm. Phage M13 has ssDNA. The phage λ DNA shows following characteristic structural features indicating that they were originally formed as circles or that they change into circles before replication:

- **Cohesive ends:** The 5′ends for these duplex DNAs project as a single strand beyond 3′ends of antiparallel strands. These single strands are formed of about 12 nucleotides. They are complementary to and pair with the nucleotides at the 3′ ends forming circles.
- **Terminal repetition:** The 3′end of DNAs bears a base sequence that is repeated at the 5′ end of strand. When such sequences are acted upon by a ***nuclease***, they lead to the formation of complementary cohesive ends leading to covalent circles.
- **Circular permutations:** The viral DNAs contain a population of molecules whose sequences are circular permutations of each other (Fig. 7.1). These circular permutated linear DNAs form identical circular molecules, whenever complementary cohesive ends are exposed by the action of ***exonuclease***.

```
1   A—B—C—D—E—F—G—G—H
    |  |  |  |  |  |  |  |  |
    A′—B′—C′—D′—E′—F′—G′—H′—H′

2   C—D—E—F—G—H—I—J—A—B
    |  |  |  |  |  |  |  |  |  |
    C′—D′—E′—F′—G′—H′—I′—J′—A′—B′
```

Fig. 7.1 Diagram to illustrate circular permutation in a linear chromosome of viruses

- **Modified nitrogenous bases:** Some of the nitrogenous bases of viral DNA are modified. For example, in T-even phages **hydroxy-methyl-cytosine** occurs in place of cytosine. Moreover, the hydroxyl group of hydroxy-methyl-cytosine is often glucosylated. These modifications protect viral DNA from degeneration by the action of host endonuclease.

2. Circular DNA: In some viruses and also in bacteriophage $\phi \times 174$, the DNA is single-stranded and circular. It is one of the smallest viral DNA with molecular weight 1.7 megadaltons and contour length of 1.77 μm. It has 5,375 nucleotides organised into nine genes. Some of the nucleotides of these genes overlap. Other examples of phages with ssDNA are G4 and S13 phages.

2. RNA-Containing Viral Chromosomes

The RNA containing viral chromosomes are linear and often composed of a single-stranded RNA molecule. These occur in plant viruses and some animal viruses.

Supercoiling of Viral Chromosome

The circular duplex DNAs when isolated in native form from viruses, are often supercoiled or supertwisted. This is caused when number of base pairs per unit length is more than normal. This produces torsional stress and as a consequence, the twisting of DNA circle.

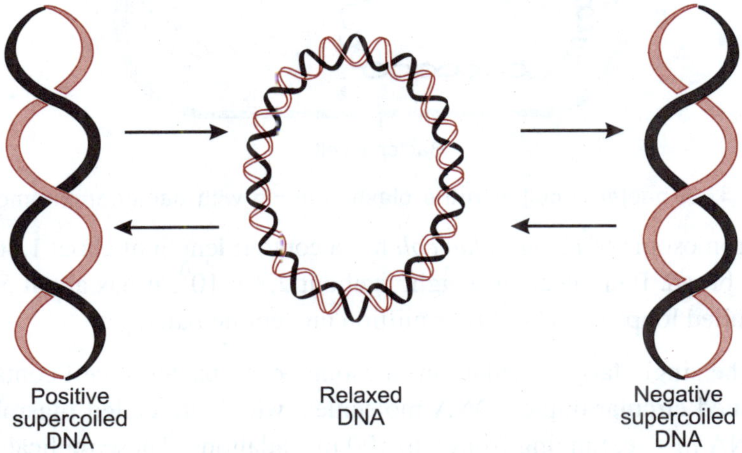

Fig. 7.2 Double-stranded circular DNA of virus and its positive and negative supercoiling

II. Prokaryotic Genome: Bacterial DNA

In bacteria and blue-green algae, the hereditary material is organised into a single large circular chromosome composed of a circular molecule of double-stranded DNA. It is known as **bacterial chromosome** or **nucleoid**. It lies free in the cytoplasm in the nuclear zone and has no protein around the DNA molecule, as is found in eukaryotic chromosomes. However, some RNA is found associated with DNA and forms its backbone or core. DNA molecule is supercoiled. Nucleoid proteins and RNA stablise the supercoiled state of DNA.

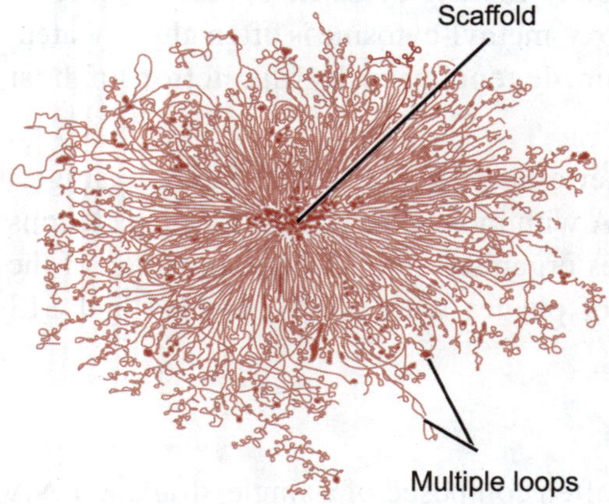

Fig. 7.3 Electron micrograph of bacterial chromosome of *Escherichia coli* showing multiple loops

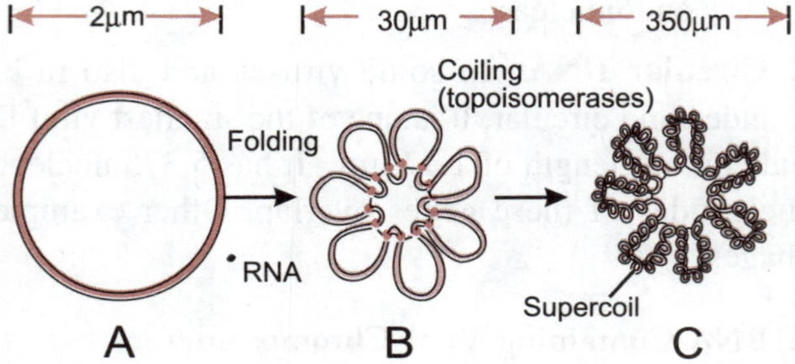

Fig. 7.4 Steps in coiling of bacterial chromosome

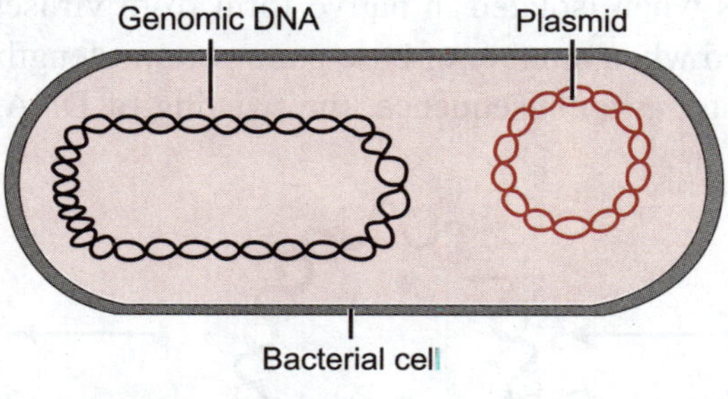

Fig. 7.5 A bacterial cell with the plasmid along with bacterial chromosome

The circular chromosome of *Escherichia coli* has a contour length of about 1,3608 μm (1.36 mμ) and is about 20Å broad. Its molecular weight is about 2.8×10^9. It has about 50 or more highly twisted or supercoiled loops and about four million nucleotide pairs.

In addition to the single large circular chromosome, each bacterial cell contains from 1 to 20 much small uncoiled circular duplex DNA molecules, which are called **plasmids.**The plasmids resemble viral DNA in size, ranging from 5 to 100 megadaltons. These replicate in fixed number along with the chromosome. Their function is not clearly known. Some of them become incorporated into the host chromosome and are called **episomes**. These are sometimes transferred from one bacterial cell to another during conjugation and give them new characteristics.

III. Mitochondrial Genome

Mitochondria of eukaryotic cells contain double-stranded, circular DNA with a molecular weight of about 10 million. It makes up less than 1 per cent of total circular DNA. A small fraction of DNA occurs in the form of interlocked circles called **catenated dimers**. These are presumably formed during replication of mitochondrial DNA.

IV. Eukaryotic Genome

In resting non-dividing eukaryotic cells, the genome is in the form of nucleoprotein complex, called **chromatin**. It is amorphous and is randomly dispersed in the nuclear matrix as interwoven network of fine chromatin threads. When a cell prepares to divide, its chromatin condenses into a species-specific number of well defined **chromosomes**.

ORGANISATION OF CHROMATIN

The chromatin is formed of about 60 per cent **proteins**, 35 per cent **DNA** and 5 per cent **RNA**. The **DNA** in the largest chromosome of *Drosophila* is about 4.0 cm long with a molecular weight 80×10^9. The DNA molecule of individual human chromosomes ranges from 1.7 to 8.5 cm in length when uncoiled. It contains $50 \times 10^6 - 250 \times 10^6$ nucleotide pairs. The information stored in DNA is organised, replicated and read with the help of a variety of **DNA-binding proteins.** These fall into two categories:

- Structural proteins – **Histones**
- Functional proteins or control proteins – **Nonhistones**

Structural proteins or **packaging proteins** are non-specifically binding proteins. These are bound to DNA along most of its length and help to package it without preventing the access of other DNA-binding proteins. They are basic proteins and are called **histones**. The **nonhistones** or **functional proteins** are associated with gene regulation and other functions of chromatin.

Histones

Histones are the main structural proteins found in eukaryotic cells. These are low molecular weight proteins with high proportion of positively charged amino acids (basic amino acids, the **arginine** and **lysine**). The positive charge helps histones to bind to DNA and play a crucial role in the packaging of long DNA molecules.

Types of Histones

There are five different types of histones that fall into two categories:

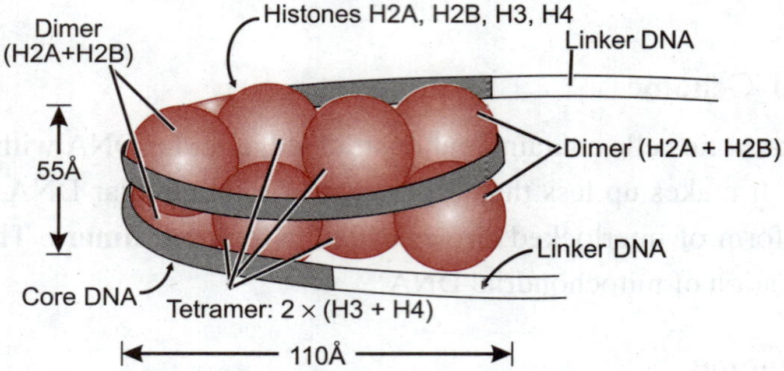

Fig. 7.6 Histone octamer in a nucleosome

1. **Nucleosomal Histones:** The nucleosomal proteins are eight molecules of four types of small proteins. They are responsible for coiling of DNA into nucleosomes. They are represented by two molecules of each H2A, H2B, H3 and H4 proteins. Each of them is formed of about 102–135 amino acids. These form the inner core of nucleosome. Nucleosomal histones exhibit great similarity because they have not changed during evolution. These are present in equimolar amounts, two of each type being present after every 200 base pairs, thus forming a **histone octamer** or **core particle.**

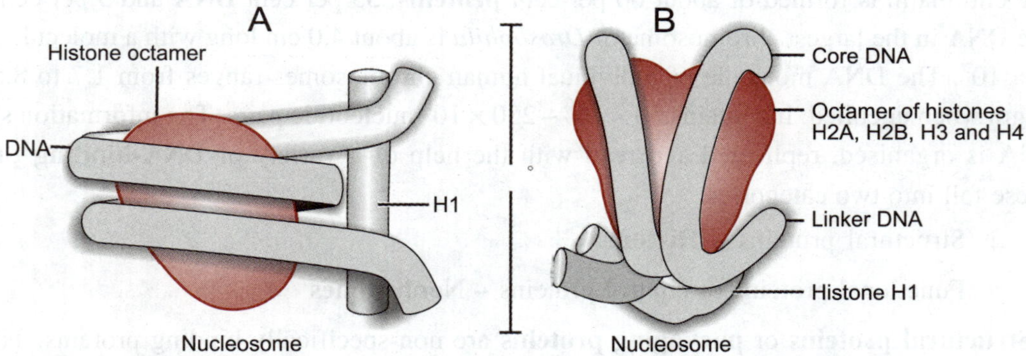

Fig. 7.7 Schematic diagram of a nucleosome or core particle showing structure of histones with DNA helix; **A.** In side view; **B.** in front view. The protein H-1 is not a member of inner circle and is technically not a part of core particle

2. **H1-histones:** These are large (of about 200 amino acids) proteins and are tissue specific. These are present one molecule per 200 base pairs and are loosely associated with DNA. Only one H1 molecule is present with each nucleosome. H1 histones are responsible for packing of nucleosomes into 30 nm nucleoprotein fibre.

Functions of Histones

Histones in eukaryotic chromosomes serve two functions:

- These serve as structural elements and help in coiling and packing of long DNA molecules.

- These cover or repress specific segments of DNA so that these segments are unable to transcribe. Their transcription is possible only when histones are removed in response to certain molecular signals.

Nucleosome Concept or Oudet Concept of Chromatin Structure

Nucleosomes are the fundamental packing units of chromatin and give chromatin the **'beads on a string'** appearance. Each nucleosome is disc-shaped about 11nm in diameter. It consists of a **core particle** and a smaller **spacer** or **linker** DNA.

1. Core Particle: The core particle consists of the octamer of histones having two copies each of **H2A, H2B, H3** and **H4.** It is about 11nm in diameter and 6nm in height. A strand of DNA having 146 base pairs is tightly wrapped around this core forming two circles (73 nucleotides per turn).

2. Spacer DNA or Linker DNA: It is a small segment of DNA having just four base pairs. One unit of histone H1 is associated with it. There is considerable variation in the length of spacer region in different species varying from one base pair to about 80 base pairs in sea-urchin sperm.

On an average, nucleosomes repeat at an interval of about 200 nucleotide pairs. Thus each human cell with 6×10^9 DNA nucleotide pairs contains 3×10^7 nucleosomes.

Histone H1 molecule has globular central region with two arms, one ending with amino-terminal and other with carboxyl-terminal. H1 molecule binds through its globular portion to a unique site on a nucleosome. The arms extend to contact other sites on histone cores of adjacent nucleosomes. Thus nucleosomes are pulled together into a closely packed regular repeating array.

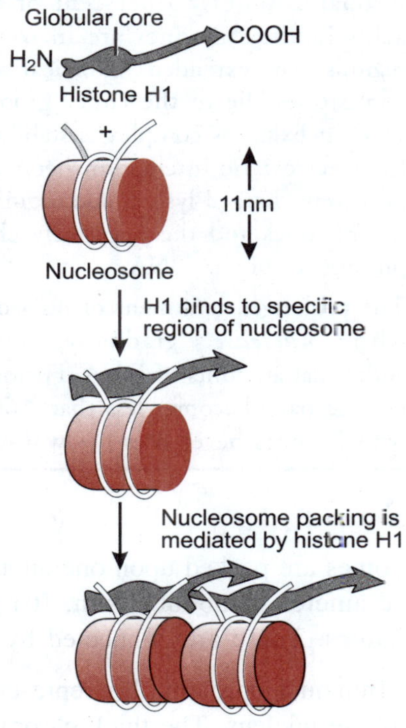

Fig. 7.8 Role of H_1 histone in nucleosomes packing

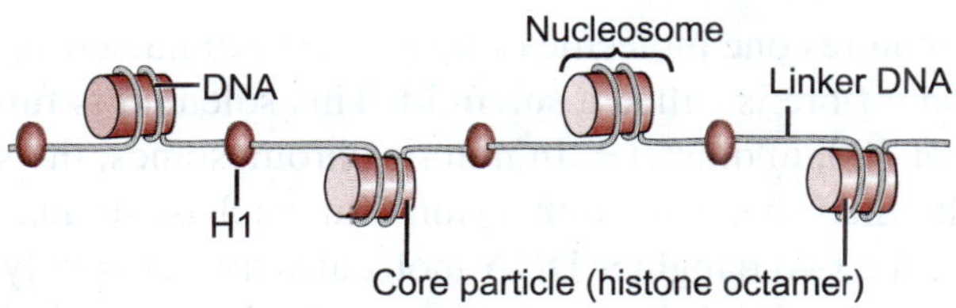

Fig. 7.9 Nucleoprotein fibre showing beads on a string appearance

Extension: Nucleosome Model

R. D. Kornberg and **J. O. Thomas** (1974) proposed a model for chromatin structure. This was called **nucleosome model** by **Oudet**, **Gross-Bellard** and **Chambon** in 1975. Nucleosomes are bead-like nucleoprotein particles formed of an octameric protein disc around which the 140bp segment of DNA is wrapped like a ribbon.

The binding of DNA and histones is intimate. About 80 per cent of amino acids in the histones are in α-helical regions. The extended α-helical regions of histones lie in the larger grooves of DNA helix. This complex is stabilised by the electrostatic interaction between the positively charged lysines and arginines of the histones and the negatively charged phosphates of the DNA.

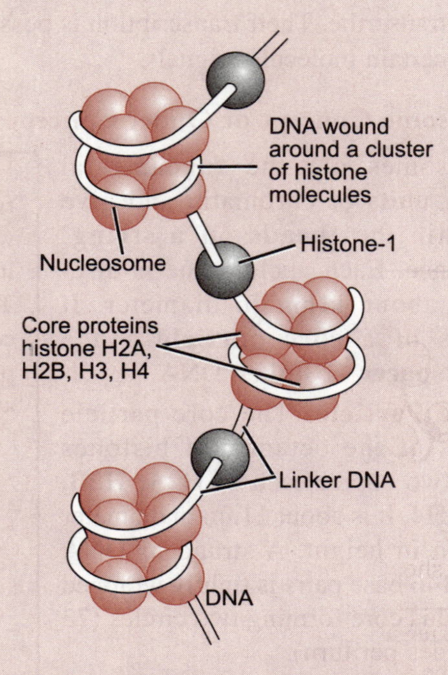

Fig. 7.10 'Beads on a string appearance' of nucleoprotein fibre shown in model

The prolonged treatment of nucleosomes with *nuclease* gradually removes additional amounts of DNA. Histones remain associated with the DNA till the number of base pairs becomes less than 160 and H1 is lost. The number of nitrogenous bases with histones never goes below 140 base pairs.

Nucleosome Packing or Formation of Chromatin Fibre

Nucleosomes are packed upon one another to form a **chromatin fibre** or **nucleoprotein fibre** with a diameter of about 30nm. It appears to have a beaded appearance. The beads are **nucleosomes**. These are connected by linker DNA.

The 11nm nucleoprotein fibre represents the first level of organisation of chromatin and is seen in interphase nucleus. The thick chromatin fibre of 30nm arises during early prophase of cell division by spiral coiling of thin 11nm chromatin fibre. The 30nm fibre has a solenoid type of ultrastructure and has 6 to 7 nucleosomes per turn.

This type of packing requires one molecule of histone H1 per nucleosome. H1 gives polarity to chromatin fibre. The 30nm fibre is called a **solenoid**. This solenoid is further folded and packed into the **supersolenoid** of 300nm diameter. In mitotic chromosomes, the supersolenoid is further condensed to produce the final shape and dimension of metaphase or anaphase choromosome. In metaphase chromosome, the two daughter DNA molecules are separately folded to produce two sister chromatids, held together by a single centromere. Each chromatid is about 700 nm thick.

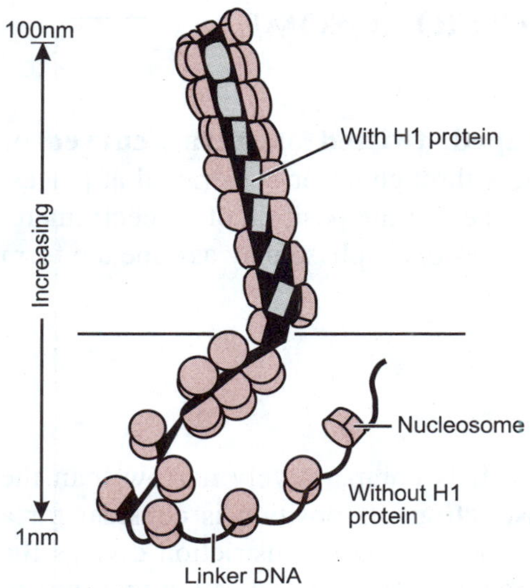

Fig. 7.11 Coiling of nucleosome fibre showing solenoid structure

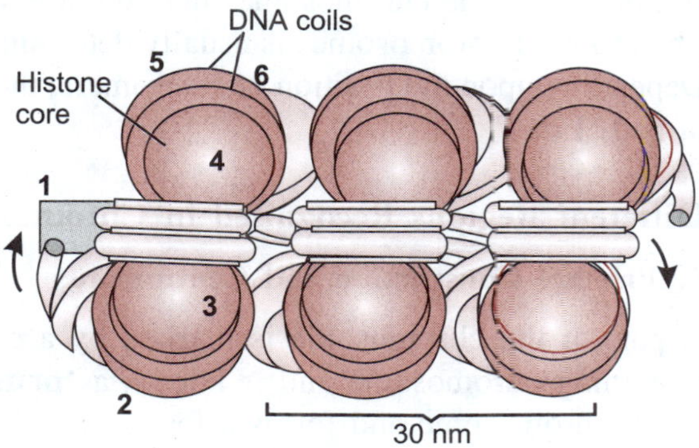

Fig. 7.12 A model showing nucleosome packing in 30nm nucleoprotein fibre

Extension: Nucleosomes in a Human Chromosome

The nucleus in human cells is typically about 5μm (5×10^{-4} cm) in diameter. The 30nm chromatin fibre of a typical human chromosome measures about 0.1cm in length and could span the nucleus more than 100 times. Thus 30nm fibre exhibits further folding and forms a series of **looped domains.** In a typical human chromosome of about 0.5nm length and of 30nm fibre, the number of looped domains may be 2,000 and each loop may contain about 20,000 to 10,00,000 nucleotides (i.e about 0.5nm of 30nm fibre).

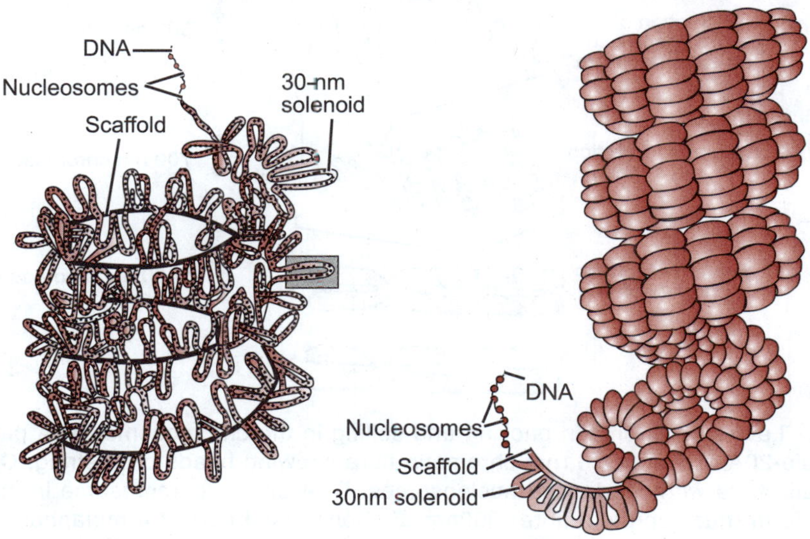

Fig. 7.13 Supercoiling of 30nm fibre into a supersolenoid around a central core or scaffold

STRUCTURE OF ANAPHASE CHROMOSOME

Shape

The anaphase chromosomes may appear as **rod-shaped, twisted** or **spiral, curved** or **filamentous**. The chromosomes may be of equal thickness throughout or constricted at places. The shape of chromosomes is usually determined by the type and the position of its centromere. Depending upon the position of centromere, the chromosomes in anaphase may assume the form of **rod, J** or **V**.

Different Regions Recognised in Chromosomes

1. Primary Constriction and Centromere

A part of the chromosome is marked by a constriction. It is comparatively narrow than the remaining chromosome and is known as **primary constriction**. Its position is constant for a given chromosome and forms a feature of identification. The primary constriction divides the chromosome into two **arms.** It shows a faintly positive Feulgen reaction, indicating presence of DNA of repetitive type. This DNA is called **centromeric heterochromatin**.

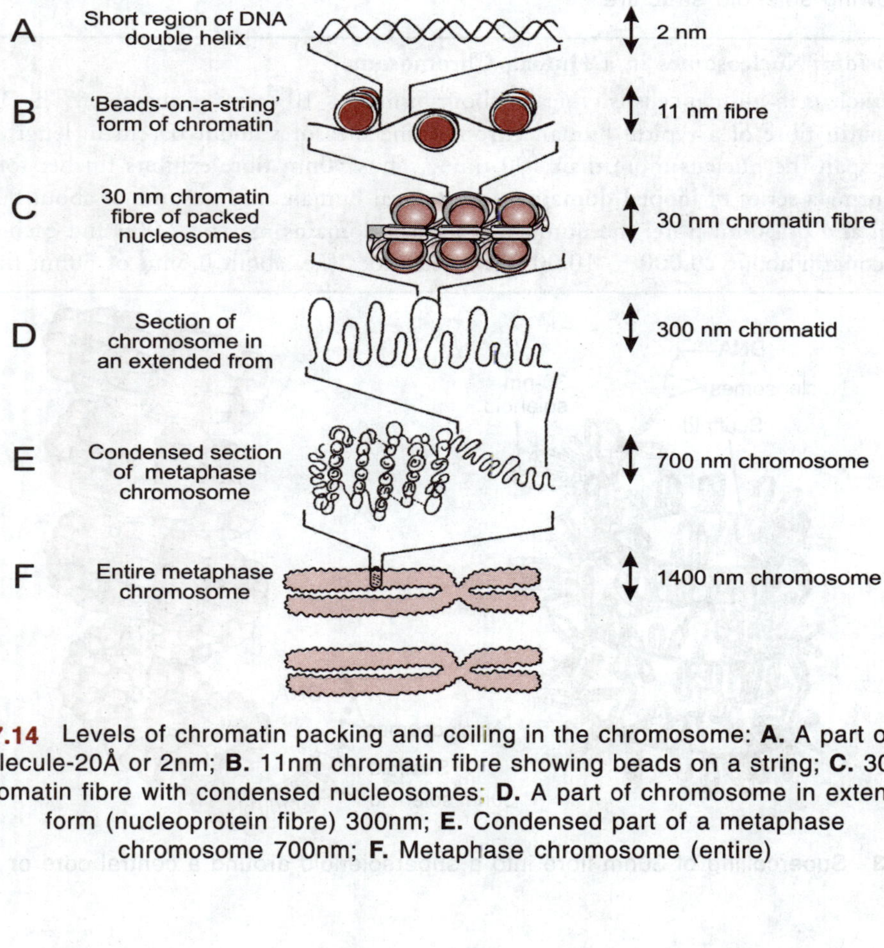

Fig. 7.14 Levels of chromatin packing and coiling in the chromosome: **A.** A part of DNA molecule-20Å or 2nm; **B.** 11nm chromatin fibre showing beads on a string; **C.** 30nm chromatin fibre with condensed nucleosomes; **D.** A part of chromosome in extended form (nucleoprotein fibre) 300nm; **E.** Condensed part of a metaphase chromosome 700nm; **F.** Metaphase chromosome (entire)

Centromere or kinetochore lies in the region of primary constriction. The microtubules of the chromosomal spindle fibres get attached to the centromere during metaphase and anaphase. Therefore, centromere is associated with the chromosomal movement during cell division.

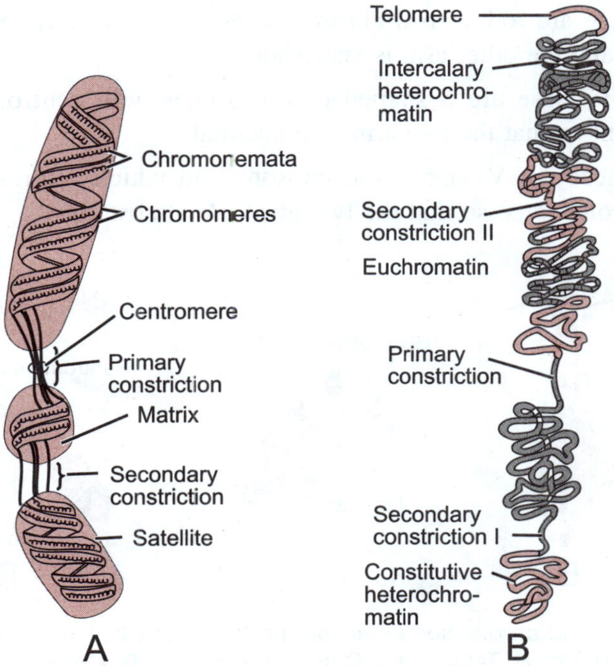

Fig. 7.15 Structure of a Chromosome. **A.** In metaphase; **B.** In anaphase

A. Types of Chromosomes Based on the Number of Centromeres

Usually, there are two centromeres one on each chromatid but there may be more or none. Depending upon the number of centromeres, the chromosomes may be:

- **Monocentric** with only one centromere.
- **Dicentric** with two centromeres.
- **Polycentric** with more than two centromeres.
- **Acentric** without a centromere. Such chromosomes represent freshly broken segments of chromosomes which do not survive for long and are digested into their constituent nucleotides.
- **Diffused** or **nonlocated** are with indistinct centromere which remains diffused throughout the length of chromosome. The microtubules of spindle fibres are attached to the chromosome arms at many points. The diffused centromeres are found in insects, some algae and some groups of plants, (e.g. *Luzula*).

B. Types of Chromosomes Based on the Position of Centromere

Based on the location of centromere the chromosomes are categorised as follows:

- **Acrocentric:** These are rod-shaped chromosomes with centromere occupying a terminal position. They have only one arm, the other is absent.
- **Telocentric:** These are rod-shaped chromosomes having subterminal centromere. One arm is very long and the other arm is very short.
- **Submetacentric:** These are S-shaped chromosomes with centromere slightly away from the mid-point so that the two arms are unequal.
- **Metacentric:** These are V-shaped chromosomes in which centromere lies almost in the middle of chromosome so that the two arms are almost equal.

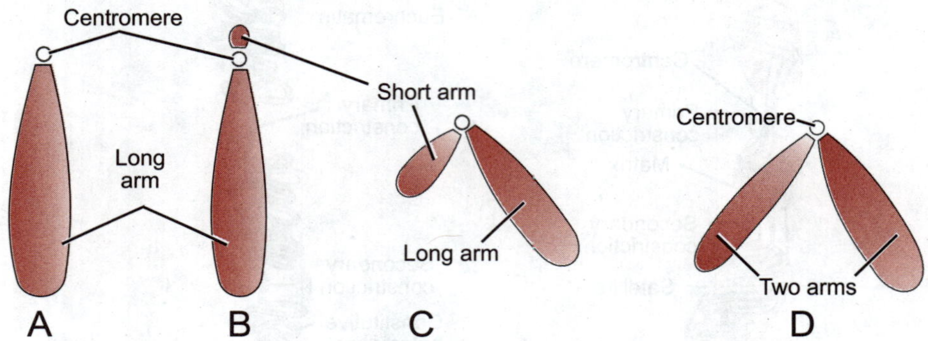

Fig. 7.16 Types of chromosomes according to the position of centromere. **A.** Acrocentric; **B.** Telocentric; **C.** Submetacentric; **D.** Metacentric

Ultrastructure of Centromere

Under electron microscope the centromere appears as a plate-like or cup-like disc, plastered upon the primary constriction. It is about 0.20 – 0.25μ in diameter and is formed of some nonchromatin material. In cross section it usually appears to consist of:

- **An electron dense layer:** It is 30 to 40nm thick with convex outer surface. Microtubules of spindle fibres get attached to it and penetrate it reaching the chromatin fibres.
- **An inner less dense layer:** It is 15 to 30nm thick which lies between the electron dense layer and the underlying chromatin fibres.
- **An outer fibrillar material:** It is on the convex face of centromere and forms a kind of **electron dense corona.**

Functions of Centromere

The centromere is considered to perform following two functions:

- It provides attachment to the microtubules of chromosomal spindle fibres and helps in chromosomal movement during cell division.

- The centromere serves as a nucleation centre for the polymerisation of tubulin, the protein used in the formation of microtubules. Thus it helps in the formation of spindle fibres during prometaphase and metaphase.

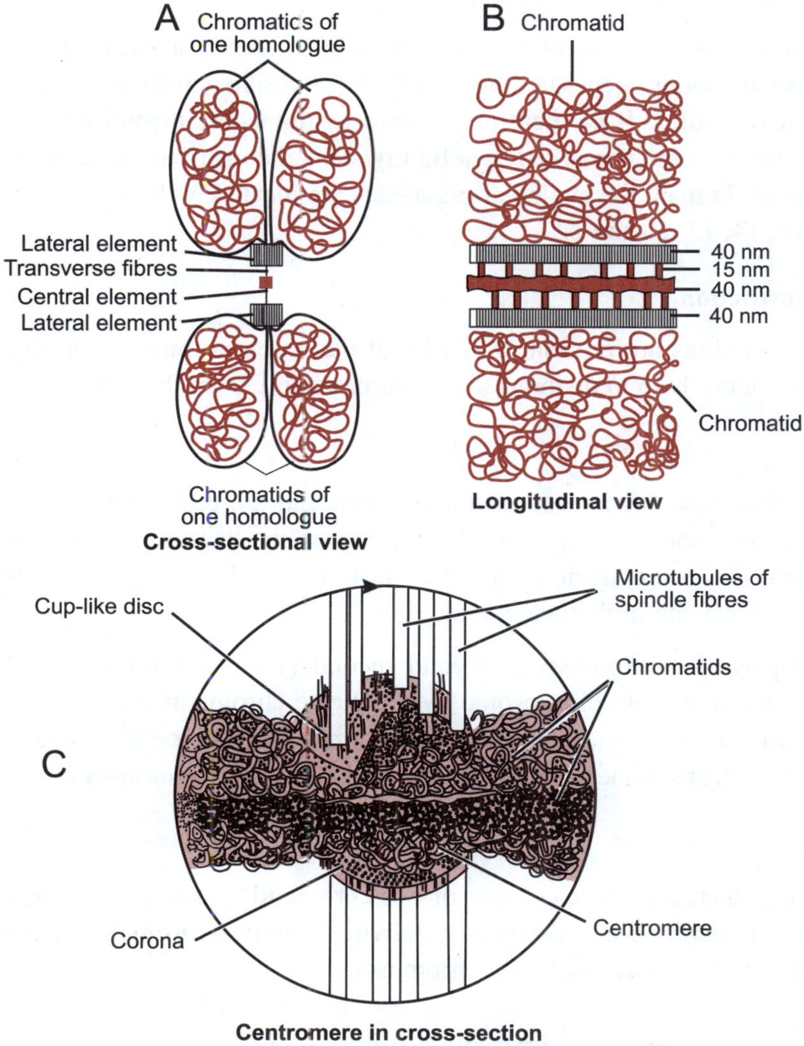

Fig. 7.17 Diagram showing location and ultrastructure of a centromere in a metaphase chromosome: **A.** Location of centromere; **B.** Structure of centromere in longitudinal section; **C.** Cross section of centromere

2. Secondary Constriction or Nucleolar Organiser

Sometimes one or both the arms of a chromosome are marked by a constriction other than the primary constriction. During interphase this area is associated with the nucleolus and is found to participate in the formation of nucleolus. It is, therefore, known as **nucleolar organiser region** or the **secondary constriction.** Its location on the chromosome is marked with a lightly stained constricted region.

3. Nucleolar Organiser Region (NOR)

In certain chromosomes, the secondary constriction is intimately associated with the nucleolus during interphase. It contains genes coding for 18S and 28S ribosomal RNA and is responsible for the formation of nucleolus. It is, therefore, known as **nucleolar organiser region (NOR).** Its location in the chromosome is marked by a lightly stained constricted area and is constant for a given chromosome. In man, the nucleolar organisers are located in the secondary constriction of chromosomes 13, 14, 15, 20 and 22.

4. Tertiary Constriction

The tertiary constrictions are present in nearly all the chromosomes. Their significance is not known. However, these help to distinguish one chromosome from the other.

5. Telomeres

The tips of the chromosomes are rounded and sealed and are called **telomeres**. These provide stability to the chromosomes and protect their individuality, because the telomeric ends do not form any permanent association with other parts of the homologous or nonhomologous chromosomes, whereas the broken ends may join.

The terminal part of a chromosome beyond secondary constriction is called **satellite**. It is attached to the main body of chromosome by a delicate chromatin filament. The satellite may appear as a rounded or elongated knob. It has a constant shape and size for a particular chromosome. The chromosome with satellite is known as **sat-chromosome**.

6. Chromatids

At metaphase stage a chromosome consists of two chromatids joined at the common centromere. In the beginning of anaphase when centromere divides, the two chromatids acquire independent centromere and each one changes into a chromosome.

SPECIAL TYPES OF CHROMOSOMES

1. Lampbrush Chromosomes

In the oocytic nuclei of those animals which have large yolky eggs, the prophase of first meiotic division is extremely extended. During this phase the oocytes grow and synthesise nutrients for the future embryo. In them, the chromosomes become greatly enlarged and assume unusual configuration. A large number of loops project out from the chromatid axis, giving it a lampbrush appearance. Hence, these chromosomes are called **lampbrush chromosomes.**

The lampbrush chromosomes are bivalents, each consisting of two chromatids. These persist during the prolonged diplotene phase of first meiotic prophase.

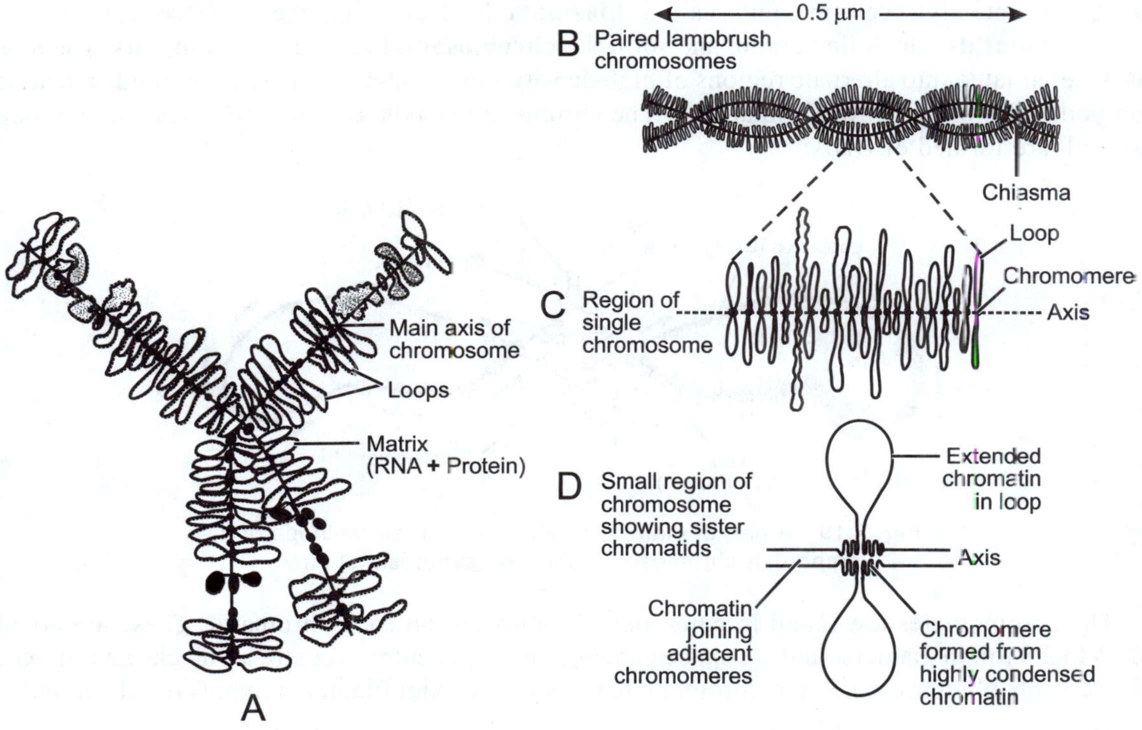

Fig. 7.18 **A.** Structure of a lampbrush chromosome; **B.** Gross structure showing two homologous chromosomes in diplotene phase; **C.** Enlarged view of a part of B; **D.** One loop of a lampbrush chromosome

History

Lampbrush chromosomes were first observed by **Flemming** (1882) in amphibian oocyte. A detailed study was made by **J. Ruckert** (1892) in the oocytes of sharks.

Occurrence

Lampbrush chromosomes are found in the oocytes of insects, sharks, amphibians, reptiles and birds which produce large and yolky eggs. These have also been found in plants and some invertebrates like *Sagitta, Sepia* and *Echinaster.*

Size

Lampbrush chromosomes are large enough to be seen under light microscope. They may be as long as 1,000 μ or more and about 20 μ in width. In salamander oocyte, they may attain a length of about 5,900 μ.

Structure

A lampbrush chromosome (in diplotene stage) consists of two homologous chromosomes which are in contact only at certain points, called **chiasmata**. Each chromosome of the pair is formed of two **chromatids** which lie parallel and form the **chromosomal axis** or the **main axis**. The axis is differentiated into alternate regions of high density loops which are lightly coloured, and arise on both sides of the chromosomal axis. The chromosomal axis, the chromomeres and the loop axis all are formed of DNA.

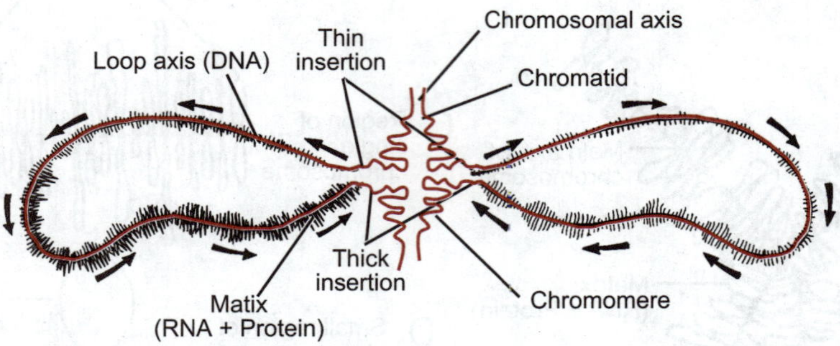

Fig. 7.19 A part of main axis with a pair of lateral loops of a lampbrush chromosome showing synthesis of RNA

The chromomeres are found in pairs, one chromomere on each chromatid. These are about 0.25 to 2.0 μm in diameter and are spaced about 2μm from centre to centre along chromatid axis. These probably represent heterochromatic regions where axial filament remains tightly coiled.

The lateral loops arise from the chromomeres either in two or in multiples of two. These extend on either side of the chromosomal axis about 550μm and are about 30 – 50Å (3–5 nm) in diameter. Each loop consists of an **axial fibre** formed of DNA. It is surrounded with the matrix composed of RNA and proteins. This gives fuzzy appearance to lateral loops.

Electron Microscopic Structure of Lampbrush Chromosome

Electron microscopic studies by **Miller** and **Beaty** (1969) on lampbrush chromosomes of salamander oocyte have shown the presence of dense granules on the loop axis of DNA. These dense granules represent large molecules of enzyme ***RNA polymerase.*** On getting attached to DNA, these initiate RNA synthesis. Arising from the ***RNA polymerase*** molecules are seen fine fibrils of RNA which increase in length.

Each loop is considered to be one long operon consisting of a series of identical copies of the same structural genes **(cistrons)** separated by **spacer DNA**. Each gene locus probably produces a very long RNA molecule. This interacts with protein to form **ribonucleoprotein** for the ribosome. There are two views regarding the formation of lampbrush chromosomes:

1. According to **Callen** and **Lyod** (1960), a chromomere is the **master gene** with solenoid supercoiling which produces several identical copies of its own. Each extends out as a lateral loop formed of linear strand of **nucleosomes,** representing the transcriptionally active stage. These are called **salvage gene copies.**

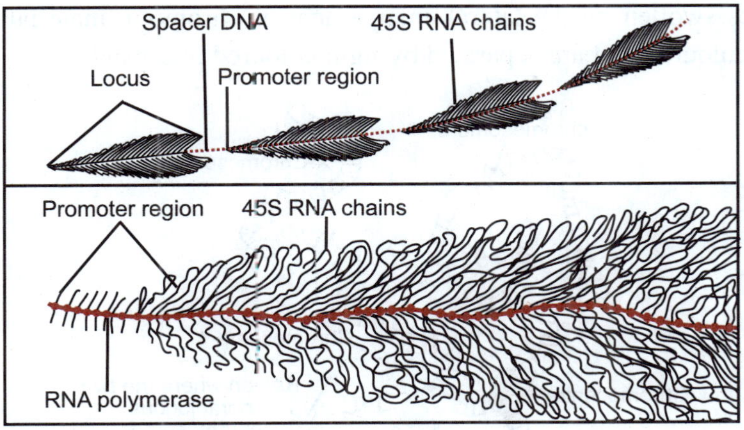

Fig. 7.20 A part of loop showing DNA molecule with RNA and RNA polymerase

2. According to **spinning out and retraction hypothesis,** a chromomere is a fully transcribed gene from end to end by spinning out a transient loop. The new loop material spins out on one side of a chromomere at the thin end of loop and returns to a condensed stage on the other side after completing the synthesis of RNA. These are associated with the rapid synthesis of yolk and protein in the maturing ovum. These disappear by the end of first prophase when chromosomes become thick and more condensed.

2. Polytene Chromosomes

Occurrence

Polytene chromosomes are special types of interphase chromosomes that are large and visible with the naked eye. These giant chromosomes are found in the cells of salivary glands of *Drosophila* and *Chironomus* larvae and in the cells of fat bodies of larval stages of certain dipterans. They were first observed in the salivary glands of *Chironomus* and hence are called **salivary gland chromosomes.**

Nature

Polytene chromosomes are **multistranded**, being composed of a large number of chromonemata. Their number has been estimated to be 1024 by **Painter** and 16,000 by **Bearmann**.

The polytene chromosomes are formed during early larval development when mitotic divisions come to an end in these tissues but DNA replication continues. DNA undergoes about 10 consecutive duplications producing 2^{10} copies of DNA fibres. They remain together forming a multistranded chromosome.

But, **Kodani** (1942) has presented a totally different interpretation. He believes that the paired homologoues consist of four chromonemata (similar to somatic chromosomes), which have become structures swollen by the lateral enlargement of chromotin material and form bands. They are darkly coloured and are separated by light coloured interbands.

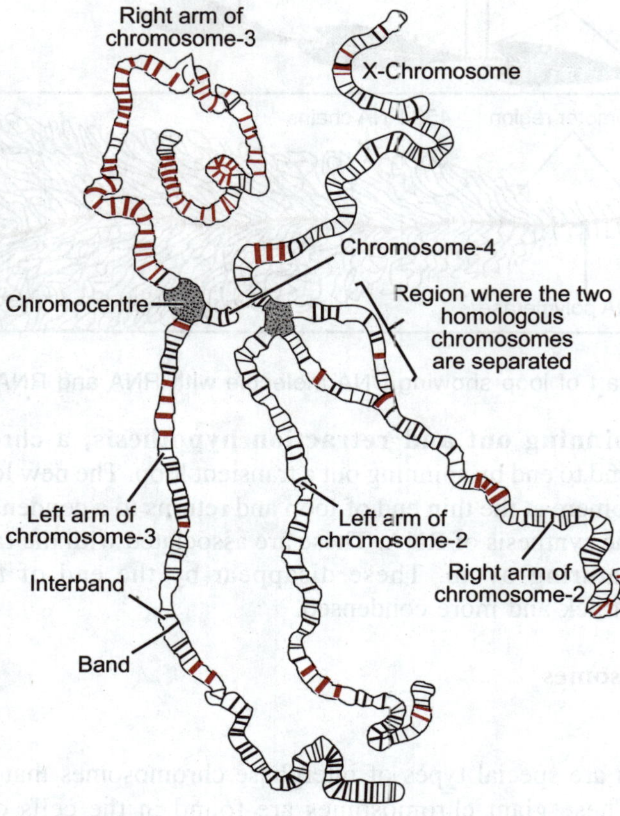

Fig. 7.21 Polytene chromosomes from salivary glands of *Drosophila*

1. Bands, Interbands and Genes

A polytene chromosome presents a distinct pattern of transverse banding. It consists of dark coloured **bands** alternating with the light coloured **interbands**.

The bands are Feulgen positive, rich in DNA and, therefore, darkly stained. They are regarded to be chromomeres where DNA fibre is supercoiled. Adjacent chromatid fibres are aligned by their chromomeres forming dark bands. The bands are specific for a particular chromosome and banding pattern is almost similar in both the homologous chromosomes of a pair. Therefore, during meiosis the pairing between homologous chromosomes is band to band.

Initially, it was believed that each band represented a single gene. Now it is estimated that an average band contains about 30,000 base pairs or 20 times the amount of DNA required to code for an average protein. It means each band is **polygenic**.

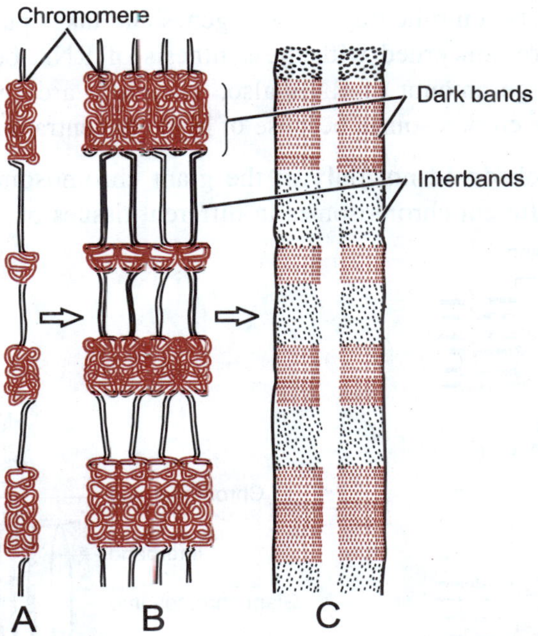

Fig. 7.22 Mechanism of chromosome folding that accounts for banding pattern in polytene chromosomes. **A.** A single chromosome; **B.** Chromatids align laterally; **C.** Dark bands and interbands

2. Chromosome Puffs and Gene Expression (Balbiani Rings)

While studying the relationship between the genes and the chromosomes, it was observed in 1881 that the structure of giant chromosomes is modified in certain regions due to enlargement and extension of bands into lateral loops. Such enlargements or puffs are known as **Balbiani rings** or **chromosomal puffs**.

The puffs form large rings around the chromosomes and are presumed to be formed by the unwinding or uncoiling of chromonemata, which are otherwise closely folded or coiled in the bands. These then project out into a series of loops. These loops increase the thickness or diameter of the chromosome and provide fuzzy appearance.

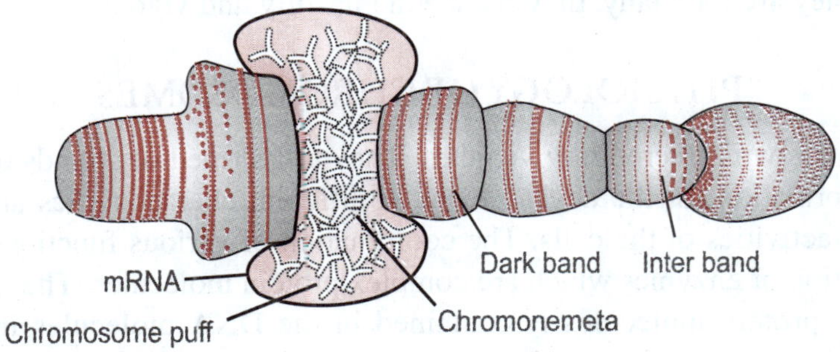

Fig. 7.23 Detailed structure of a polytene chromosome of chironomus larva

The formation of puffs is controlled by specific genes and takes place at definite time and in specific regions. They are concerned with the synthesis of RNA because besides DNA and protein, puffs contain large amount of RNA also. The puffs are chiefly associated with the metabolic activities of the chromosomes because of high concentration of RNA.

The puffs have been observed not only in the giant chromosomes of salivary glands of *Drosophila* but also in different chromosomes in different tissues of dipterans.

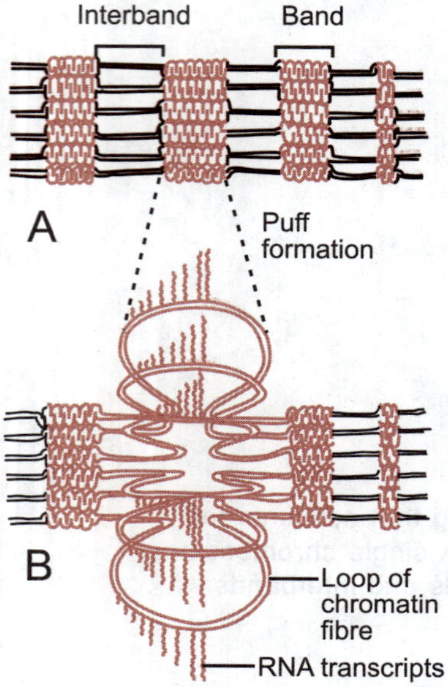

Fig. 7.24 A part of a giant polytene chromosome of *Drosophila* showing **A.** Bands and interbands; **B.** Puff formation

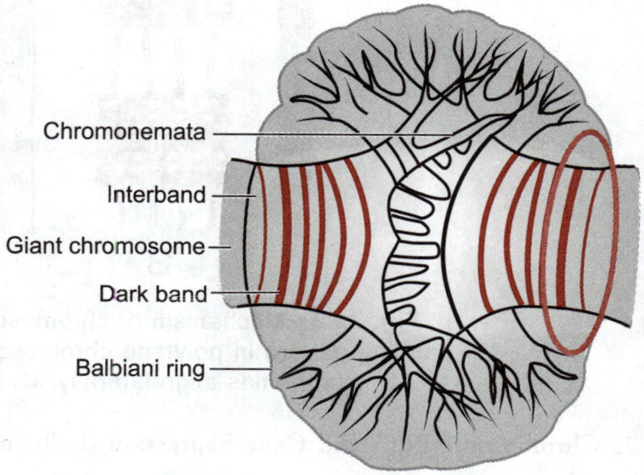

Fig. 7.25 A part of a giant chromosome of *Drosophila* showing Balbiani rings or chromosomal puffs

3. Supernumerary Chromosomes

In certain plants and animals one or more additional chromosomes were observed in addition to the normal number. Such accessory chromosomes are very small and generally inert, being composed of heterochromatin. Normally, their presence in the nucleus does not affect the phenotype but if they are too many, they reduce the fertility and vigour.

PHYSIOLOGY OF CHROMOSOMES

Chromosomes are known as hereditary vehicles. They are formed of strands of DNA molecules which contain information for the development of different characteristics and performance of various metabolic activities of the cells. The coordination of various functions is brought about through the formation of enzymes which are complex protein molecules. The information for the synthesis of these protein molecules is contained in the DNA molecules in the sequence of nitrogen bases. The sequence of three nitrogen bases, that codes for one molecule of a particular amino acid, constitutes a **triplet codon**. DNA of a particular cistron transcribes its **mRNA**. It carries genetic code similar to its DNA. It moves out of the nucleus and gets attached to the ribosome. The tRNA molecules pick specific activated amino acids and recognise the codons on the mRNA by their specific anticodons. The amino acids are thus linked together forming a polypeptide chain, which is finally changed into functional protein either singly or by the association of several polypeptide chains.

TEXT QUESTIONS

1. Describe the structure and functions of chromosomes.
2. Describe the chemical nature of a chromosome.
3. How do special types of chromosomes help in understanding the functioning of chromosomes?
4. Describe the structure and replication of chromosomes.
5. Write short notes on:
   (a) Lampbrush chromosomes (b) Polytene chromosomes
   (c) Secondary constriction (d) Nucleosome
   (e) Histones.
6. Write a brief note on the structure of viral chromosome.
7. What is the difference between bacterial and eukaryotic chromosomes?
8. Describe nucleosome concept of chromation structure.

8
Chapter

DNA Replication

IN THIS CHAPTER

Overview of DNA Replication: Basic Rules for DNA Replication; Basic Requirements for DNA Replication; DNA Replication is Semiconservative: Meselson and Stahl's Experiment; Cairans Autoradiography Experiment; Taylor's Experiment on *Vicia Faba*; DNA Strands are Synthesised in 5′→3′ Direction; Semidiscontinuous Replication of DNA; Origins of Replication; Unidirectional and Bidirectional Replication of DNA; Accuracy of DNA Replication; Editing or Proof Reading during Replication; Mechanism of DNA Replication in Prokaryotes: Activation of Deoxyribonucleotides; Initiation; Synthesis of New Strand; Termination; Helix Formation; Mechanism of DNA Replication in Eukaryotes: Differences in DNA Replication in Prokaryotes and Eukaryotes; Enzymology of DNA Replication: Proteins of DNA Replication; Initiation Proteins; Helicases; Swivelases or Topoisomerases; Primases; Replication Models: Theta (θ) Replication; Rolling Circle Model; D-loop Model; Inhibitors of Replication.

Replication is an essential property of genetic material. DNA replication is the process whereby DNA makes exact copies of itself. Approximately twenty or more enzymes and proteins are needed during replication. Replication occurs at the rate of between 50 nucleotides per second (in mammals) and 500 nucleotides per second (in bacteria). The hydrogen bonds between the complementary bases on the two strands of the parent DNA molecule break and the strands unwind, each strand acts as a template for the synthesis of a new strand complementary to itself. DNA polymerases move down the two single strands linking free nucleotides to their complementary bases on the templates. The process continues until all the nucleotides on the templates have joined with appropriate free nucleotides and two identical molecules of DNA are formed. Every time a cell divides it needs to make a duplicate copy of all its genes. It means, at each cell division the cell has to carry out replication of its entire DNA for transferring genetic information from parent cell to daughter cells.

AN OVERVIEW OF DNA REPLICATION

Watson and **Crick** suggested a very simple mechanism of DNA replication on the basis of its double helical structure. During replication, the weak hydrogen bonds between the nitrogenous bases of the complementary nucleotides break so that the two polynucleotide strands of DNA separate and uncoil. The strands thus separated are complementary to each other. Each separated strand acts as a template for the synthesis or polymerisation of other strand and two identical daughter DNA molecules are formed. Because of the specificity of base pairing, each nucleotide of the separated chains attracts its complementary nucleotide from the cell cytoplasm. Once the nucleotides are attached by the hydrogen bonds, their sugar radicals join through their phosphate components, completing the formation of a new polynucleotide chain.

Thus, each strand of the double helix DNA serves as a template or model on which its complementary strand is synthesised. This method of DNA replication is described as **semiconservative,** because each daughter DNA molecule is a hybrid conserving one parental polynucleotide strand and the other one newly synthesised strand.

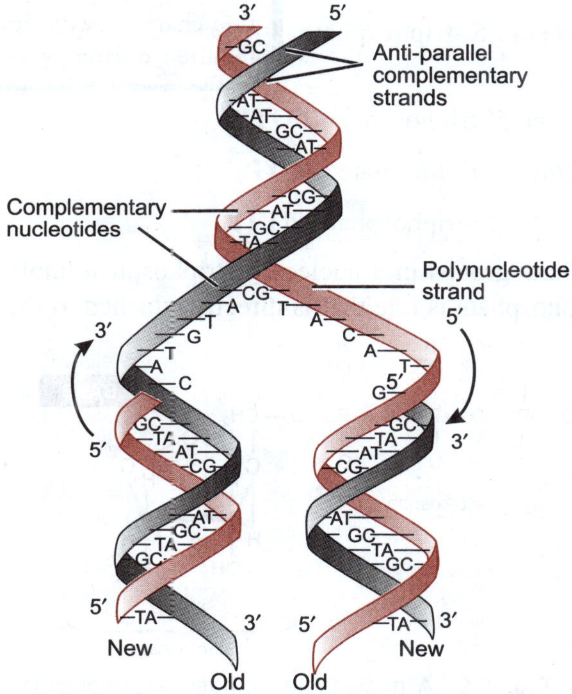

Fig. 8.1 Generalised model of semi-conservative replication of DNA

Basic Rules for DNA Replication

The basic rules that are followed strictly during DNA replication can be summarised as follows:

1. There is specificity of base pairing during DNA replication, i.e., adenine pairs with thymine and guanine with cytosine.
2. Nucleotide monomers are added one by one to the 3′ end of the growing strand by the enzyme ***DNA polymerase.***
3. The sequence of nitrogenous bases in each daughter strand being formed is complementary to the base sequence in the template strand.
4. C–3′ carbon of deoxyribose present on the 3′ end of the new polynucleotide chain of DNA has OH group and is free to bind to another nucleotide. C–5′ carbon of deoxyribose on the 5′ end of polynucleotide chain has a phosphate. Therefore, the new polynucleotide chain is always synthesised in 5′→3′ direction.

Basic Requirements for DNA Replication

The cells ready to undergo DNA replication are required to possess following precursors, primers, enzymes and proteins:

> **Extension: Energy for polymerisation**
>
> 5′-monophosphates and 5′-diphosphates of these nitrogenous bases can not be polymerised. Only 5′-triphosphates can be polymerised by ***polymerase*** enzyme into polynucleotide chain. These triphosphates provide energy required during polymerisation.

1. **Precursor Nucleotide Molecules:** Cells need all the four deoxyribonucleotide 5′-triphosphates (dNTPs) namely dATP, dGTP, dCTP and dTTP.
   - 2′-Deoxyadenosine 5′-triphosphate (**dATP**)
   - 2′-Deoxguanosine 5′-triphosphate (**dGTP**)
   - 2′-Deoxycytidine 5′-triphosphate (**dCTP**)
   - 2′-Deoxythymidine 5′-triphosphate (**dTTP**)

   The three phosphate groups in a nucleoside triphosphate molecule are designated α, β and γ, where α-phosphate is one that is directly attached to the deoxyribose.

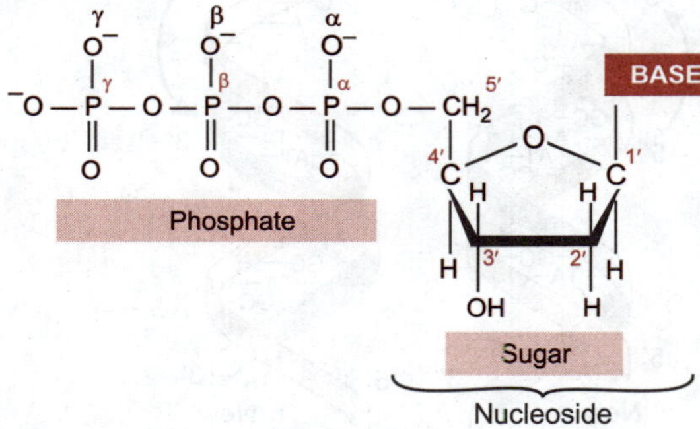

Fig. 8.2 A nucleoside triphosphate molecule

2. **Template DNA:** A single strand of DNA is needed which determines the sequence of nucleotides during polymerisation of new strand.
3. **RNA Primer:** It is needed for the initiation of polymerisation of new polynucleotide strand.

> **Extension: Need for a Primer**
>
> Enzyme ***DNA polymerase III*** that carries out DNA replication can not initiate polymerisation, i.e. it can not add the first nucleotide to start the synthesis of daughter DNA strand. It can add nucleotides only to a chain of nucleotides that is already paired with the template DNA. Therefore, another enzyme, ***RNA polymerase,*** also called ***primase*** is needed. It synthesises a short RNA oligonucleotide segment formed of a sequence of about 10 RNA nucleotides complementary to parental strand of DNA. It is called primer RNA. ***DNA polymerase III*** recognises the primer and adds DNA nucleotides to its 3′ end to construct new DNA strand. RNA primer is then cleaved off and replaced by DNA nucleotides.

Table 8.1 Major enzymes and factors needed during DNA replication

Enzymes or Factors	Activity
1. Helicase proteins	Unwind DNA helix.
2. Binding or strand stabilising proteins	Stabilise DNA strands that are isolated by proteins.
3. DNA topo-isomerases or DNA gyrases	Relieve supercoils in front of replication fork.
4. Primase	Synthesis of RNA primer.
5. DNA polymerases	Polymerisation of a complementary strand of DNA by the assembly of nucleotides on RNA primer; fill the gaps left after the removal of RNA primer.
6. DNA ligase	Seals nicks left after DNA primases and fills gaps left by the removal of RNA primer.

4. **Enzymes:** Enzymes required for DNA replication are ***DNA polymerases***, ***polynucleotide ligases*** and ***helicases***, etc. They are shown in Table 8.1, but are discussed in detail separately under the heading enzymology of DNA replication.
5. **Proteins:** A number of proteins are associated with DNA replication. They help in identifying origin of replication, recognising initiation point, unzipping of nucleotide strands, stabilising the unzipped single DNA strands and recoiling of new or daughter DNA molecules. The name and function of these enzymes are also discussed under enzymology.

DNA Replication is Semi-conservative

There can be three possible modes of DNA replication:

1. Semi-conservative Replication: In semi-conservative mode of replication, each replicated DNA molecule consists of one old and one new strand. During this replication, the double helix of DNA unzips, its two polynucleotide strands separate and new complementary strand is synthesised on each of the two parental template strands. This mechanism ensures the faithful copying of genetic information at each cell division.

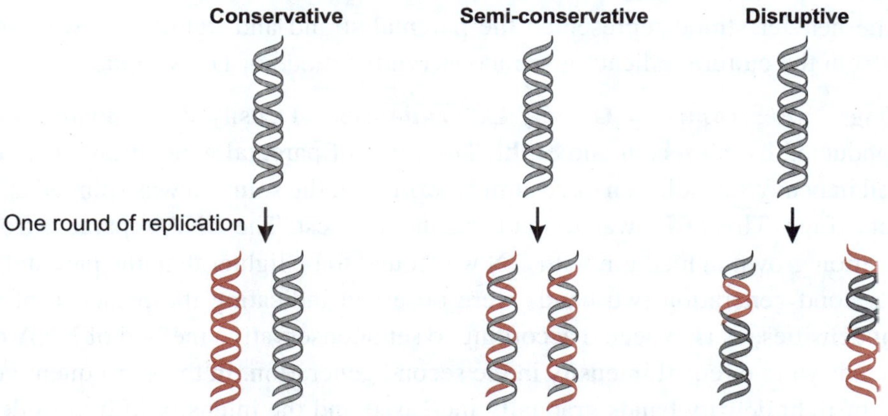

Fig. 8.3 Results of one round of replication of DNA for each of the three possible modes by which replication could be accomplished

2. Conservative Replication: In this method of replication, the two newly synthesised polynucleotide chains or strands are brought together. These join to form a new DNA molecule, while the parental strands reassociate to form the parent DNA molecule. It means original DNA helix retains itself while the new one is formed as its copy.

3. Disruptive Replication: In this case parental strands are dispersed in the two daughter DNA helices so that each strand is formed of some parts of parental DNA and some parts of new DNA. This would involve cleavage of parental strands during replication.

I Meselson and Stahl's Experiment in Support of Semi-conservative Replication of DNA

1. Using Heavy Isotopes of Nitrogen: The hypothesis of semiconservative replication of DNA was tested by **Matthew Meselson** and **Franklin Stahl** in 1958. They cultured a species of bacteria (*Escherichia coli*) in a culture medium containing heavy isotopes of nitrogen (^{15}N). After these bacteria had replicated for several generations in that medium, radioactive ^{15}N became incorporated in both the strands of the DNA. This DNA was denser and heavier than the DNA present in bacteria grown in ^{14}N medium. When these bacteria with ^{15}N were transferred in cultural medium containing ^{14}N isotopes, it was found that DNA separated from fresh generation of bacteria possessed one strand heavier than the other. The heavier strand represented the parental strand and lighter one was the new one synthesised from the culture indicating semiconservative mode of DNA replication.

Matthew Meselson
(May 24, 1930)

The experimental evidence that DNA replication is semiconser-vative was presented by **Matthew Meselson** and **Franklin Stahl** in 1958.

Meselson was also associated with the first identification of mRNA with **Brenner** and **Jacob**. He also contributed to the understanding of process of recombination between DNA molecules. This led to establishing the **recombinant DNA technology**. In 1968 with **R. Yuan**, he successfully isolated a *restriction endonuclease* enzyme.

2. Separating DNA Strands by Density Determination: Density determination experiments were also conducted by **Meselson** and **Stahl**. The DNA of parental generation labelled with ^{15}N was dissolved in heavy salt solution of cesium chloride and the solution was spinned at high speed in an ultracentrifuge. This DNA was found to be the heaviest. The DNA separated from bacteria of first generation grown in medium with ^{14}N was found to be lighter than the parental DNA with ^{15}N. In the second generation two bands were observed indicating the presence of DNA with two different densities, as is expected according to semiconservative method of DNA replication. These two bands were of equal intensity in the second generation. In the subsequent generations, the intensity of light density bands gradually increased and the intensity of hybrid density band gradually decreased.

3. Interpreting the Results: Meselson and **Stahl** interpreted their results as follows:

- After first replication, each daughter DNA of generation I was a hybrid possessing one heavy strand received from the parent DNA molecule and one light strand newly synthesised.
- When this hybrid DNA molecule replicated, it contributed its heavy strand (with ^{15}N) to one daughter DNA molecule and its light strand to another daughter DNA. Therefore, DNA molecules of second generation were 50 per cent hybrid DNA with $^{15}N/^{14}N$ nitrogen and other 50 per cent were light DNA molecules with $^{14}N/^{14}N$ nitrogen.
- DNA molecules of third generation had only one-fourth or 25 per cent hybrid DNA and remaining 75 per cent were normal DNA molecules with $^{14}N/^{14}N$ nitrogen.

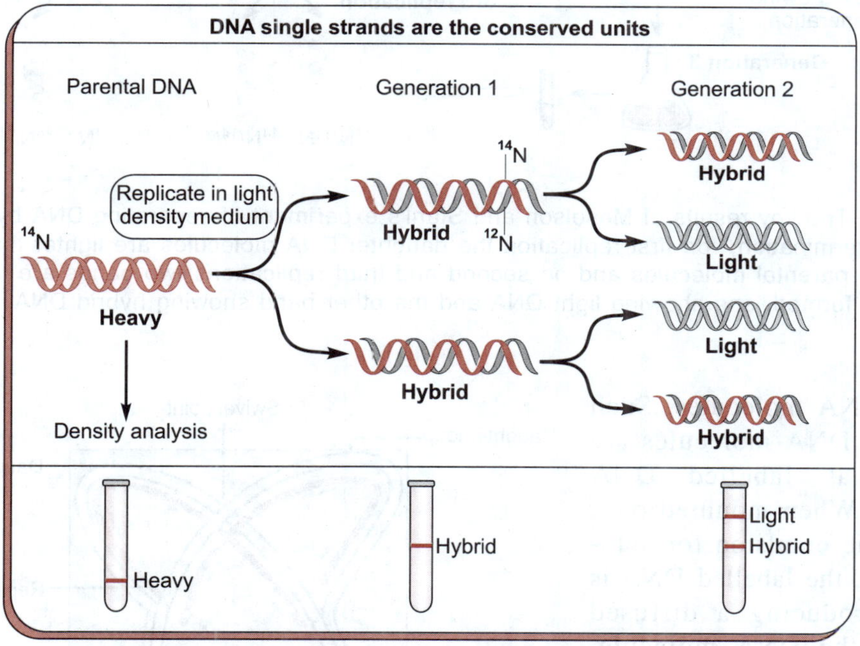

Fig. 8.4 Replication of DNA is semiconservative

II Cairans Autoradiography Experiment to Support Semiconservative Replication of DNA

J. Cairans demonstrated the semiconservative mode of DNA replication in the chromosome of bacterium, *Escherichia coli*, by autoradiography using radioactive thymine. Radioactive thymine or tritiated thymidine is obtained by using heavy isotopes of hydrogen (^3H). By growing *E. coli* in the culture medium containing tritiated thymidine, the radioactivity was incorporated in the

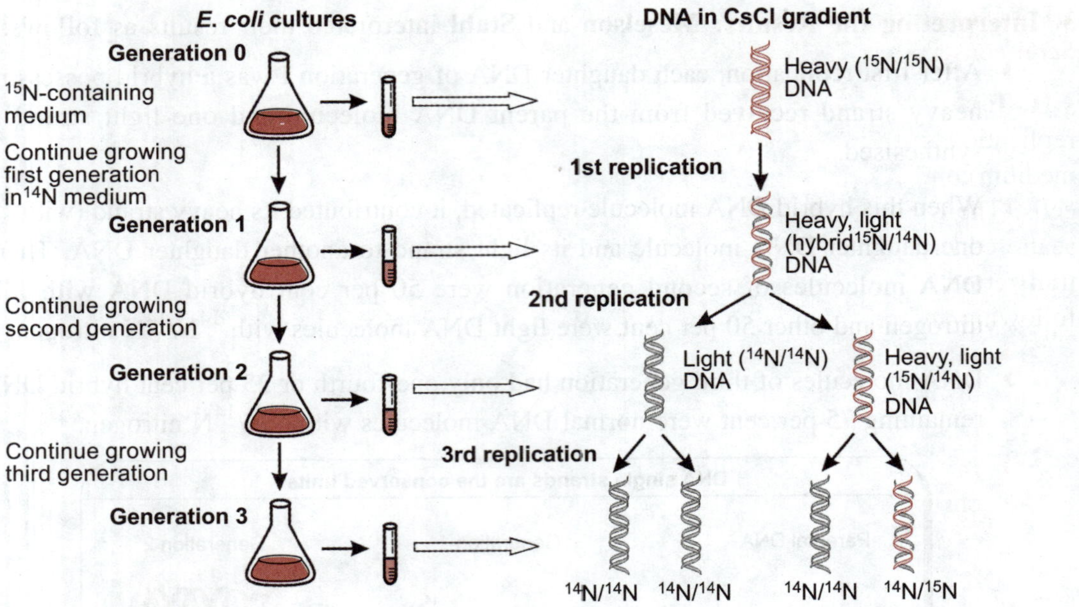

Fig. 8.5 The key results of Meselson and Stahl's experiment of separating DNA by density determination; on first replication the daughter DNA molecules are lighter than parental molecules and on second and third replications two bands are formed, one showing light DNA and the other band showing hybrid DNA

daughter DNA molecules. Such radioactive DNA molecules are described as labelled DNA molecules. When mounted on a photographic emulsion for autoradiography, the labelled DNA is exposed producing a diffused profile of the DNA molecule. The duplicating DNA molecule shows a **replicating fork**, the point at which two chains of DNA become four. After first replication, the radioactivity is seen to be incorporated to only one of the two strands of DNA. Therefore, both daughter DNA molecules are found to be labelled after first replication. This supports the semiconservative mode of DNA replication.

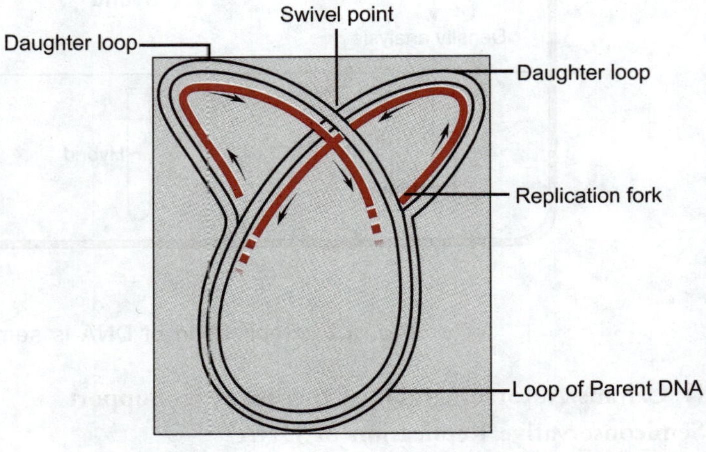

Fig. 8.6 Autoradiography of the single circular chromosome of *E. coli* labelled with tritiated thymidine–Cairans experiment

III Taylor's Experiments on Vicia faba Root Tips to Support Semi-conservative DNA replication in Eukaryotes

J.H. Taylor (1957) and his co-workers demonstrated the semi-conservative method of DNA replication in the root tip cells of *Vicia faba* by autoradiography. The roots were grown in a medium containing radioactive thymidine so that radioactivity is incorporated in the DNA of these cells. The outline of this labelled chromosome on a photographic film appears in the form of scattered black dots of silver grains. When these root tips with labelled chromosomes were transferred to the unlabelled medium containing colchicine and studied for radioactivity, the following observations were made:

- In the chromosomes of first generation the radioactivity was found to be uniformly distributed in both the chromatids, because in them the original or template strand of DNA double helix was labelled with radioactivity and the new one was non-labelled.
- In the chromosomes of second generation only one of the two chromatids in each chromosome was radioactive.

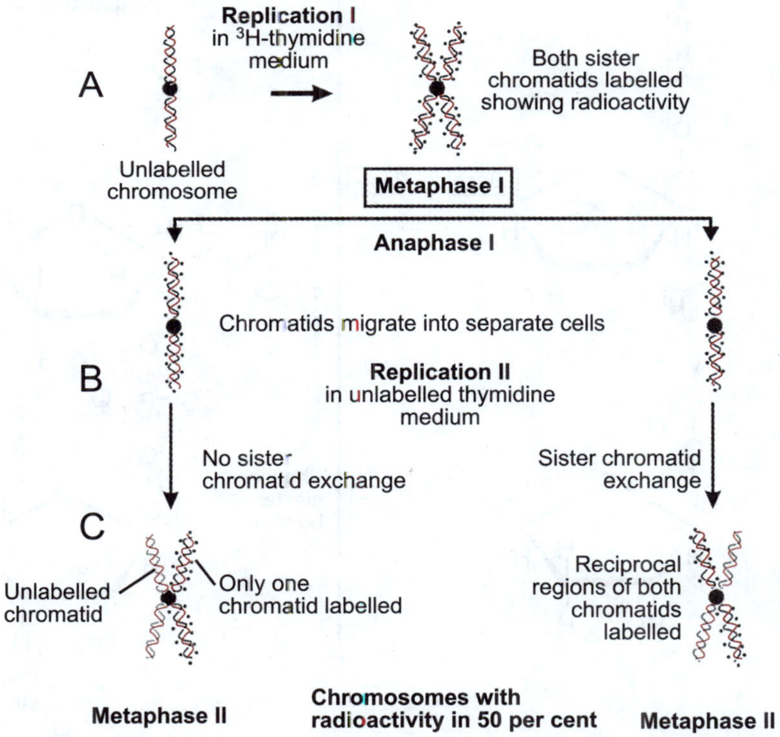

Fig. 8.7 Taylor's experiment demonstrating the semiconservative mode of replication of DNA in root tips of *Vicia faba;* **A.** Chromosome of parental generation having uniformly distributed radioactivity; **B.** Chromosomes of first filial generation showing uniform radioactivity in both daughter chromosomes; **C.** Chromosomes of second generation where radioactivity is seen in 50 per cent daughter chromosomes, the other 50 per cent are nonradioactive

Taylor's experiment demonstrates the semiconservative method of chromosome replication. But as it is known that each chromatid is composed of only double helical molecule of DNA, this represents the replication of DNA.

Synthesis of New DNA Strands Always Proceed in 5′→ ′

During DNA replication, the polymerisation of new strands on both the parental strands (3′→5′ strand and 5′→3′ strand) proceeds only in 5′→3′ direction. This is because of the following two reasons:

1. ***DNA polymerase*** enzyme can carry out polymerisation in 5′→3′ direction only.
2. When DNA is synthesised in 5′→3′ direction, the energy required for polymerisation is derived from the hydrolysis of triphosphate group of the free dNTP when it is being added to 3′ hydroxyl group (3′OH) of the growing polynucleotide chain. If the nucleotides

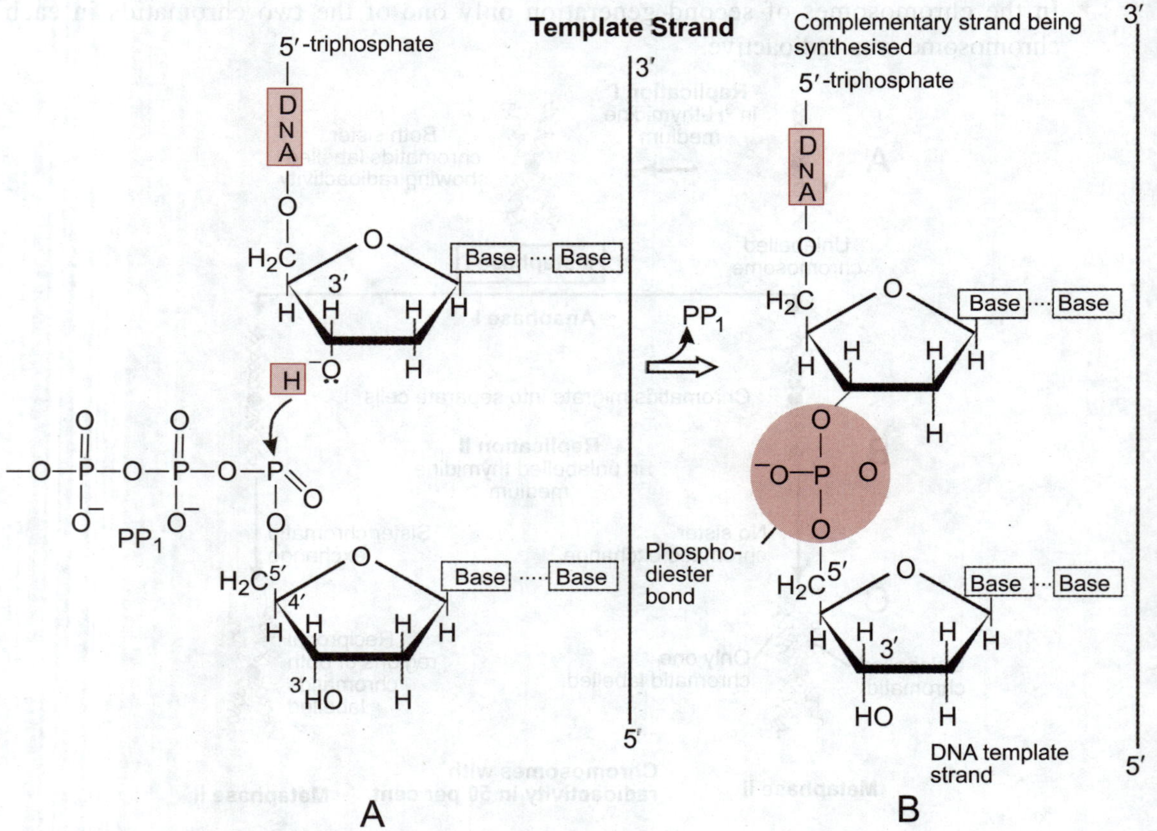

Fig. 8.8 **A.** Joining of a nucleotide to the nucleotide of template strand of parental DNA and **B.** Linking of two nucleotides in the complementary strand through phosphodiester bond between C3′ carbon of deoxyribose and phosphate group attached to C5′ carbon of previous nucleotide. *Note* the new nucleotide is joined to C5′ carbon leaving C3′–OH group free, showing 5′→3′ direction of synthesis

are added in the 3′→5′ direction, the energy of polymerisation would be derived from the hydrolysis of 5′-triphosphate group of the terminal nucleotide already added to the growing strand of DNA.This would eliminate the possibility of proof reading, because removal of a mismatched terminal nucleotide would also remove the 5′-triphosphate group needed for polymerisation and further chain elongation.

The Two Strands of DNA are Assembled in Different Ways (Semi-discontinuous Replication of DNA)

Anti-parallel strands of DNA exhibit continuous and discontinuous DNA synthesis. This is called **semi-discontinuous replication**. The polymerisation of DNA strands always proceeds in 5′ → 3′ direction. Since the two parent strands of a DNA molecule are antiparallel, the new strands at the replication fork are oriented in opposite directions along the antiparallel parent templates. It means their replication on 3′→5′ strand and on 5′→3′ strand occur differently and includes following two procedures:

1. Leading Strand Synthesis or Continuous Strand Synthesis

Continuous replication occurs in 5′→3′ direction using 3′→5′ strand of parental DNA as template. DNA strand synthesised in 5′→3′ direction is called **leading strand.** The leading strand is synthesised as one piece adding nucleotides to its 3′ end. The process allows rapid synthesis with about 1,000 nucleotides being added per second to each new strand.

2. Lagging Strand Synthesis

Lagging strand is synthesised on 5′→3′strand of template DNA in short segments of 1,000 to 2,000 nucleotides in 5′→3′ direction. These segments are called **Okazaki fragments**. This was described by Japanese scientist **Reiji Okazaki** and colleagues. Synthesis of each Okazaki fragment begins with RNA primer, synthesised by ***primase*** and is carried out by ***polymeraseIII*** as in leading strand. Thus, synthesis of lagging strand of DNA is discontinuous and needs a series of RNA primers, one for each Okazaki fragment.

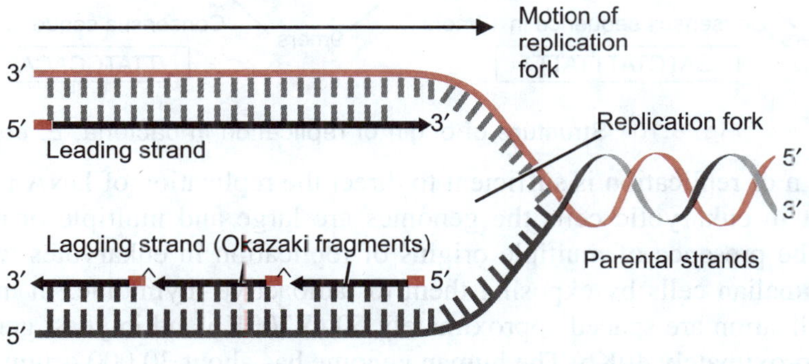

Fig. 8.9 Synthesis of leading and lagging DNA strands during DNA replication. *Note* that the leading strand grows towards the replication fork, while the segments of lagging strand grow away from the fork

Table 8.2 Differences between leading and lagging strands of replicating DNA

Leading Strand	Lagging Strand
1. It grows continuously as a single piece.	1. It is formed initially as short segments, called Okazaki fragments.
2. It needs a single RNA primer to start its growth.	2. Each segment needs a separate RNA primer to start a new Okazaki segment.
3. It does not need *DNA ligase* for its growth.	3. *DNA ligase* is needed to join Okazaki fragments.
4. Direction of growth is in 5′→3′ direction.	4. Direction of growth of complete strand is 3′→5′, but each Okazaki fragment is synthesised in 5′→3′ direction.
5. It is formed on 3′→5′ strand.	5. The template DNA strand for lagging strand is 5′→3′.

Origins of Replication

The replication of DNA in both prokaryotes and eukaryotes starts at a specific or unique sequence of nitrogenous bases in the DNA molecule. This is called **origin of replication**. In *E. coli*, the origin of replication consists of 245 base pairs, four repeats of 9 base pairs and three repeats of 13 base pairs called **9mers** and **13mers**. These serve as binding sites for proteins that initiate replication process. The initiator protein **DnaA** is responsible for the initial step in unwinding of DNA helix. A number of subunits of **DnaA** protein bind to each of the several 9mers. This is followed by the binding of **DnaB** and **DnaC** proteins responsible for further opening of DNA helix. These proteins are called ***helicases***. The energy required to breakdown hydrogen bonds between two DNA strands is supplied by the hydrolysis of ATP.

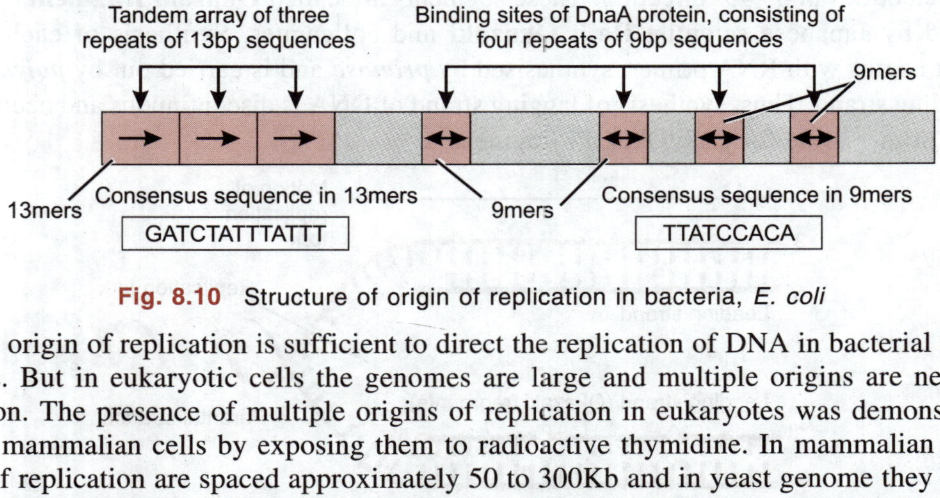

Fig. 8.10 Structure of origin of replication in bacteria, *E. coli*

Single origin of replication is sufficient to direct the replication of DNA in bacterial and viral genomes. But in eukaryotic cells the genomes are large and multiple origins are needed for replication. The presence of multiple origins of replication in eukaryotes was demonstrated in cultured mammalian cells by exposing them to radioactive thymidine. In mammalian cells the origins of replication are spaced approximately 50 to 300Kb and in yeast genome they lie at an interval of approximately 40Kb. The human genome has about 30,000 origins of replication.

In yeast cells, the origins of replication are called **autonomously replicating sequences (ARS).** They are formed of about 100 base pairs each. Many of them have a **core sequence** or

consensus sequence of 11 base pairs. This core sequence binds to a protein complex, called **origin of replication complex** (ORC). They initiate DNA replication by unwinding DNA at the origins in the chromosomes.

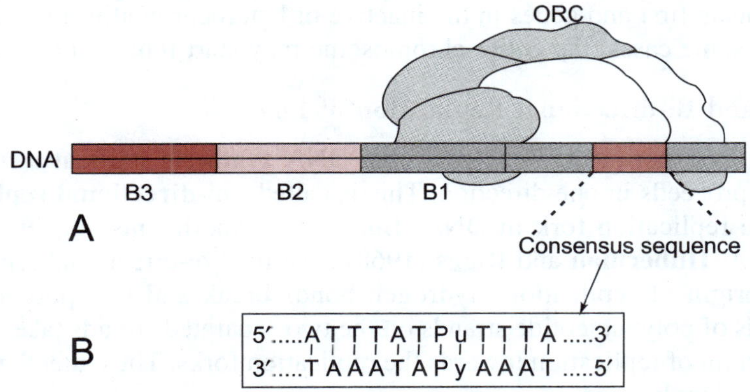

Fig. 8.11 **A.** ARS segment containing core sequence or consensus sequence and associated origin of replication complex (ORC); **B.** Core sequence or consensus sequence in ARS

> A and T richness of consensus sequence is in accordance with the function of origin of replication, because separation of two hydrogen bonds between A = T is easier and will need less energy than the disruption of three hydrogen bonds between G≡C base pairs.

Replicon and Replication Fork

In eukaryotic chromosomes replication proceeds from multiple sites. A length of DNA whose replication is initiated at one origin of replication is described as a **replicon**. The origin of replication lies near the centre of replicon. When origin of replication is activated, it produces two **replication forks** that move in opposite directions. Finally they meet replication forks from other replicons (Fig. 8.11).

Table 8.3 A summary of replicons in a few organisms

Organism	No. of Replicons	Average Length	Fork Movement
1. Bacterium *(E.coli)*	1	4200 kb	50,000 bp
2. Yeast *(Saccharomyces cerevisiae)*	500	40 kb	3,600 bp
3. Fruit Fly *(Drosophila melanogaster)*	3,500	40 kb	2,600 bp
4. Toad *(Xenopus laevis)*	15,000	2,000 kb	500 bp
5. Mouse *(Mus musculus)*	25,000	150 kb	2,200 bp

Replicons vary in length. They may be as short as comprising of 13,000 base pairs (about 4μm long) to as long as of 9,00,000 base pairs (*i.e.* about 280 μm). In higher eukaryotes the number of replicons may vary from 20,000 to 1,00,000 or more per haploid genome. A eukaryotic chromosome during DNA replication appears to have a number of bubbles, called **replication bubbles.**

In prokaryotes, there is only one origin of replication, one replication fork and only one replicon. In eukaryotes, replication of DNA at all the origins or of all the replicons does not occur simultaneously. It is initiated in specific clusters. Typically, genes in active segments of the chromosome replicate first and genes in the inactive or heterochromatin segments replicate later in S–substage. In some cases, the entire chromosome may start replication early or late.

Uni-directional and Bi-directional Replication of DNA

J. Cairans from his experiments concluded that DNA synthesis starts at a fixed point on the chromosome and proceeds in one direction. This is called **uni-directional replication of DNA**. There is only one replication fork in DNA. But recent experiments suggest of bi-directional replication of DNA. **Huberman** and **Riggs** (1968) described bi-directional replication of DNA. At the point of origin of replication, hydrogen bonds break and two polynucleotide strands separate. Synthesis of polynucleotide strands on the two separated strands takes place in both the directions from origin of replication towards the replication forks. Thus, there are two replication forks in bidirectional replication.

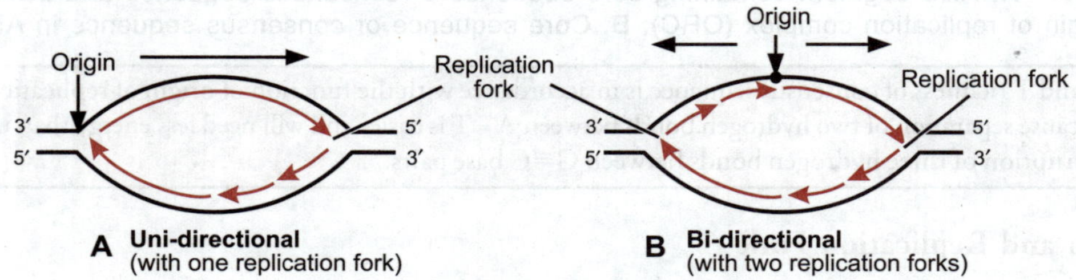

Fig. 8.12 Uni-directional and bi-directional replication of DNA

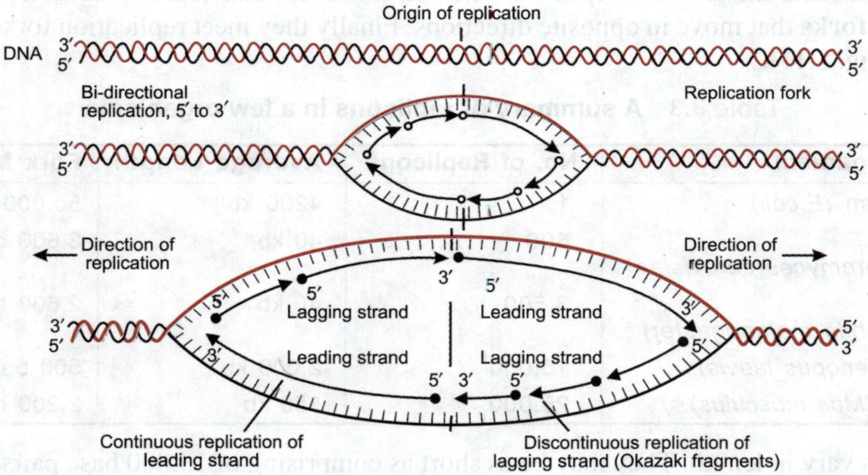

Fig. 8.13 Bi-directional DNA replication in a linear eukaryotic chromosome. Synthesis of DNA is initiated at the origin of replication and proceeds in $5' \rightarrow 3'$ direction in both the strands. The two replication forks migrate in opposite directions

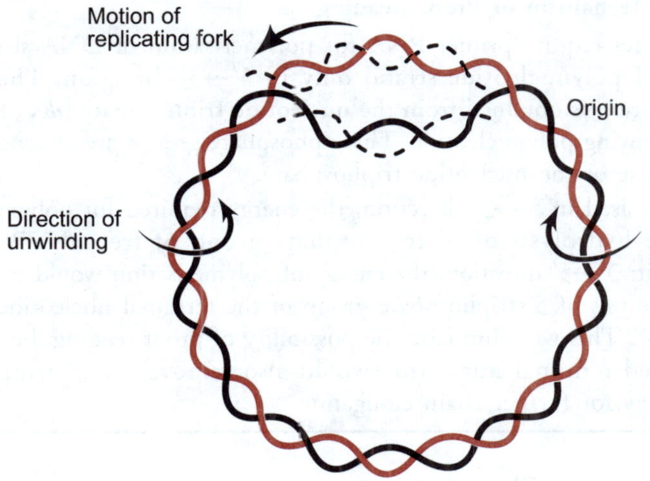

Fig. 8.14 Bi-directional replication of circular DNA in prokaryotes showing two Y-shaped junctions or two replication forks

Levinthal and **Cairans** proposed that during replication the two strands do not separate completely. Instead, when DNA strands start unzipping at one end, simultaneously the unzipped segments start attracting their respective nucleotide pairs. In this way the unzipping of the original DNA strands and synthesis of fresh DNA strands go side by side. It means in a replicating DNA, a Y-shaped growing point must be visible. **Cairans** has demonstrated two Y-shaped regions in the circular replicating DNA of viruses and bacteria. One of these points is known as **growing point** and the other **initiation point.** This represents **uni-directional replication.** In case of **bi-directional replication** two replication forks are formed and new DNA strand on each parental strand grows in both directions from the point of origin of replication.

Bi-directional replication has been widely observed in phage, bacterial and plasmid DNA except a few where DNA replication is unidirectional. Eukaroytic DNA shows bi-directional replication.

Accuracy of DNA Replication or Fidelity of DNA Replication

DNA replication is the synthesis of a new strand complementary to the template strand at each nucleotide position. Although, the functioning of ***DNA polymerase*** is very accurate, synthesis of complementary strand is not perfect and a non-complementary nucleotide is occasionally inserted by mistake. It has been estimated that the frequency of errors during DNA replication is roughly one incorrect base pairing per 10^9 to 10^{10} nucleotides incorporated. This error frequency is much lower than predicted on the basis of complementary base pairing. This accuracy or fidelity in DNA replication is attributed to the following activities of ***DNA polymerase I***:

Extension: Probable Mechanism of Proof Reading

DNA polymerase enzymes require primer RNA for polymerisation of DNA strand and catalyse the growth of polynucleotide strand only in 5′→3′ direction. The energy required for polymerisation is obtained from the **nucleotide triphosphate** (*d*NTP) which enters the 3′ end of growing polynucleotide. Three phosphate groups are attached to the 5′ carbon of deoxyribose of the nucleotide triphosphate.

When DNA is synthesised in 5′→3′ direction, the energy required for polymerisation is obtained from the hydrolysis of 5′ triphosphate group of free *d*NTP. In case polymerisation occurs in 3′→5′ direction, the energy of polymerisation would have to be derived from the hydrolysis of 5′ triphosphate group of the terminal nucleotide already incorporated into DNA. This will eliminate the possibility of proof reading, because the removal of mismatched terminal nucleotide would also remove the 5′ triphosphate group needed for energy for further chain elongation.

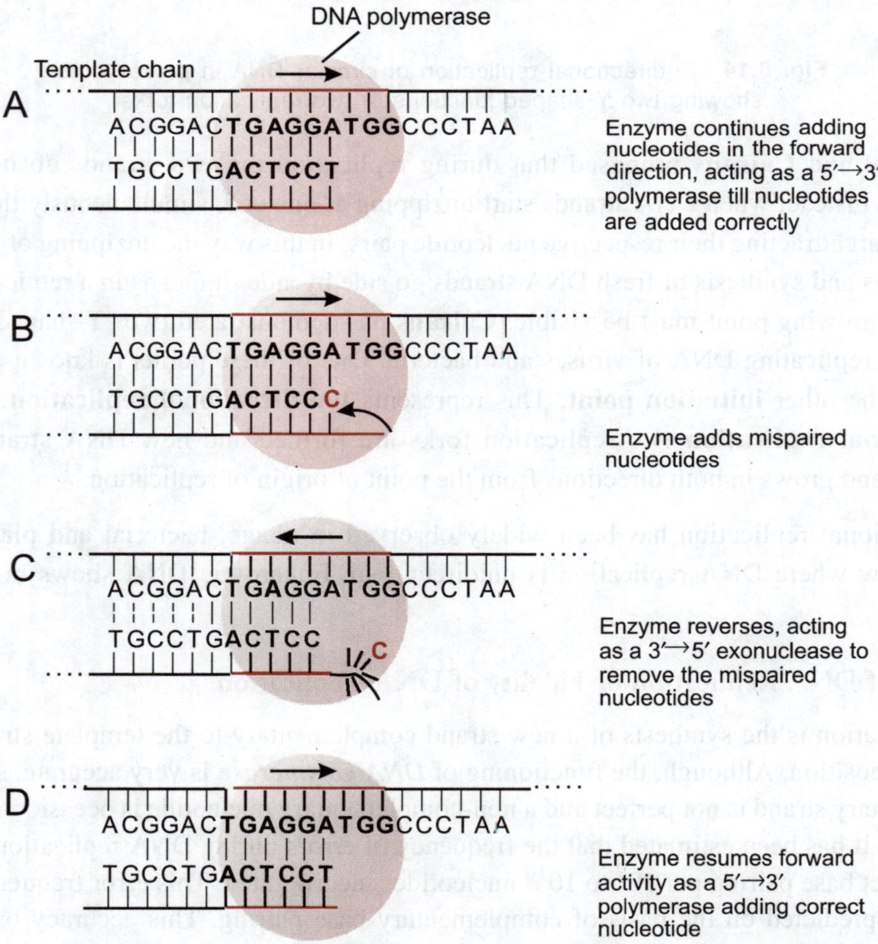

Fig. 8.15 3′→5′ exonuclease proof reading activity of *DNA polymerase*

1. Exonuclease Proof Reading Activity of DNA Polymerase I

DNA polymerases have two types of activities: 3′→5′ and 5′→3′ exonuclease activities. The 5′→3′ exonuclease activity helps in removing RNA primers from Okazaki fragments. The 3′→5′ exonuclease activity is associated with proof reading the DNA strand under synthesis and to detect and excise the mismatched nucleotide incorporated. This 3′→5′ exonuclease proof reading activity is found in ***DNA polymerase I*** of *E.coli* and ***DNA polymerase*** δ and ε of eukaryotes. They remove the mismatched base incorporated at the 3′ end of growing polypeptide chain. This increases the accuracy of replication by hundreds to thousands times.

2. Selection of Correct Base

DNA polymerase selects the correct base for insertion into the new polynucleotide strand during replication. It discriminates against incorporation of a mismatched base and selectively increases the accuracy of DNA replication about hundred folds and reduces the expected error frequency from 10^{-4} to 10^{-6}.

MOLECULAR MECHANISM OF DNA REPLICATION IN PROKARYOTES

In *E. coli* the entire process of DNA replication involves following four steps. Each step is governed by specific enzymes. Thus, more than a dozen enzymes and proteins are involved in initiation and completion of DNA replication.

I Activation of Deoxyribonucleotides

The deoxyribonucleoside monophosphates (AMP, GMP, CMP, TMP) occur in the nucleoplasm. These are activated into triphosphates (ATP, GTP, CTP and TTP) by utilising ATP. This process is known as **phosphorylation** and is catalysed by the enzyme ***phosphorylase***.

II Initiation

DNA replication is initiated at a definite sequence of nucleotides, called **initiation point** or **origin of replication** (*oriC*). This consists of a specific sequence of 100 to 200 or more base pairs. In prokaryotic chromosomes, there is only one origin of replication per chromosome. Each point of origin along with the part of DNA to be replicated from it forms one **replicon.**

In *E. coli*, the origin of replication is formed of 245 base pairs, consisting of three repeats of 13 base pairs and four repeats of 9 base pairs. These are called **13mers** and **9mers** respectively. The process of initiation involves action of nine different proteins and can be separated into following steps:

1. Recognition of Initiation Point

Specific initiator protein, **DnaA**, recognises the initiation point on DNA. A single complex of about 30 molecules of DnaA protein binds to the four repeats of 9 base pairs sequence in one of the strands of the duplex and initiates the separation of intertwined polynucleotide chains at all the three mers of 13 base pairs.

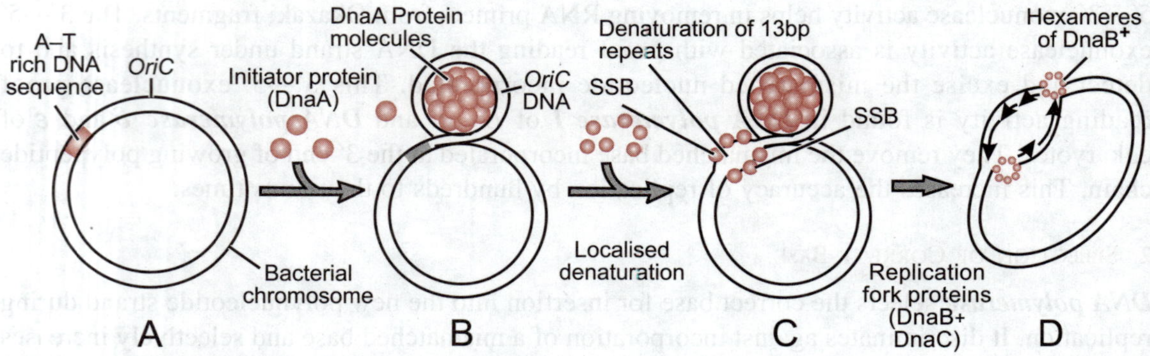

Fig. 8.16 Model for initiation of DNA replication on the origin of replication in *E. coli*. **A.** A chromosome of *E.coli* showing origin of replication (OriC), and A = T rich sequence; **B.** About 20 DnaA protein molecules with ATP bind at four 9 base pair repeats; **C.** Denaturation of three 13 base pairs repeats; **D.** Hexamers of DnaB protein bind to each strand with the aid of DnaC protein and unwind DNA

2. Unwinding or Opening of DNA Double Helix

Initiator protein, **DnaA,** after binding to the origin of replication, denatures DNA in the region of three repeats of 13 base pairs sequence. This process requires ATP and the bacterial histone-like protein (HU). About 45bp are melted at the origin to open up the two strands. This initiates the formation of **pre-priming complex**, which is formed by five proteins, namely DnaA, DnaB, DnaC, SSB and gyrase or topoisomerase. Firstly, the two hexamers of ***helicase*** enzyme along with **DnaB**, bind into each strand of unwounded DNA close to DnaA. Each DnaB is loaded onto DNA strand with the help of six DnaC proteins. Two molecules of protein **DnaB** unwind DNA bidirectionally and create two potential replication forks. A topoisomerase enzyme, called ***DNA gyrase***, promotes separation of DNA strands by inducing negative supercoiling in advance of helicase.

Strand separation is maintained by several molecules of single strand binding proteins **(SSB)** by preventing the renaturation of separated strands and stabilises them so that these can serve as templates.

III. Synthesis of New Strands of DNA

The synthesis of new strands on the separated strands of template DNA involves following six steps:

1. RNA Priming or Formation of RNA Primer

The ***primase*** enzyme (also called ***DNA directed RNA polymerase***) binds to the first priming sequence on the leading strand and synthesises a short **primer RNA** segment consisting of 10 – 60 nucleotides only at the origin of replication complementary to the separated strand of DNA.

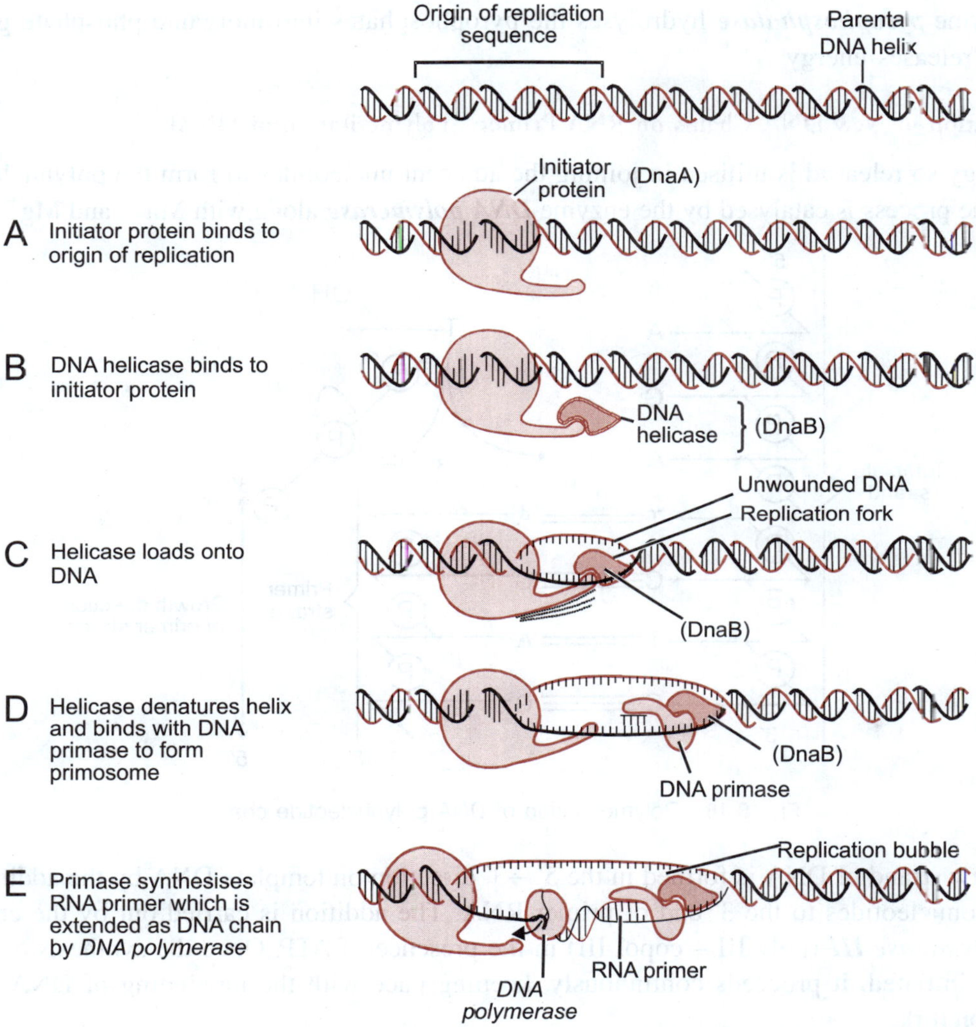

Fig. 8.17 Schematic representation of the formation of replication bubble at origin of replication in *E. coli* and initiation of RNA primer and then new DNA strand

Synthesis of RNA primer is essential because ***DNA polymerase*** can not initiate synthesis of new DNA strand. It can only polymerise growth of DNA chain.

2. Assembling of Complementary Strands

Deoxyribonucleoside triphosphates pair with the appropriate nitrogenous bases of template DNA strands according to base pairing rule.

3. Conversion of Deoxyribonucleoside Triphosphates to Monophosphates

The deoxyribonucleoside triphosphate molecules on pairing with the nitrogenous bases of template strand set free pyrophosphate (p ~ p) molecules and change to deoxyribonucleotides.

The enzyme ***pyrophosphatase*** hydrolyses the pyrophosphates into inorganic phosphate groups (Pi) and releases energy.

4. Formation of New DNA Chains on RNA Primers (Polymerisation of DNA)

The energy so released is utilised in joining the adjacent nucleotides to form the polynucleotide chain. The process is catalysed by the enzyme ***DNA polymerase*** along with Mn^{++} and Mg^{++} ions.

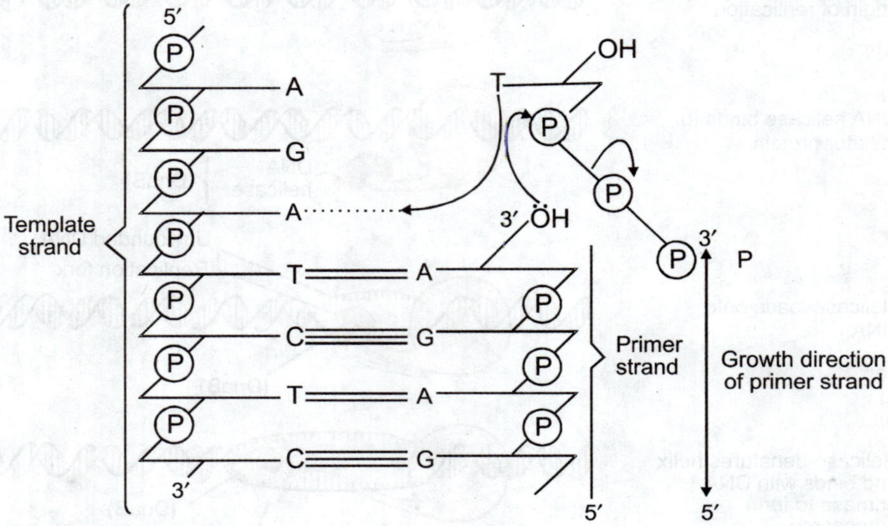

Fig. 8.18 Polymerisation of DNA polynucleotide chain

The new strand of DNA is formed in the 5′→3′ direction on template DNA by the addition of deoxyribonucleotides to the 3′ end of primer RNA. The addition is carried out by the enzyme ***DNA polymerase III*** (poly III – copol III) in the presence of ATP. Once the synthesis of DNA strand is initiated, it proceeds continuously, keeping pace with the unwinding of DNA at the replication fork.

Therefore, the new strand on 3′→5′ template parent strand grows towards the replication fork in 5′→3′ direction by the addition of nucleotides to the growing 3′ end. It means the 5′→3′ strand grows as one piece and is called **leading strand**. Its synthesis is described as **leading strand synthesis**. In contrast, the new strand on 5′→3′ template strand elongates away from the replication fork. It is synthesised discontinuously as a series of short segments of nucleotides, called **Okazaki fragments**, which are joined by the enzyme ***DNA ligase***. Each Okazaki fragment is formed of 100 – 200 nucleotides in eukaryotes and 1,000 – 2,000 nucleotides in prokaryotes and needs its own RNA primer for its polymerisation. This 3′→5′ new strand synthesised discontinuously on 5′→3′ template parent strand is called **lagging strand**.

Since, the synthesis of leading strand is continuous and that of lagging strand is discontinuous, the overall replication of DNA is said to be **semi-discontinuous**.

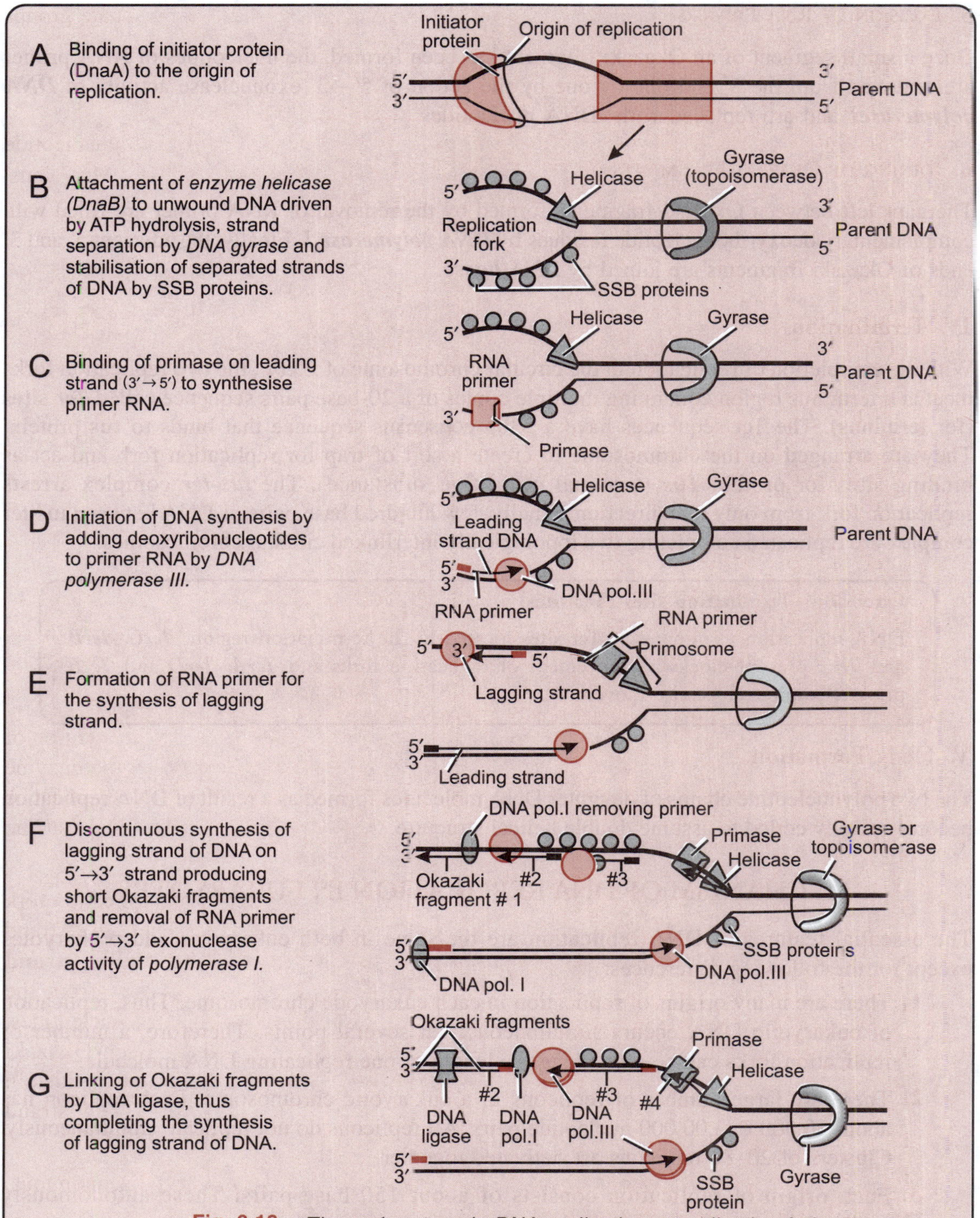

Fig. 8.19 The major steps in DNA replication at replication fork

5. Excision of RNA Primers

Once a small segment of an Okazaki fragment has been formed, the nucleotides of RNA primer are removed from the 5′ end one by one by the action of 5′→3′ exonuclease activity of ***DNA polymeraseI*** and are replaced with DNA nucleotides.

6. Joining of Okazaki Fragments

The gaps left between Okazaki fragments formed by the removal of RNA primer are filled with complementary deoxyribonucleotide residues by ***DNA polymerase I***. Finally, the adjacent 5′ and 3′ ends of Okazaki fragments are joined by ***DNA ligase***.

IV. Termination

With the completion of replication in the circular chromosome of *E.coli*, the two replication forks meet at a terminus region containing multiple copies of a 20-base pairs sequence called ***Ter* sites** (for terminus). The Ter sequences have a 23bp consensus sequence that binds to **tus** protein. They are arranged on the chromosome to create a sort of trap for replication fork and act as binding sites for protein ***Tus*** (terminal utilisation substance). The ***tus-ter*** complex arrests replication fork from only one direction. Finally, few hundred base pairs of DNA between tus-ter complex are replicated completing two topologically interlinked circular chromosomes.

> **Extension: Termination sites (Ter-sites)**
>
> DNA replication terminates at **Ter-sites** located in the termination region. *TerC*, *TerB* and *TerF* prevent clockwise movement of replication forks and *TerA*, *TerD* and *TerE* prevent counter clockwise movement.

V. Helix Formation

The two polynucleotide chains of daughter DNA molecules formed as a result of DNA replication become spirally coiled to assume double helical structure.

MECHANISM OF DNA REPLICATION IN EUKARYOTES

The essential features of DNA replication are the same in both eukaryotes and prokaryotes except for the following differences:

1. There are many origins of replication in each eukaryotic chromosome. Thus, replication of eukaryotic DNA occurs simultaneously at several points. Therefore, a number of replication forks can be seen along the length of one replicating DNA molecule.
2. There are large number of replicons in a eukaryotic chromosome. Each replicon has about 50,000 to 3,00,000 nucleotide pairs. All replicons do not replicate simultaneously. Clusters of 20–80 replicons are activated together.
3. Each origin of replication consists of about 150 base pairs. These autonomously replicating sequences are called **replicators.**

4. Initiation of replication in all eukaryotes requires a multisubunit protein, the **origin recognition complex (ORC)**.

5. **Replication bubbles** are formed at the origins due to separation of two strands of parent DNA at origins of replication.

6. In eukaryotic cells, ***DNA polymerase-α*** is a multisubunit enzyme with similar structure in all eukaryotic cells. It functions only in the initiation of a new strand. Two small proteins associated with it help in the synthesis of short primers containing RNA and a short segment of DNA for Okazaki fragments of the lagging strand. This short segment of DNA is 4 nucleotide long and is called **iDNA** or **initiator DNA**. ***DNA polymerase α*** lacks proof reading 3′→5′ exonuclease activity. These primers are extended by ***DNA polymerase δ***, which is also a multisubunit enzyme. Along with specific protein, it functions analogous to the β–subunit of ***polymeraseIII*** of *E. coli*. Therefore, ***DNA polymerase δ*** has 3′→5′ proof reading exonuclease activity and also carries out synthesis of both lagging and leading strands.

   DNA polymerase ε functions during DNA repair, acts analogous to ***polymeraseI*** at the replication fork and also removes primers of Okazaki fragments on lagging strand.

7. Termination of replication on linear chromosomes of eukaryotes involves synthesis of special structures called **telomeres** at the end of chromosomes with the help of enzyme ***telomerase.***

Replication of the Ends of Eukaryotic Chromosomes (Synthesis of Telomeric DNA by Enzyme Telomerase)

The eukaryotic chromosomes are formed of linear double-stranded DNA molecules that extend from one end to the other. These ends are called **telomeres** or **telomeric ends**. During replication, these telomeric ends face special problem. ***DNA polymerase*** can carry out polymerisation on RNA primer by the addition of nucleotides. After replication, the newly synthesised strands are still with RNA primer and are bonded to the parental strand. The gaps are created between Okazaki fragments once the RNA primers are removed. These gaps in the discontinuous strand created by the removal of RNA primers are filled with the addition of nucleotides to the 3′– OH end by ***DNA polymerase*** and their ends are sealed by ***DNA ligase***. However, the gap created by the removal of RNA primer from the 5′-end of each newly synthesised strand, remains because ***DNA polymerase*** can not initiate new DNA synthesis and there is no strand to provide the upstream 3′– OH group to receive the nucleotides. As a result, at each successive round of DNA replication, the end of chromosome will get shortened and shortened.

The mechanism of replication of telomeric ends of eukaryotic chromosomes was studied in protozoan, *Tetrahymena*, by **Elizabeth Blackburn, C. W. Greider** and **J. W. Szostak**. They found that an enzyme, ***telomerase***, adds several repeats of the sequence present at the telomeric end of the parental DNA. In *Tetrahymena*, the telomeres at each end of a chromosome have repeated sequence 5′–TTGGGG–3′. The enzyme ***telomerase*** can add several repeats of this sequence to the 3′–end of the complementary strand using 5′→3′ synthesis. These repeats then

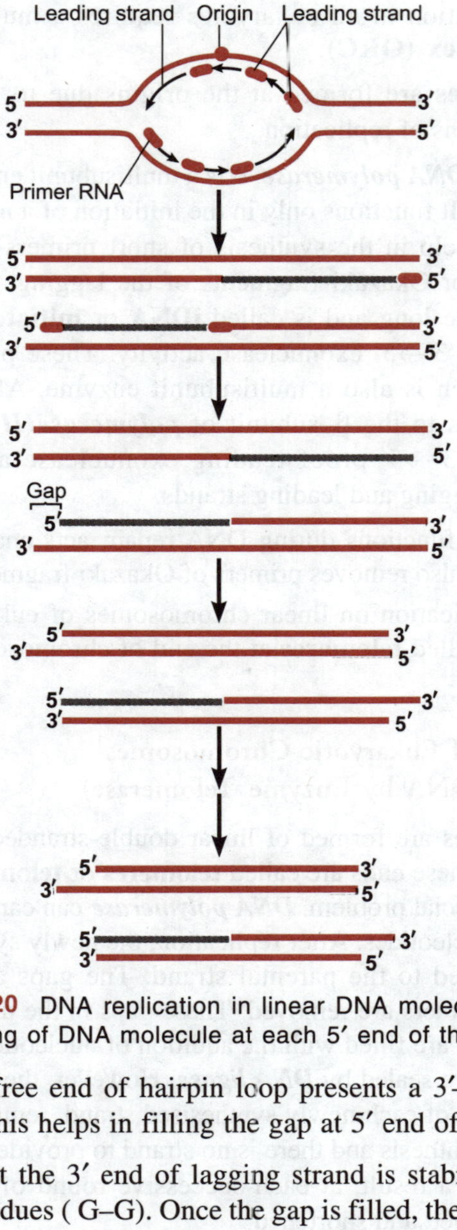

Fig. 8.20 DNA replication in linear DNA molecule causing shortening of DNA molecule at each 5′ end of the new strand

form a **'hairpin loop'**. The free end of hairpin loop presents a 3′–OH group that serves as a substrate for ***polymerase I***. This helps in filling the gap at 5′ end of the daughter strand.

The hairpin loop formed at the 3′ end of lagging strand is stabilised by hydrogen bonding between opposite guanine residues (G–G). Once the gap is filled, the hairpin loop is then cleaved off at its terminus.

Extension: Role of Telomerase Enzyme

Special enzymes, called **telomerases**, are used for the replication of telomeric ends of eukaryotic chromosomes. ***Telomerase*** enzyme is a complex of proteins and RNA. The RNA can pair with a telomere repeat and acts as a template for the synthesis of complementary telomere repeats.

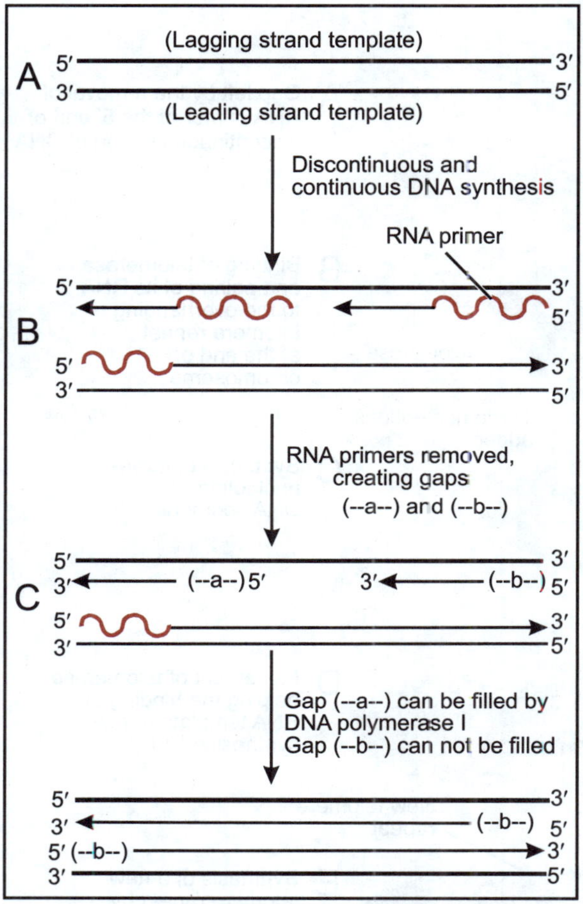

Fig. 8.21 Diagram illustrating the problem of replicating a linear chromosome in eukaryotes at the ends, because a gap is left following the removal of RNA primer at the 5′ end of newly synthesised strand from 3′→5′ lagging strand

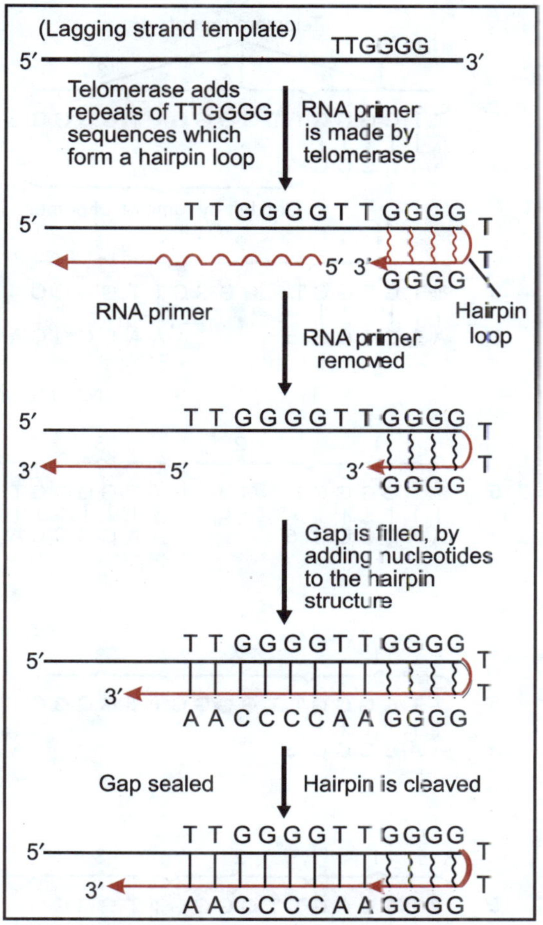

Fig. 8.22 Replication of telomeric end by enzyme telomerase in protozoan. *Tetrahymena*

Telomerase Enzyme

This enzyme is made up of both protein and RNA. The RNA component of ***telomerase*** has a base sequence complementary to 6-base pair telomere repeat unit, *i.e.* 3′···AACCCC···5′. Because of this, ***telomerase*** binds to the overhanging telomere repeat at the end of parent DNA strand and sticks out beyond the chromosome to be used as a template for the synthesis of new chromosomal telomere repeats. The enzyme ***telomerase*** first catalyses the addition of three nucleotides, TTG to new DNA on RNA template. Later, synthesis of TTGGGG telomere repeat is completed, followed by the addition of more telomere repeats.

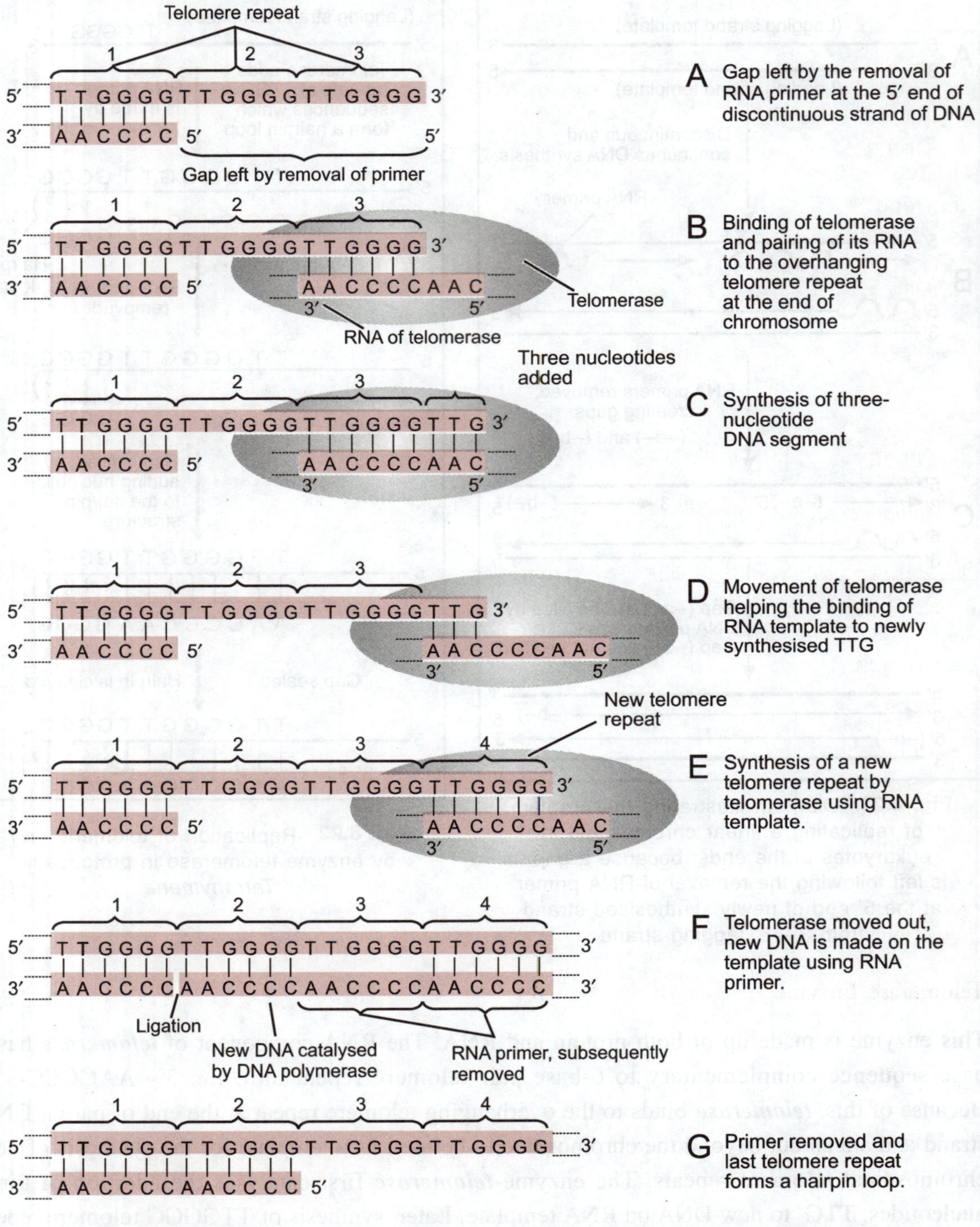

Fig. 8.23 Role of telomerase in the synthesis of telomeric DNA in *Tetrahymena*

Table 8.4 Differences in the replication of DNA in prokaryotes and eukaryotes

Replication in prokaryotes	Replication in eukaryotes
1. There is only one origin of replication per DNA molecule.	1. Origins of replication are many (over 1,000) in each eukaryotic chromosome.
2. Origin of replication is formed of about 100 – 200 or more nucleotides.	2. Each origin of replication is formed of about 150 nucleotides.
3. Replication of DNA occurs at one point in each prokaryotic chromosome.	3. Replication of DNA occurs at several points simultaneously in each chromosome.
4. Only one replication fork is formed in each replicating prokaryotic chromosome.	4. A number of replication forks are formed simultaneously in each replicating DNA.
5. Prokaryotic chromosome has one replicon.	5. Eukaryotic DNA molecules have large number of replicons (50,000 and above), but replication does not occur simultaneously on all replicons.
6. No replication bubble is formed during DNA replication.	6. A number of replication bubbles are formed in one replicating DNA molecule.
7. Initiation of DNA replication in prokaryotes is carried out by proteins DnaA and DnaB.	7. Initiation of DNA replication is carried out by a multisubunit protein, origin recognition complex (ORC).

Table 8.5 Differences in the synthesis of leading and lagging strand in prokaryote and mammalian cells.

DNA replication in *E. coli*	DNA replication in Mammalian cells
1. Leading strand is synthesised by *polymerase III*.	Leading strand is synthesised by *polymerase* δ.
2. Synthesis of lagging strand is initiated by *primase*.	Synthesis of lagging strand is initiated by a complex of primase and *polymerase* α (pol.α).
3. Synthesis or extension of Okazaki fragments on RNA primer occurs by *polymerase III*.	Extension of RNA primer into Okazaki fragments occurs by *polymerase* δ.
4. Removal of RNA primers and filling of gaps	Removal of primers and filling of gaps between Okazaki fragments is carried out by *polymerase* δ.

ENZYMOLOGY OF DNA REPLICATION

Studies on enzymology of DNA replication were first carried out by **Arthur Kornberg** and colleagues in 1957. They isolated an enzyme from lysates of *E. coli* that was able to direct DNA synthesis *in vitro*. This enzyme is now called ***DNA polymerase I***.

About 20 or more different proteins and enzymes are required during DNA replication. They collectively form a **DNA replicase system** or **replisome**.

Enzymes Associated with DNA Replication

The enzymes associated with DNA replication are:

- DNA polymerase enzymes
- Polynucleotide ligase
- Exonuclease.

DNA Polymerase Enzymes in Prokaryotes

Enzymes that add successive nucleotides to a growing DNA strand are called **DNA polymerases.** These have three sites for attachment:

Arthur Kornberg (March 3, 1918)

Kornberg isolated *DNA-polymerase I*, for which he was awarded Nobel Prize in 1959. It was shared with Severo Ochoa. He has contributed to the mechanism and concept of DNA replication and gene expression.

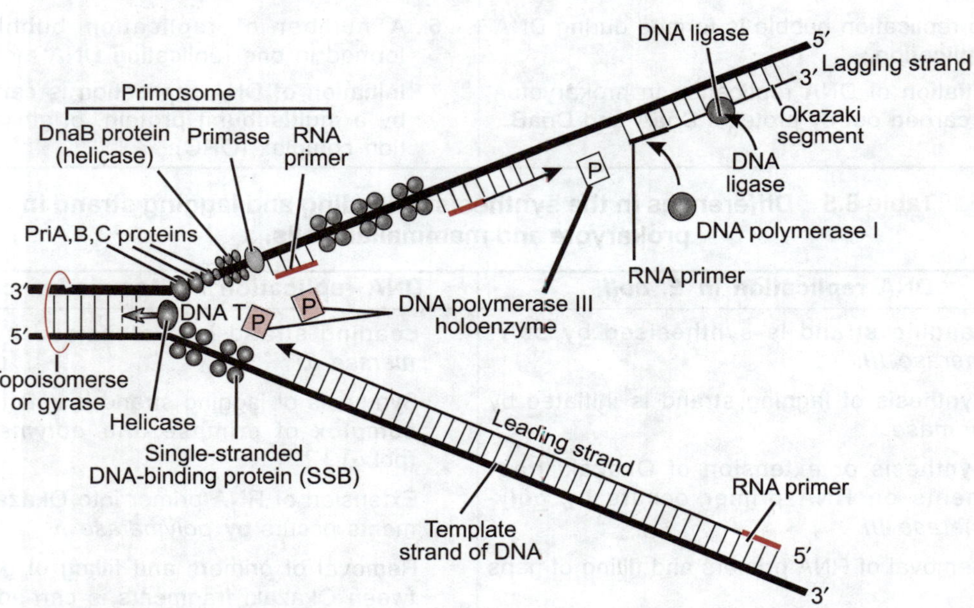

Fig. 8.24 Diagram to show different enzymes and proteins required in DNA synthesis and composition of primosome

1. **Template site** for the attachment to the template DNA.
2. **5′- triphosphate site** for attachment of deoxyribonucleoside 5′–triphosphate, and
3. **Primer terminus site** for the attachment to the 3′–OH end of the RNA primer. DNA polymerases are associated with the addition of deoxyribonucleoside triphosphate nucleotides to primer RNA in the direction from 5′–PO_4 end to the 3′–OH end.

There are three ***DNA polymerase*** enzymes that participate in the process of DNA replication and play distinct roles in the replication and repair of DNA. These enzymes are:

1. ***DNA polymerase I (Pol.I)***
2. ***DNA polymerase II (Pol.II)***
3. ***DNA polymerase III (Pol.III)***

All the DNA polymerases show following fundamental properties:

1. All polymerases synthesise DNA in 5′→3′ direction adding a dNTP to the 3′–hydroxyl group of the growing chain.
2. DNA polymerases can simply add a new dNTP to a preformed primer strand that is hydrogen bonded to template.
3. They can not initiate DNA synthesis *de novo* by catalysing polymerisation of free dNTPs.

1. DNA Polymerase I

It was the first enzyme discovered that could catalyse DNA synthesis. It was isolated from *E. coli* by **Arthur Kornberg** and his colleagues in 1957. It is roughly spherical with a diameter of about 6.5nm. It has a molecular weight of 1,90,000 and is formed of a single polypeptide chain of about 1,000 amino acid residues. It possesses a sulphydroxyl group, single interchain disulphide and one zinc molecule at the active site.

A ***DNA polymerase I*** molecule in reality is a complex structure being formed of:

(a) ***DNA polymerase 3′→5′ exonuclease***

(b) ***5′→3′ exonuclease***

(c) ***3′→5′*** proof reading ***exonuclease***.

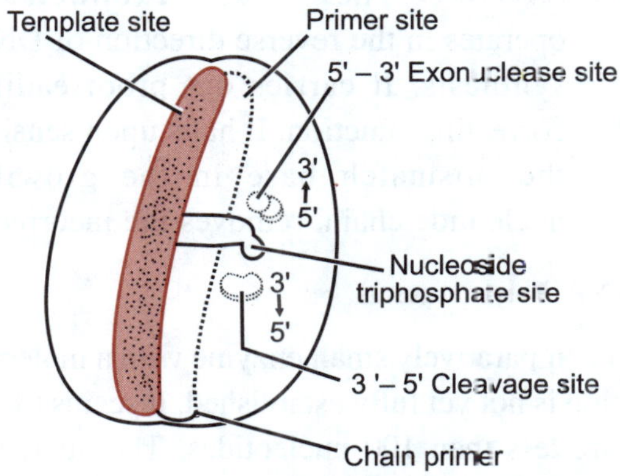

Fig. 8.25 A model of ***DNA polymerase I*** enzyme, showing its five different active sites

There are five specific binding sites on the spherical molecule of ***DNA polymerase I***

- **Template site** for binding the template DNA.
- **Primer site** for binding primer strand of RNA.
- **Primer terminus site** for 3′–hydroxyl terminus of primer.
- **5′–triphosphate site,** a locus for incoming deoxyribonucleoside 5′–triphosphate group.
- **5′–3′ exonuclease site,** a locus for 5′→3′ ***exonuclease*** activity situated in the path of growing chain.

Function: Earlier ***DNA polymerase I*** was considered to carry out DNA replication. It is now known that ***DNA polymerase I*** is not the primary enzyme of replication. Instead, it carries out a number of clean up functions during replication. ***DNA polymerase I*** is a multifunctional enzyme. It is associated with following activities:

- **5′→3′ Polymerase Activity:** It catalyses the addition of mononucleotide units(deoxyribonucleotide triphosphate residues) to the free 3′–hydroxyl end of polynucleotide chain of DNA. This helps in the growth of polynucleotide chain in 5′→3′ direction. A pure ***DNA polymerase I*** can add about 1,000 nucleotide residues per minute per molecule of enzyme at 37°C.
- **5′→3′ Exonuclease Activity:** The 5′→3′ exonuclease operates in the direction of DNA synthesis. It catalyses the activity of DNA repair by removing and replacing damaged base pairs. It also catalyses removal of pieces of RNA primers used in DNA replication from the 5′ end of each Okazaki fragment and replaces them with DNA nucleotides.

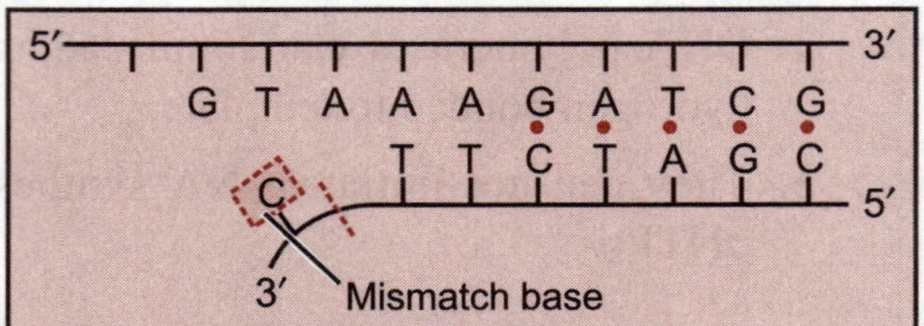

Diagrammatic representation of 3′→ 5′ exonuclease activity of ***polymerase I*** enzyme during proofreading or editing

- **3′→5′ Proofreading Exonuclease Activity:** The 3′→5′ ***exonuclease*** operates in the reverse direction of DNA synthesis. It carries out proofreading correcting function. It halts upon sensing the mismatch base in the growing nucleotide chain, removes the incorrect nucleotide and places the correct one.

Polymerase I removes the mismatched last nucleotide from the 3′ end of the growing DNA strand. Therefore, this enzyme activity of ***polymerase I*** is referred as *3′- **exonuclease*** activity.

2. DNA Polymerase II

It is a comparatively small enzyme with a molecular weight of about 1,20,000 daltons. Its biological function is not yet fully established. It seems to repair DNA by filling gaps in polynucleotide chains that are less than 100 nucleotides. The nucleotide residues are added in 5′→3′ direction. Thus, ***polymerase II*** can not replicate long strands but can fill gaps in DNA duplex.

3. DNA Polymerase III

DNA polymerase III enzyme is a **holoenzyme.** It was discovered by **T. Kornberg** and **M. L. Gefter** in 1972. It is the largest, most complex and most active enzyme out of the three polymerases. It has a molecular weight of about 9,00,000 daltons and consists of 10 – 20 subunits. Because of its primary role in replication, ***polymerase III*** is also called ***replicase***.

DNA polymerase III is a dimer with two similar multisubunits complexes. Each complex catalyses the replication of one DNA strand. Each unit of enzyme contains 10 different subunits.

Table 8.6 Comparison of three types of DNA polymerases from *E. coli*

Characteristics	DNA polymerase I	DNA polymerase II	DNA polymerase III
1. Molecular weight (In daltons)	1,90,000	12,0,000	9,00,000
2. Polypeptide subunits	1	4	10
3. Molecules per cell	400	Not known	10–20
4. Genes	Pol. A	Pol. B	Pol. C (dnaE), dnaN, DnaQ and DnaT.
5. Affinity for triphosphate (TPNs) precursors	Low	Low	High
6. Polymerisation rate	16 – 20/second.	7/second.	250 –1,000/sec.
7. Use of primer (single strand)	+	–	–
8. 5′→3′ Polymerisation activity	+	+	+
9. 3′ → 5′ Exonuclease activity	+	+	+
10. 5′→ 3′ Exonuclease activity	+	–	–
11. Main function	Polymerisation, proof-reading	Fills gaps in DNA duplex.	Polymerisation of long DNA strands.

These subunits are alpha (α), beta (β), beta2 (β2), epsilon (ε), theta (θ), tau (τ), gamma (γ), delta(δ), delta prime (δ′), chi (χ), and psi (ψ). All the ten subunits are needed in DNA replication. Subunits α, β, and θ form the core enzyme, the remaining seven subunits increase the processivity of core enzyme:

- ***Subunit α*** of ***polymerase III*** is the largest subunit. It catalyses 5′→3′ addition of nucleotides to the growing polynucleotide chain.
- The **β subunit** of ***polymerase III*** recognises and binds to the template strand of parental DNA. It is also called ***copolymerase III***. It is a dimer subunit consisting of two polypeptide chains (β2) in each subunit. The four polypeptides of the two subunits associate in pairs to form a ring or donut-shaped structure that encircles the DNA template and acts like a clamp. Each dimer along with core enzyme acts as a **sliding clamp** that allows holoenzyme ***polymerase III*** to slide from one nucleotide to the next on the template strands of DNA till it reaches the end. The speed of sliding is very fast about 1,000 nucleotides per second. This increases the processivity of ***polymerase III*** Presumably, it also helps in the repair of damaged DNA. β-subunit is active in both dividing and non-divding cells.
- The ***ε*-subunit** proof-reads 3′→5′ for mistakes.

- The ***θ* subunit** of ***polymerase III*** is associated with α and ε subunits to form the core polymerase enzyme.
- The **subunit *γ*** of ***polymerase III*** is found in mitochondria. It carries out replication of mitochondrial DNA only.

Enzymes γ, δ, χ, ψ form **clamp-loading complex.** A single clamp-loading complex consists of six subunits of five types γ, 2, δ, δ′, γ, ψ. The two core polymerases are linked together by a dimer of τ(tau) subunits. This dimeric polymerase then associates with a single clamp-loading complex, completing **DNA *polymerase III***. ***Polymerase α***, ***β*** and ε are most active in the dividing cells and are, therefore, associated with DNA replication.

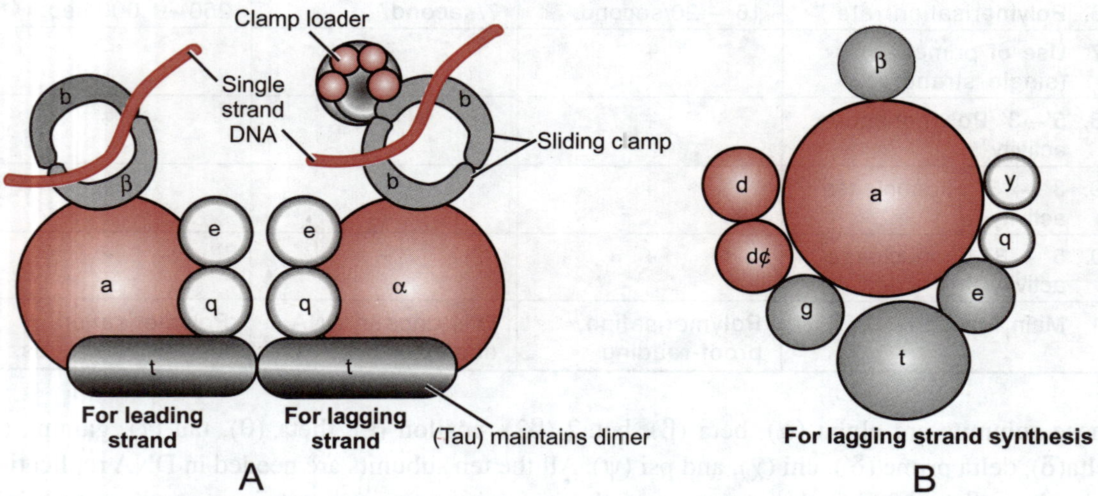

Fig. 8.26 A molecule of DNA polymerase III holoenzyme. It is formed of 10 kinds of protein chains and each is represented twice. **A.** Arrangement of different polypeptides in two subunits of polymerase III; **B.** Polypeptide subunits

DNA Polymerases in Eukaryotes

In eukaryotes, DNA polymerases have been isolated from yeasts, rat liver cells and human tumour cells. They have following five types of DNA polymerases:

1. **DNA polymerase α:** ***Alpha polymerase*** is a large, high molecular weight protein. It is found in both nucleus and cytoplasm and is, therefore, known as ***cytoplasmic polymerase***. It is associated with the replication of mitochondrial DNA.
2. **DNA polymerase β:** The ***beta polymerase*** is found only in the nucleus and is called ***nuclear polymerase*** or ***small polymerase***. It occurs only in vertebrate cells.
3. **DNA polymerase γ:** The ***gamma polymerase*** is encoded in the nucleus of mammalian cells.

Table 8.7 Subunits of DNA polymerase III of *E. coli*

Subunits	No. of subunits per holoenzyme	Molecular wt. of subunits	Functions of subunits
α	2	1,32,000	→ Polymerisation activity. It synthesises short primers containing RNA and DNA for Okazaki segments on lagging strand.
ε	2	27,000	→ $3' \rightarrow 5'$ **exonuclease** activity, carries out proof reading.
θ	2	10,000	
τ	2	71,000	→ Associated with stable template binding, core enzyme dimerisation.
γ	2	52,000	→ Clamp-loading complex that loads β subunits on lagging strand at each Okazaki fragment.
δ	1	35,000	
δ′	1	33,000	
χ	1	15,000	
ψ	1	12,000	
β	4	37,000	→ DNA clamp that slides over DNA, required for optimal processivity,

4. **DNA polymerase δ:** The ***delta polymerase*** enzyme is found in mammalian cells. Its action is assisted by proliferating cell nuclear antigen (PCNA). This enzyme along with ***polymerase*** α is associated with DNA replication.
5. **DNA polymerase ε:** The ***epsilon polymerase*** enzyme is PCNA-protein independent enzyme. It occurs in mammalian Hela cells and in budding yeast cells. It was previously known as ***DNA polymerase δ II.***

2. Polynucleotide Ligase or DNA ligase Enzyme

DNA ligase can join the ends of two fragments of DNA strand by catalysing the synthesis of a phosphodiester bond between a 3′– hydroxyl group at the end of one chain and 5′– phosphate group at the end of other chain. ***DNA ligase*** obtained from *E.coli* is formed of a single polypeptide chain having molecular weight about 74,000 to 77,000. About 200 to 400 copies of this enzyme are present in each *E. coli* cell.

DNA ligase performs following functions:

- Joins the Okazaki segments of lagging strand of DNA in the process of DNA replication.
- Repairs single-strand nicks in duplex DNA.
- Links the ends of linear DNA duplexes to yield circles.
- Joins segments of DNA in the process of recombination, which occurs during meiosis, transduction or in genetic transformation.

Proteins Associated with DNA Replication

At least nine different proteins participate in the initiation of DNA replication. They recognise origin of replication (the site from where replication is to start), open or unwind DNA helix at the origin, stabilise the separated single strands and establish a prepriming complex for subsequent reactions. These proteins are:

1. Initiator Proteins

These proteins recognise origin of replication in the two strands of DNA. **DnaA** protein recognises sequence of nucleotides in the origin of replication site (four repeats of nine nucleotides) and opens the duplex at specific site.

2. DNA Unwinding Proteins or Helicases

The unwinding proteins or **rep proteins** were discovered by **B. Alberts** and his colleagues in phage T_4 bacteriophage. A number of them are found both in prokaryotic and eukaryotic cells. Their molecular weight varies from 10,000 to 75,000. They possess specific binding sites for short segments of about 9 nucleotide residues. One particular protein called **DnaA** identifies origin of replication and initiates unwinding of the helix and separation of the two strands of DNA by breaking hydrogen bonding between base pairs of antiparallel strands of DNA. A number of molecules of this unwinding protein bind to each of the unit of binding site in advance of the formation of replicating fork. Other unwinding proteins like **DnaB, DnaC,** etc., help in opening up and destabilising the DNA helix. These unwinding proteins are called **helicases,** because they open and destabilise DNA helix.

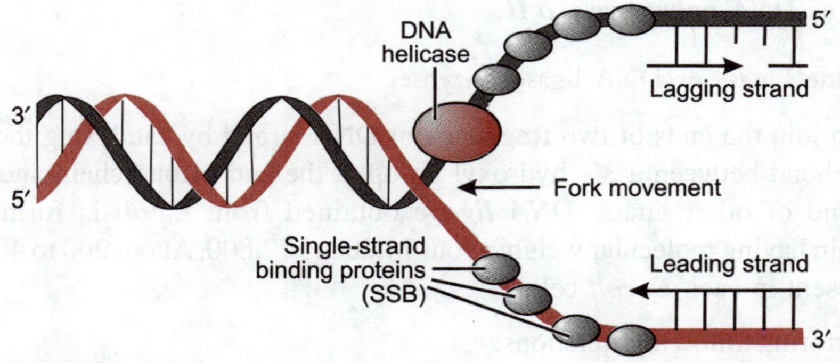

Fig. 8.27 Proteins associated with unwinding of DNA at replication fork

These enzymes use energy from ATP to unwind DNA in advance of replication fork. In *E. coli* two different helicases are found. One helicase attaches to the lagging strand template and moves in 5′→3′ direction and the other attaches to the leading strand template and moves in 3′→5′ direction.

3. Accessory Proteins

Two types of proteins, namely the **sliding clamp proteins** and **brace proteins** are required for **DNA replication**. These are also known as **clamp-loading proteins.** These load the ***polymerase III*** in *E. coli* and ***polymerase*** δ in eukaryotes onto the primer RNA and maintain a stable association with the template.

1. The **brace proteins** are called **γ-complex** in prokaryotes and **replication factor C (RFC)** in eukaryotes. They recognise binding site at the junction of primer and template and bind to it.

2. The **sliding clamp proteins** are β proteins in *E. coli* and proliferating cell nuclear antigens (PCNA) in eukaryotes. These proteins bind adjacent to the brace proteins and form a ring around the template DNA. The ring formed by the sliding clamp maintains association of polymerase with its template as replication proceeds.

4. Untwisting Proteins or Swivelases or Topoisomerases

DNA strands separation creates topological stress in helical DNA. This stress is relieved by *swivelases* or *topoisomerases*. These enzymes cause reversible breakage and rejoining of DNA strands. There are two types of topoisomerases:

1. **Type I topoisomerases** break just one strand of DNA.
2. **Type II topoisomerases** introduce simultaneous break in both the strands. Earlier they were named gyrases.

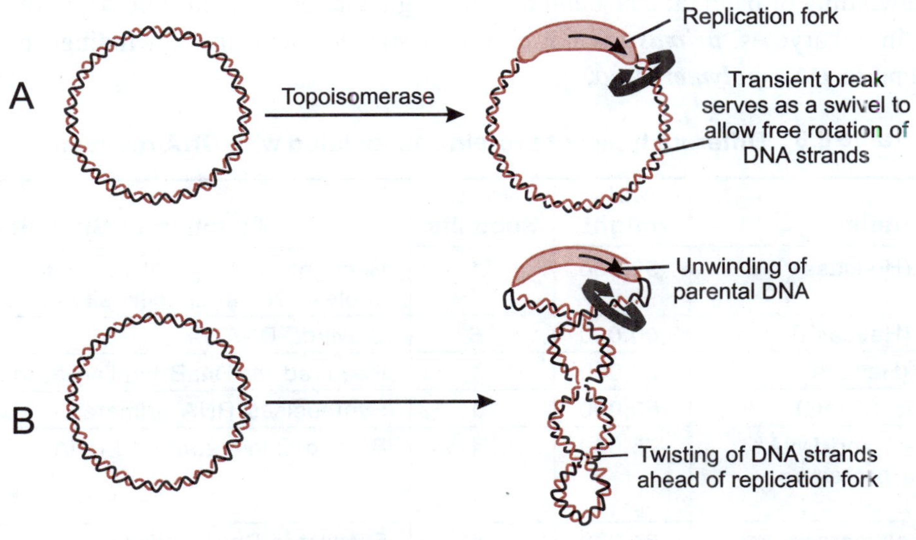

Fig. 8.28 Role of topoisomerase during DNA replication: **A.** The enzyme opens a single chain break in one of the two strands of a DNA double helix; **B.** When the two strands of template DNA unwind, the DNA ahead of replication fork rotates through one full turn in opposite direction, causing circular molecules to become twisted. The problem is solved by topoisomerases which catalyse the reversible transient breakage and rejoining

The breaks introduced by both types of these enzymes serve as **swivels.** These allow the two strands of template DNA to rotate freely around each other and prevent overwinding of DNA. These enzymes are essential in the replication of circular DNA so that its replication can proceed without twisting the DNA ahead of the fork. In eukaryotes also, these enzymes are required, otherwise the complete chromosome would rotate during DNA replication.

5. Single Strand DNA Binding Proteins (SSBP)

The separated strands of DNA helix are stabilised and protected from the hydrolytic cleavage of phosphodiester bonds by **DNA-binding proteins**. These are also called **single strand binding proteins (SSBPs).**

6. Primases

Primases are specific kind of DNA dependent RNA polymerases. They function only during DNA replication. They synthesise a small piece of RNA, called **RNA primer,** which serves as a primer on which deoxyribonucleotides are added and a new polynucleotide strand complementary to template strand of DNA is synthesised. It means primases can initiate formation of polynucleotide strand which DNA polymerases can not.

In *E.coli*, each primase functions only in association with six other proteins. These proteins are collectively called a **primosome.** The other proteins of primosome are believed to be associated with the unwinding of parental DNA and recognising the target sequence or origin of replication sequence. In eukaryotes, ***primase*** is not closely associated with the unwinding proteins. It is tightly bound to ***DNA polymerase*** α.

Table 8.8 Different types of proteins associated with DNA replication.

Protein	Mol. weight	No. of Subunits	Function of Subunits
1. DnaA (Helicase)	52,000	1	Recognises origin of replication, opens duplex DNA at specific sites in origin.
2. DnaB (Helicase)	3,00,000	6	Unwinds DNA.
3. DnaC (Helicase)	29,000	1	Required for DnaB binding at the origin.
4. Primase (DnaG)	60,000	1	Synthesises RNA primers.
5. Single-strand DNA binding proteins (SSBP)	75,000	4	Bind to single strand of DNA.
6. RNA polymerase	4,54,000	5	Facilitates DnaA activity.
7. DNA gyrase (DNA topoisomerase II)	40,0000	4	Relieves torsional strain generated by unwinding of DNA.
8. DNA ligase	74,000	1	Ligases Okazaki segments and nucleotides.

REPLICATION MODELS

Following models of DNA replications have been identified in prokaryotes and eukaryotes:

1. Theta (θ) Replication of DNA

This model of DNA replication was studied in the circular DNA of bacteria *E. coli*. Replication begins with the unwinding of DNA double helix at initiation point, also called **origin of replication**. An **initiator protein** binds at this point. Using energy from ATP, the initiator protein slightly unwinds the double helix at the initiation point, separating the two strands. This produces two forks, called **replication forks** or **Y-junctions**. DNA replication proceeds simultaneously on both the strands and in both directions from the initiation point in a semi-conservative fashion. The replication forks are at the ends of a replication bubble that grows in size as replication continues. The unwinding and replication move along the circle leading to the formation of two new circles. Each circle has one maternal or template strand and other newly formed strand.

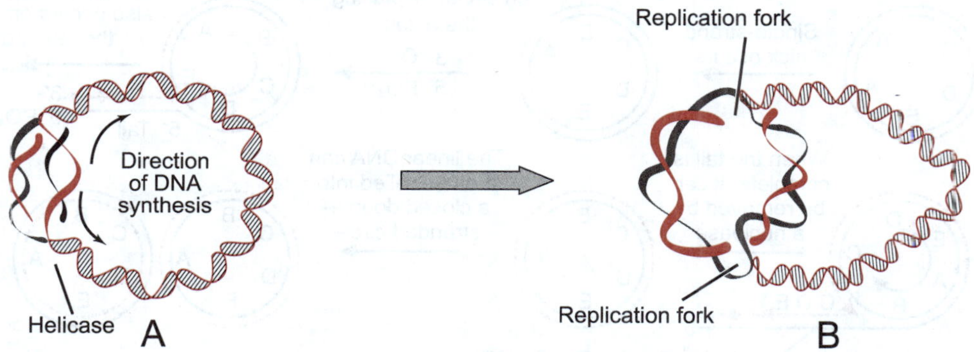

Fig. 8.29 *θ* replication of bacterial DNA in *E. coli*, showing two replication forks or Y junctions

In this replication, the circular DNA molecule does not break. It maintains its circular form throughout the replication process. However, in the replication of circular DNA an intermediate **theta structure** is formed, similar in shape to Greek letter theta (θ). Therefore, DNA replication of circular chromosome in prokaryotes is called **θ replication**. It is also seen in circular DNA of plasmids, mitochondria, chloroplasts and some viruses. Each point of origin and the area of DNA that is subsequently replicated forms one **replicon**.

2. Rolling Circle Model

Rolling circle model of DNA replication occurs in F^+ or Hfr cells of bacteria, in F-plasmids and in some bacteriophages. To start with, a nick or break is formed in one of the strands by a site-specific endonuclease. This opens up one of the two strands of DNA helix. The open strand has a free 3′– OH end and a free 5′– PO_4 end. The complementary closed strand of circular DNA serves as a template and ***DNA polymerase*** adds nucleotides to the free 3′ end of the open strand. There is no need of a primer, because the original break produces a primer configuration

(3′ – OH). As nucleotides are added to 3′ end of the open strand, the unbroken circular template strand rolls and the 5′– PO_4 end of open strand is displaced as **tail**. Since the molecular shape created in this replication resembles Greek letter sigma (σ), the rolling circle replication has been called **σ–mode of DNA replication.**

On completion, the resulting double helix separates from the tail by enzyme ***nuclease***. The enzyme ***DNA ligase*** then joins the ends of replicated circular strand, completing circular double helical DNA. Simultaneously, the –PO_4 tail is replicated in discontinuous fashion forming a double-stranded molecule. Such a structure is called **concatemer**. The two ends of replicated tail can also be joined to form a circle by enzyme ***DNA ligase*** or the linear molecule is packed into phage head.

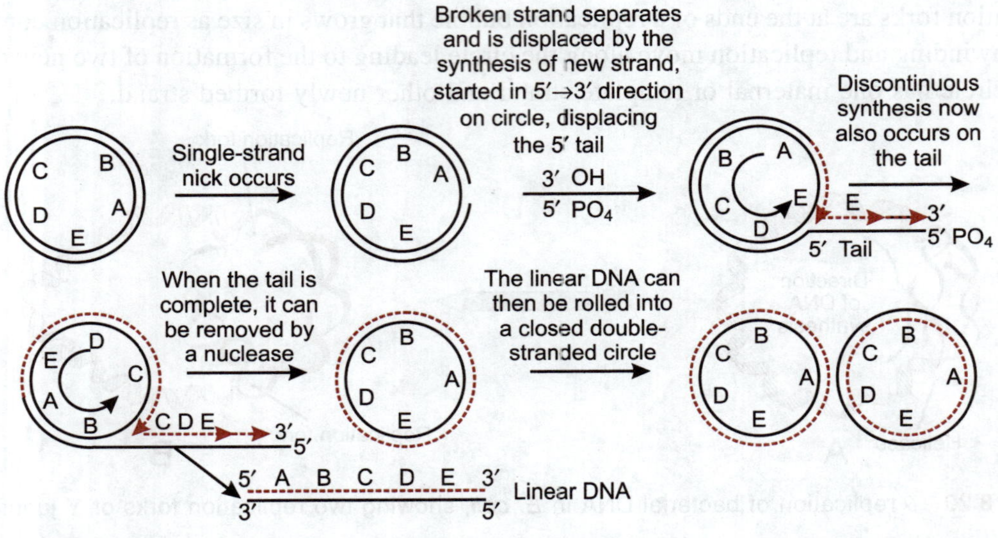

Fig. 8.30 Rolling circle model or σ-mode of replication of circular DNA in a bacteriophage

The concatemers are essential intermediates in phage production. During bacterial mating, a linear DNA molecule is formed during replication process from donor cell. It is passed on to recipient cell to complete conjugation.

3. D-Loop Model

This type of replication of DNA is found in circular DNA molecules of chloroplasts and mitochondria. In them, origin of replication is located at different points on each of the two parental template strands. Replication begins on one strand at its origin of replication, displacing the other. The displaced strand forms a **displacement loop** or **D-loop.** Replication continues until it passes the origin of replication on the other strand. Replication is then initiated on the second strand in the opposite direction. This results in the formation of two circular DNA molecules. In some species mitochondrial and chloroplast DNAs have multiple D-loops.

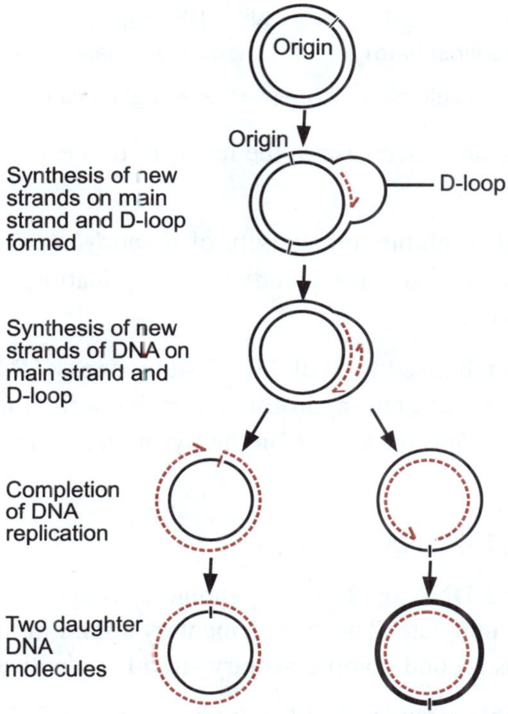

Fig. 8.31 Displacement loop (D-loop) model of DNA replication in mitochondria and chloroplasts

Inhibitors of DNA Replication

A variety of substances inhibit DNA replication. These act at various levels during DNA replication.

> **Extension: DNA inhibitors**
> - Some inhibitors interfere with nucleotide biosynthesis.
> - Some inhibitors interact with DNA template.
> - Nucleotide analogs also act as inhibitors in DNA replication.
> - Some inhibitors bind to replication proteins.
> - Some drugs and antibiotics interfere with DNA replication.

Some inhibitors interfere with nuclectide biosynthesis. Substances like **methotrexate** and **fluorodeoxyuridylate** interfere with nucleotide biosynthesis. **Methotrexate** is used for the the treatment of cancers. It inhibits actions of dihydrofolate reductase causing accumulation of dihydrofolate in the cell. High levels of dihydrofolate inhibit following two reactions:

- dUMP (deoxyuridine monophosphate) + N^5, N^{10}–methylene tetrahydrofolate ⟶ dTMP + Dihydrofolate
- PRPP + L-Glutamine ⟶ PRA + L-glutamate + PP_i

Therefore, methatrexate causes decrease in the levels of dTTP, dATP and dGTP and is used in treating cancer.

5-fluorouracial (5FU) also inhibits the activity of ***thymidylate synthetase*** causing depletion of dTMP and dTTP to the cells and consequently DNA replication. Therefore, 5FU is also used as a powerful anticancer drug.

Inside the cells, FU is metabolised to 5FdUMP (5 fluorodeoxy UMP). It binds to thymidylate synthetase and N^5, N^{10}–methylene tetrahydrofolate forming a tight tertiary complex. This bond can not be broken because of 5–fluro groups on the pyrimidine ring and thymidylate synthetase get inactivated.

Replication of Single-stranded DNA and RNA

Viruses with single-stranded DNA or RNA as genome also synthesize complementary strands using their single strand as template. The complementary strand in turn is used to synthesize its complementary strand. This second complementary strand is identical with the parent strand.

During replication of single-stranded DNA a double-stranded DNA molecule is formed as an **intermediate** or as a **transient** molecule.

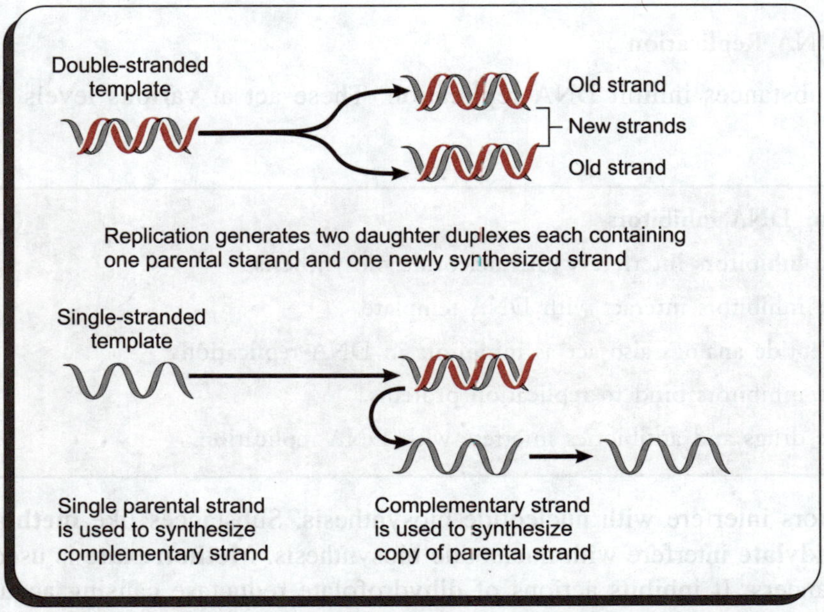

Fig. 8.32 Replication of a double-stranded and a single-stranded DNA molecule

In **retroviruses** single-stranded RNA synthesizes single-stranded DNA with the help of enzyme ***reverse transcriptase.*** This process of synthesis of single-stranded DNA from RNA by complementary base pairing is called **reverse transcription.**

The single-stranded DNA is then converted into a double-stranded DNA. This duplex DNA gets incorporated into host DNA and becomes its integral part. This is seen in parasitic RNA viruses which produce various diseases in the host.

TEXT QUESTIONS

1. Describe the mechanism of DNA replication. What enzymes are necessary during this process?
2. Describe the role of different forms of enzyme DNA polymerase in DNA replication.
3. Give the enzymology of the origin, continuation and termination of DNA replication in *E. coli*.
4. Discuss the role of various enzymes and proteins that participate in DNA replication.
5. Distinguish between the three possible ways in which the double helix could be replicated.
6. Describe an experiment that determines semi-conservative replication in *E. coli.*
7. Differentiate between the roles of ***DNA polymerase I*** and ***DNA polymerase III*** during DNA replication.
8. Explain why replication of the lagging strand is discontinuous.
9. Draw the density gradient banding pattern that Meselson and Stahl had obtained if DNA replication had turned out to be conservative or disruptive.
10. How is the leading strand of a DNA duplex replicated? How is the lagging strand replicated? Give differences in the synthesis of two types of strands.
11. Describe the basis for the requirement that the leading and lagging strands be replaced by different mechanisms.
12. Explain the following:
    (a) Semi-conservtive replication of DNA (b) Replication origin
    (c) Replication fork (d) Okazaki fragment
    (e) Concatamer (f) Lagging strand
    (g) Leading strand (h) Replication
    (i) DNA polymerase.
13. Describe experiment conducted by Taylor on *Vicia faba* root tips. What does it signify?

14. Describe quantitative changes in the DNA contents during cell cycle.
15. Describe role of DNA unwinding and untwisting proteins.
16. Name the specific binding sites on the molecule of polymerase–I enzyme.
17. What is the difference between continuous and discontinuous replication?
18. Give reason for discontinuous replication of DNA.
19. What is a primosome and a replisome? Give relation between them.
20. Describe synthesis of Okazaki fragment.
21. Why lagging strand in DNA replication is formed in small pieces?
22. Describe the role of following enzymes and proteins in the process of DNA replication:
    (a) Helicase (b) Primase
    (c) SSB Proteins (d) DNA polymerase I
    (e) DNA polymerase III (f) DNA ligase.
23. What is the function of DNA ligase or polynucleotide ligase enzyme?
24. What is copolymerase III?
25. What is the difference between DNA polymerase and RNA polymerase? Give one main difference.
26. What is replisome?
27. Give the role of polymerase δ.
28. What is D–loop?
29. What is rolling circle model?
30. What is the role of topoisomerase I in DNA replication?
31. What are primosomes?
32. What is the role of single-stranded binding proteins (ssb proteins)?
33. How many subunits constitute polymerase III holoenzyme?
34. Which enzyme in DNA replication has both activities of polymerase and exonuclease?
35. What are endonucleases?
36. What are Okazaki fragment?
37. Define semiconservative method of replication.
38. What do you mean by theta structure?

39. Define a replicon.

40. What is replication fork?

41. Describe the role of following enzymes and proteins in DNA replication:

(a) DNA polymerase I
(b) DNA polymerase II
(c) DNA gyrase
(d) DNA topoisomerase
(e) Primase
(f) SSBP
(g) Swivelase
(h) Helicase
(i) DNA ligase

42. Name the enzyme of DNA replicase system which has proof reading activity.

43. What is D-loop model of DNA replication and where is it seen?

9

Chapter

DNA Repair

IN THIS CHAPTER

Necessity of DNA Repair, Mistakes in DNA, Agents Causing Mistakes and Damage in DNA, Biochemical Mechanism of DNA Repair, Mismatch Repair, Thymine Dimer Repair, Photoreactivation, Base Excision Repair, Recombinational Repair, SOS Repair or Error Prone Repair, Extensive Damage Repair by Homologous Recombination.

NECESSITY OF DNA REPAIR

DNA molecules are most vital for the cell and for the survival of individual organisms. But DNA molecules are quite fragile and are constantly subjected to radiations and intracellular physical and chemical stresses. Oxygen, UV light, X-rays, alkylating agents and free radicals all can cause damage and introduce changes in nucleotide sequences. These changes can result from breaks in polynucleotide strands, chemical modification of nitrogenous bases in the strands and incorporation of mismatched bases during replication. The commonly occurring changes in DNA can be due to formation of pyrimidine dimers, alkylation or deamination of nitrogenous bases. Both spontaneous and induced damages in DNA, if left uncorrected, may result in mutations that disrupt the function of essential enzymes or structural proteins and pose a significant risk to the existence of the cell. For correcting such occasional mistakes in DNA structure, several efficient correcting mechanisms have evolved in order to protect organisms against the potential deleterious effects of mutations.

MISTAKES IN DNA

Types of Mistakes

The various mistakes that are often introduced in the nucleotide sequences in DNA molecules can be classified into following categories:

- Single strand breaks in DNA
- Double strand breaks in DNA
- Alteration of bases such as intrastrand dimer formed by two adjacent pyrimidines (thymine dimer by UV radiation).

Table 9.1 Types and causes of DNA mistakes

Mistakes	Causes
1. Insertion of incorrect base in one strand (*i.e.* replacement of purine by pyrimidine or vice versa)	Replication error Spontaneous deamination of pyrimidine C to U Spontaneous deamination of purine A to hypoxanthine.
2. Alteration of bases	Introduced by ionising radiations and UV rays. They lead to the formation of thymine dimers.
3. Loss of bases	Introduced by alkylating agents that cause depurination by breaking N-glycosidic bond.
4. Single strand breaks	Caused due to the breakage of phosphodiester bonds by chemicals (peroxides, sulphhydryl containing compounds) by ionising radiation and by free radicals in water.
5. Double strand breaks	Are single strand breaks introduced in separate strands of DNA at the same or at different levels.
6. Cross-linking of two chains of the helix	Are antibody-mediated covalent linkages between a base of one strand and an opposite base in complementary DNA strand.

Agents Causing Mistakes and Damage

The various agents causing damage to DNA or introducing mistakes in nucleotides sequence can be placed under following categories:

1. Radiation

Ionising radiation, for example, X-rays, naturally occurring cosmic rays, electron beams, γ-rays and radiation beams of various kinds used in radiotherapy are associated with breaks in polynucleotide strands of DNA.

The strand breaks are caused by the action of secondary electrons produced by a β-particle or an X-ray photon or by the production of free radicals in water that can attack the phosphodiester bonds.

2. Chemicals

Strong chemicals like alkytaling agents such as methyl ethanesulphonate (**MES**), ethyl ethanesulphonate (**EES**) and nitrosoguanidine, etc., react primarily with guanine by adding an alkyl group to N-7 of purine ring. This weakens the N-glycosidic bond and causes **depurination**.

Extension: Depurination

The loss of purine from DNA strand is called depurination. Depurination occurs by the breakage of N-glycosidic bond of a purine nucleotide. The rate of spontaneous depurination is about one purine removed per 300 purines per day at pH 7 and 37°C, *i.e.* 10^4 depurinations per day in a mammalian cell.

Depurination can be induced by alkylating agents which add an alkyl group to N-7 of purine ring. Alkylation weakens the N-glycosidic bond and causes loss of a purine from the strand.

Aflutoxin, benzopyrene, nitrosamines, nitrosourea, psoralen and cisplatin are some of the chemical agents or drugs which cause damage to DNA and introduce gene mutations.

BIOCHEMICAL MECHANISM OF DNA REPAIR

Cells possess a variety of mechanisms to repair the damages in DNA structure caused by various mutagenic agents. These are:

1. Repair of Damage Caused by Incorrect Base Insertion (Mismatch Repair)

During DNA replication, both the enzymes ***polymerase I*** and ***polymerase III*** occasionally incorporate wrong or incorrect base in the daughter strand under synthesis. A wrong base fails to form a hydrogen bond with the template base in the parental or template strand. Such errors are usually corrected by the editing function of these enzymes.

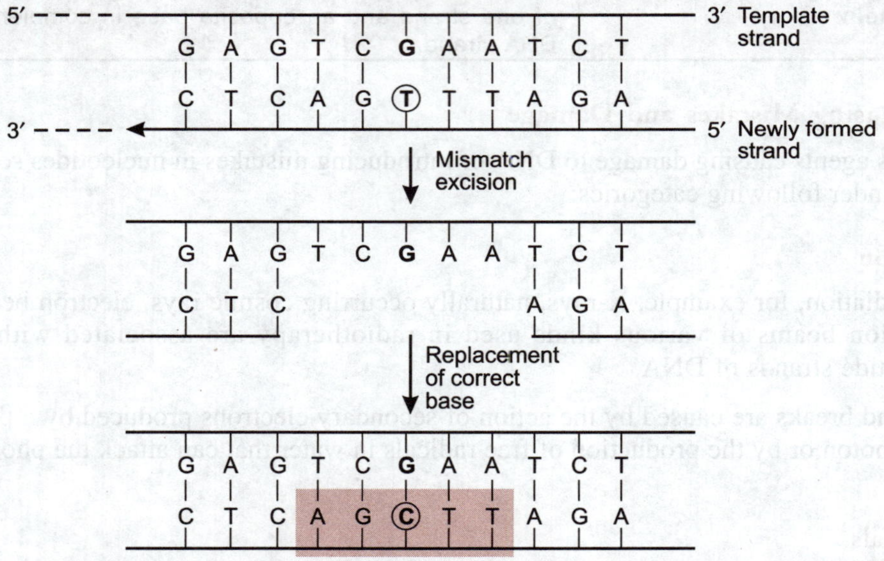

Fig. 9.1 Excision of a short segment from newly synthesised strand

The errors missed by the editing function of these enzymes are corrected by a second correcting system, called **mismatch repair**. In mismatch repair a pair of non-hydrogen-bonded base is recognised as incorrect and a polynucleotide segment with unmatched pair is excised, thereby removing the unmatched pair. The gap is filled by ***polymerase I*** (pol. I). The mismatch repair system also counteracts the mutagenic effects of tautomeric bases. Such molecules in rare or tautomeric form are misincorporated into DNA because tautomeric A pairs with C and tautomeric G pairs with T. Such pairs are correctly hydrogen bonded to template strand and are not recognised by editing enzyme. But when tautomeric base returns to its original form, it forms a mismatch base pair and is corrected by mismatch repair system.

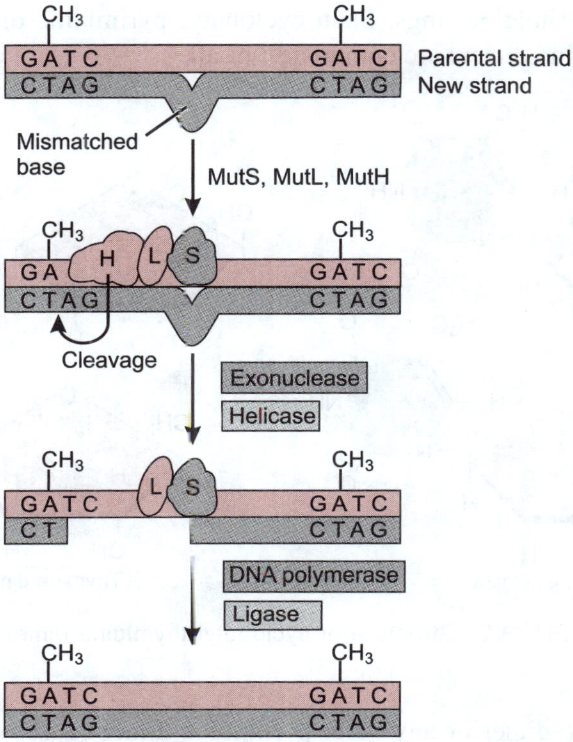

Fig. 9.2 Mismatch repair mechanism in *E. coli*

The mismatch repair system requires a complex of three proteins, MutS, MutH and MutL. The protein **MutS** recognises and binds to the mismatched base. **MutH** binds to DNA at the hemimethylated GATC sites and cleaves the unmethylated (newly systhesised) strand. **MutL** binds to MutS and MutH and helps in final repair steps.

> **Extension: Mismatch repair genes**
>
> Mismatch repair is a highly conserved mechanism. It functions in the similar fashion in bacteria as well as in eukaryotes. Several homologs of bacterial *mutS*, *mutH* and *mutL* genes have been identified in yeast and mammals. This type of mismatch repair system is similar in all eukaryotic cells.
>
> It is believed that families of eukaryotic genes, homologous to *mutS*, *mutH* and *mutL* arose before the evolutionary divergence of animals, plants and fungi groups.

2. Repair of Thymidine Dimers

This can be discussed under following heads:

1. Formation of Pyrimidine Dimers

UV radiations produce intrastrand photodimers between two adjacent pyrimidine bases of the same strand. The dimers result from the formation of **cyclobutyl ring** by the joining of 5th and 6th carbon atoms of two pyrimidine rings. Such cyclobutyl pyrimidine dimers are most common photodimers and thymidine dimers are commonest of all.

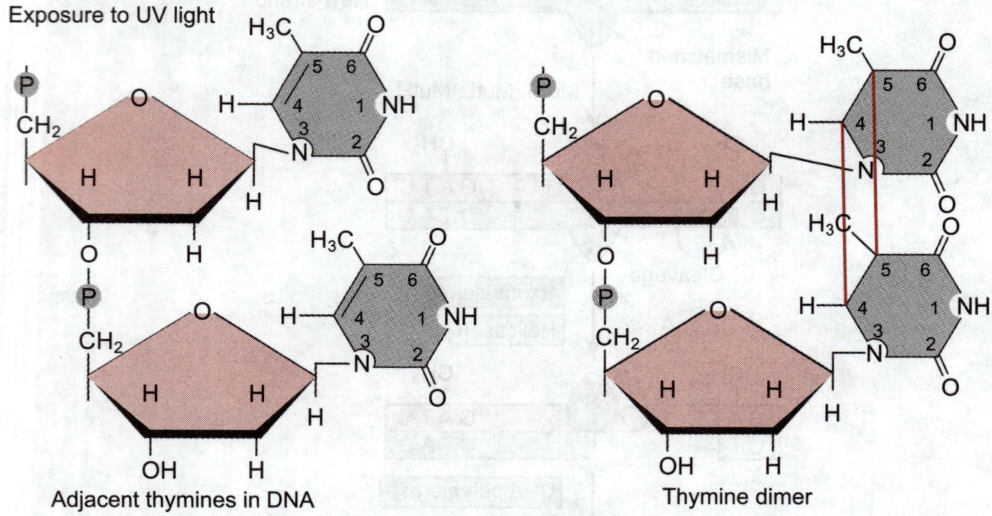

Fig. 9.3 Structure of cyclobutyl thymidine dimer

2. Effect of Dimers

The presence of thymine dimer or any other pyrimidine dimer causes distortion of DNA helix because the two thymines are pulled closer. This produces a '**kink**' in the polynucleotide strand in which dimer is formed. The kink causes inhibition of the advancement of replication fork and thus interferes with DNA replication.

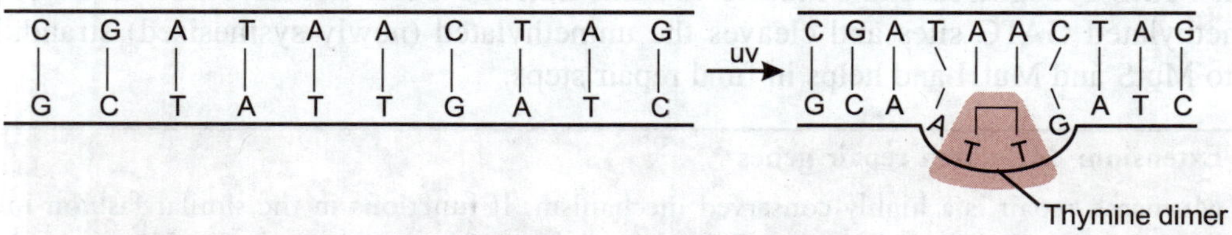

Fig. 9.4 Distortion of DNA helix and formation of 'Kink' caused by thymine dimer

3. Mechanism of Dimer Repair

There are four major pathways for the repair of damage in DNA caused by thymidine dimer. These are separated into two categories:

1. Photo-reactivation or light induced repair
2. Light independent repair
   - Excision repair
   - Recombinational repair
   - SOS repair

1. Photoreactivation or Light Induced Repair : Photoreactivation is light induced enzymatic cleavage of pyrimidine dimer (thymidine dimer). It is catalysed by enzymes called ***photolyases*** or **PR enzymes**. They absorb radiant energy of wavelengths 300 to 600 nm and split the cyclobutyl ring between two pyrimidine molecules of the dimer. As a result the cyclobutyl thymidine dimers are converted to pyrimidine monomers.

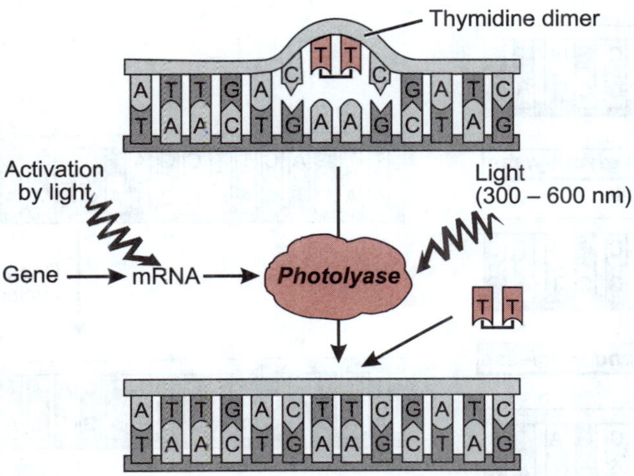

Fig. 9.5 Mechanism of enzymatic photoreactivation repair of thymidine dimer

2. Excision Repair or Dark Repair : The excision repair mechanism is a multistep enzymatic process that can remove and replace individual nitrogenous bases (**base excision repair**) and also the longer nucleotide chains (**nucleotide repair**). This repair mechanism does not require sunlight. So it is also called **dark repair**.

- **Base excision repair** includes enzymatic removal of chemically modified base (such as uracil, 5-methylcytosine or 3-methyladenine), followed by removal of the nucleotide. The gap thus created is filled by a complementary base pair. It is carried out in the following steps with the aid of several enzymes. **1.** The enzyme ***DNA glycosylase*** removes the chemically modified base, causing apurination or apyrimidation. **2.** The enzyme ***endonuclease*** cleaves the phosphodiester bond in the sugar phosphate backbone at the apurinic or apyrimidinic site, forming a nick. **3.** Enzyme ***deoxyribose phosphodiesterase*** removes the deoxyribose phosphate molecule. **4.** The repair polymerase (***DNA polymerase*** β) adds the single nucleotide to the 3′ end of the nicked strand at the gap. **5.** Enzyme ***DNA-ligase*** seals the nick by joining the two parts of the polynucleotide strand.

- **Nucleotide excision repair** operates to remove pyrimidine dimer (or thymidine dimer) and fills the gap by complementary nucleotides. It requires the entire DNA replication machinery. The enzyme ***repair-endonuclease*** recognises the distortion in the polynucleotide strand produced by a thymidine dimer. It cleaves the sugar-phosphate backbone of the DNA strand on both the sides of the kink. This exposes 3′-OH group at the 5′ incision site. The oligonucleotide consisting of about 30 nucleotides is removed by ***DNA helicase***. The resulting gap is repaired by ***DNA polymerase*** δ or ε and is closed by ***DNA ligase***.

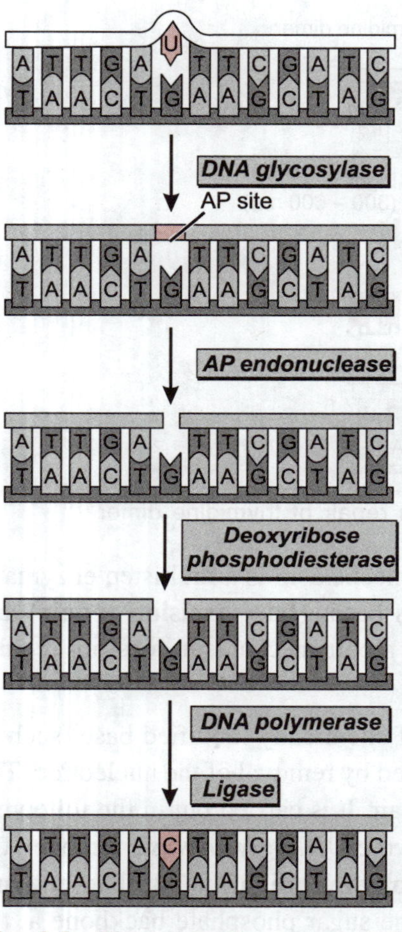

Fig. 9.6 Diagrammatic representation of base excision repair when methylated cytosine, i.e. uracil enters DNA strand during replication

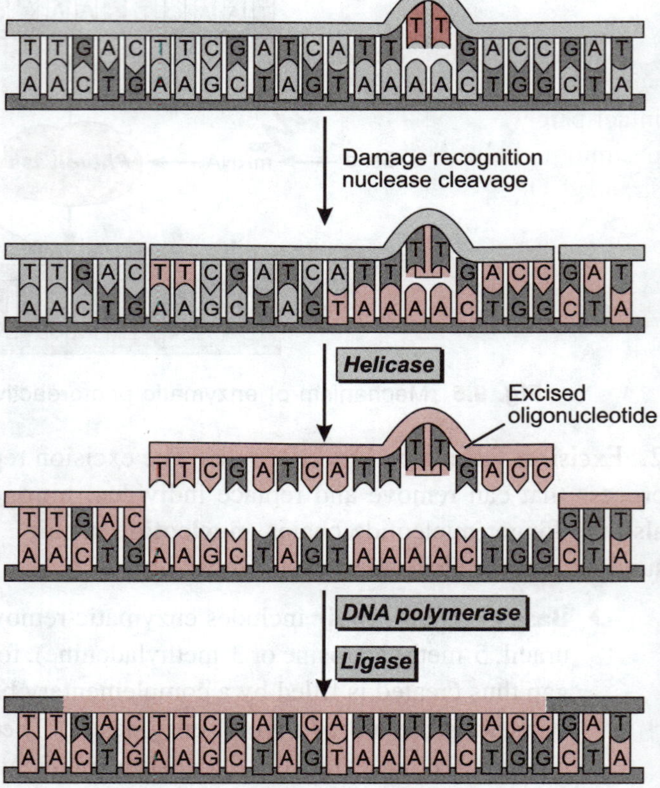

Fig. 9.7 Nucleotide excision repair of a pyrimidine dimer utilising DNA replication machinery

3. Recombinational Repair : In recombinational repair the undamaged strand of parental DNA is copied during replication to produce a normal daughter molecule. During DNA replication, ***polymerase III*** is stalled at the site of thymine dimer. The enzyme passes over the dimer block and initiates chain growth beyond the block. Therefore, the daughter strands have large gaps, one for each thymidine dimer. Complete normal double-stranded DNA molecules are formed from such fragmented and kinked DNA molecules by sister-strand exchange mechanism.

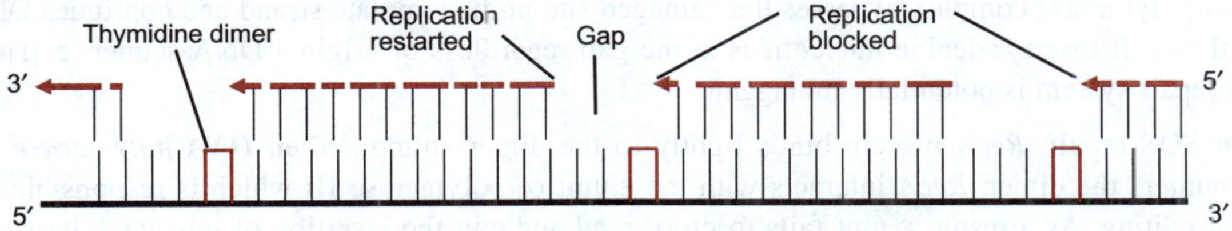

Fig. 9.8 Blockage of replication due to thymidine dimers

In single strand exchange, the gap produced in the daughter strand is filled by recombination with the undamaged parental strand. Due to this exchange, a gap is formed in the previously intact parental strand. This is filled by the action of DNA polymerase and ligase enzymes. The pyrimidine dimer is later on removed by excision repair to produce two intact normal double stranded DNA molecules.

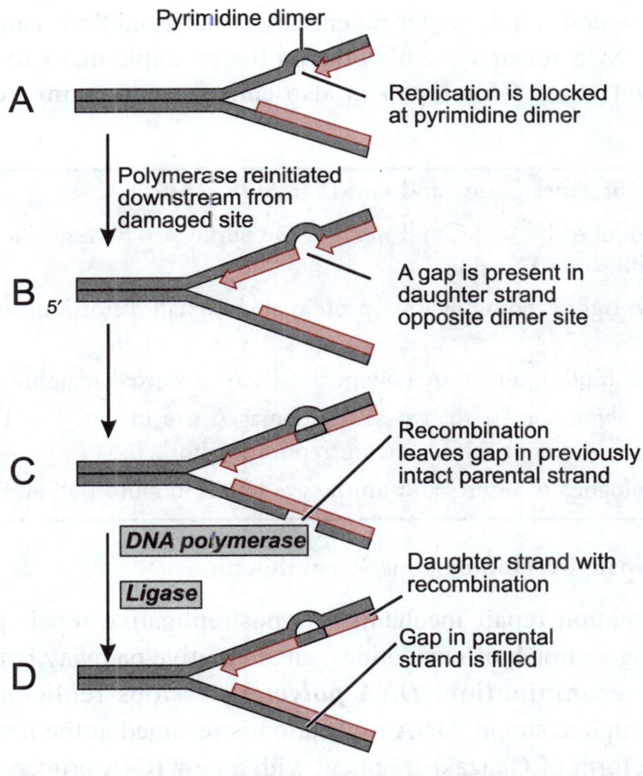

Fig. 9.9 Post-replication repair by recombinations. **A.** Replicating DNA molecule with a pyrimidine dimer; **B.** DNA polymerase bypasses the dimer and starts postdimer synthesis of new strand; **C.** The resultant gap in the newly synthesised DNA strand is repaired by recombination with undamaged parental strand; **D.** Enzymes DNA polymerase and DNA ligase fill the resultant gap in the previously intact parental strand

4. SOS Repair or Error Prone Repair : SOS repair is a **bypass repair system** that allows polymerisation to proceed across a dimer despite the distortion of the helix. During replication, the DNA polymerase complex bypasses the damaged site in the template strand and continues DNA synthesis. It inserts adenine nucleotides at the gap regardless of original DNA sequence. Thus, this repair system is potentially mutagenic.

In SOS repair, *RecA* protein binds tightly to the dimer strand. When ***DNA polymerase III*** encounters the dimer, *RecA* interacts with the ε unit of polymerase III which is responsible for DNA editing. As a result, ε unit fails to proof read and edit the insertion of mismatch base and allows replication to proceed overlooking the insertion of wrong base. In *E. coli* proteins UmuC and UmuD bind to DNA polymerase and alter the stringency of nucleotide incorporation. Adenines are inserted into the newly synthesised strand at the lesion site irrespective of base sequence in the parental template strand. If the damaged site is thymidine dimer, the newly synthesised DNA is not mutated. But if the damaged site has cytosine dimer, the newly synthesised strand is mutated.

SOS repair system operates only under potentially lethal conditions caused by extensive UV irradiation. Therefore, SOS repair system works on the principle that survival with mutation is better than no survival at all. This repair is also called **error prone repair** or **translation replication**.

Extension: Role of genes umuC and umuD in SOS repair

The proteins produced by *umuC* and *umuD* gene family are necessary for dimer bypass. They perform following functions:

- They help in tight binding of *RecA* protein at the small distortion caused by pyrimidine dimer.
- They help in binding of DNA polymerase III to distorted region.
- They help polymerase III to bypass the damaged site in template DNA and move forward in the direction of advancing replication fork.

Genes homologous to umuC and umuD are found in almost all animals and plants.

Extensive Damage Repair by Homologous Recombination

Homologous recombination repair mechanism is postreplicative repair process. When DNA damage includes a longer stretch of nucleotides, an alternative pathway is found to operate. It is called **homologous recombination**. ***DNA polymerase*** stops replication on reaching the damaged site in the template strand. DNA replication is resumed at the next priming site on the template strand in the form of Okazaki fragment with a new RNA primer.

The gap left in the daughter strand or newly synthesised strand may be of several hundred nucleotides. This gap is then repaired by homologous DNAs recombination with the complementary parental strand using specific recombination enzymes. Enzymes ***DNA polymerases*** and ***DNA ligases*** repair the resulting gap in the complementary parental strand.

TEXT QUESTIONS

1. Describe the various types of damages that occur in DNA. Discuss the agents responsible for such DNA damages.
2. Describe in brief the photoreactivation repair mechanism. What types of damages are repaired by photoreactivation?
3. Explain dark repair mechanism or excision repair mechanism.
4. Describe the mismatch repair mechanism in *E. coli*. Discuss the role of three proteins that are required during mismatch repair.
5. Discuss various mechanisms used in pyrimidine dimer or thymidine dimer repair.
6. Write short notes on:
   (a) Error-prone repair
   (b) Mismatch repair
   (c) Role of methylation in DNA repair
   (d) Post replication repair
   (e) Recombinational repair.
7. Write in brief how pyrimidine dimers or thymidine dimers are formed.
8. What is a Kink? How is it formed?

10

Chapter

Ribonucleic Acid

IN THIS CHAPTER

Occurrence; Structure; Differences between DNA and RNA; Genetic RNA; Nongenetic RNAs; Messenger RNA or Nuclear RNA (mRNA) ; Differences between Prokaryotic and Eukaryotic mRNA; Informosomes; Ribosomal RNA (rRNA); Transfer RNA (tRNA); Functions of Different Types of RNAs.

Ribonucleic acid (**RNA**) is single-stranded nucleic acid found in all living cells. Its different forms are associated with the transmission of information from nucleus into the cytoplasm and with the synthesis of proteins for the regulation of cell activities. However, RNA is the only macromolecule which also functions for the storage of information and also as a catalyst. The catalytic RNAs are called **ribozymes**. RNA is also presumed to have played an important role as an essential chemical intermediate in the evolution of life on this planet.

OCCURRENCE

RNA is found chiefly in the cytoplasm but also in the nucleolus. Inside the cytoplasm it occurs freely as well as in the ribosomes. RNA can also be detected from mitochondria, chloroplasts and associated with the eukaryotic chromosomes. In some plant and animal viruses RNA acts as hereditary material.

STRUCTURE

The structure of RNA can be easily understood by comparing it with the structure of DNA.

1. More commonly RNA is a single-stranded structure consisting of an unbranched polynucleotide chain, but it is often folded back on itself forming helices. DNA is a double-stranded structure and its two polynucleotide chains are wounded spirally around the main axis.
2. RNA like DNA is formed of several hundreds or thousands of nucleotides arranged in a linear sequence and connected together by 3′–5′ phosphodiester bonds.
3. The sugar found in nucleotides of RNA is **ribose**, whereas it is deoxyribose in DNA. The nucleotides of RNA are called **ribonucleotides**.

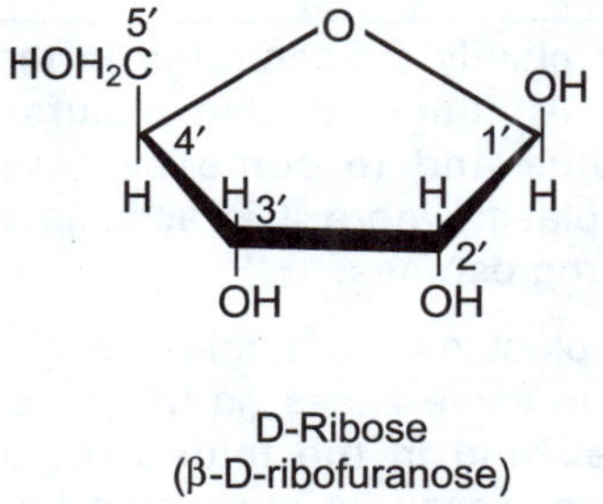

Fig. 10.1 A molecule of ribose sugar

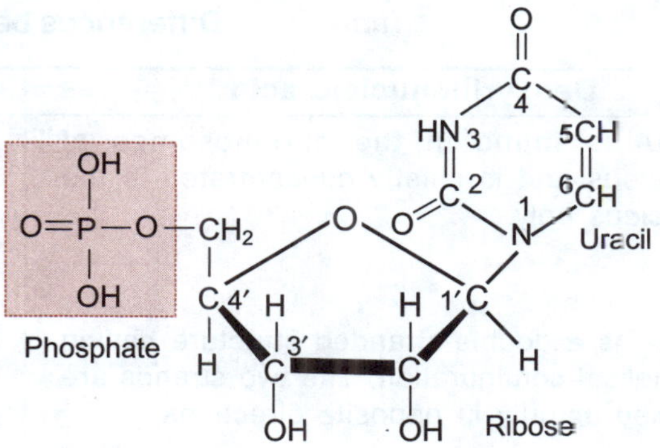

Fig. 10.2 A nucleotide of RNA (Uridylic acid)

4. The four nitrogenous bases found in RNA are adenine, cytosine, guanine, and uracil, whereas those in DNA are adenine, cytosine, guanine and thymine. Therefore, in RNA **thymine of DNA is substituted by uracil**. Some unusual nitrogenous bases are also found in RNA.
5. The base composition of RNA does not agree to the **A::U = G:::C = 1**.
6. Intramolecular pairing between the nucleotides of the single strand of RNA provides stability to RNA. In DNA, nucleotides of two polynucleotide strands pair through hydrogen bonds.
7. DNA is the hereditary material, whereas RNAs are of different types for performing different functions during protein synthesis. However, in most plant viruses and some animal viruses RNA also acts as hereditary or genetic material.

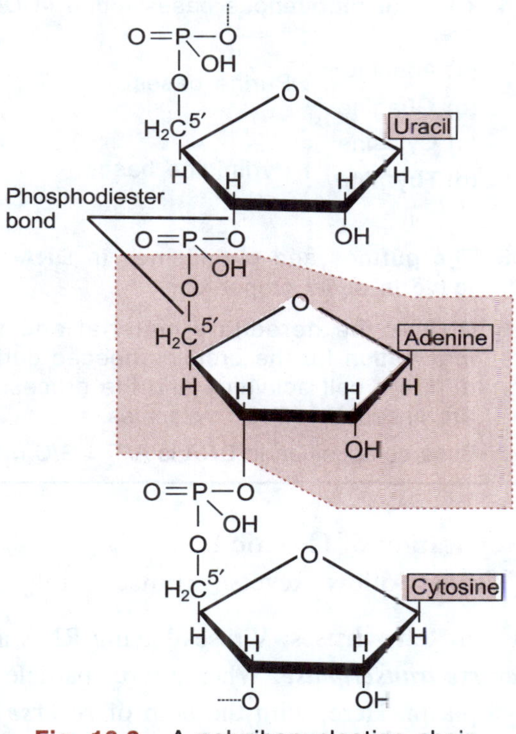

Fig. 10.3 A polyribonucleotide chain showing ribonucleotides and phosphodiester bond

THE GENETIC RNA

In most plant viruses, some animal viruses and in many bacteriophages, DNA is not found and RNA acts as hereditary material. This RNA may be single-stranded or double-stranded. When it is double-stranded it follows the same rules of base pairing as in the case of DNA. The Table 10.2 given on page 240 provides a list of viruses whose genetic material is RNA.

Table 10.1 Differences between DNA and RNA

Deoxyribonucleic acid	Ribonucleic acid
1. DNA is found in the chromosomes of nucleus and is chiefly concentrated in the nucleus only.	1. RNA is chiefly concentrated in the cytoplasm. Although, it also occurs in the nucleolus and to some extent in the nucleoplasm where it is associated with the chromosomes.
2. DNA is a double-stranded structure having a helical configuration. The two strands are coiled spirally in opposite directions.	2. RNA molecules are single-stranded. The strand in some cases (in tRNA) is folded on itself and in the folded region, the nucleotides may be connected by hydrogen bonds.
3. The sugar molecule in DNA is deoxy-ribose.	3. The sugar molecule in RNA is ribose.
4. The four nitrogenous bases found in DNA are: (a) Adenine, (b) Guanine } Purine bases (c) Cytosine, (d) **Thymine** } Pyrimidine bases.	4. The four nitrogenous bases found in RNA are: (a) Adenine, (b) Guanine } Purine bases (c) Cytosine, (d) **Uracil** } Pyrimidine bases. *i.e.* in RNA thymine is replaced by uracil.
5. The purines and pyrimidines in DNA are found in equal proportion.	5. It is not necessary that purines and pyrimidines occur in equal proportion.
6. DNA is the hereditary material and the information for the proteins needed during different cell activities and life processes are encoded in DNA molecules.	6. RNA participates in protein synthesis.
7. Base composition in DNA is A/T = G/C = 1.	7. Not so in RNA.

Replication of Genetic RNA (Viruses follow Reverse Transcription)

1. In Retroviruses: Viruses having RNA as the hereditary material possess an enzyme called ***reverse transcriptase***. When a virus particle enters the host cell, its RNA is released in the host cytoplasm. Here, with the help of ***reverse transcriptase*** a strand of complementary DNA is synthesised using RNA as template. This single DNA strand synthesises its complementary antiparallel strand completing the formation of double-stranded DNA. This viral DNA then joins the DNA of host cell becoming its integral part. The integrated DNA copy of host DNA is called **provirus** and this process of DNA transcription is called **reverse transcription**. Utilising synthetic machinery of host cell, this DNA produces several copies of RNA and protein molecules to be used in the formation of viral protein coat.

RNA → ssDNA → dsDNA → Viral mRNA → Viral proteins

Table 10.2 Genetic RNA in different viruses

Types of Viruses	Names of Viruses	Types of RNA
1. Plant viruses	1. TMV	1. Single-stranded
	2. Wound tumour	2. Double-stranded
2. Animal viruses	3. Influenza virus	3. Single-stranded
	4. Poliomyelitis	4. Single-stranded
	5. Reo virus	5. Double-stranded
3. Bacteriophages	6. MS2	6. Single-stranded
	7. F2	7. Single-stranded
	8. r17	8. Single-stranded

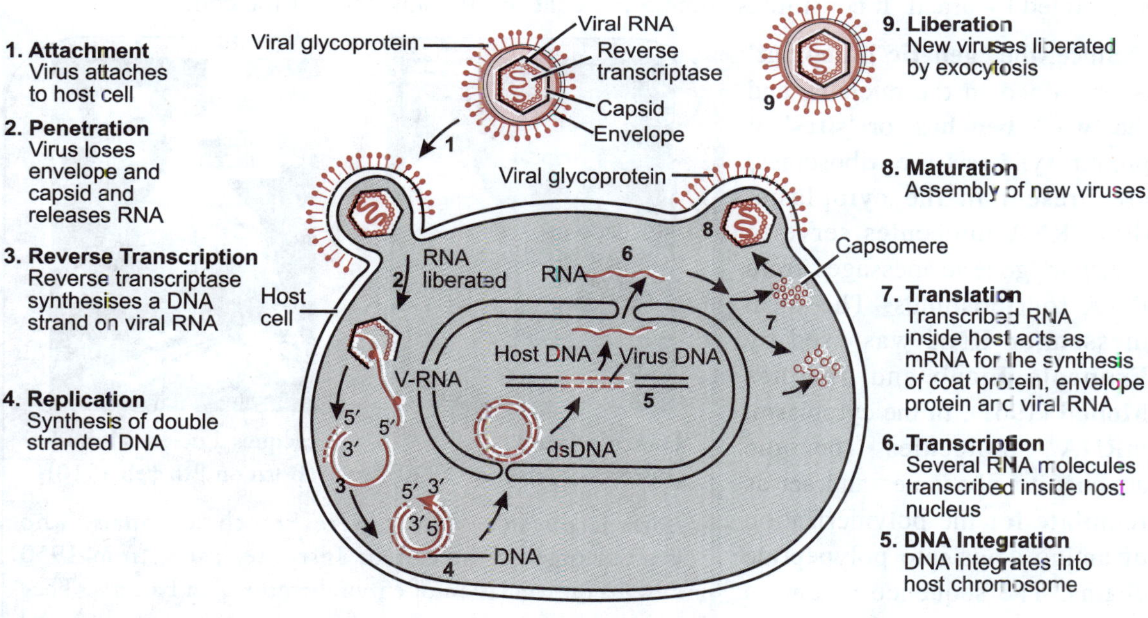

Fig. 10.4 Replication of viral particles of retrovirus by reverse transcription inside host cell

2. In RNA-RNA Viruses: These viruses synthesise new genomic RNA on RNA template with the help of an enzyme called ***replicase***. These new RNA molecules then synthesise mRNAs which guide synthesis of viral proteins using host's ribosomes, tRNA, amino acids and required enzymes.

$$\text{RNA} \rightarrow \text{RNA} \rightarrow \text{mRNA} \rightarrow \text{Polypeptide}$$

NON-GENETIC RNA

In all other organisms, where DNA is the hereditary material all types of RNAs are non-genetic. The non-genetic RNAs are synthesised from DNA template. In general, three types of RNAs have been distinguished:

1. Messenger RNA or nuclear RNA (mRNA)
2. Ribosomal RNA (rRNA)
3. Transfer RNA (tRNA)

1. Messenger RNA or Nuclear RNA

Messenger RNA carries genetic information from chromosomal DNA to the cytoplasm, where it acts as a template for protein synthesis. It is complementary to DNA and carries the copy of the same base sequence as found in that part of DNA from which it is copied, except that thymine is substituted by uracil. It constitutes about 5 per cent of the total RNA of the cell.

Since, the **'genetic blueprint'** is contained in the nucleus, and the **'work benches'** or **'sites'** of protein synthesis (i.e. ribosomes) are present in the cytoplasm, these RNA molecules serve to transport genetic messages from DNA to ribosomes. The term **messenger RNA** was used by **Francois Jacob** and **Jacques Monod** (1961). In the cytoplasm, mRNA molecules become attached to ribosomes and act as **template** for the polymerisation of amino acids into polypeptide chains. The sequence of amino acids in each polypeptide chain is determined by the sequence of codons in its mRNA molecule.

Francois Jacob (Born on 17th June 1920)

Jacques Lucien Monod (Born on 9th Feb 1910)

Both Jacob and Monod were French nationals, who worked together in Pasteur Institute, Paris, from 1950 onward on regulation of enzyme synthesis in bacteria. They advanced operon concept of gene regulation and shared Nobel Prize (1965) with Andre Lwoff. They also proposed the existence of RNA messenger in 1961.

Characteristics of mRNA

mRNA has following characteristics:

1. It is formed as a complementary strand to one of the two strands of a DNA.
2. It, therefore, carries the same sequence of base arrangement as found in that part of DNA from which it is copied except that the thymine of DNA is substituted by uracil in mRNA. mRNA, therefore, contains the same information as coded in that part of DNA.

3. After synthesis, it immediately diffuses out of the nucleus into the cytoplasm, where it gets associated with the ribosomes.
4. Here mRNA acts as a template for protein synthesis.
5. mRNA has a short life span and withers away after few translations.
6. It has a high turnover.
7. mRNA from different organisms may be alike in gross base content, but they usually differ in the arrangement of bases.
8. The molecules of mRNA are linear and longest of all the three types of RNAs. Their length is related to the size of polypeptide chains for which these have codec information.
9. There is one mRNA for each polypeptide chain.
10. Because of great variation in the size of mRNA molecules, mRNA is also called **heterogeneous nuclear RNA** or **hnRNA**.

Structure of mRNA

The main part of mRNA is the sequence of nucleotides that codes for a polypeptide. However, a fully processed mRNA molecule has at either end some sequences that are not translated. Because of these sequences the two ends of each mRNA molecule are distinctly marked. Each fully processed mRNA molecule has following parts:

1. A cap of methylated guanine (**G-cap**) at its proximal or 5′ end.
2. A **start** or **initiation codon** (**AUG** or **GUG**) next to G-cap.

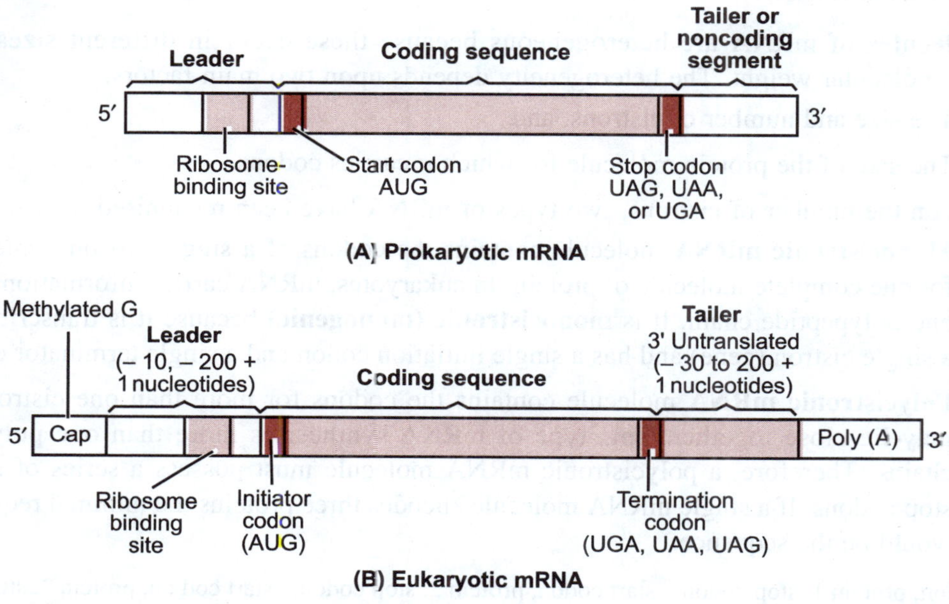

Fig. 10.5 A fully processed mRNA with marked ends: **A.** Prokaryotic mRNA; **B.** Eukaryotic mRNA

3. A long **coding region.**
4. A **termination codon** or **stop codon (UAA, UAG** or **UGA)** at the end of coding region at the distal end (i.e. 3′ end).
5. A **poly-A tail** of many adenine nucleotides.
6. A small **non-translated** or **non-coding segment I** may be present after the G-cap and before poly-A tail. The non-translated sequence at 5′ of mRNA is called the **leader**, because it precedes the start of initiation codon. The leader segement has ribosome-binding site, where mRNA binds to ribosome. The nontranslated sequence at 3′ end is the **tailer** or **noncoding segment II**. It follows the termination codon. Though, the leader and tailer sequences are not translated, their presence is essential for the translation of message coded in the coding region.

Table 10.3 Differences in prokaryotic and eukaryotic mRNA

Character	Prokaryotic mRNA	Eukaryotic mRNA
1. Nature	May be polycistronic or monocistronic	Only monocistronic
2. 5′ Methylated G-cap	Absent	Present
3. Ribosome binding site or Shine Dalgarno sequence	Present	Absent
4. Poly-A, tail at 3′ end	Absent	Present

Heterogeneity of mRNA

The molecules of mRNA are heterogeneous because these occur in different sizes having different molecular weight. The heterogeneity depends upon two main factors.

1. The size and number of cistrons, and
2. The size of the protein molecule for which it carries codons.

Based on the number of cistrons, two types of mRNA have been recognised.

- **Monocistronic mRNA** molecule contains the codons of a single cistron, which codes for one complete molecule of protein. In eukaryotes, mRNA carries information only for one polypeptide chain. It is **monocistronic (monogenic)** because it is transcribed from a single cistron (gene) and has a single initiation codon and a single terminator codon.
- **Polycistronic mRNA** molecule contains the codons for more than one cistron which may lie close together. This type of mRNA synthesises more than one polypeptide chains. Therefore, a polycistronic mRNA molecule must possess a series of start and stop codons. If a single mRNA molecule encodes three proteins the minimal requirement would be the sequence:

Start codon, protein 1, stop codon – start codon, protein 2, stop codon,– start cod on, protein 3, stop codon

1st sequence | 2nd sequence | 3rd sequence

In such mRNA molecules, the leader sequence preceding the first start signal may be several hundred bases long and there is usually a sequence called a **spacer** of 5 to 20 bases between one stop codon and the next start codon. The structure of a tricistronic mRNA is shown in Fig. 10.6. The mRNA molecule for the metabolism of histidine codes for the synthesis of ten specific enzymes.

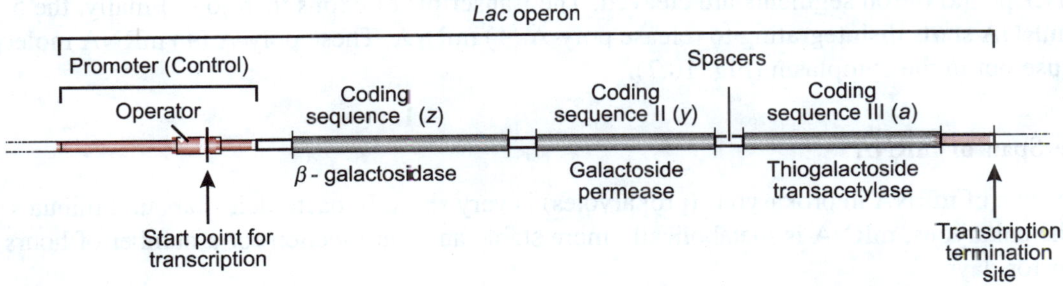

Fig. 10.6 Arrangement of control elements and coding sequences or cistrons in a typical polycistronic *lac* operon gene of *E. coli*

Biosynthesis of mRNA

Synthesis of mRNA is accomplished by using one of the two DNA strands. It is carried out from the 5′ end towards 3′ end on 3′ → 5′ polynucleotide strand. ***RNA polymerase*** attaches to the **initiation site** or **promoter end** of the structure gene of DNA cistron and catalyses RNA synthesis. The phenomenon of synthesis of mRNA from DNA is known as **transcription**.

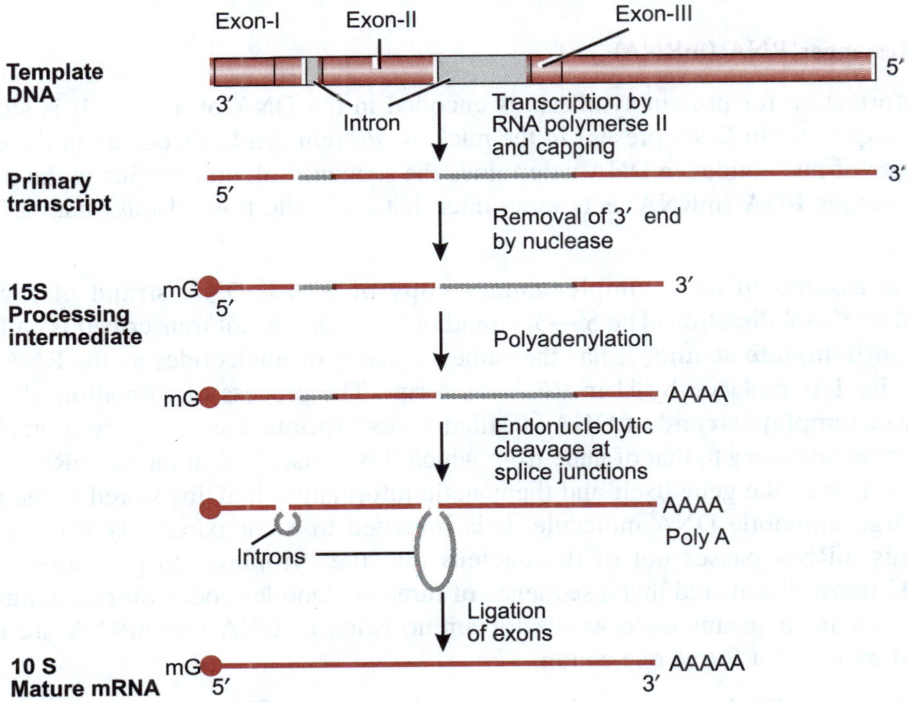

Fig. 10.7 Formation of mRNA from primary transcript in eukaryotic cells

In eukaryotes, mRNA is synthesised as heterogeneous nuclear RNA (hnRNA) inside the nucleus. These molecules are much bigger in size than the functional mRNA molecules because these are copies of both introns and exons. They are called **primary transcripts**. A sequence of about 210 nucleotides of polyadenylic acid (**poly-A**) is added to the 3′ end of each primary transcript and intron segments are cleaved. The transcripts of exons then join. Finally, the 5′ end of hnRNA starts disintegrating to release poly-A (+) mRNA. These poly-A (+) mRNA molecules diffuse out in the cytoplasm (Fig. 10.7).

Life Span of mRNA

Life span of mRNA in prokaryons (prokaryotes) is very short. In bacteria it is about 2 minutes. But in the eukaryons, mRNA is metabolically more stable and can function for a number of hours and even for days.

Informosomes

In eukaryons mRNA is associated with proteins, forming ribonucleoprotein complexes. Some of these complexes remain free in the cytoplasm without being attached to the polyribosomes. These were named **informosomes** by **Spirin**. They are very stable and can remain in the cell cytoplasm for several days. Their stability is attributed to the presence of protein sheath around each mRNA molecule. In informosomes, proteins and mRNA are found in the ratio of 4: 1. These occur in the form of particles of 80S or less.

Role of Messenger RNA (mRNA)

Genetic information for protein synthesis is encoded in the DNA of a gene. It is stored in the nucleotide sequences in DNA present in the nucleus. Protein synthesis occurs in the cytoplasm. The sequence of nucleotides in DNA determines the sequence of amino acids in the polypeptide chain. **Messenger RNA** (mRNA) acts as an intermediate in the flow of information from DNA to protein.

mRNA is assembled as a complementary copy of 3′→5′ DNA strand of a gene. It is synthesised in 5′→3′ direction. The 5′→3′ strand of DNA that is not transcribed is called **coding strand** or **antitemplate strand.** It has the same sequence of nucleotides as the RNA transcript except that its T is replaced by U in RNA transcript. The process of formation of RNA from **noncoding** or **template strand** of DNA is called **transcription**. The sequence of nucleotides in mRNA is complementary to that of gene from which it is transcribed. It means mRNA retains the same information as the gene itself and the genetic information remains stored in the nucleus as a part of large immobile DNA molecule. It is imparted to a comparatively small and mobile mRNA. This mRNA passes out of the nucleus into the cytoplasm to participate in protein synthesis. **Gamow** discovered that a sequence of three nucleotides codes for one amino acid in a polypeptide chain. It means code words for amino acids in DNA and mRNA are nucleotide triplets, and each triplet forms one codon.

One molecule of DNA serves as the template for many mRNA molecules, each of which serves as a template for the formation of a large number of polypeptide chains.

2. Ribosomal RNA (rRNA)

Occurrence: The ribosomal RNA (rRNA) constitutes bulk for the cellular RNA (over 80 per cent of the weight of total RNA of the cell). It occurs in ribosomes which are nucleoprotein molecules.

Types of rRNA

Inside the ribosome of eukaryotic cells, rRNA occurs in the form of particles of three different dimensions. These are designated 28S, 18S and 5.8S. The 28S and 5.8S molecules occur in large subunit (60S subunit) of ribosome, whereas 18S molecule is present in the small subunit (50S subunit) of ribosome. The 28S molecules of rRNA have a molecular weight $1.5 - 1.8 \times 10^6$ daltons, whereas 18S particles are only 0.7×10^6 daltons. In prokaryotic cells there are only 23S, 5S and 16S rRNAs with a molecular weight 1.2×10^6, 3.5×10^4 and 0.6×10^6 daltons respectively. Inside the ribosomes, rRNA remains folded as shown in Fig. 10.8.

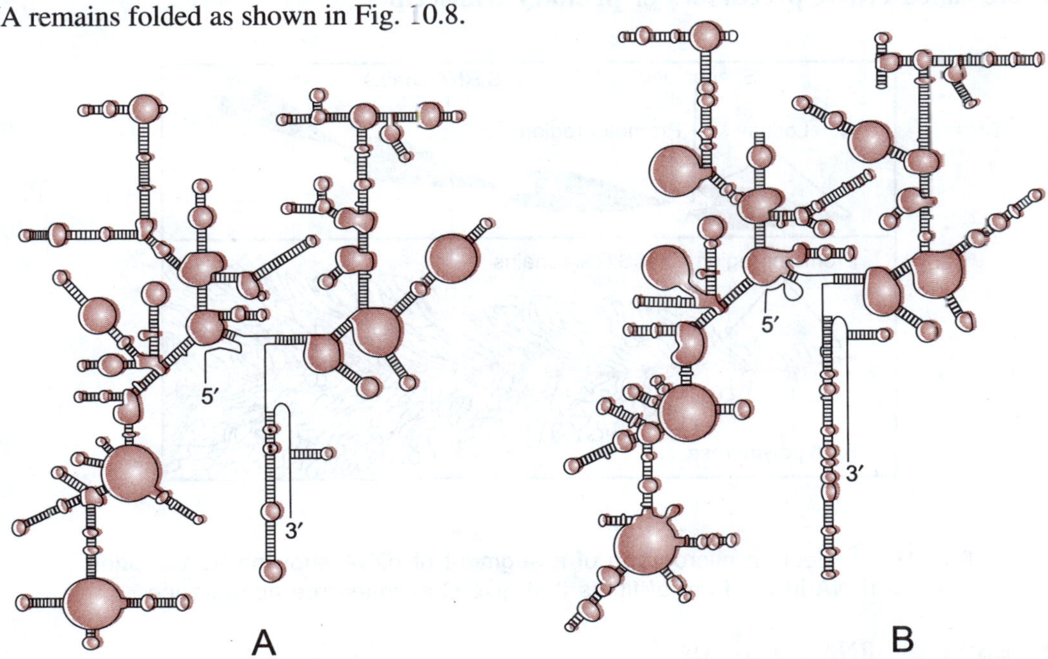

Fig. 10.8 Diagram to show folding in one molecule of rRNA **A.** in prokaryotes 16S rRNA of *E.coli*; **B.** in eukaryotes – 18S rRNA of yeast cell

Base Composition of rRNA: rRNA differs in base content from tRNA and mRNA. It is relatively rich in guanine and cytosine. The base components in rRNA of *E. coli* have a molar ratio of adenine 21: uracil 19: guanine 37: cytosine 23.

Biogenesis of Ribosomal RNA

Cells contain millions of ribosomes, and each ribosome contains several molecules of rRNA. The DNA sequences encoding rRNA are normally repeated hundreds of times showing **tandem arrangement**. This DNA is called **rDNA**. It is clustered in one or more **nucleoli**. Thus, nucleolus is a ribosome-producing organelle. The part of chromosome containing rDNA is called **nucleolar organiser**. In bacteria about 10-20 cistrons are associated with rRNA synthesis, but in eukaryotes 210–2,100 tightly clustered rDNA cistrons are involved in rRNA synthesis. The process of rRNA synthesis involves following steps:

1. Synthesis of rRNA Precursor

rRNA precursor is synthesised on rDNA cistrons as short fibrils with the help of ***RNA polymerase*** enzyme. Under electron microscope, these fibres appear as a chain of 'Chrismas tree', because about 100 or so fibrils show gradual increase in their length from one end of the 'Christmas tree trunk' to the other. The shorter fibrils are located closer to initiation point of transcription and the longer fibrils are closed to completion of transcription. The completed RNA fibrils are called **rRNA precursors** or **primary transcripts**.

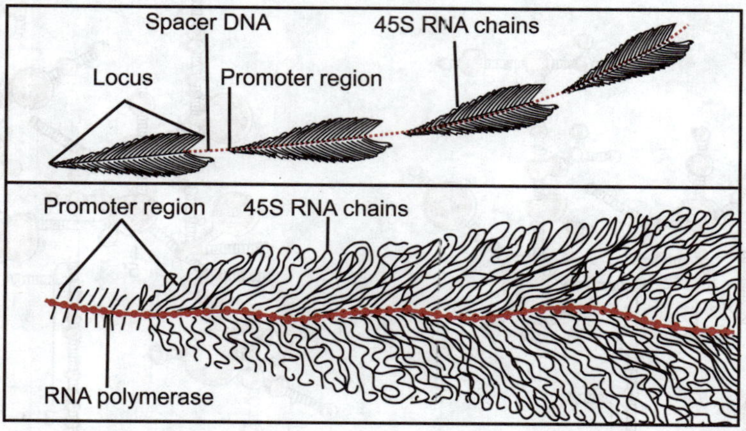

Fig. 10.9 Electron micrograph of a segment of rDNA showing transcription of rRNA in the form of fibrils that give Christmas tree appearance

2. Processing of rRNA Precursor

Of the four rRNA molecules, three (28S, 18S and 5.8S) are synthesised as a single primary transcript, also called the **pre-rRNA**. 5.8S rRNA is synthesised from separate RNA precursor outside the nucleolus. The primary transcript for rRNA is a 45S molecule of about 13 kilobases (i.e. consisting of 13,000 nucleotides). The processing of 45S rRNA occurs inside the nucleolus and when it is tightly associated with protein forming **ribonucleoprotein particles** (RNP). 45S rRNA undergoes four principal cleavage events as follows:

- **First cleavage** of primary transcript (45S RNA) occurs at site I and removes the 5′ terminal leader sequence and produces **41S intermediate**.
- **Second cleavage** occurs in 41S intermediate at site 2 or site 3. Cleavage at site 3 generates **32S intermediate**.

- **Final processing steps** involve separation of 32 intermediate into 28S and 5.8S sections of rRNA. The 28S and 5.8S with proteins form large ribosomal subunit.

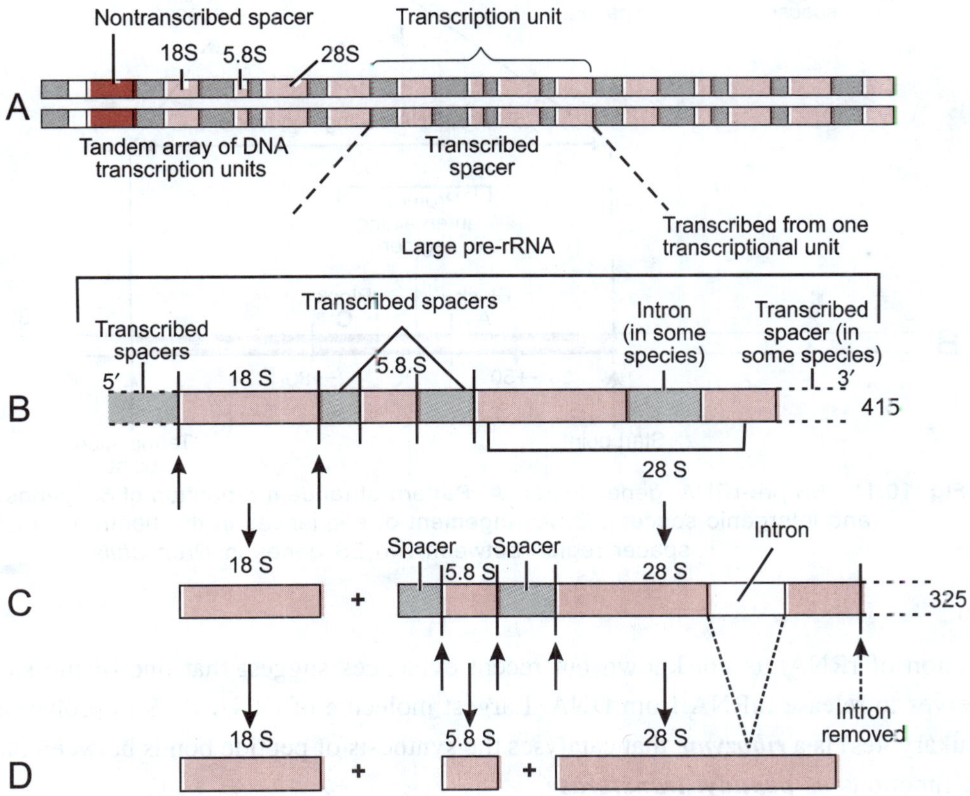

Fig. 10.10 Processing of ribosomal RNAs: **A.** The eukaryotic transcription unit that includes the genes for the three largest rRNAs occurring in multiple copies, arranged in tandem arrays. Non-transcribed spacers (black) separate the units. Each transcription unit includes the genes for the three rRNAs and four transcribed spacers; **B.** Single long primary transcript of large pre-rRNA with a sedimentation coefficient of about 45S; **C and D.** RNA processing yields mature 18S, 5.8S and 28S rRNA molecules. RNA cleavage actually occurs in a series of steps and position of cleavages is shown by small vertical arrows

3. Synthesis and Processing of 5S rRNA

5S rRNA is about 120 nucleotides long. It has extranucleolar origin. The rDNA for 5S rRNA is located outside the nucleolar organiser. It also exhibits tandem arrangement of nucleotides or genes. The process of transcription of 5S rRNA requires ***RNA polymerase-III*** enzyme and three factors namely TFIIIA, TFIIIB and TFIIIC. Factor TFIIIA is specifically required for initiation of rRNA.

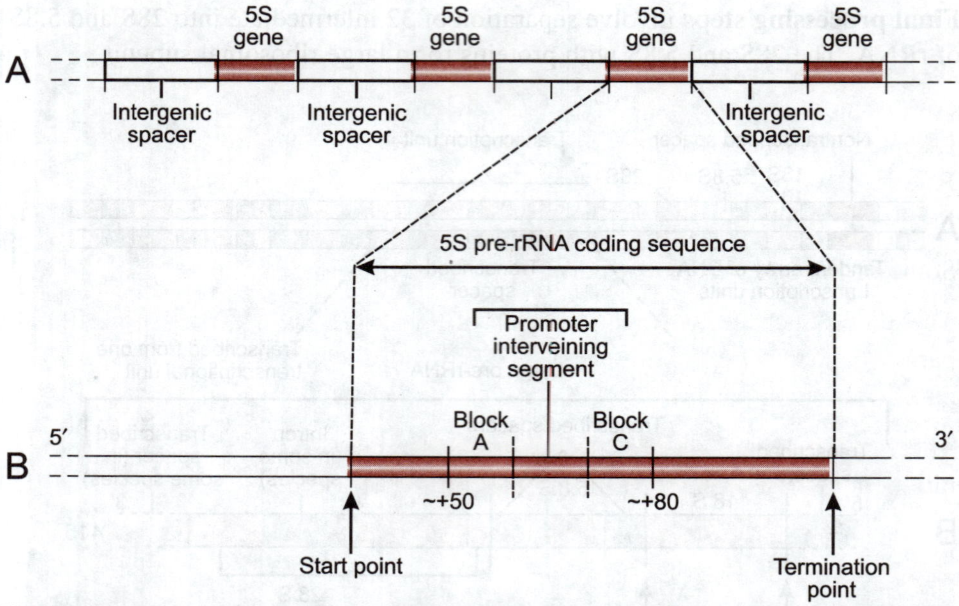

Fig. 10.11 5S pre-rRNA gene cluster: **A.** Pattern of tandem repetition of 5S genes and intergenic spacers; **B.** Arrangement of sequences in the nontranscribed spacer region between two 5S genes in *Drosophila*

Function

The function of rRNAs is not known but recent evidences suggest that one of the subunit of rRNA serves to release mRNA from DNA. Largest molecule of rRNA (23S in prokaryotes and 28S in eukaryotes) is a ***ribozyme*** that catalyses the synthesis of peptide bonds between the amino acids. It functions as ***peptidyl transferase***.

3. Transfer RNA (tRNA)

The transfer RNA is a family of about 60 small-sized ribonucleic acids which can recognise the codons on mRNA and exhibit high affinity for 21 activated amino acids, combine with them and carry them to the site of protein synthesis. tRNA molecules have been variously termed as **soluble RNA** or **supernatant RNA** or **adaptor RNA**. tRNA is about 10-15 per cent of total weight of RNA of the cell.

Characteristics of tRNA

tRNA molecules have following characteristics:

- tRNA molecules are smallest, containing 75 to 93 nucleotides.
- Each tRNA molecule has a **3′- OH terminus** and a **5′-monophosphate terminus** (not 5′– triphosphate as in other RNA molecules).

- Their sedimentation coefficient is 4S and molecular weight 24,000 to 31,000 daltons.
- Its polynucleotide chain undergoes secondary and tertiary foldings because of internal complementary base pairing. As a result, a tRNA molecule acquires a precise L-shaped **three-dimensional configuration,** resembling a twisted clover leaf. In 1965, **R.W. Holley** proposed a **Clover leaf model** of tRNA molecule. For this he got Nobel Prize with **Khorana** and **Nirenberg** in 1968.
- Some of the bases of the two arms are coiled over one another.
- The 3′ end of the polynucleotide chain of tRNA is called acceptor end. It ends in CCA base sequence. This represents site for the attachment of activated amino acid. The 5′ end of the chain terminates with guanine base.
- The band in chain of each tRNA molecule contains a definite sequence of three nitrogenous bases which constitute the **anticodon**. It recognises the codon on mRNA.

Clover-Leaf Model of tRNA

According to Clover leaf model four different regions or special sites can be recognised in the molecule of tRNA. These are:

1. Amino Acid Arm or Acceptor Arm (AA-arm) or Acceptor Stem: It is double helical and stem-like. It possesses both 5′ and 3′ ends of the molecule. It has seven base pairs and the unpaired 3′ terminal always has a base triplet CCA with – OH at the tip. The – COOH group of a specific amino acid joins with the – OH group of adenosine base of CCA in the presence of ATP, forming **amino acyl tRNA**. It is common to all the tRNA molecules. Therefore, 3′ end of acceptor arm is called **carrier end**.

2. Anticodon Arm: It is a loop-like structure consisting of 7 to 11 unpaired nucleotides. It lies opposite to the AA arm. Three of the 11 nitrogenous bases of these ribonucleotides are complementary to one of the triplet codon of mRNA molecule. Since the base triplet on mRNA chain is called **codon,** its complementary base triplet on tRNA molecule is termed **anticodon.** Anticodon reads its appropriate codon on mRNA and temporarily binds to it. Therefore, the terminal end of this arm is called **recognition end.**

3. DHU Loop (Dihydrouridine Loop): It is also a loop-like arm. It has enzyme recognition site that binds to specific amino acid activating enzyme which catalyses the union of a specific amino acid to tRNA molecule.

4. TψC loop: It is also a loop-like or hairpin arm of the tRNA with a site for attachment to a ribosome. This site is common to all the molecules of tRNA. It is formed of five base pairs and a terminal loop with seven nitrogenous bases.

5. Variable Arm: It contains from 4 to 21 nucleotides.

6. The TψC loop, which is thought to interact with ribosomal RNA, is at the corner of the L.

7. All bases except five (16, 17, 20, 47, and 76) are stacked, even though half of the bases are not in double-stranded helical regions. Even the bases of the single-stranded 3′-terminal segment (except for the terminal adenine) are stacked.

8. There are nine base pairs that are not shown in the clover leaf model. These are called **tertiary base pairs** because they are responsible for the three-dimensional folding. Only one of these, a G·C pair is a standard Watson-Crick pair.

9. In addition to the tertiary base pairs described in item 6, the tertiary structure is stabilized by hydrogen bonds between bases and different ribose units and between ribose units in separate nucleotides.

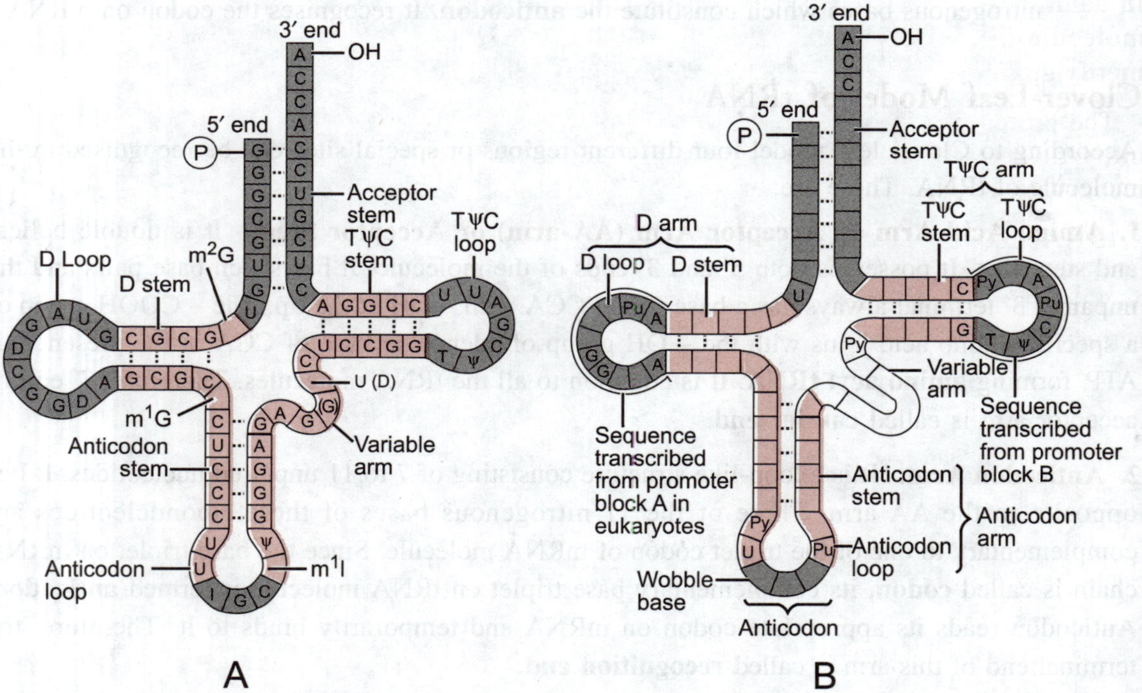

Fig. 10.12 **A.** Clover leaf model of a eukaryotic alanine tRNA molecule from yeast as determined by Holley. (The modified nitrogenous bases present in tRNA are 9-inosine ϕ-pseudouridine, T-ribothymidine, MeI–Methylinosine, MeG-1-methylguanosine, DiMeG-N^2-dimethyl guanosine, DiHu-5, 6-dihydrouridine. The usual bases are adenine (A), uridine (U), cytosine (C) and guanine (G)); **B.** tRNA structure; each tRNA molecule possesses four special sites represented in the figure. (i) Acceptor arm of tRNA represents site for the attachment of amino acid; (ii) D-loop represents specific site for the recognition of amino acid; (iii) anticodon loop bears codon recognition site (anticodon) and (iv) TψC bears ribosome recognition site

Extension: Structural similarity in tRNA molecules

Two observations make it likely that all tRNA molecules are structurally similar:

1. Most of the tertiary base pairs are formed from bases that are conserved in all tRNA molecules.
2. Different tRNA molecules can co-crystallize, forming uniform crystals.

A low-resolution X-ray crystallographic analysis of these mixed crystals does not indicate that two types of molecules are present and yield molecular dimensions that are the same as those of a single kind of molecules. This suggests that all tRNA molecules are structurally similar.

Unusual Bases in tRNA

In addition to the usual bases of RNA (cytosine, guanine, adenine and uracil), each tRNA molecule has several unusual bases. Some of them are **pseudouridine, inosinic acid, methylguanine, methyl aminopurine**, etc.

The presence of these rare nucleotides (unusual) does not affect the pairing of tRNA with mRNA. These probably prevent intramolecular base pairing in the open tRNA loop or help in the recognition of ***aminoacyl tRNA synthetase enzyme***.

tRNA molecules occur in both active and inactive forms. The inactive molecules of tRNA lack the C–C–A sequence of nitrogenous bases at 3′ end of the chain either in full or in part. By the addition of this nucleotide sequence an inactive tRNA molecule is activated. The process is governed by cytidine triphosphate **(CTP)** and adenine triphosphate **(ATP)**.

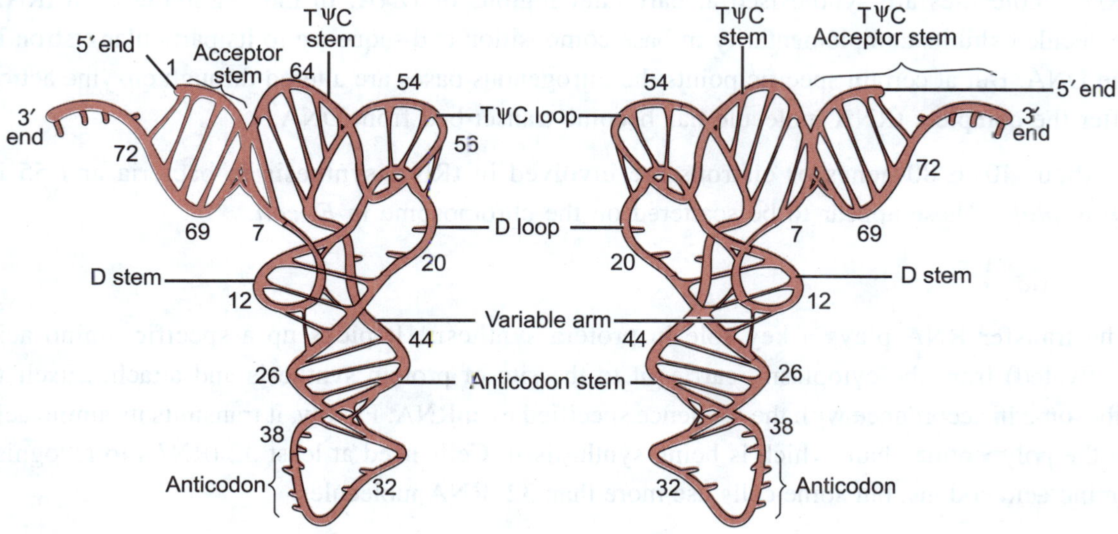

Fig. 10.13 Three-dimensional structure of yeast tRNA in opposite side views as deduced from X-ray diffraction

Table 10.4 Differences between RNA types

Features	Ribosomal RNA (rRNA)	Messenger RNA (mRNA)	Transfer RNA (tRNA)
1. Percentage of cell's Total RNA	About 80 per cent	About 5 per cent	About 15 per cent
2. Length of Molecule	Variable	Longest	Shortest
3. Shape of Molecule	Greatly coiled	Linear	Clover leaf-like, folded into L-shape
4. Types	Six	Numerous	About 60
5. Life Span	Long, used again and again in translation	Very short, 2 minutes to 4 hours, degraded after translation	Long, used again and again in translation
6. Function	Forms part of ribosomes and helps in the attachment of mRNA to ribosome during protein synthesis	It carries coded instructions from DNA to cytoplasm for protein synthesis	Acts as amino acid carrier; transports activated amino acids from cytoplasm to mRNA on ribosome

Transcription and Processing of tRNA

tRNA molecules are synthesised at particular regions of DNA. In the beginning each tRNA molecule exhibits complementarity in base composition and sequence to its particular cistron in the DNA. But at certain specific points the nitrogenous bases are altered through enzyme action after the complete tRNA molecule has become transcribed from DNA.

About 40 to 80 genes or cistrons are involved in tRNA synthesis in bacteria and 55 in *Drosophila*. These appear to be scattered on the chromosome in *E. coli*.

Function

The transfer RNA plays a key role in protein synthesis. It picks up a specific amino acid (activated) from the cytoplasm, carries it to the site of protein synthesis and attaches itself to ribosome in accordance with the sequence specified by mRNA. Finally, it transmits its amino acid to the polypeptide chain which is being synthesised. Cells need at least 32 tRNAs to recognise amino acid codons, but some cells use more than 32 tRNA molecules.

Extension: Ribozymes

American biochemist, **Thomas Cech**, and his colleagues in 1983 discovered that certain pieces of RNA acted as enzymes or biocatalysts to cut selected pieces out of strands of viral RNA. These were named **ribozymes. Thomas Cech** got Nobel Prize for Chemistry for this discovery in 1989. Like enzymes, ribozymes are not used up during reaction. Therefore, they can continue to snip up one virus after the next until all the viruses in the culture are destroyed.

Functionally, there are two types of ribozymes:

1. Ribozymes that carry out **intra-molecular catalysis** and catalyse reactions on themselves. They have folded structure.
2. Ribozymes that carry out **inter-molecular catalysis** and act on other molecules without themselves being changed.

Many important cellular reactions are catalysed by ribozymes:

1. The maturation of hnRNAs (heterogeneous nuclear RNA).
2. The chipping out unnecessary sections from primary RNA transcripts of ribosomal RNA synthesising genes.
3. 23S and 28S rRNAs of ribosomes catalyse the formation of peptide bonds during synthesis of polypeptide chains in prokaryotic and eukaryotic cells. Catalytic role of ribosomal RNA was established by **Cech** in early 1980s while studying ***Tetrahymena*** ribosome and by **Sidney Altman** from the study of RNase P. The final support for catalytic role of rRNA came in 1992 from **Harry Noller** and his colleagues.

Scientists are working to use ribozymes therapeutically.

Other Types of RNAs

In addition to these three types of RNAs, following types of RNAs are also found:

1. Small Nuclear RNAs (snRNAs): These are small RNA molecules formed of 100 to 215 nucleotides. These are found in the nucleus of eukaryotic cells. These are important for DNA replication and processing of long, newly transcribed heterogeneous nuclear RNA molecules.

2. Genetic RNAs: In some viruses, RNA is the genetic material. It carries the hereditary blue-print. In some viruses hereditary RNA is single-stranded but in others it is double-stranded.

3. Enzymatic RNAs or Ribozymes: Thomas Cech and **Sidney Altman** (1983 and 1985) independently discovered that certain RNAs act as biological catalysts. These are called **ribozymes.** These are associated with the maturation of **hnRNAs** (heterogeneous nuclear RNA). Even 23S and 28S ribosomal RNAs of ribosomes are found to catalyse the formation of peptide bonds during synthesis of polypeptide chains in prokaryotic and eukaryotic cells.

Functions of Different types of RNAs

1. RNAs are essential for protein synthesis.
   - mRNA receives genetic information from DNA by transcription, comes out of the

nucleus into the cytoplasm and translates the coded information into a protein with a determined sequence of amino acids.

- tRNA picks up amino acid from cytosol of the cell and brings it to mRNA template at a specific position for polymerisation.
- rRNAs with proteins form structural component of ribosomes and provide surface for attachment for mRNA and tRNA.

2. In TMV, HIV and influenza virus, RNA acts as genetic material.
3. **Small nuclear RNAs (snRNAs)** are formed during DNA replication and are essential for DNA replication.
4. **Ribozymes** act as biological catalysts. These are needed during maturation of heterogeneous nuclear RNAs into functional mRNA molecules. Some ribozymes control peptide formation during synthesis of polypeptide chain.

TEXT QUESTIONS

1. What is RNA? Discuss its structure. How does it differ from DNA?
2. Give a brief account of different types of RNAs found in the cell. What role do these RNAs play in the process of protein synthesis?
3. Name those organisms in which RNA is the hereditary material. How it can be proved that RNA carries genetic information in these organisms?
4. Describe structure of mRNA. Write a note on its biogenesis.
5. Discribe replication of genetic material in retroviruses.
6. Describe clover leaf model of tRNA.
7. Why newly formed RNAs (primary transcripts) have to undergo processing before becoming functional? Discuss the process of its maturation.
8. Discuss the role of mRNA. Write a note on informosomes.
9. Discribe the structure and role of different types of ribosomal RNAs.
10. What is the difference between polycistronic and monocistronic mRNA?
11. Give one example of each of the following:
    (a) RNA is hereditary material (b) DNA is single-stranded
    (c) DNA is nongenetic.
12. Write short notes on:
    (a) RNP particles (b) SnRNA
    (c) Ribozymes (d) RNA processing
    (e) Informosomes (f) Clover leaf model.

13. What are ribozymes? Who discovered them? Describe their role in the cells.
14. Give differences between different types of RNAs found in the cell.
15. Why mRNA is called heterogeneous RNA?
16. Where will you find the following:
    (a) 5′ cap
    (b) Abnormal base pairs
    (c) Anticodon loop
    (d) Polycistronic mRNA
    (e) Reverse transcriptase.
17. Describe differences in the structure of prokaryotic mRNA and eukoryotic mRNA.
18. Give one important function of each of the following:
    (a) Leader segment
    (b) Anticodon
    (c) Amino acid arm
    (d) TψC arm
    (e) 45S RNA
    (f) Polymerase III
    (g) Replicase
    (h) Methylated G cap
    (i) RNA polymerase.
19. What is the difference between coding and noncoding strand of DNA? Which of the two acts as a template strand for the transcription of RNA?
20. Who described the catalytic role of RNA?

11

Chapter

Transcription and Processing of RNA

IN THIS CHAPTER

Transcription; Site and Time of Occurrence; Sense and Antisense Strands; Mechanism of RNA Transcription; Material Required; RNA Polymerase in Prokaryotes; Transcription Unit; Promoter Site; Functional or Consensus Sequences in Promoter Region; Molecular Mechanism of Transcription in Prokaryotic Cells; Molecular Mechanism of Transcription in Eukaryotic Cells; Eukaryotic Polymerases; Functional Sequences in Promoter Region in Eukaryotes; RNA-directed RNA Synthesis; Post-transcriptional Modification of RNA or RNA Processing; Processing of tRNA and rRNA; Processing of mRNA; Spliceosomes; RNA Synthesis *in vitro*; Comparison between Replication and Transcription.

TRANSCRIPTION

DNA contains information for the synthesis of cell's specific proteins. DNA is located in the nucleoid (in prokaryotes) or nucleus (in eukaryotes) and protein synthesis occurs in the cytoplasm. DNA does not move to the site of protein synthesis (ribosomes) to directly guide the process. Instead, it transfers its information to single-stranded mRNA molecules which move to the ribosomes to direct protein synthesis. The process of the formation of all the three families of RNA, *i.e.* mRNA, tRNA and rRNA from DNA templates is called **transcription** (meaning: written across). It involves rewriting the genetic message coded in DNA into RNA molecule. The transcribed mRNA moves out of the nucleus to the ribosomes in the cytoplasm and directs protein synthesis.

Site and Time of Occurrence

Transcription occurs in the nucleus during the G_1 and G_2 phases of the interphase. RNA synthesis does not occur when cell is dividing. DNA has promoter and terminator sites. Transcription starts at the initiation point close to the promoter site and stops at the terminator site. Enzyme ***RNA polymerase*** recognises the promoter sequence of the gene and binds to it to initiate transcription.

During transcription only one of the two strands of DNA duplex is transcribed. As in DNA, enzyme polymerase can add nucleotides to the 3′–OH end of the polynucleotide chain, RNA chain is also synthesised in 5′→3′ direction. It is transcribed on 3′→5′ DNA strand. Thus 3′→5′ DNA strand (from which RNA is transcribed) is called **template strand** or **antisense strand** or **noncoding strand.** Its complementary strand or 5′→3′ strand is called **antitemplate strand** or **sense strand** or **coding strand** because it has the same sequence of nitrogenous bases as the RNA transcript (except T of coding strand is substituted for U) and it also has the same polarity.

> **Extension: Sense and Antisense Strands**
>
> - The transcribed RNA has 5′→3′ polarity and is called sense RNA. It is transcribed in 5′→3′ direction on the 3′→5′ strand of DNA, which is called **template strand** because it acts as a template for RNA synthesis.
> - 3′→5′ strand or template strand is also called **antisense strand** because its polarity is just opposite to RNA strand and its nucleotide sequence is complementary to sense RNA.
> - The 5′→3′ strand or complementary strand of DNA is called **sense strand** because the sequence of nitrogenous bases in this strand is similar to that of RNA (except T is substituted by U).
> - Since RNA is the coding instrument for protein, the nontemplate or sense strand is also called coding strand and template strand as **noncoding** or **anticoding strand.**

MECHANISM OF RNA TRANSCRIPTION IN PROKARYOTES

The mechanism of RNA synthesis was worked out in late 1950s by the American investigators, **Jerard Horowitz, Samuel B. Weiss** and **Audery Stevens** in *in vitro* experiments. The existence of mRNA was first shown by **Spiegelman** and coworkers and ***RNA polymerase*** activity was discovered by **Weiss** in 1960.

Material Required

RNA transcription requires:

1. The enzyme ***RNA polymerase*** or ***DNA directed RNA polymerase***.
2. DNA template—the transcription unit.
3. All the four types of ribonucleoside triphosphates (ATP, CTP, GTP and UTP).
4. Divalent metal ions Mg^{2+} or Mn^{2+} as a cofactor.

No primer is needed for RNA synthesis.

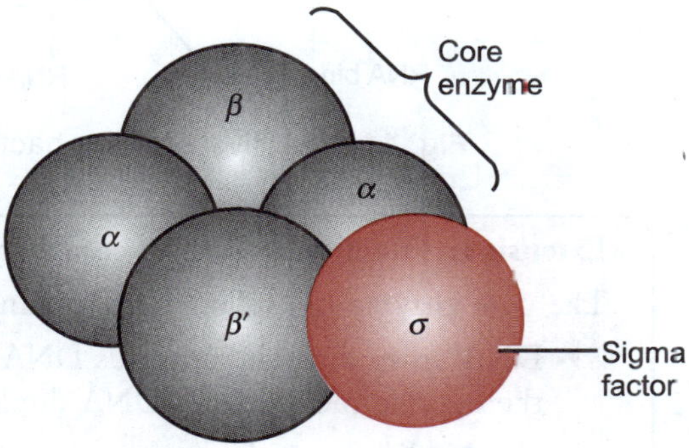

Fig. 11.1 RNA polymerase enzyme of bacterium *E. coli* showing association of its 5 polypeptides ααββ′ and σ

RNA Polymerase in Prokaryotes

Bacterial ***RNA polymerase*** is a complex **holoenzyme** that catalyses transcription of RNA. In prokaryotes, single ***RNA polymerase*** **(RNAP)** enzyme carries out synthesis of all the three types of RNAs. But in eukaryotes three distinct RNA polymerase enzymes, namely ***RNA polymerase***

I, RNA polymerase II and ***RNA polymerase III*** catalyse synthesis of rRNA, mRNA and tRNA respectively. In addition, mitochondria and chloroplasts have their own RNA polymerases, which resemble bacterial enzyme.

In prokaryotes single RNA polymerase catalyses synthesis of all the three types of RNAs. In *E. coli*, ***RNA polymerase*** enzyme is a large protein with a molecular weight of 4,80,000. It is a holoenzyme and consists of **core enzyme** and **sigma subunit**.

1. Core Enzyme

Core enzyme is formed of four tightly associated protein chains namely ββ′αα with molecular weights of 1,60,000, 1,50,000, 90,000, and 4,000 respectively. It binds non-specifically to DNA template and migrates downstream till its sigma factor recognises the promoter region.

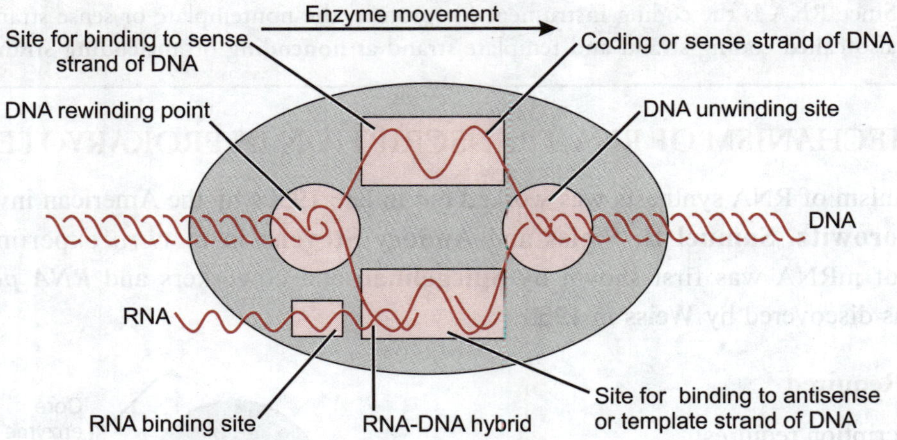

Fig. 11.2 Active sites on bacterial core enzyme of RNA polymerase

> **Extension: Functional sites on core enzyme of RNA polymerase**
>
> The core enzyme has following four functional sites:
>
> - DNA unwinding site unwinds DNA duplex and separates two nucleotide strands as the enzyme moves along DNA duplex to be transcribed.
> - Site binding to antisense strand or template strand of DNA molecule.
> - Site binding to the sense strand or coding strand of DNA molecule.
> - DNA rewinding site binds to site where separated sense and antisense strands of DNA start rewinding after the transcription in that segment of DNA is complete.

1. The α Subunits: The core enzyme has two molecules of α-polypeptide. This is coded by gene ***rpoA***. The α subunits are needed for the assembly of core enzyme and probably help in the recognition of promoter.

2. The β Subunit: Only one copy of β subunit is present in one molecule of core enzyme of RNA polymerase. It is encoded by gene ***rpoB***. The β subunit binds with the incoming nucleotide to be added to RNA chain and helps in the formation of the first phosphodiester bond.

3. The β′ Subunit: It is encoded by ***rpoC*** gene and binds with the template strand or antisense strand of DNA. Therefore, both β and β′ are catalytic subunits of core enzyme.

2. Sigma Subunit (σ)

Sigma subunit is formed of a single polypeptide chain, loosely attached to the core enzyme. It is encoded by the gene ***rpoD.*** It recognises the start signal on DNA molecule and directs the core enzyme of ***RNA polymerase*** to bind to the promoter region upstream of initiation codon. The σ subunit recognises two special sequences of bases in the promoter region of the coding strand (i.e. nontemplate strand) of DNA. These are **–10 sequence** and **–35 sequence**.

Therefore, ability of holoenzyme ***RNA polymerase*** to recognise promoter sequence is due to σ subunit, which also facilitates opening or melting of DNA helix. The σ subunit separates from core enzyme once about 10 nucleotides are joined to initiate RNA transcription.

The core enzyme without sigma factor binds nonspecifically to any region of DNA and can transcribe DNA there, because it fails to recognise specific promoters on DNA template. It is also unable to discriminate between the two strands of DNA helix to be used as a template. **Sigma factor** recognises specific transcription initiation sites where RNA synthesis can begin. These sites are called **promoter sites.** Thus, there are different sigma factors, one each for every promoter site.

Functions of RNA Polymerase

1. ***RNA polymerase*** unwinds about 15 bases of DNA around the initiation site to form an open promoter – DNAcomplex and provides single-strand of DNA to act as template for transcription.
2. ***RNA polymerase*** basically catalyses the formation of phosphodiester bonds between successive nucleotides of a polynucleotide chain during synthesis of both DNA and RNA. With sigma factor, it contributes to the synthesis of RNA, because sigma factor recognises promoter site.

RNA polymerase lacks proof reading 3′→5′ exonuclease activity. Therefore, one error for every 10^4 to 10^5 ribonucleotides incorporated is introduced during RNA transcription. But the mistake during RNA transcription is not serious because of its high turnover and Wobble pairing during translation.

Transcription Unit

The transcription unit is a stretch of nucleotides in DNA that is transcribed into a single functional RNA molecule. It is called an **operon** in prokaryotes and a **gene** in eukaryotes. A typical transcription unit has following essential parts:

- A promoter region
- A start point or initiation sequence
- A coding segment
- A terminator sequence.

RNA polymerase binds to the promoter. Transcription begins at the start point, progresses along the length of coding sequence and terminates at the terminator sequence or at termination codon. The transcription of mRNA starts in 5′→3′ direction on 3′→5′ strand of DNA.

Promoter Site or Promoter Region and Functional or Consensus Sequences in Promoter Region in Prokaryotes

Promoter site is a segment of regulatory DNA located just preceding the initiation codon of a cistron or gene, beyond the 5′end of coding DNA. Enzyme ***RNA polymerase*** binds to this site for the initiation of transcription. This stretch of DNA in front of structure gene is also called **upstream region** and the sequence of bases in this region is referred as **upstream sequence**. It is a highly variable site having specific sequence of 20–200 base pairs. In *E. coli*, it consists of about 70 base pairs located before the transcription start site.The promoter region in prokaryotes has two or three important **functional sequences.** These are called **consensus sequences,** i.e. these sequences from different systems or from promoters of different cistrons differ only by one or two bases. These functional sequences are:

1. –10 Sequence or Pribnow Box

It is a consensus sequence of 6–10 bases upstream to the initiation site for mRNA. In *E. coli*, it occupies – 10 region of promoter. All sequences in Pribnow boxes of different genes are variants of **5′....TATAAT.... 3′**. In this segment at the position 6 **'T'** is present in all promoters and is called **'conserved T'**. This sequence of bases in – 10 region of promoter is named **Pribnow box** after its discoverer **David Pribnow**.

Pribnow box sequence orients RNA polymerase on promoter and helps it in locating the precise bases at which transcription will begin.

> **Extension: Consensus and Conserved Sequence**
>
> A **consensus sequence** is the base sequence in a DNA strand that is present in all promoters or in other regions of DNA which perform the same function but with some variation. It means consensus sequence from different sources are similar in structure and function but not identical. They vary from one group of organisms to another.
>
> In contrast, a **conserved sequence** remains unchanged in all DNA sequences having the same function.

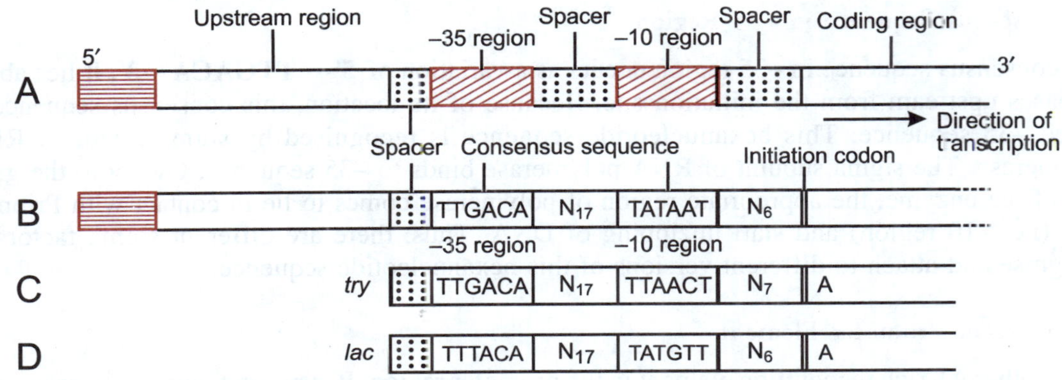

Fig. 11.3 Base sequence of promoter site in coding or nontemplate DNA of *E. coli*. **A.** Different regions of a promoter site; **B.** Consensus sequences in promoter region; **C.** Promoter region of tryptophan gene; **D.** Promoter region of *lac* gene

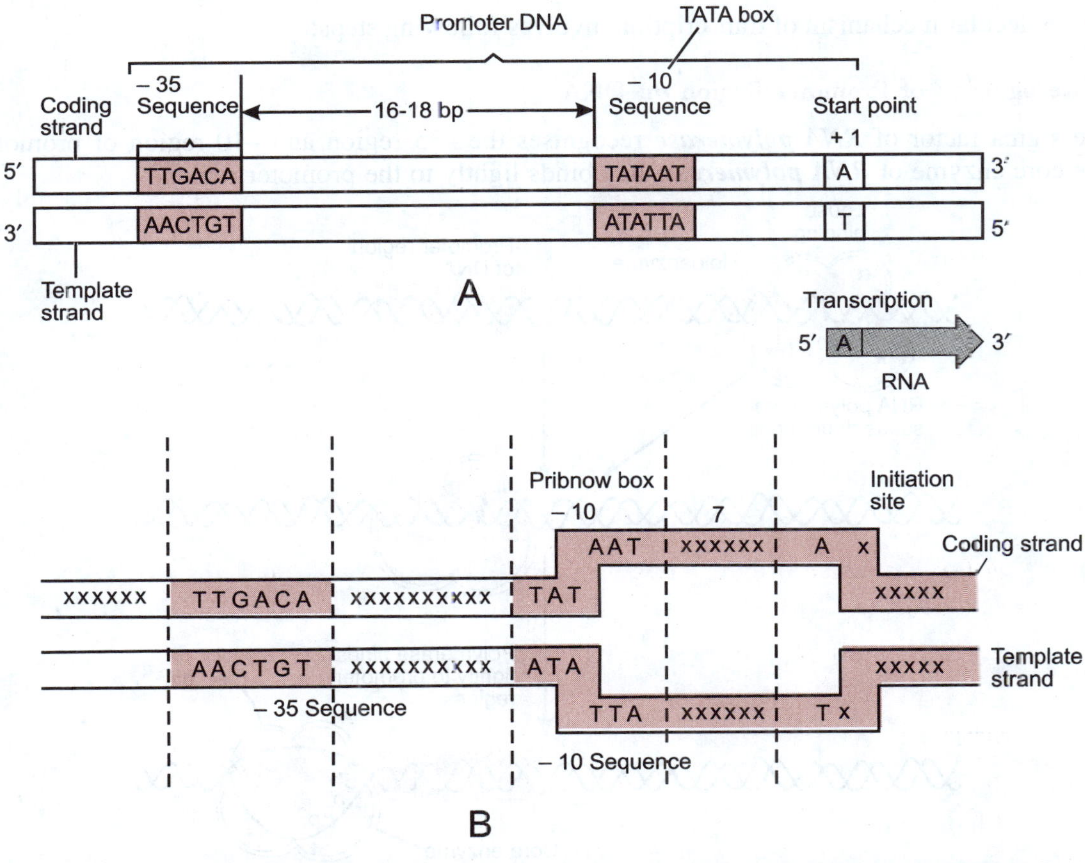

Fig. 11.4 **A.** Promoter region in prokaryote, *E. coli;* **B.** Base sequence of promoter site in DNA of *E. coli*. The key regulatory sequence for the initiation of transcription are found in this region, located at –35 and –10 base pairs upstream from the initiation site

2. Consensus Sequence in –35 Region

The consensus sequence in –35 region occurs as a variation of **5′··· TTGACA···3′.** It lies about 35 bases upstream from the initiation site. Because of its location, this consensus sequence is called –35 sequence. This hexanucleotide sequence is recognised by sigma factor of RNA polymerase. The sigma subunit of RNA polymerase binds to –35 sequence. Owing to the great size of the enzyme, the appropriate region of polymerase comes to lie in contact with Pribnow box, (i.e. –10 region) and start unzipping of DNA. Thus, there are different sigma factors to recognise and attach to different versions of this hexanucleotide sequence.

3. Upstream Promoter Element

It is another AT-rich recognition element in the promoter region. It is present only in the promoters of certain highly expressed genes and is located between – 40 and – 60 base sequences. This upstream promoter sequence binds to the α subunit of RNA polymerase.

Molecular Mechanism of Transcription in Prokaryotic Cells

The molecular mechanism of transcription involves following steps:

1. Recognition of Promoter Region on DNA

The sigma factor of ***RNA polymerase*** recognises the –35 region and –10 region of promoter. The core enzyme of ***RNA polymerase*** then binds lightly to the promoter region.

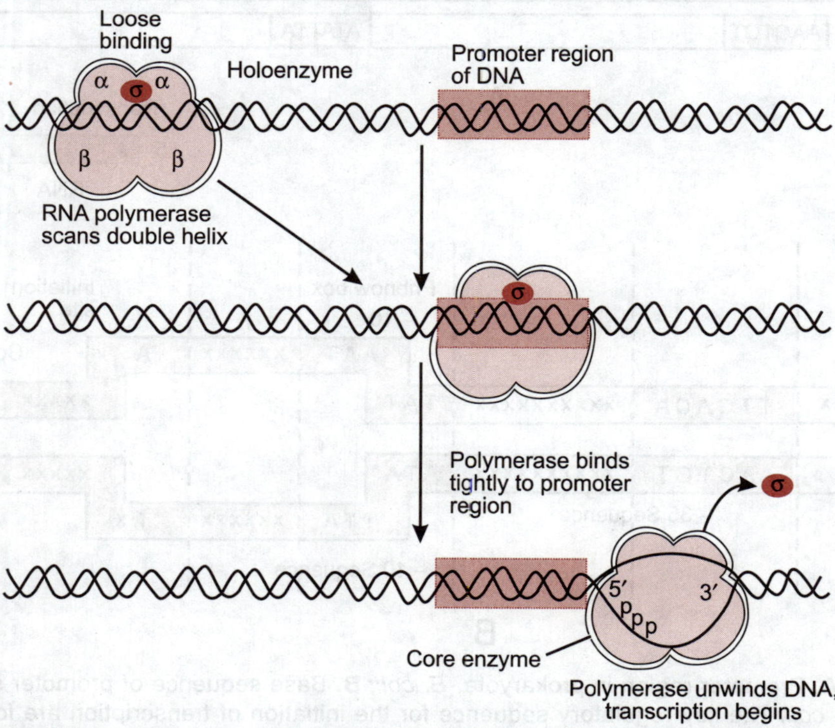

Fig. 11.5 Recognition of promoter region on DNA by RNA polymerase and unwinding of DNA helix

2. Binding of RNA Polymerase to DNA Duplex

The core enzyme ***RNA polymerase*** binds to a specific **promoter** in the DNA double helix. This site is located on the 5′ end upstream of the initiation site of the gene to be transcribed. It intiates separation of DNA strands and thus determines where to start RNA synthesis. The promoter also determines which DNA strand is to be transcribed.

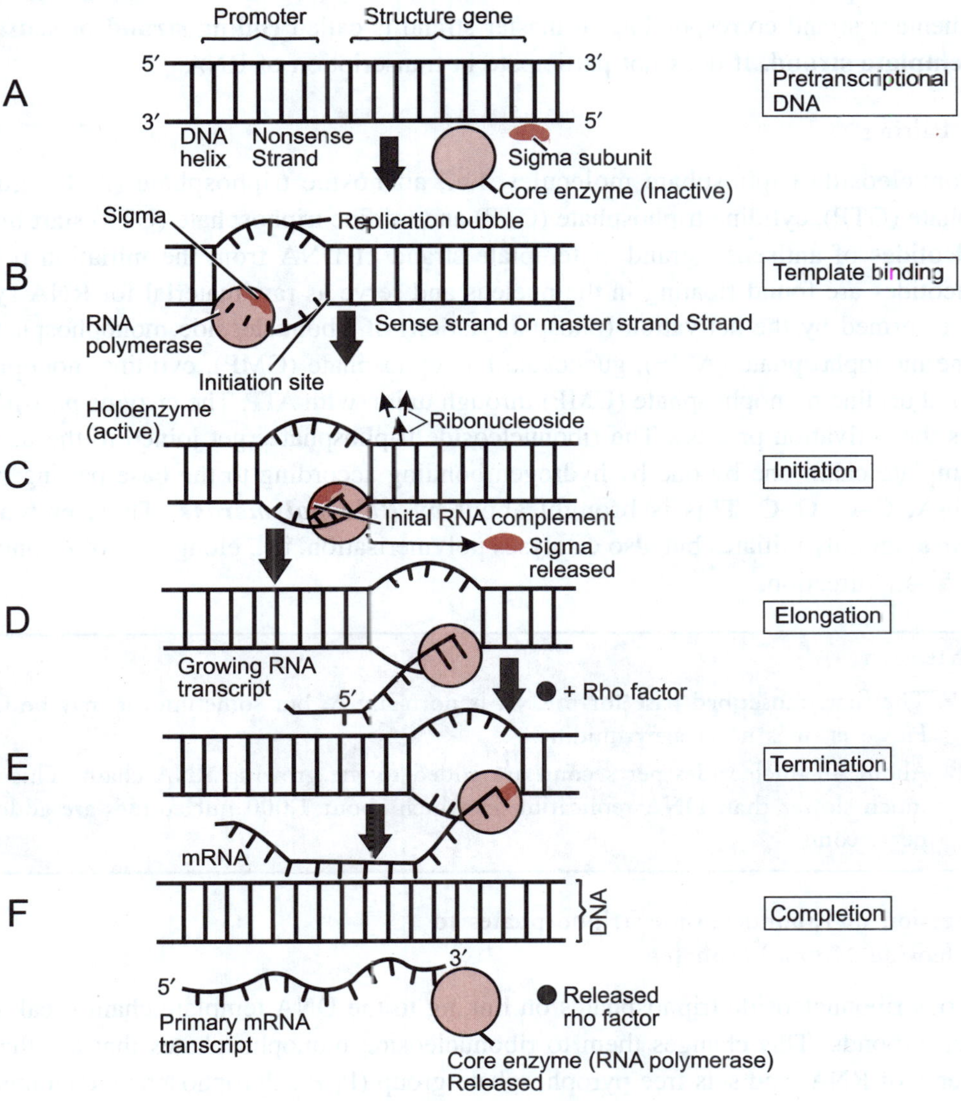

Fig. 11.6 Different steps in transcription

3. Exposure of DNA Bases and Initiation of Polynucleotide Chain

RNA polymerase initiates local unwinding and separation of two strands of DNA. The separation begins from about the middle of Pribnow box which is 6 – 10 base long segment of DNA prior to the initiation point of RNA synthesis. Approximately 10 – 15 base pairs are separated, leading to the formation of **transcription bubble.**

The separation of DNA strands exposes the bases of the DNA for the initiation of synthesis of polynucleotide chain of mRNA. Out of the two DNA strands only one, i.e., 3′→5′ strand functions as template. It is called **master strand** or **nonsense strand** or **template strand.** The complementary strand corresponding to master strand is called **coding strand** or **sense strand** or **nontemplate strand.** It does not participate in transcription of RNA.

4. Base Pairing

The ribonucleoside triphosphate molecules, i.e., adenosine triphosphate (ATP), guanosine triphosphate (GTP), cytidine triphosphate (CTP) and uridine triphosphate (UTP) start binding to the nucleotides of antisense strand or template strand of DNA from the initiation point. The ribonucleotides are found floating in the nucleus and serve as raw material for RNA synthesis. These are formed by the activation (phosphorylation) of ribonucleoside monophosphates, *viz.*, adenosine monophosphate (AMP), guanosine monophosphate (GMP), cytidine monophosphate (CMP) and uridine monophosphate (UMP) through union with ATP. The enzyme ***phosphorylase*** catalyses the activation process. The ribonucleoside triphosphates get joined to the base of the DNA template chain one by one by hydrogen bonding according to the base pairing rule, i.e., A–U, U–A, C–G, G–C. This is brought about by ***RNA polymerase***. Thus, enzyme ***RNA polymerase*** not only initiates but also catalyses polymerisation, i.e., elongation of ribonucleotide chain in 5′→3′ direction.

Extension

- The first transcribed base for mRNA is normally A, but sometimes it may be G. However, it is never a pyrimidine.
- About 40 nucleotides per second are added to the growing RNA chain. This is much slower than DNA replication in which about 1,000 nucleotides are added per second.

5. Conversion of Ribonucleoside Triphosphates to Ribonucleoside Monophosphates

The various ribonucleoside triphosphates on linking to the DNA template chain break off their high-energy bonds. This changes them to ribonucleoside monophosphates that are the normal components of RNA, and sets free pyrophosphate group (P~P). Pyrophosphate contains a high-energy bond (~). It undergoes hydrolysis by the enzyme ***pyrophosphatase.*** This releases energy and sets free inorganic phosphate P_i. The first ribonucleotide triphosphate retains all the three phosphates and is, thus, chemically distinct from the other nucleotides added after it.

$$P \sim P + H_2O \xrightarrow{\text{Pyrophosphatase}} 2\,P_i + \text{Energy}$$

6. Elongation of RNA Polynucleotide Chain

With the energy so released, each ribonucleoside monophosphate joined to DNA template chain then joins the ribonucleotide arrived earlier, making the RNA chain longer. The enzyme ***RNA polymerase*** catalyses the formation of **phosphodiester bonds** between successive nucleotides and requires a divalent ion Mg^{++} and Mn^{++}.

When a few nucleotides (about 10) are added, the sigma factor is released from the enzyme ***RNA polymerase.*** About 10–12 nucleotides of the growing RNA chain are bonded to template DNA at any one time. As the polynucleotide chain grows beyond 12 nucleotides, it begins peeling off DNA template.

mRNA synthesis on DNA template takes place in 5′→3′ direction so that the successive nucleotides are attached at 3′ end of growing RNA strand. Core enzyme moves along DNA, unwinds the two strands exposing the nucleotides of template DNA. As a result complementary copy of RNA is transcribed on 3′→5′ strand of DNA. The transcribed RNA molecule is called **RNA transcript** or **primary RNA transcript**.

7. Termination of Transcription and Separation of RNA Chain

As transcription proceeds, the hybrid DNA-RNA molecule dissociates, partly freeing the RNA molecule under synthesis. When the polymerase core enzyme reaches a **terminator sequence** on the DNA, it leaves the DNA. The terminator sequence lies in the template strand and consists of two inverted repeats separated by about six bases and followed by a sequence of A′s. The string of A′s in template gives rise to a sequence of U′s at the 3′ end of mRNA. It is the terminator sequence in mRNA.

The inverted repeats in mRNA are complementary, and can pair to generate a 'hairpin' structure. A long RNA molecule can have many hairpin structures. They provide an opportunity of termination but termination occurs at any position within the sequence of U′s. The fully formed RNA chain is now released and polymerase moves to start transcription of new RNA transcript. One gene forms several molecules of RNA, which are released one after the other on completion from the DNA template. In *E. coli,* two methods of termination have been identified:

1. Without termination factor, and
2. With termination factor (rho).

1. Rho-protein Dependent Termination: The chain terminating protein ***rho*** (ρ) acts as an accessary termination factor. It acts by disrupting the RNA-DNA hybrid. It dissociates the RNA transcript from template and terminates transcription.

Extension: Rho Protein

Rho protein is a specialised helicase. It consists of a hexamer of six identical subunits that recognise and bind to a sequence of 50 to 90 bases located upstream of the terminator in mRNA. The rho-binding sequence is rich in C and poor in G. Rho binds only to the growing mRNA.

2. Rho-protein Independent Termination: This type of RNA transcript termination depends on a specific DNA sequence that ends RNA transcript with a stable hairpin structure. Such termination sites have two or more stretches of G – C pairs arranged in the form of inverted repeat followed by a string of adenine (poly A) in the master strand. The released mRNA is called **primary transcript.** The self complementary residues in the transcript fold into a hairpin loop, ending in poly U sequence at 3′ end. It undergoes processing to become functional. From nucleus, it comes out into the cytoplasm through nuclear pores.

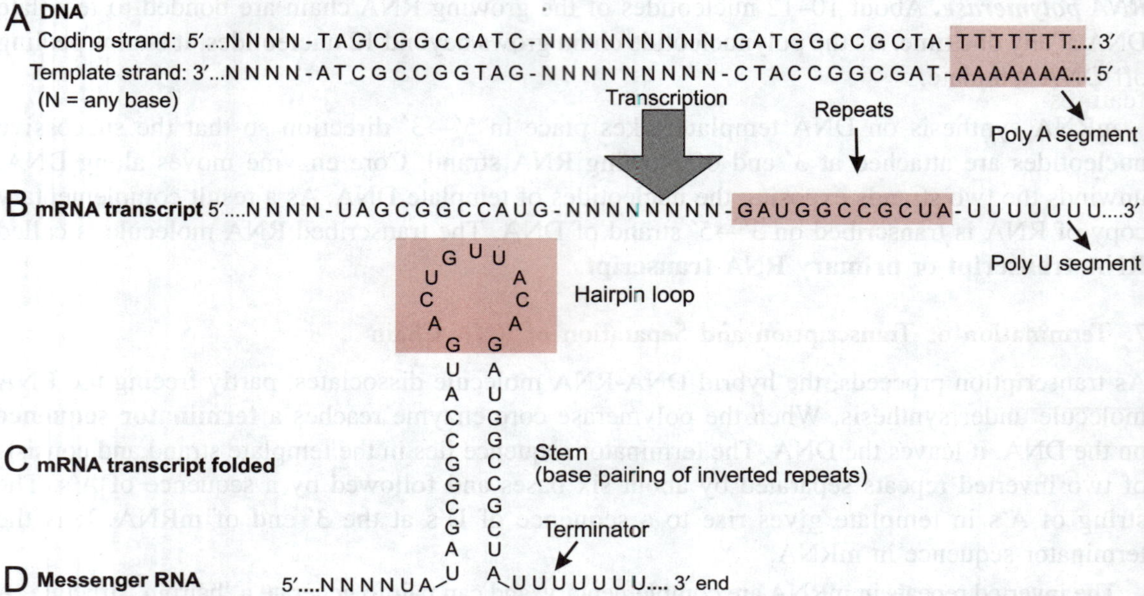

Fig. 11.7 Termination of RNA transcript from DNA template without rho factor in prokaryotes. The termination region contains two stretches of G – C pairs arranged in the form of inverted repeat.

8. Reformation of DNA Helix

As the RNA chain grows, the transcribed region of the DNA molecule gets hydrogen bonded to the sense strand and the two are spirally coiled to assume the original double helical form. When the last ribonucleotide is added, the RNA polymerase and RNA chain are fully released from the DNA, which also completes its winding into a double helix. The protective protein coat is added again to the DNA duplex.

Extension: Differences between Replication and Transcription

The differences in replication and transcription are:

- In replication new DNA molecules are formed, while in transcription RNA molecules are synthesised.
- Uracil base is inducted in RNA in place of thymine base.
- RNA transcripts do not remain connected with DNA template, but separate out as single strands and move out of the nucleus in the cytoplasm and participate in the synthesis of protein.
- Several copies of RNA transcripts are released from each DNA template.

9. Processing of RNAs

The originally transcribed RNA molecules are called **primary transcripts.** In both prokaryotes and eukaryotes, they undergo extensive changes to become functional. This is called **processing of RNAs** or **post-transcriptional modifications of RNAs.**

MOLECULAR MECHANISM OF TRANSCRIPTION IN EUKARYOTIC CELLS

Transcription in eukaryotic cells is more complex and differs from that in prokaryotes in following features:

1. Three different ***RNA polymerases*** are required in the transcription of three different types of RNAs.
2. **Eukaryotic promoters** are more varied than prokaryotic promoters. There are three different types of promoters for three different polymerases. Some eukaryotic promoters are located downstream from the transcription start point while some are present upstream.

Table 11.1 Differences between prokaryotic and eukaryotic transcription

Prokaryotic Transcription	Eukaryotic Transcription
1. Occurs in the cytoplasm.	1. Occurs in the nucleus.
2. There is no definite phase for its occurrence.	2. Takes place in the G_1 and G_2 phases of cell cycle.
3. A single RNA polymerase synthesises all the three types of RNA (mRNA, tRNA, rRNA).	3. Three RNA polymerases I, II and III synthesise rRNA, mRNA, and tRNA respectively.
4. Coupled transcription-translation is the rule.	4. Coupled transcription-translation is not possible.
5. Initiation of transcription does not need any proteins or initiation factors.	5. Initiation of transcription requires proteins called **transcription factors**. These are TFIIA, TFIIB, TFIID, TFIIE, TFIIF and TFIIH. These recognise TATA box.
6. Pre-RNA molecules are released and processed in the cytoplasm.	6. Pre-RNAs are released and processed in the nucleus.
7. RNA polymerase is a complex of five polypeptides.	7. RNA polymerases are complexes of 10 to 15 polypeptides.
8. Transcriptional unit has more than one genes (**Polycistronic**).	8. Transcriptional unit has only one gene. (**Monocistronic**).
9. The mRNA primary transcript has fewer surplus nucleotides.	9. The mRNA primary transcript has a large number of surplus nucleotides.
10. The 23S, 16S, 5S rRNAs are formed from a single primary transcript.	10. The 28S, 18S, 5.8S and 5S rRNAs are formed from two primary transcripts.

3. **Transcription factors:** Many transcription factors are involved in binding of eukaryotic RNA polymerase to DNA. Unlike prokaryotic sigma factor, the transcription factors are not part of RNA polymerase and bind to promoter even before the polymerase binds to initiate transcription.

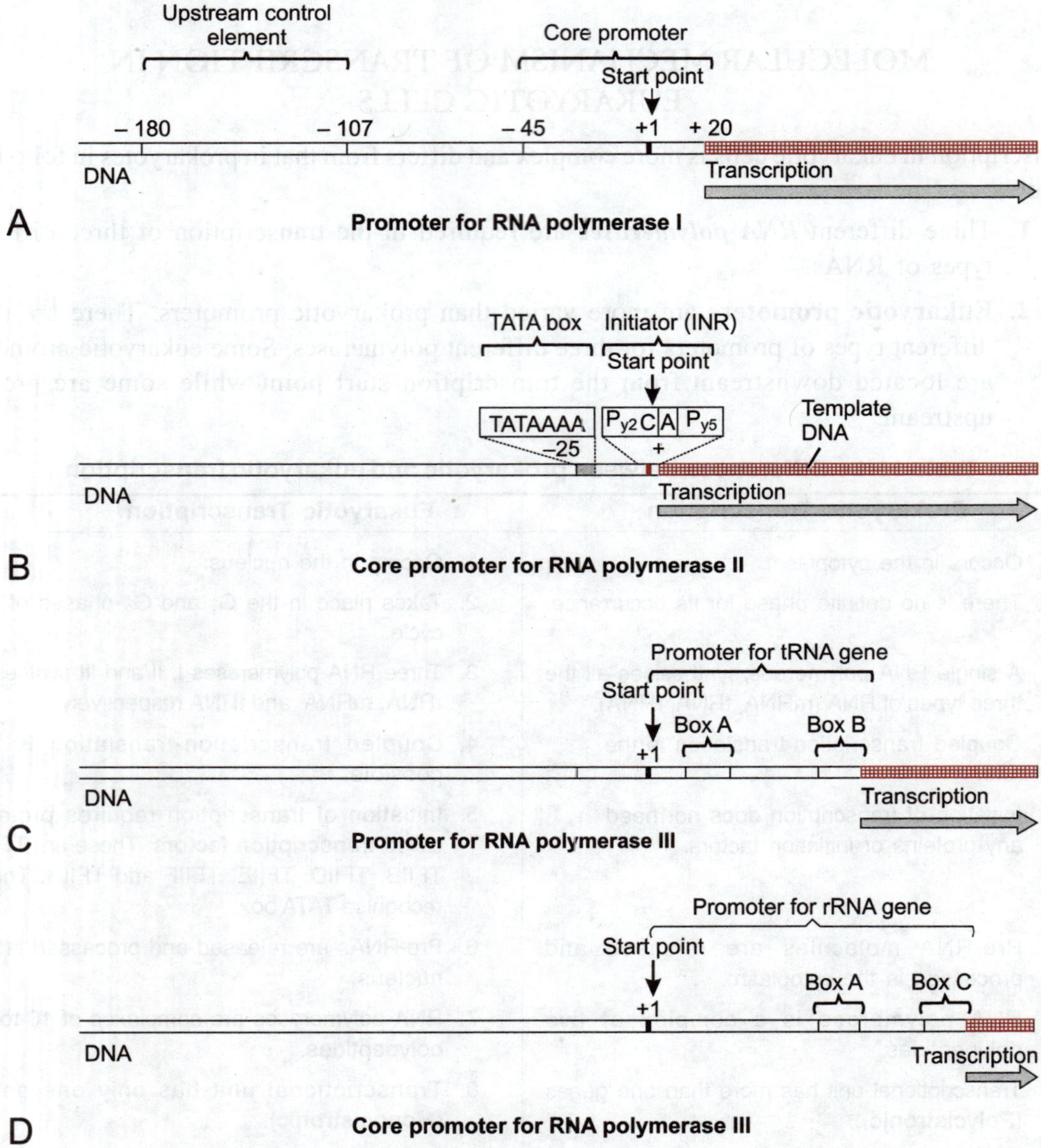

Fig. 11.8 Different promoters for three different types of RNA polymerases I, II and III in eukaryotes. Promoter for mRNA is upstream but for tRNA and rRNA is downstream of start point

The three nuclear RNA polymerases are:

1. RNA Polymerase I

It is found in the nucleolus and is responsible for the synthesis of ribosomal RNA. The RNA molecule synthesised in the nucleolus is precursor to three ribosomal RNAs (28S, 18S, 5.8S rRNA).

2. RNA Polymerase II

It is found in the nucleoplasm and transcribes protein-encoding gene. It synthesises precursors of mRNA. It also synthesises most small nuclear RNAs (snRNAs) which are needed in post-transcriptional processing of primer RNA.

3. RNA Polymerase III

It is also found in nucleoplasm. It is responsible for the synthesis of a variety of small RNAs including tRNA (Fig. 11.8C) and 5S rRNA (Fig. 11.8D).

The three eukaryotic polymerase enzymes are large and formed of many polypeptide subunits. Their molecular weight is about 5,00,000. For example, eukaryotic RNA polymerase II has more than 10 subunits and three of them are also present in RNA polymerase I and III. The largest subunit of RNA polymerase II is related to β subunit of prokaryotes.

Table 11.2 List of different types of RNA polymerases and RNA transcribed by them

Types of RNA Polymerases	Types of RNA synthesised
Type I polymerase	28S rRNA, 18S rRNA, 5.8 rRNA
Type II polymerase	mRNA, SnRNA, ScRNA
Type III polymerase	tRNA, 5S rRNA and other small RNAs
Mitochondrial RNA polymerase	Mitochondrial RNAs inside mitochondria
Chloroplast RNA polymerase	Chloroplast RNAs inside chloroplast

These are specific proteins present in the nucleoplasm and are required for the correct functioning of RNA polymerases during initiation of RNA transcription. They bind to TATA box and are also called **TATA binding proteins (TBP).** These are classified into two categories:

1. General Transcription Factors or Basal Transcription Factors

These are needed for the transcription of all the genes transcribed by a particular RNA polymerase. They are designated as TFI, TFII and TFIII and correspond to the three RNA polymerases.

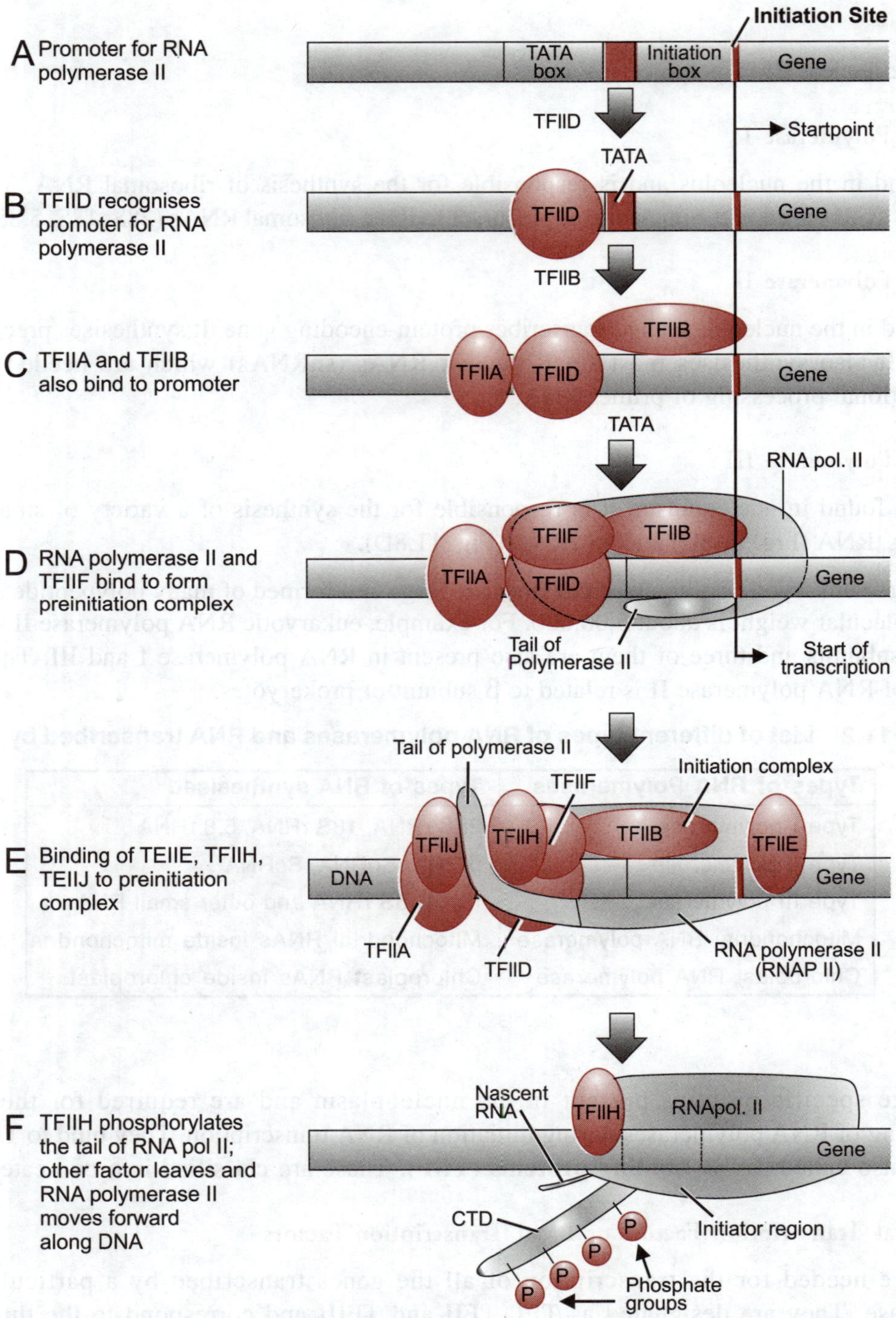

Fig. 11.9 Formation of preinitiation complex during initiation of transcription in eukaryotes. The steps in the formation of transcription apparatus and movement of RNA polymerase II forward from the binding site

2. Additional Transcription Factors or Specific Transcription Factors

These are needed for the transcription of particular specific genes under specific circumstances. These are responsible for regulating gene expression and, therefore, control the expression of individual genes.

Transcription Factors needed for RNA Polymerase II

Transcription factors TFIIB, TFIID, TFIIE, TFIIF and TFIIH are general transcription factors, required for ***RNA polymerase II*** to start RNA transcription. TFIID has a TATA binding protein (TBP) which binds specifically to the TATAA consensus sequence. TFIID also binds to the TBFIIB forming TATAA-TFIID-TBIIB complex at the promoter. With the help of TFIIF factor ***RNA polymerase*** binds to this complex.

The release of RNA polymerase II from promoter and its movement along template DNA require other three TFII complexes, TFIIE, TFIIH and TFIIJ. TFIIH unwinds DNA around the transcription site (**helicase activity**) and phosphorytales the tail of largest subunit of RNA polymerase at the C-terminal. This end is also **called carboxy-terminal domain** (CTD). It consists of seven amino acids repeated approximately 50 times. On being phosphorylated at serine or threonine residues, ***polymerase*** is released from promoter region to slide over the template strand of DNA. All other TFII complexes except for TFIIH are left behind as RNA polymerase moves forward.

Table 11.3 Transcription factors associated with RNA polymerase II

TBP	binds to TATA box, a part of TFIID
TFIID	includes TBP; recognises Polymerase II-specific promoter; first to bind to TATA box
TFIIA	binds upstream of TATA box; required for binding of RNA Pol II to promoter
TFIIB	binds downstream of TATA box; required for binding of RNA Pol II to promoter
TFIIF	accompanies RNA Pol II as it binds to promoter
TFIIE	required for promoter clearance and elongation of ribonucleotide chain
TFIIH	phosphorylates the tail of RNA Pol II, retained by polymerase during elongation
TFIIJ	required for promoter clearance and elongation.

Functional or Consensus Sequences in Promoter Region in Eukaryotes

Sequence of nitrogenous bases similar to that found in prokaryotes are present in the promoter region of eukaryotes also. There are different promoters for all the RNA polymerases.

Promoters for Polymerase II

The promoter region for RNA polymerase II is formed of three regions: the **initiation box, TATA box** and a variety of **upstream elements**.

TATA BOX OR GOLDBURG HOGNESS BOX

The consensus sequence in TATA box is a hexanucleotide sequence, consisting of **5′...TATAAA...3′**. It is usually represented by **TATXAX,** where X may be T or A. It is comparable to Pribnow sequence of prokaryotic promoter. It is the site of assembly of a preinitiation complex and determines the site of initiation of transcription. It lies about 25bp upstream of initiation box. It is rich in AT sequences.

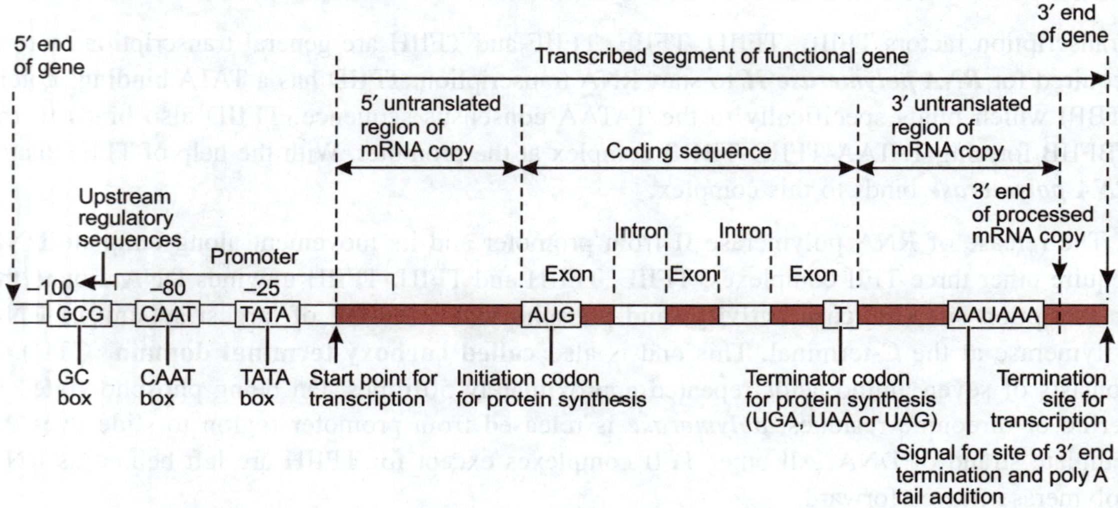

Fig. 11.10 A eukaryotic gene with its promoter region

Table 11.4 Similarity between Pribnow box and TATA box sequences

Type of Sequence	General Group	Organism	Sequence
Pribnow Sequence	In Viruses	$\phi \times 174$	5′ ----- TACAGTA ----- 3′
		SV40	5′ ----- TATAATG ------ 3′
	In Prokaryotes	*E. coli*	5′ ----- TATAATG ------ 3′
TATA Box	In Eukaryotes	Mouse	5′ ----- TATAAAG ------ 3′
		Chicken	5′ ----- TATATAT ------ 3′

The assembly of pre-initiation complex (transcription factors and polymerase) is essential for initiation of transcription.

In addition, eukaryotic genome has proximal control elements, named **GC box** with base sequence GGGCGG at –100 site and a **CAAT box** with base sequence **GCCCAATCT.**

Promoter for RNA Polymerase I

The promoter for RNA polymerase I has an **upstream control element** and a **core promoter.** The core promoter site is rich in GC sequences. The protein (**UBFI upstream binding factor for pol I**) recognises and binds to both the upstream control element and the core promoter. Later, **SLI (selectivity factor for pol I)** binds to DNA in association with UBFI. Finally, RNA polymerase I binds and transcribes rRNA genes.

Promoter for RNA Polymerase III

The promoter for RNA polymerase III for both 5S RNA and tRNA are located within the gene itself. The recognition sites for them are downstream of the start site. Two factors TFIIIC and TFIIIB are needed for the binding. Firstly, TFIIIC (or TFIIIA) binds to both sites and induces TFIIIB to bind to the promoter near the start sites and TFIIIB binds RNA polymerase III to promoter in DNA.

RNA-DIRECTED RNA SYNTHESIS

In RNA viruses whose genome is RNA, RNA is transcribed from RNA. These viruses induce the formation of an enzyme ***RNA-directed RNA polymerase*** or ***RNA replicase*** in the host cell. ***RNA replicase*** acts like DNA-directed RNA polymerase and exhibits the following characteristics:

- RNA replicase directs the synthesis of RNA on RNA in $5' \rightarrow 3'$ direction.
- Ribonucleotides are added to the RNA chain in the same fashion as in DNA directed-RNA polymerase.
- Ribonucleotide triphosphates enter the chain on $3'$ end and pyrophosphate is released.
- The RNA transcript is complementary to RNA template.
- Editing, proof reading or repair are not associated with RNA replicase.
- RNA replicase works on single-stranded template RNA.

RNA EDITING

RNA editing is the post-transcriptional modification of primary mRNA transcript by the addition or deletion or both addition and deletion of one or more nucleotides in pre-mRNA. This phenomenon was observed in mid 1980s in the production of certain mRNAs in the mitochondria of some protozoa.

Example

The conserved protein, **subunit III of cytochrome oxidase** (CO III) is encoded by mt DNA. The sequence of CO III gene and its mRNA transcripts were compared in three protozoans, *Trypanosoma brucei* (*Tb*), *Crithidia fasiculata* (*Cf*) and *Leishmania tarrentolae* (*Lt*). It was found the mt DNA sequences of this gene in *Cf* and *Lt* were collinear with mRNAs but not in *Tb*.

The mRNA of *Tb* had several U nucleotides inserted which were not encoded in the DNA, and the DNA of *Tb* had several T nucleotides which were not represented in the transcript.

Tb DNA G GTTTTTGG AGG G G

Tb RNA UU G U G UUUUUGG UUU AGG UUUUUUU G UU G

Fig. 11.11 DNA sequences of cytochrome oxidase subunit II (CO III) in protozoan *Trypanosoma brucei* and its conserved mRNA

There is extensive insertion of U nucleotides in mRNA and deletions of several T nucleotides. This editing is carried out by a special RNA molecule, called **guide RNA** (gRNA). It pairs with mRNA transcript and causes the cleavage of the transcript, templating the missing U nucleotides and ligating the transcript.

Editing may involve addition of single C nucleotides at many positions or substitution of C to U or U to C. The C to U editing is found to result in the formation of an AUG initiation codon from ACG codon in some chloroplast mRNAs in some higher plants. It is assumed that RNA editing process is an additional regulatory control process and results in the generation of tissue specific initiation or stop codons. C to U editing is found in mRNA transcribed from a nuclear gene that encodes for cytoplasmic apo-lipoprotein B in mammals.

POST-TRANSCRIPTIONAL MODIFICATION OF RNA OR RNA PROCESSING

The immediate product of transcription is called **primary transcript**. It is biologically inactive. The process of chemical modifications in the primary transcript to generate the final functional RNA product is called **RNA processing**. In RNA processing:

- Larger RNA precursors are cut into smaller RNAs by ***ribonuclease-P enzymes*** **(cleaving).**
- Unwanted nucleotides are removed by enzymes named ***nucleases*** **(splicing).**
- Useful regions are rejoined by ***ligase*** enzyme **(union).**
- Certain nucleotides are added at the end enzymatically **(terminal addition),**
- Molecule may fold on itself to assume proper shape **(folding),** and
- Some nucleotides may be modified **(nucleotide modification).**

Extension: Primary Transcript and Ribonuclease P Enzyme

Primary transcript is RNA molecule formed immediately after transcription. The primary transcripts of three different types of RNAs before undergoing processing are known as precursors or **pre-mRNA, pre-tRNA** and **pre-rRNA**. In eukaryotes, pre-mRNA was also referred as **hnRNA** or **heterogeneous nuclear RNA**.

Ribonuclease P Enzyme

The enzyme ***ribonuclease P*** is a ribozyme. It is formed of both a protein and a small RNA molecule. The catalytic activity of enzyme is due to RNA present in the large sub-unit of ribosome (23S rRNA in prokaryotes and 28S rRNA in eukaryotes). The protein component merely modulates the activity of RNA component.

Processing of Pre-tRNA Transcript

Eukaryotic pre-tRNA molecules are biochemically altered. The process has been studied in detail in yeast, where following four processes are needed for functional tRNA molecules to be formed:

- Trimming of ends by the cleavage of phospho-ester bonds.
- Splicing for removing an intron.

- Addition of some terminal sequences.
- Heterocyclic base modification, usually by **methylation.**

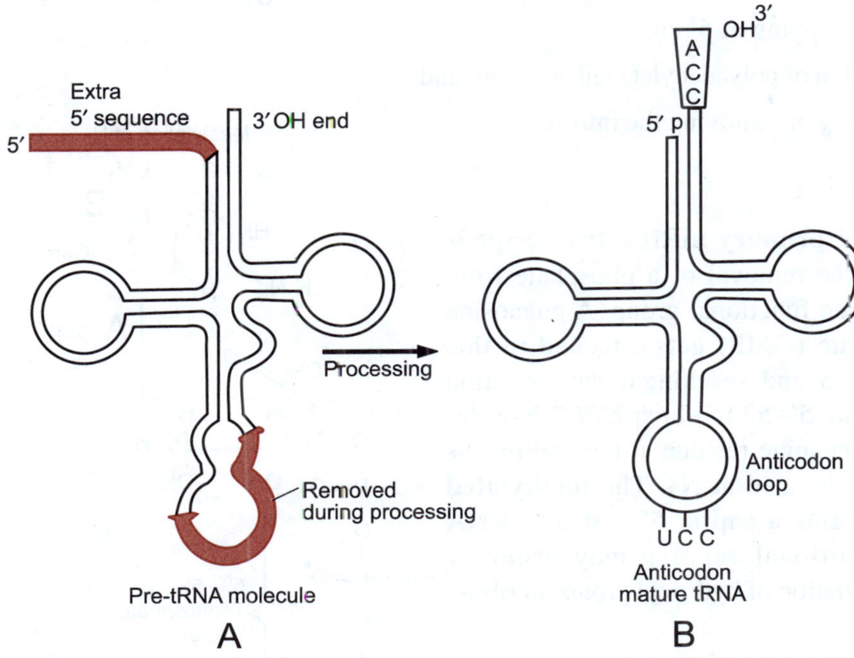

Fig. 11.12 Post-transcriptional modification or processing of a pre-tRNA molecule n *E. coli.* **A.** Pre-tRNA molecule; **B.** Mature tRNA. Processing involves removal of extra segment at 5′end and of intron sequence from anticodon loop and the addition of CCA sequence at the 3′ OH end

In prokaryotes pre-tRNA molecule undergoes only a few of the above modification steps. In some prokaryotic tRNAs some introns are removed, in some terminal sequences are added and in some others, a few heterocyclic and ribose rings are modified. The endonuclease enzyme, ***ribonuclease P***, catalyses removal of a segment from 5′ end of pre-tRNA by hydrolytic cleavage.

Processing of rRNA-Transcript

The primary transcript of rRNA contains precursors for all the three rRNAs (23S, 16S and 5S). The processing involves its trimming to smaller segments of appropriate size with the help of ribonucleases. It has already been discussed in biogenesis of rRNA.

Processing of Pre-mRNA Transcript

Most prokaryotic mRNA transcripts require very little or no post-transcriptional modification to be changed into functional mRNA. In prokaryotes, often translation of mRNA begins while it is being transcribed. This is called coupled **transcription-translation**. But in eukaryotes, the primary-mRNA transcript molecules have copy of both exons and introns. Therefore, they are spliced and chemically modified in the nucleus before being transported to the ribosomes for translation. Following three biochemical processes occur during mRNA processing in eukaryotes:

- RNA capping at 5′ end
- Addition of polyadenylate tail at 3′ end, and
- Splicing or removing the introns.

1. RNA Capping

The **5′ end of primary mRNA transcript** is modified by the removal of a phosphate from the triphosphate functional group. A guanosine monophosphate (GMP) gets attached to this diphosphate at 5′ end, resulting in the formation of an unusual 5′–5′ triphosphate covalent linkage. The guanine residue at the end of this linkage is methylated at N_7. The **methylated guanosine** forms a cap at 5′ end of mRNA (Cap 0). Additional capping may occur by another methylation of hydroxyl groups of ribose

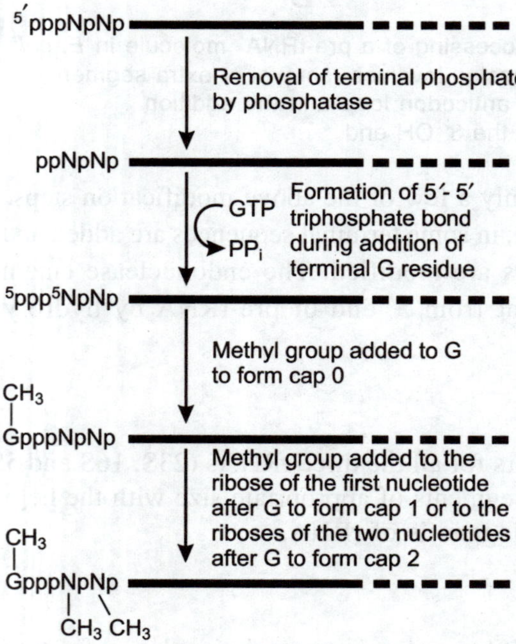

Fig. 11.13 Steps involved in the formation of 5′G cap

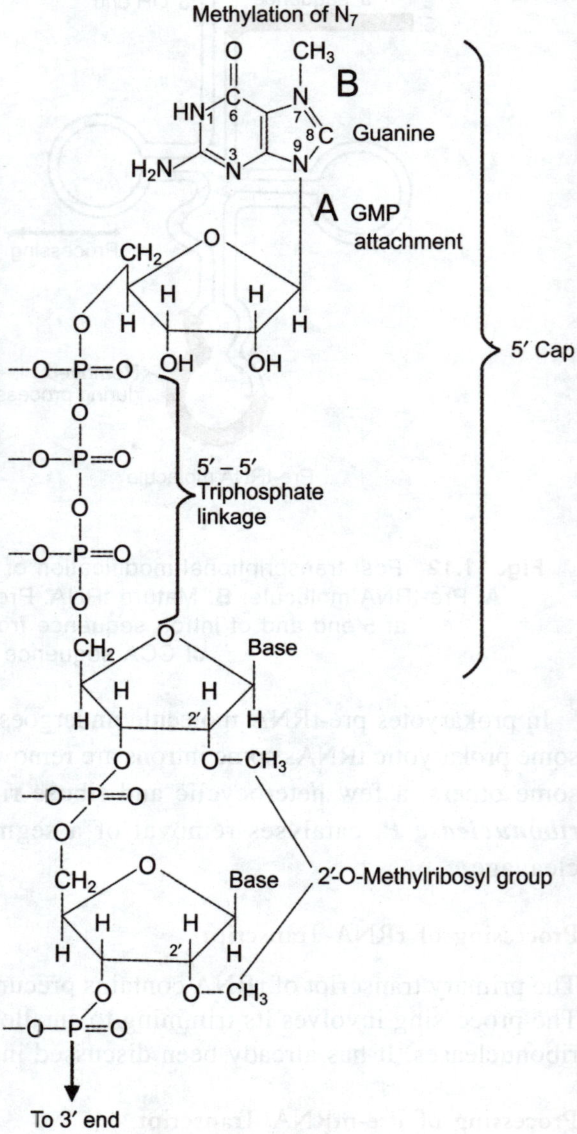

Fig. 11.14 Capping of mRNA involving: A. Attachment of a GMP at 5′ end and B. Methylation of guanine of GMP

of first one or two nucleotides in some higher eukaryotes (Cap1 or Cap 2). Only cap 0 is formed in lower eukaryotes.

Capping of RNA is probably necessary for marking mature mRNA and to protect it against cleavage by exonucleases.

2. Poly A Addition

The **3′end of primary mRNA transcript** is modified by the addition of a polyadenylate **(poly A)** tail. It is catalysed by ***polyadenylate polymerase (poly A polymerase)*** enzyme. The poly A tail is formed of 20 to 250 nucleotide residues. Before the addition of poly A tail, an endonuclease removes a few nucleotides from the 3′ end of the transcript. The cleavage site has tail recognition sequence –AAUAA– and GU rich sites and is recognised by a polyadenylation complex.

The **polyadenylation complex** is formed by the joining of poly A adenylate and poly A binding protein (PABP) with cleavage-polyadenylation specificity factor (CPSF) and cleavage stimulation factor (CST). This complex binds to the tail recognition sequence –AAUAAA– and GU-rich sites and stimulates the process of cleavage of nucleotides and addition of poly A tail.

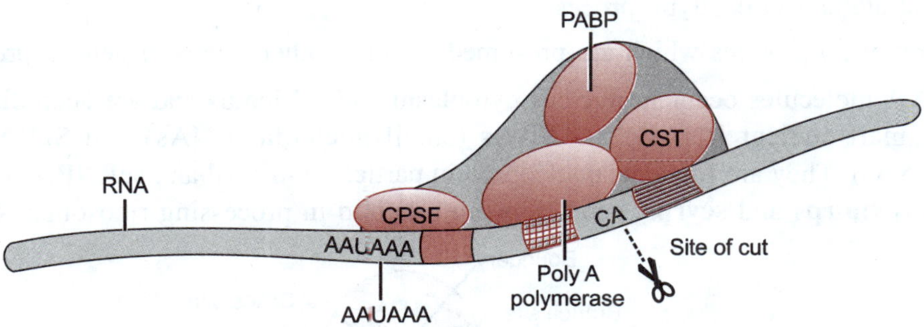

Fig. 11.15 Structure of polyadenylation complex

Most eukaryotic mRNA molecules have a 3′ poly A tail. It is thought to stabilise mRNA molecule by increasing resistance to cellular nucleases and helps it to cross the nuclear membrane.

3. RNA Splicing

The process of cutting out introns from primary mRNA transcript and rejoining the exons to form functional mRNA is called **RNA splicing.** It occurs in the nucleus of only eukaryotic cells which have split genes.

Mechanism of RNA Splicing

The mechanism of RNA splicing depends on specific base sequences at the junctions between introns and exons. The consensus sequence at the 5′ end of mRNA intron, called **5′ splice site**, starts with consensus sequence 5′-**GU** and 3′ **splice site** at the 3′ end of intron ends with consensus sequence 3′-**AG.**

In the middle of the intron is present a special adenine residue. It is used as a **branch site** during splicing. The conenus recognition sequences for these sites are:

1. 5′ splice site: 5′...AG↓**GU**AAGU....3′
2. 3′ splice site: 5′...YYYYYYNC**AG**↓....3′
3. Branch site: 5′...UACUA**A**C....3′

Spliceosome or Splicing Apparatus

The 5′ and 3′ splice sites and the branch sequence are recognised by a complex formed of proteins and small segments of RNA. These RNA-protein complexed together form a large splicing apparatus, called **spliceosome.** This complex brings 5′ and 3′ splice sites in intron together.

Spliceosome is a large body formed of:

1. SnRNAs—5 in number.
2. 41 associated proteins that add to its mass.
3. 70 other proteins that act as splicing factors and help in binding the spliceosome to RNA substrate and in catalytic process.
4. Another 30 proteins which are presumed to act at other stages of gene expression.

Small RNA molecules occur in nucleus cytoplasm and nucleolus and are accordingly called **SnRNAs** (small nuclear RNAs), **SnoRNAs** (small nucleolar RNAs) and **ScRNAs** (small nucleolar RNAs). They are found as nucleoprotein particles (**snRNP** and **scRNP**). They are also designated as **snurps** and **scyrps.** sno RNAs are required in processing ribosomal RNA.

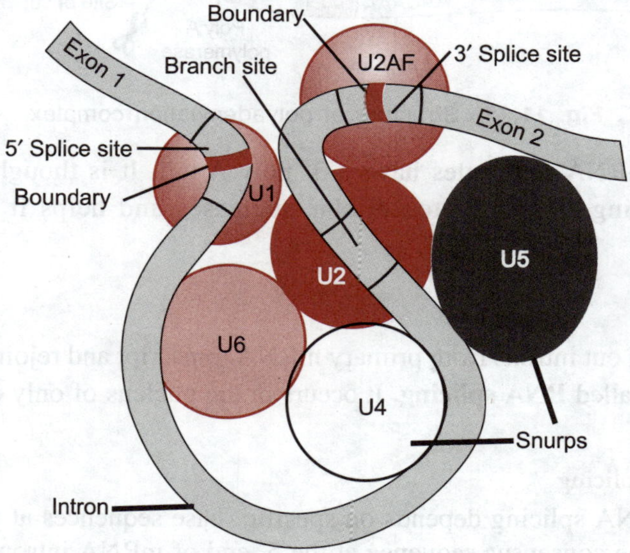

Fig. 11.16 Structure of a spliceosome showing six ribonucleoproteins (U1 to U6) that are assembled at the splice sites of intron/exon boundaries

The five snRNAs present in the spliceosome and associated with splicing are U1, U2, U5, U4 and U6. They are named according to the snRNAs. Each snRNP contains a single snRNA and several proteins. The U4 and U6 snRNPs are found as a single particle (U4/U6).

- Binding of U1 snRNP to 5′ splice site of intron is the first step in splicing.
- U2 + auxilliary factor (U2AF) binds to the region between the branch site and 3′ splice site. It has two units: large and small subunits. The complex thus formed is called **E complex.**
- Splice factor SF_1 or BBP connects U2AF to U1 snRNP and form **complex A.**
- Binding of U5 and U4/U6 snRNPs to this complex convert it to **B1 complex** or **B1 spliceosome.**
- Release of U1 snRNP allows U6 snRNA to interact with the 5′ splice site. By this B1 spliceosome is converted to **B2 spliceosome.**
- Dissociation of U4 from U6 snRNP allows U6 snRNA to pair with U2 snRNA and to form catalytic active site.

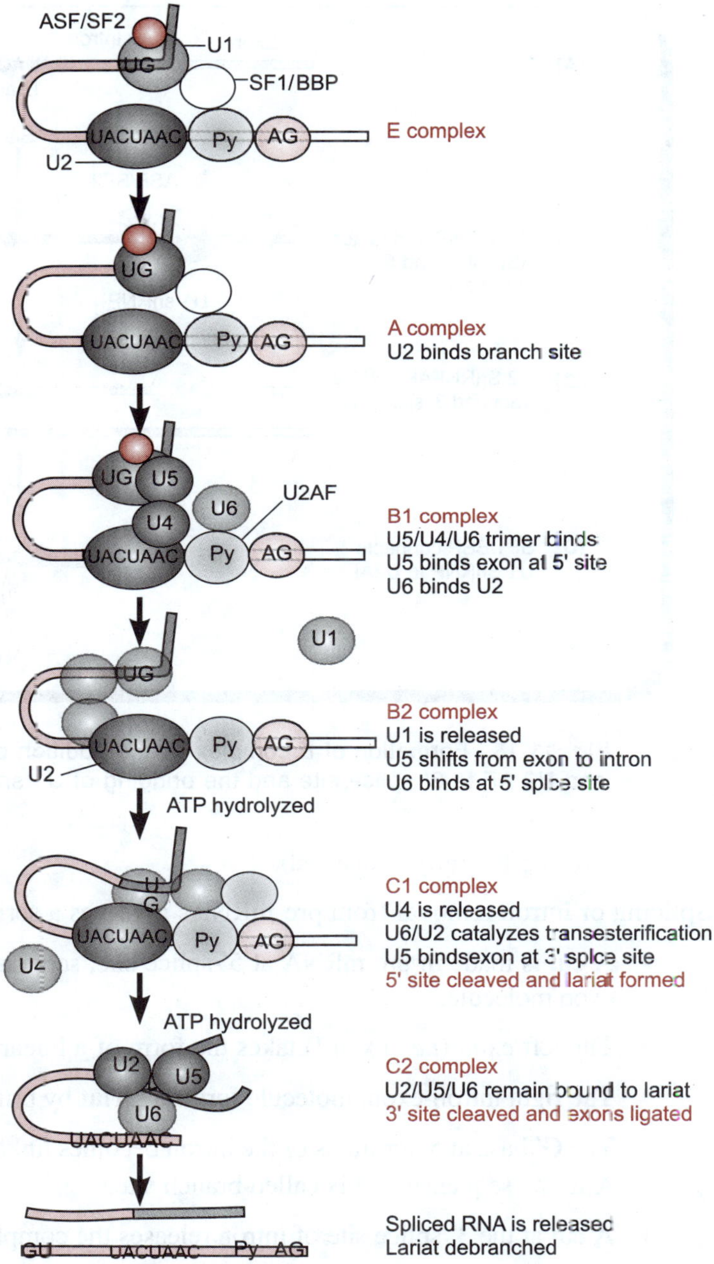

Fig. 11.17 Different steps in the splicing reaction leading to spliceosome formation and splicing of exon 1 and exon 2 and separation of lariat

- U2–U6 interaction results in the removal of intron segment.
- This brings 3′ end of one exon close to 5′ end of other exon both of which join together.

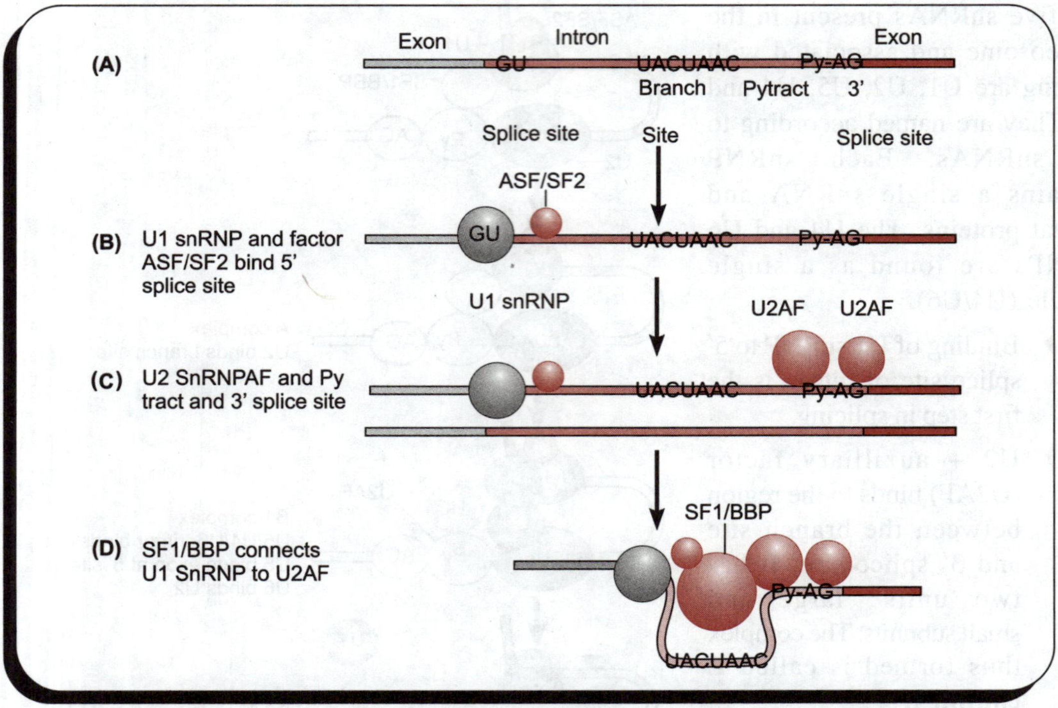

Fig. 11.18 Formation of E complex by the addition of U1 snRNP to 5′ splice site and U2 AF to 3′ splice site and the bridging of U1 snRNP and U2AF by sF_1/BBP

Steps in Splicing (or Intron Removal)

Splicing or intron removal from pre-mRNAs involves a series of well-defined steps:

- A cut is made in pre mRNA at 5′-splice site, separating the left exon and the right intron-exon molecule.
- The left exon (*i.e.,* exon I) takes the form of a linear molecule.
- The right intron-exon, molecule forms a **lariat** by bringing its 5′ and 3′ splice sites together.
- The G′ base at 5′ terminus of the intron becomes linked to a base within the intron with base A in the sequence that is called branch site.
- A cut at the 3′ splice site of intron releases the complete intron in lariat form.
- The 5′ end of right exon (2) is now ligated or spliced to 3′ end of left exon (*i.e.,* exon 1) by covalent bond.
- Lariat is debranched and the excised intron becomes linear. It is rapidly degraded.
- The 5′ end of exon 1 is free to attack the other intron-exon junction.

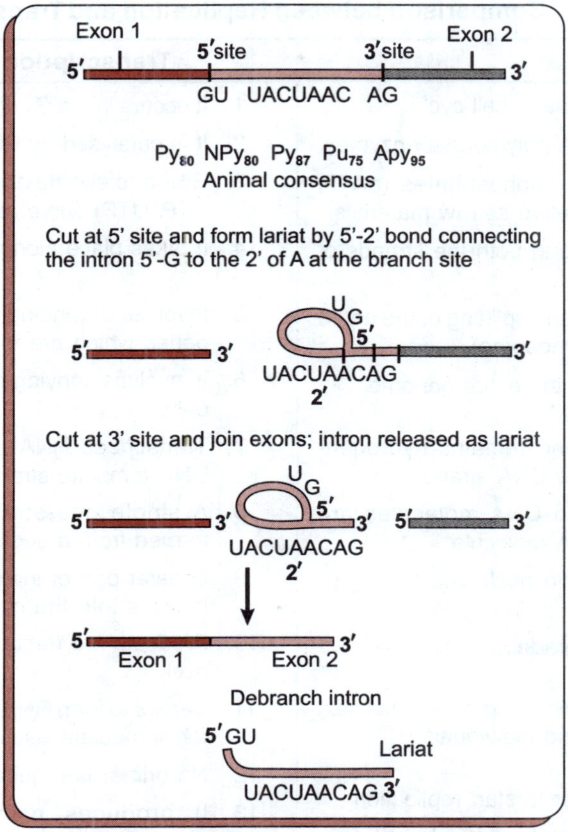

Fig. 11.19 splicing occurs in two stages. First the 5′ exon is cleaved off, and then it is joined to the 3′ exon

The **branch site** helps in identifying 3′ splice site. It lies 18 to 40 nucleotides upstream of 3′ splice site. In eukaryotes, additional cryptic sites are present. In case branch site is lost due to mutation, the cryptic sites help in splicing. In yeast cells the consensus sequence of branch site is UACUAAC and it is highly conserved. The removal of introns from eukaryotic pre-mRNA occurs in the nucleus in complexes called **spliceosomes.**

Self-Splicing RNA Introns

Certain genes have **self-splicing RNA introns.** The primary transcript of such a gene carries out RNA splicing in the absence of snRNPs. The intron RNA catalyses the process of splicing. These **catalytic RNA** molecules are also called **ribozymes**. The process of splicing involves transesterification reaction.

Table 11.5 Comparison between Replication and Transcription

Replication	Transcription
1. It occurs in the S phase of cell cycle.	1. It occurs in the G_1 and G_2 phases of cell cycle.
2. It is catalysed by DNA polymerase enzymes.	2. It is catalysed by RNA polymerase enzymes.
3. Deoxyribonucleoside triphosphates (dATP, dGTP, dCTP, dTTP) serve as raw materials.	3. Ribonucleoside triphosphates (ATP, GTP, CTP, UTP) serve as raw materials.
4. Replication occurs along both the strands of DNA.	4. It takes place along one strand of DNA.
5. It involves unwinding and splitting of the entire DNA molecule (chromosome).	5. Involves unwinding and splitting of only those genes which are to be transcribed.
6. It involves copying of the entire genome.	6. It involves copying of certain individual genes only.
7. Replicated DNA strand remains hydrogen-bonded to its template DNA strand.	7. Transcribed RNA strand separates from its DNA template strand.
8. Two double-stranded DNA molecules are formed from one DNA molecule	8. A single one-stranded RNA molecule is formed from a segment of one DNA strand.
9. Products remain within nucleus.	9. Greater part of the products passes from the nucleus into the cytoplasm.
10. Products are not degraded.	10. Products are degraded after their function is over.
11. Serves to conserve the genome for the next generation of cells and individuals.	11. Serves to form RNA copies of individual genes for immediate use in protein synthesis.
12. It requires RNA primer to start replication	12. No primer is required to start.
13. It produces normal DNA molecules that do not need any processing.	13. It produces **primary RNA transcript** molecules which need processing to acquire final form and size.

RNA SYNTHESIS IN VITRO

Two years before **Kornberg's** isolation of ***DNA polymerase*****, Grunbergmango** and **Ochada** isolated an enzyme specifically involved in RNA metabolism. This enzyme was named ***RNA polymerase***. It can link different nucleotides into a long RNA chain or can break the chain into smaller sections. In the presence of this enzyme RNA synthesis can occur even in the absence of primer, in any proportion of nucleotides and even in the presence of only one kind of nucleotides. This means RNA chain can be obtained consisting solely of a single repeating base (polyadenylic acid "poly-A" formed of adenine ribonucleotides) or of polyU. **Weiss, Horwitz** and **Stevens (1950, 60)** isolated another ***RNA polymerase*** which could work only in the presence of DNA as a template for RNA *in vitro* synthesis.

The substances needed for the synthesis are nucleoside diphosphate, adenosine diphosphate (ADP), uridine diphosphate (UDP), guanine diphosphate (GDP) and cytosine disphosphate (CDP), but all the four substrates need not be present at the same time.

TEXT QUESTIONS

1. Describe various steps in the transcription of RNA in prokaryotes.
2. Describe differences in the prokaryotic and eukaryotic transcription.
3. Discuss the structure of polymerase I and function of its various sub-units.
4. Discuss with suitable diagrams the mechanism of post-transcriptional processing of mRNA. Why is processing essential?
5. Give details of various transcription factors needed for RNA transcription in eukaryotes.
6. What is termination? Discuss various methods of termination of gene transcription in *E. coli.*
7. Discuss nature of factor Rho. What role does it play in RNA transcription and how?
8. Explain role of σ factor in RNA transcription.
9. Give differences in the structure of promoter region in prokaryotes and eukaryotes.
10. How are introns removed from RNA by splicing?
11. How is poly-A tail added to eukaryotic mRNA?
12. Discuss *in vivo* synthesis of mRNA in eukaryotes and pokaryotes. What is the basic difference in the two cases?
13. Write short notes on:
    (a) RNP particles
    (b) snRNP
    (c) RNA Splicing
    (d) RNA processing
    (e) Heterogeneous nuclear RNA (HnRNA)
    (f) Biogenesis of mRNA
    (g) Spliceosome
    (h) Informosomes
    (i) Heterogeneity of mRNA
    (j) Transcription factor
    (k) Polycistronic mRNA.
14. What do you mean by coding and non-coding DNA?
15. Why does messenger RNA need a methyl G cap and a poly-A tail?
16. Compare DNA-directed RNA synthesis and RNA-directed RNA synthesis in terms of the following characteristics:
    (a) Major elongation enzyme
    (b) Direction for new chain elongation
    (c) Form of entering nucleotides
    (d) Mechanism of phosphodiester bond formation
    (e) Proof reading
    (f) Template.

17. Describe in brief the role of each of the following proteins in the transcription of RNA:

(a) σ submit (b) ρ-protein

(c) RNA polymerase (d) Rho factor.

18. What is the function of tRNA?

19. Which group of cells always have monocistronic RNA?

20. What is primary transcript?

21. What is rDNA?

22. Define heterogeneous nuclear RNAs.

23. Define the following terms:

(a) 3′ terminus (b) Polymer

(c) 5′ cap (d) Antisense strand

(e) Complementary (f) Hogness box

(g) Primary transcript (h) Initiation box.

24. Expand the following abbreviations:

(a) TF (b) UBF

(c) TBP (d) CTD

(e) 5′ UTR (f) SLT.

12
Chapter

Genetic Code

IN THIS CHAPTER

The linear arrangement of nitrogenous bases in DNA is said to determine the sequence of amino acids in a protein molecule. There are only four nitrogenous bases in a DNA molecule, namely adenine (A), cytosine (C), guanine (G), and thymine (T). It means that the precise sequence of these four nitrogenous bases on the DNA strand somehow direct the proper amino acids to their proper places through the intervention of mRNA.

Genetic Code

Several theories were proposed to explain the mechanism by which a particular sequence of nitrogenous bases in DNA, by transcribing complementary bases in mRNA, determines the position of specific amino acid in the protein molecule. The theory which is widely accepted till now was proposed by **F. H. C. Crick.** The theory holds the existence of a genetic code and its smallest unit which codes for one amino acid is known as **codon.**

A codon (code word) is the nucleotide or nucleotide sequence in mRNA which codes for a particular amino acid, whereas the genetic code is the sequence of nitrogenous bases in mRNA molecule which encloses information for the synthesis of protein molecules.

Genetic Code is a Triplet Code

The main problem of genetic code was to determine the exact number of nucleotides in a codon which codes for one amino acid. Since there are only four nitrogenous bases in mRNA for 20 amino acids, combination of only one or two nitrogenous bases can not provide sufficient code words for 20 amino acids. A **singlet code** consisting of only one nucleotide provides just four codons A, C, G and U. These are insufficient to code for 20 amino acids. Similarly a combination of two nitrogenous bases **(doublet code)** provides $4 \times 4 = 16$ codons still insufficient for 20 amino acids.

A
C
G
U

Codons

AA	AC	AG	AU
CC	CA	CG	CU
GG	GA	GC	GU
UU	UA	UG	UC

Singlet Code: 4 × 1= 4 codons **Doublet Code:** 4 × 4 = 16 codons.

Fig. 12.1 Codons of singlet and doublet code of nitrogenous bases

Table 12.1 Triplet codons of mRNA for amino acids

	U	C	A	G	
U	UUU } Phe UUC UUA } Leu UUG	UCU } Ser UCC UCA UCG	UAU } Tyr UAC UAA Ochre (Terminator) UAG Amber (Terminator)	UGU } Cys UGC UGA } Terminator UGG} Trp	U C A G
C	CUU } Leu CUC CUA CUG	CCU } Pro CCC CCA CCG	CAU } His CAC CAA } Gln CAG	CGU } Arg CGC CGA CGG	U C A G
A	AUU } Ile AUC AUA AUG } Met	ACU } Thr ACC ACA ACG	AAU } Asn AAC AAA } Lys AAG	AGU } Ser AGC AGA } Arg AGG	U C A G
G	GUU } Val GUC GUA GUG	GCU } Ala GCC GCA GCG	GAU } Asp GAC GAA } Glu GAG	GGU } Gly GGC GGA GGG	U C A G

Gamow (1954) pointed out the possibility of three letter code, *i.e.* each codon consisting of three nitrogenous bases.

This will give 4 × 4 × 4 = 64 code words or codons, which are more than enough to code for twenty amino acids. The Table 12.1 provides the list of 64 triplet codons for amino acids.

It is evident from the table that several of the triplets have the same letters but in different sequences and these code for different amino acids. It means that the sequence of letters in the triplets is most important in determining what amino acid is to be coded.

Although, informations are coded in the form of nitrogenous base sequences in DNA molecule, it is customary to represent the code letters of mRNA because the message from DNA is carried into the cytoplasm by mRNA and the code on mRNA is translated into the sequence of amino acids in the polypeptide chain.

Deciphering of Genetic Code

By the process of *in vitro* synthesis, scientists have found out codons for all the twenty amino acids. These codons have been listed in Table 12.2.

Table 12.2 Triplet codons for amino acids

	Amino acid	Abbreviation	DNA code	mRNA codon
1.	Alanine	ala	CGA, CGG CGT, CGC	GCU, GCC, GCA, GCG
2.	Arginine	arg	GCA, GCT, GCC TCT, GCG, TCC	CGU, CGA, CGG CGC, AGA, AGG
3.	Asparagine	asn	TTA, TTG	AAU, AAC
4.	Aspartic acid	asp	CTA, CTG	GAU, GAC
5.	Cysteine	cys	ACA, ACG	UGU, UGC
6.	Glutamine	gln	GTT, GTC	CAA, CAG
7.	Glutamic acid	glu	CTT, CTC	GAA, GAG
8.	Glycine	gly	CCA, CCG, CCT, CCC	GGU, GGC, GGA, GGG
9.	Histidine	his	GTA, GTG	CAU, CAC
10.	Isoleucine	ile	TAT, TAA, TAG	AUA, AUU, AUC
11.	Leucine	leu	AAT, AAC, GAA, GAG, GAT, GAC	UUA, UUG, CUU, CUC, CUA, CUG
12.	Lysine	lys	TTT, TTC	AAA, AAG
13.	Methionine	met	TAC	AUG
14.	Phenylalanine	phe	AAA, AAG	UUU, UUC
15.	Proline	pro	GGA, GGG, GGT, GGC	CCU, CCC, CCA, CCG
16.	Serine	ser	AGA, AGG, AGT, AGC, TCA, TCG	UCU, UCC, UCA, UCG, AGU, AGC
17.	Threonine	thr	TGA, TGG, TGT, TGC	ACU, ACC, ACA, ACG
18.	Tryptophan	trp	ACC	UGG
19.	Tyrosine	tyr	ATA, ATG	UAU, UAC
20.	Valine	val	CAA, CAG, CAT, CAC	GUU, GUC, GUA, GUG
21.	Terminating triplets	–	ATT, ATC, ACT	UAA, UAG, UGA

NATURE AND CHARACTERISTICS OF GENETIC CODE

The characteristics of genetic code are as follows:

1. Genetic Code is a Triplet Code

A codon of the present day genetic code that specifies one amino acid in a polypeptide chain comprises of a sequence of three nitrogenous bases on mRNA in a specific sequence.

2. Genetic Code is Commaless

Genetic code on mRNA is read continuously without internal punctuation along the reading frame. It means there is no comma between the adjacent codons and each codon is immediately followed by the next codon with no intervening spaces for comma and no skipping of any nucleotide in the message.

3. Genetic Code is Non-overlapping

The codons in mRNA are read in successive groups of three nucleotides in 5′→3′ direction. A message of AAGAAGAAG .... in mRNA will code for lysine–lysine–lysine .... because AAG codes for lysine. Since, there is no comma theoretically the reading of message can begin as AAG or AGA or as GAA depending upon where the reading begins as shown below:

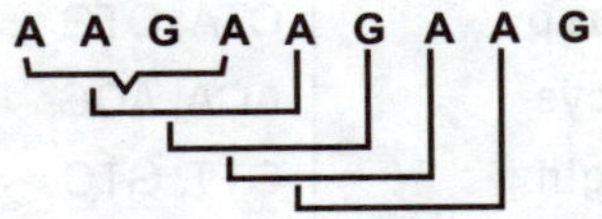

This type of reading of message is described as **overlapping sequence reading frame**. But evidences are available to show that translation of genetic code in an mRNA begins at the correct point and there is no overlapping of codons. i.e., genetic code is **nonoverlapping**.

4. Genetic Code is Unambiguous

The genetic code inside the cell medium (*in vivo*) is said to be nonambiguous because a particular codon always codes for the same amino acid. No doubt the same amino acid may be coded by more than one codons **(degeneracy)**, but one codon never codes for two different amino acids.

In certain rare cases the genetic code is found to be **ambiguous** (same codons coding for different amino acids under different conditions). For example, in streptomycin sensitive strain of *E. coli*, UUU codon, normally codes for phenylalanine, but may also code for isoleucine, leucine or serine when ribosomes are treated with streptomycin. Another example of ambiguity is seen when GUG, sometimes, functions as a start codon in protein synthesis. As a start codon GUG specifies amino acid methionine, whereas GUG within the coding sequence codes for valine. This is called **site specific variation in codon translation** and is also described as **context effect in codon translation**. This ambiguity is enhanced at high Mg-ion concentration, low temperature and in the presence of ethyl alcohol.

5. Genetic Code is almost Universal with Few Exceptions

The same genetic code is said to be present in all kinds of living organisms including viruses, bacteria, unicellular and multicellular organisms. Thus, for example, lysine is coded by AAA or AAG in the mRNA of all organisms, arginine by CGU, CGC, CGA and so on. Hence, scientists can isolate mRNA from one organism, translate it by using synthetic machinery isolated from another organism and obtains proteins as if it has been synthesised in the original organism. However, code is not absolutely universal. There are a few differences in the genetic code in mitochondria, chloroplasts and some organisms. Protein synthesis in plant mitochondria uses the universal nuclear genetic code. But, mitochondria of other organisms have differences in their codons shown as under.

1. **AUA** and **AUG** both code for methionine, instead only AUG in nuclear genetic code. In nuclear code AUA specifies isoleucine.
2. **UGA** in the nuclear code acts as a stop codon but in mitochondria it specifies tryptophan.
3. **AGA** and **AGG** are stop codons in mitochondria but act as arginine codons in the nucleus.
4. **UAA** and **UAG** do not act as stop codons in mitochondria of some ciliated Protozoa They code for glutamine.

Table 12.3 Differences between human and yeast mitochondrial genetic codes

Sl.No	Amino Acid			
	Codon	Nuclear code	Mitochondrial code in human	Mitochondrial code in yeast
1.	UGA	Termination	Tryptophan	Tryptophan
2.	AUA	Isoleucine	Methionine	Isoleucine
3.	CUN*	Leucine	Leucine	Threonine
4.	AGG, AGA	Arginine	Termination	Arginine
5.	CGN	Arginine	Arginine	Termination

*N represents all the four nitrogenous bases A, C, G, U.

6. Genetic Code is Degenerate

Since there are $4^3 = 64$ possible codons for only 20 amino acids that are found in proteins, it was concluded that in most cases a single amino acid is coded for by two to several different codons. This multiple system of coding is known as **degenerate system** or **degenerate genetic code**. Except for methionine and tryptophan, which have only one codon each, all other amino acids have two or more codons. The degeneracy of coding system provides protection to organisms against many harmful mutations, stabilises phenotypes by lessening the effect of random mutations and minimises the consequences of base pairing errors. Codons that specify the same amino acid are called **synonyms**. Most synonyms differ only in their third base of the codons. For example, GUU, GUC, GUA and GUG all code for amino acid valine. It means the major degeneracy occurs at the third position at the 3′ end of the triplet codon. When first two

Extension: Genetic Code

Genetic code contains 64 codons $4^3 = 64$. Out of these, three are stop codons or **termination codons** (UGA, UAA, UAG) and one **initiation codon** (AUG). The remaining 60 codons code for 20 amino acids. According to Crick's wobble hypothesis, only 32 types of tRNA could be sufficient to transport 20 amino acids, not the 60. Because of wobbling nature of first base of anticodon, one tRNA can recognise several different codons coding for the same amino acid.

bases are specified, the same amino acid may be coded for whether the third base is U, C, A or G. This base is described as **'Wobbly base'**.

For example, note the genetic codes for the following amino acids:

(a) **Serine:** UCU, UCC, UCA, UCG and AGU, AGC
(b) **Arginine:** CGU, CGC, CGA, CGG and AGA, AGG
(c) **Leucine:** CUU, CUC, CUA, CUG and UUA, UUG
(d) **Valine:** GUU, GUC, GUA, GUG.

The following table represents the occurrence of multiple codons for different amino acids and clearly illustrates the degeneracy of genetic code:

S. No.	Amino Acids	No. of codons for each amino acid	Total no. of codons
1.	Arginine, Leucine, Serine	6 each	18
2.	Alanine, Glycine, Proline, Threonine, Valine	4 each	20
3.	Isoleucine, Stop codons	3 each	6
4.	Asparagine, Aspartic acid, Cysteine, Glutamic acid, Glutamine, Histidine, Lysine, Phenylalanine and Tyrosine	2 each	18
5.	Methionine and Tryptophan	1 each	02
			64

Reasons for Degeneracy: Wobble Hypothesis (Wobble Occurs in the Anticodon)

Wobble or Fluctuating Base: Crick (1966) proposed **Wobble hypothesis**. According to this hypothesis, the base in first position of anticodon on tRNA is usually an abnormal base, like inosine, pseudouridine, tyrosine, etc. These abnormal bases are able to pair with more than one type of nitrogenous bases in the third position of the codon on mRNA. For example, inosine **(I)** can pair with A, C or U. This base is called a **Wobble base** or **fluctuating base**.

- Wobbling between anticodons and codons allows some tRNA molecules to read more than one codon.
- The codon/anticodon base pairing rules are known as Wobble rules.

The last two nucleotides of an anticodon on tRNA are strictly complementary to first two nucleotides of codon in mRNA as per classical rule of pairing. But the pairing of third base on mRNA with first base of anticodon of tRNA is less stringent and may occur in a nonclassical fashion, so that in some cases a single tRNA may base pair with more than one codon. This is called **Pairing Wobble** or **Wobble Base Pairing**.

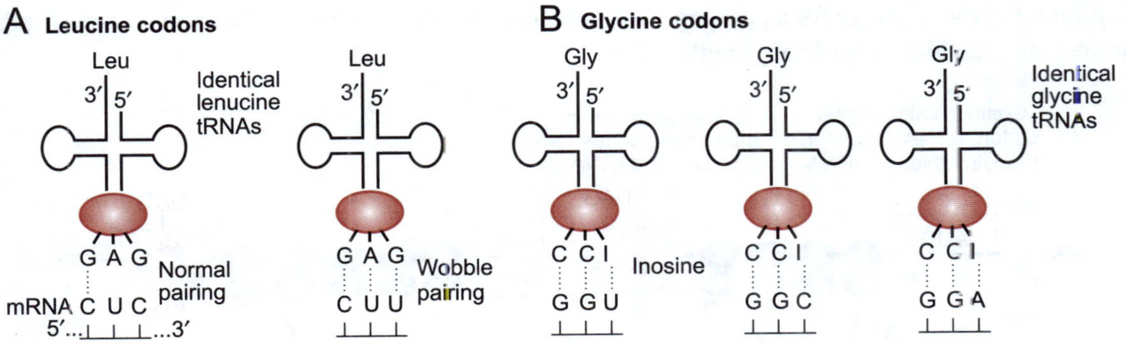

Fig. 12.2 Two examples of wobble base pairing. **A.** Two different leucine codons CUC and CUU can be read by the same anticodon of leucine tRNA; **B.** Three different glycine codons GGU, GGC, and GGA can be read by the same anticodon of glycine tRNA molecule having abnormal or wobble base (I)

Table 12.4 Anticodon – codon base pairing rules and Wobble pairing

Nucleotide at first position at the 5′ end of tRNA anticodon	Normal pairing with nucleotide at 3′ position of codon at 3′ end of mRNA	Wobble base pairing with nucleotide in mRNA
G	C	U or C
C	G	G
A	U	U
U	A	A or G
I (Inosine)	–	A, U or C

Biological Significance of Degeneracy of Genetic Code

Degeneracy of genetic code is a boon for the survival of living beings, because of the following reasons:

1. It is protective device against mutations. The synonym codons usually differ only in the base occupying third position in the codon. The first two nitrogenous bases are important in coding. Therefore, mutations that change the third base of the codons often go unnoticed and amino acid in the polypeptide chain remains unchanged.
2. Chain termination mutations usually lead to the synthesis of incomplete and nonfunctional proteins. This may lead to the appearance of recessive character. Thus occurrence of three termination codons helps in reducing the probable harmful effect of certain mutations.

7. Genetic Code has Chain Initiation (start) and Chain Termination (stop) Codons

A reading frame of the genetic message always has a specific start and stop signals for protein synthesis. The codon present in the beginning of the cistron known as **initiation codon** marks the beginning of the message for a polypeptide chain. The initiation codon is **AUG** and it codes for the amino acid, methionine or formylmethionine.

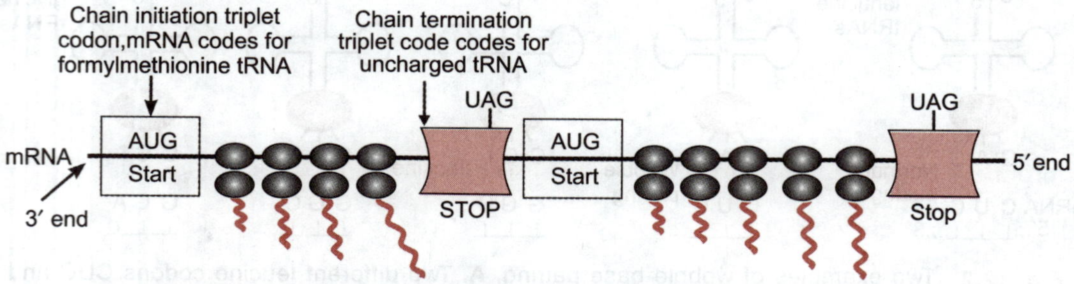

Fig. 12.3 Role of initiation and termination codons

Similarly, the last codon of a cistron helps in reading the termination of polypeptide chain. This is known as **termination codon**. There are three termination codons: **UAA, UGA** and **UAG**. Earlier when the function of these codons was not known, these were called **nonsense codons**. No tRNA molecule has anticodon for these codons. Therefore, synthesis of polypeptide chain stops here, that is why these codons are called **stop codons** or **termination codons.**

The initiator and terminator codons are known as signals and this phenomenon is known as punctuation. Punctuation helps in delimiting the different cistrons on a polycistronic mRNA.

The signals or spacers do not occur between the codons of a message but initiate and terminate larger sections of the message between functional genes. These particular triplets are recognised by two special tRNAs that carry no amino acid and prevent further growth of the protein chain.

8. Genetic Code has Collinearity

The codons in DNA and mRNA and the corresponding amino acid residues in the polypeptide chain have a linear arrangement which has been demonstrated by the studies of T_4 mutants. These produce incomplete head protein molecules. These mutants can be shown to map in linear sequence by the recombination technique. This suggests that the code is **collinear**.

DISCOVERY OF GENETIC CODE OR CRACKING OF GENETIC CODE

The existence of a triplet code was simply an assumption till **Marshall Nirenberg** (Nobel Prize winner) and **Heinrich Matthaei** in 1961 proved its existence by experiments. They were able to synthesise artificial mRNA which contained molecules of only one base uracil. It was named **polyuridylic** (poly U) molecule. The synthetic poly U mRNA was placed in a cell-free system containing protein-synthesising system, enzyme ***polynucleotide phosphorylase*** extracted from *Escherichia coli* and the twenty amino acids together with necessary ATP. After some time a small protein-like molecule was produced which was formed by the linking of phenylalanine (a homopolymer).

Nirenberg and Matthaei Initially used Homopolymers

In their initial experiments, **Nirenberg** and **Matthaei** used RNA homopolymers of each type of ribonucleotides, like homopolymers of U(UUUUUU..........), C (CCCCCC..........) and of G (GGGGG..........). They found that **poly U** produced polyphenylalanine polypeptide, poly A resulted in the synthesis of polylysine and poly C produced polyproline. On the basis of above information codons UUU, AAA and CCC were assigned for amino acids phenylalanine, lysine and proline.

Marshall Nirenberg
(April 10, 1927)

Nirenberg successfully prepared cell free protein synthesising system from *E.coli*. By adding new mRNA (poly U), Nirenberg was able to synthesise polyphenyl polypeptide chain. This was described as **cracking of genetic code**, for which he shared 1968 Nobel Prize with H.G.Khorana.

Har Gobind Khorana
(January 9, 1922)

Har Gobind Khorana was born in 1919 in Raipur (now in Pakistan). He made remarkable contribution in artificial synthesis of nucleic acids. He discovered how to synthesise triplet RNA molecules of known sequence thereby assigning the genetic code.

He also synthesised **artificial genes** (long DNA molecules).

Khorana shared in 1968 Nobel Prize with Marshall Nirenberg and Robert Holley.

Use of Mixed Copolymers

After establishing the codons of homopolymers, **Nirenberg**, **Matthaei** and **Ochoa** used RNA heteropolymers. Using all the four ribonucleotides to construct mRNA, they established the nature of codons formed by two or more nitrogenous bases, such as ACA, CCA, CAA, etc. These were described as **copolymers**. Finally, by the end of 1968, composition of all the 64 triplet codons corresponding to all 20 amino acids were established.

Triplet Binding Assay

In 1964, **Nirenberg** and **Philip Leder** developed the **triplet binding assay**. For this, triplets of known sequences were synthesised in the laboratory to serve as templates. The radioactively charged tRNA, the RNA triplets and ribosomes were incubated on a nitrocellulose filter and radioactive tRNA associated with RNA triplet, and ribosomes were isolated from the nitrocellulose filter and were identified. This revealed the specific codon and anticodon association.

Contribution of Khorana

Indian born biochemist, **Dr. H.G. Khorana**, devised an ingenious technique for artificially synthesising mRNA with repeated sequences of known nucleotides. For this valuable contribution he was awarded Nobel Prize in 1968.

By using synthetic DNA, **Khorana** and his coworkers prepared chains of polyribonucleotides with known repeating sequences of two, three and four nucleotides as follows:

- Poly CUC UCU CUC UCU ....
- Poly CUA CUA CUA CUA ....

In first case CUC and UCU are two codons arranged alternately in the polynucleotide chain. This dictates the formation of polypeptide chain having two amino acids (leucine and serine) arranged alternately. The second case is an example of homopolymer chain. The polynucleotide chain is formed of repeated linkage of codon CUA. This dictates the formation of a polypeptide chain consisting of only one amino acid leucine.

Table 12.5 List of synthetic co-polymers of RNA, codons derived from them and amino acids incorporated to form polypeptide

Copolymers	Codons formed	Coded for amino acid in polypeptide chain
UG	UGU GUG	Cysteine Valine
AC	ACA CAC	Threonine Histidine
UUC	UUC UCU CUU	Phenylalanine Serine Leucine
AUC	AUC UCA CAU	Isoleucine Serine Histidine
UAUC	UAU, UAC CUA, UUA UCA, UCA AUC, AUU	Tyrosine Leusine Serine Isoleucine
GAUA	GAU AGA UGA UAG AUA UAA	Asparitic acid None None None Isoleucine None

TEXT QUESTIONS

1. What do you mean by 'Genetic code'? Discuss in brief the special features of genetic code.
2. Describe the contribution of Dr. H.G. Khorana in understanding genetic code.
3. What is cracking of the genetic code?
4. Outline the essential features of triplet code.
5. Write short notes on:
   (a) Contribution of Kornberg
   (b) Contribution of H.G. Khorana
   (c) Ambiguity
   (d) Collinearity.
6. The triplet genetic code is degenerate. How could you justify the statement?
7. It is found that both (a) UUU and VVC code for phenylalanine, (b) but when ethyl alcohol is used in high concentration, the incorporation of leucine and isoleucine is sharply increased, while the incorporation of phenylalanine is decreased for the same codons Which of these situations (a) or (b) represents ambiguity and which degeneracy?
8. What is the role of nonsense codons in protein synthesis?
9. Who coined the term 'genetic code'?
10. What is codon?
11. What is anti-codon?
12. What is genetic code?
13. Name three nonsense codons.
14. How many bases code for one amino acid?
15. Name the amino acids which have one codon each.
16. Genetic code is non-overlapping and degenerate, why?
17. Name two codons that code for more than one amino acid.
18. Which base triplets code for the amino acid phenylalanine?
19. Are there any base triplets that code for amino acids and also for start signals?
20. What is the role of nonsense codons in protein synthesis?
21. Which RNA molecule bears codons and which RNA molecules has anti-codon?
22. What anti-codons will be required to recognise the following codons: (a) AAU (b) CGA.
23. Who cracked the genetic code?
24. Describe biological significance of degeneracy of genetic code.
25. What is the use of synthetic copolymers?
26. Why is the concept of 'universality of genetic code' threatened?

13
Chapter

Protein Synthesis

IN THIS CHAPTER

Concept About Flow of Genetic Information: One Way Flow; Circular Flow; Inverse Flow; Reverse Transcription; Raw Material Needed for Protein Synthesis; Mechanism of Protein Synthesis; Transcription; Basic Steps in Translation; Translation in Prokaryotes; Transcription coupled Translation; Polyribosome formation in Prokaryotes; Translation in Eukaryotes; Role of Ribosomes in Protein Synthesis; Role of tRNA in Protein Synthesis; Role of mRNA in Protein Synthesis; Translocation of Protein Synthesised; Inhibition of Protein Synthesis; Post-translational Processing of Proteins; Activation of Amino Acids; Attachment of Activated Amino Acids with tRNA Molecules; Stages during Translation.

Proteins are giant molecules (macromolecules) of utmost biological importance because these not only form the building material of a cell but act as enzymes controlling all the biochemical reactions going on in the living system and also control the expression of phenotypic traits.

Chemically, proteins are **polymers of amino acids**. The amino acid molecules are linked together in linear fashion forming a long **polypeptide chain.** The adjacent amino acids are joined together by **covalent peptide bond** between the carboxyl group of one amino acid and the amino group of the other.

Only 20 amino acids are biologically important and participate in the formation of polypeptide chains of proteins. As is evident that there are several thousands of proteins but only 20 amino acids, the various types of proteins are formed by different arrangement of these amino acids. The change even of a single amino acid in a polypeptide chain of more than hundreds can destroy or change the function of a protein. All the proteins which are required by the cell are synthesised by the cell itself.

Protein synthesis, therefore, includes arrangement of a definite number of amino acids in a definite sequential order. This sequence of amino acids in the polypeptide chain is determined by the sequence of nucleotides in the polynucleotide chain of DNA. A sequence of three nitrogenous bases in DNA codes for one specific amino acid of the polypeptide chain. This is known as **triplet code**.

CONCEPTS ABOUT FLOW OF GENETIC INFORMATION

1. One Way Flow of Information: **Crick** (1958) suggested that DNA transmits the genetic information for protein synthesis by transcribing mRNA **(transcription)** and other types of RNAs. This information encoded in the ribonucleotide sequences in RNA is then translated into the sequence of amino acids **(translation)**. This is known as **one way flow of information** or **central dogma**.

2. Circular Flow of Information: **Barry Commoner** (1968) suggested about circular flow of information. The DNA transcribes RNA (mRNA, tRNA, etc). RNA translates the information into protein, protein then directs the synthesis of RNA and RNA dictates the synthesis of DNA.

3. Inverse Flow of Information: **Temin** (1970) described the presence of an enzyme, ***RNA dependent DNA polymerase***, which controls the synthesis of DNA from a single strand of RNA. **D. Baltimore** (1970) described the presence of this enzyme in certain RNA tumour viruses. In tumour viruses, RNA is genetic RNA but in certain cases the nongenetic RNA synthesised under the control of DNA may also control the synthesis of DNA by inverse flow of information. This is known as **central dogma reverse or reverse transcription** (**H. Temin**, **Baltimore** along with **R. Dulbecco** were awarded Nobel Prize in Medicine in 1976).

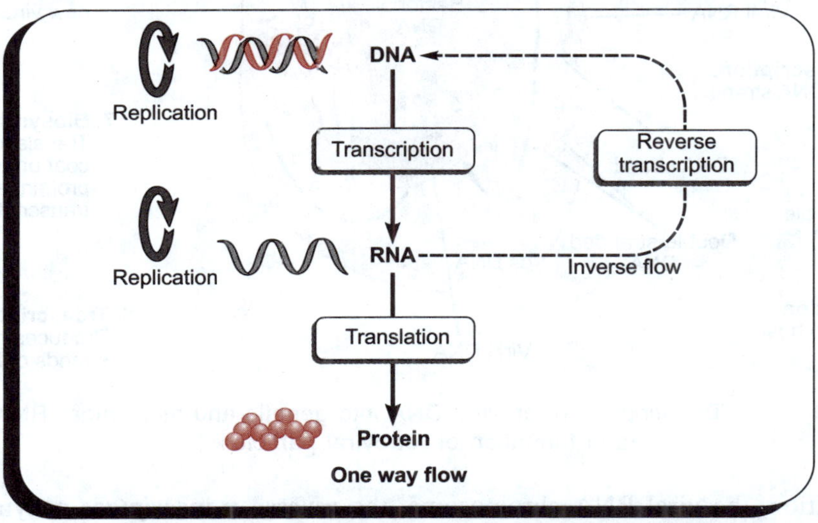

Fig. 13.1 The central dogma states that information in nucleic acid can be perpetuated or transferred, but transfer of information into protein is irreversible

REVERSE TRANSCRIPTION

In eukaryotes, DNA is the hereditary material. It can replicate its own copy and also transfer the information to messenger RNA by transcription which is then translated in the form of protein. In many RNA viruses such as polio virus, influenza virus, double-stranded RNA phages, and most of the single-stranded RNA phages, the synthesis of viral RNA is primed in the host cell directly by the viral RNA introduced by the viral infection. The synthesis is controlled by ***RNA-dependent-RNA polymerase*** enzyme, also called the ***replicase***.

In 1970, **Temin** and **Mizutani,** and also **Baltimore** discovered a phenomenon called **reverse transcription** in certain single-stranded tumour-producing RNA viruses or RNA retroviruses. *Rous sarcoma* virus and Rauscher mouse leukemia virus can synthesise DNA from RNA template because large viruses have a special enzyme, ***reverse transcriptase*** or ***RNA-directed DNA polymerase***.

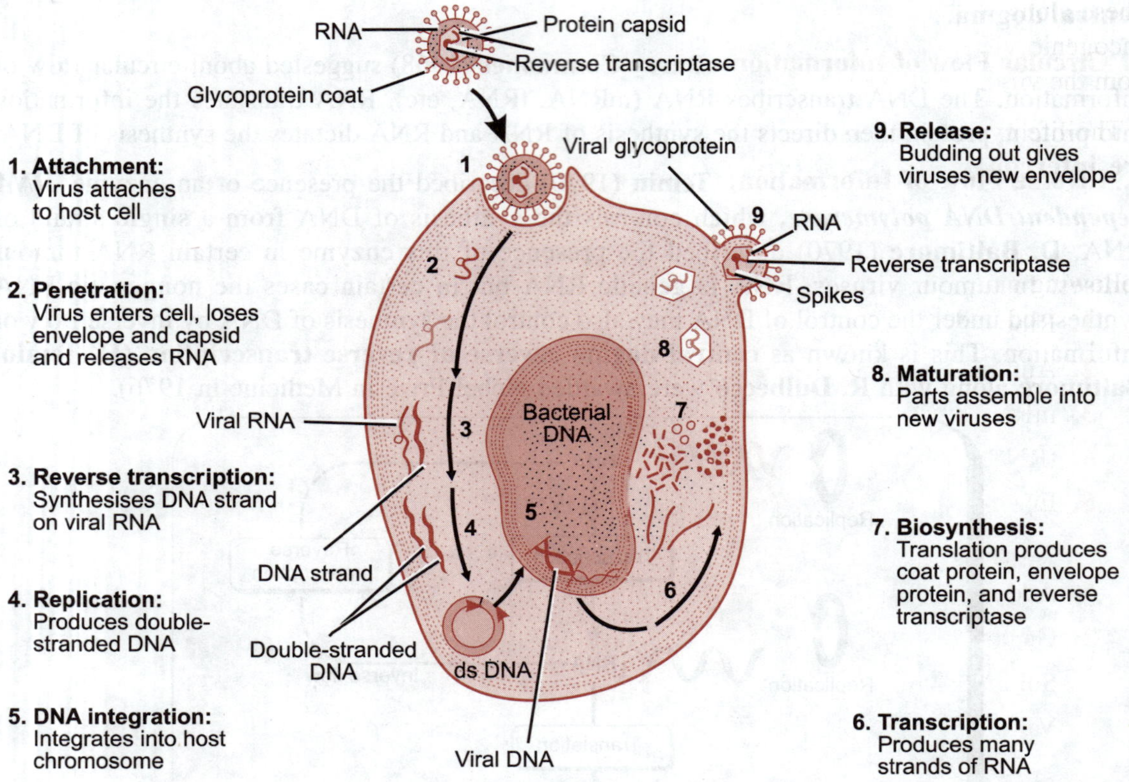

Fig. 13.2 Transcription of proviral DNA into genetic and messenger RNA and formation of retroviral particles

During infection, the viral RNA chromosome, the ***reverse transcriptase*** enzyme and some additional protein from retrovirus enter the host cell. Enzyme ***reverse transcriptase*** then synthesises a DNA strand complementary to all the viral chromosomes, forming a RNA/DNA hybrid. Synthesis occurs in 5′→3′ direction. ***Reverse transcriptase*** possesses ***RNAse*** activity and can copy DNA as well as RNA. It also degrades RNA strand of the RNA/DNA hybrid. This free DNA strand is used as a template to synthesise a DNA double helix. The double-stranded DNA is then integrated into the host chromosome and is named **provirus** or **proviral DNA**.

The provirus replicates along with the host chromosome for several cell generations. It may even become incorporated into the germinal tissue and may be passed successively through generations. Such **endogenous proviruses** have been found in large number of animals including birds, rodents, primates and man. Thus, genetic RNA may lead to the formation of genetic DNA.

Under normal conditions, these proviruses remain inactive. But, sometimes, a cellular signal can stimulate transcription of proviral DNA into RNA. Some of the RNA is messenger RNA while the rest is retroviral chromosomal RNA. This mRNA is translated into **viral proteins.** These along with retroviral RNA form retroviral particles. A byproduct of such transcription in oncogenic viruses (cancer-producing viruses) is the production of oncogenic proteins derived from the viral chromosome.

The life cycle of retroviruses confirms the statement, **'The genetic DNA and genetic RNA are interconvertible'.**

RAW MATERIAL FOR PROTEIN SYNTHESIS

Following machinery and substances are required for protein synthesis:

1. Ribosomes
2. Amino acids
3. mRNA
4. tRNA
5. Enzymes: (a) Amino-acid activating enzymes: ***amino acyl-tRNA synthetase***, and
   (b) ***Peptide polymerase*** system
6. Adenosine triphosphate (ATP) as an energy source
7. Guanosine triphosphate (GTP) for synthesis of peptide bond
8. Soluble protein and transfer factors
9. Various inorganic cations such as Mg^{2+}, K^+ and NH_4^+.

MECHANISM OF PROTEIN SYNTHESIS

The central dogma of protein synthesis comprises of following two steps:

I. Transcription II. Translation

I. Transcription of mRNA from DNA: An Overview

The synthesis of RNA on template DNA through the enzymatic action of ***DNA-directed RNA polymerase*** or simply ***RNA polymerase*** is called **transcription.** It is the first step of central dogma. It involves transfer of information from DNA to mRNA. The information encoded in the base sequence of DNA are copied into the base sequence in mRNA. Therefore, mRNA is complementary to DNA segment from which it is transcribed. Transcription of DNA into RNA proceeds by complementary base-pairing mechanism similar to DNA replication, except that only one of the two strands (3′→5′ strand) is copied. The detailed process of transcription has been discussed in the previous chapter.

II. Translation (Assembly of Polypeptide Chain): An Overview

Translation is the second step of central dogma. It involves a change in the language from nucleotide order in an mRNA molecule to amino acid sequence in a polypeptide. In translation, the order of nucleotides of mRNA is read as triplet codons and determines the order of amino acids in the growing polypeptide chain.

RNA polymerase binds to a particular binding site at the beginning of a gene. Starting from there, it moves along the strand and encounters with each DNA nucleotide and adds the corresponding complementary RNA nucleotide to the growing mRNA strand. When ***RNA polymerase*** arrives at the stop signal at the opposite end of the gene, it gets disengaged from the DNA and a newly assembled mRNA molecule is released. The transcribed mRNA moves out of the nucleus, undergoes processing and attaches to the ribosomes to direct protein synthesis. Transcription occurs only during G_1 and G_2 substages of interphase at the time of protein synthesis.

Extension: An Overview of Protein Synthesis

The hereditary information encoded in genes is expressed in two phases: 1. **Transcription,** in which enzyme ***RNA polymerase*** assembles an mRNA molecule on template DNA strand. Its nucleotides sequence is complementary to the DNA nucleotide sequence of the gene from which it is copied. 2. **Translation,** in which a ribosome assembles a polypeptide, whose amino acid sequence is specified by the nucleotide sequence in the mRNA.

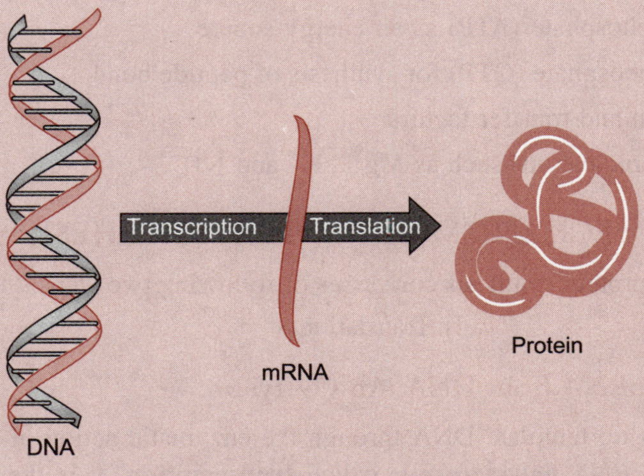

Fig. 13.3 The central dogma of gene expression. DNA is transcribed to make mRNA, which is translated to make a protein

Translation begins when an rRNA molecule within the ribosome recognises and binds to the initiation codon on mRNA. The ribosome then moves along mRNA molecule, three nucleotides (= one codon) at a time. Codon specific tRNA with its amino acid comes to join the codon on mRNA. This is then transferred to the growing polypeptide chain. The ribosome continues to move on mRNA and its polypeptide chain keeps elongating, until ribosome reaches the stop/termination codon which does not code for any amino acid. The ribosomes, the polypeptide chain and mRNA are all set free.

The process of translation involves follwing steps:

- Activation of amino acids.
- Attachment of activated amino acids with tRNA (Amino-acetylation of tRNA).
- Stages of translation: Initiation, elongation and termination of polypeptide chain.
- Modification of released polypeptide chain.

1. Activation of Amino Acids

Amino acids in the cytoplasm occur in inactive form and they can not take part in protein synthesis. Hence these are activated by giving them energy. The activation is facilitated through ATP, whose molecules unite with the amino acids forming highly reactive **amino acid phosphate-adenyl complexes** which are known as **aminoacyl adenylates.** The process of activation is governed by specific enzyme, ***aminoacyl tRNA synthetase***. Usually each amino acid has its own specific ***aminoacyl tRNA synthetase enzyme***. Hence, there are as many enzymes as the number of amino acids (*i.e.* 20). Each enzyme has double specificity. The specificity is provided by **recognition region**. It recognises its own amino acid and finds out its own tRNA. The mechanism of activation can be represented as below:

$$\text{AA} + \text{ATP} \xrightarrow{\text{AA activating enzyme}} \underset{\substack{\text{Activated} \\ \text{amino acid}}}{\text{AMP}\sim\text{AA}} + \underset{\text{Pyrophosphate}}{\text{PP}}$$

Where, AA	represents	amino acid
ATP	represents	adenosine triphosphate
AMP	represents	adenonsine monophosphate
PP	represents	pyrophosphate (inorganic)

For example, with amino acid serine following reaction occurs:

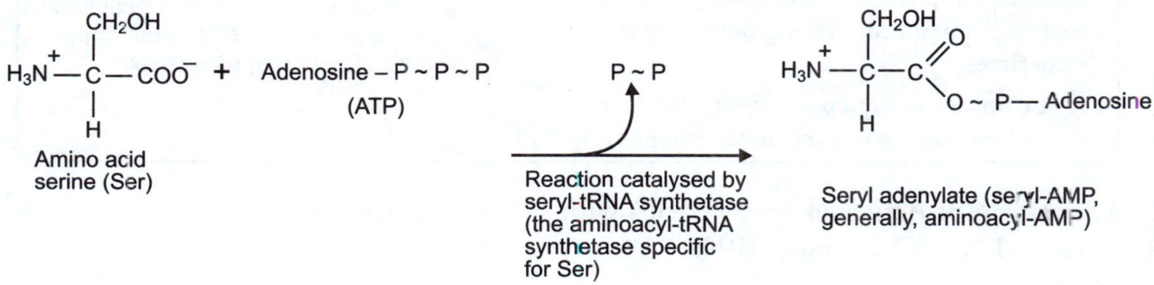

Fig. 13.4 Reaction that occurs in the activation of amino acid serine with the help of enzyme *amino-acyl tRNA synthetase*

Extension: Aminoacyl tRNA Synthetases

The enzymes **aminoacyl tRNA synthetases** are required to link amino acids to their cognate tRNA molecules. There are twenty different ***aminoacyl tRNA synthetases*** in each cell, one for each of the amino acids that take part in protein synthesis. But as many as 60 different tRNA molecules are found in the cell. It means some amino acids have more than one tRNA and only one aminoacyl tRNA synthetase for all tRNAs of that amino acid. It means:

1. The same aminoacyl tRNA synthetase enzyme can recognise all the tRNAs for the same amino acid, in case there are more than one tRNAs for one amino acid.
2. Different or more than one synthetases are present in the cell for each of those amino acids which are specified by more than one codon.

The specificity of ***aminoacyl tRNA synthetases*** ensures the joining of correct amino acid with each tRNA with a high degree of accuracy. However, wrong pairing of amino acid and tRNA does occur because each of these enzymes have to exercise two kinds of specificity, i.e. an ***aminoacyl tRNA synthetase*** has to recognise a proper tRNA and a proper amino acid. Since all tRNA molecules have a very similar primary, secondary and tertiary structures, pairing of enzyme with specific tRNA does go wrong sometimes.

The linkage of amino acid with tRNA is through an ester bond and needs energy from ATP.

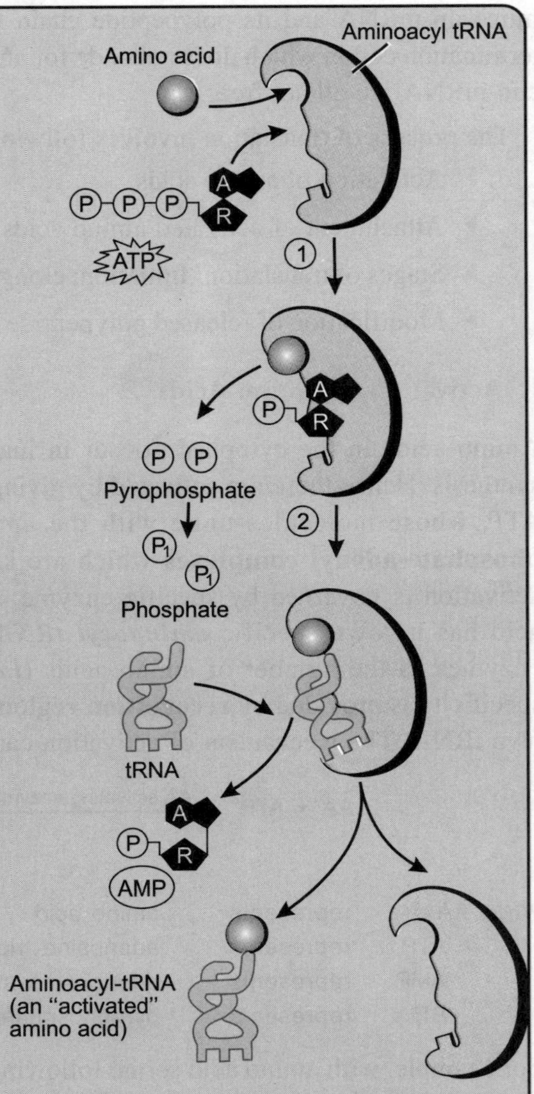

Fig. 13.5 Process of amino acid activation by enzyme *aminoacyl tRNA-synthetase* and its attachment with tRNA

2. Attachment of Activated Amino Acids with tRNA or Formation of Aminoacyl-tRNA (Charging of tRNA)

The enzyme bound activated amino acids, **aminoacyl adenylates,** become attached to the 3′ end of their respective tRNA molecules. The attachment is catalysed by the same enzymes, ***aminoacyl transfer RNA synthetases***, that catalyse activation of their amino acid. The product thus formed is known as **aminoacyl transfer RNA complex** (aminoacyl tRNA). The reaction is as under:

$$AA \sim AMP + tRNA \xrightarrow[\text{enzyme}]{\textit{AA tRNA synthetase}} AA \sim tRNA + AMP$$

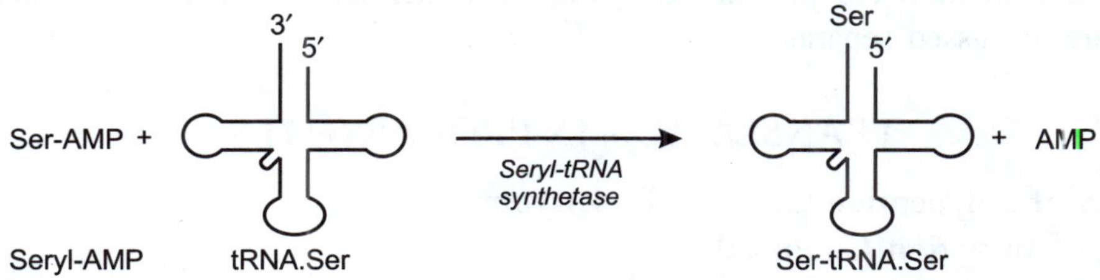

Fig. 13.6 Reaction that occurs in the binding of activated amino acid serine with ATP with the help of enzyme aminoacyl tRNA synthetase

This is important to note that particular varieties of amino acids join specific tRNA molecules. This means that for 20 amino acids at least, there are 20 different tRNA molecules and 20 different enzymes. The tRNA charged with its cognate amino acid serves as an **adaptor molecule** for decoding the information on mRNA. Therefore, tRNA is also called **adaptor RNA.** The tRNA with attached amino acid is said to be **acylated** or **charged**. The tRNA molecule without an amino acid is **uncharged tRNA** while with an incorrect amino acid, it is called **mischarged tRNA**. There are no tRNA for termination codons. Hence, synthesis of polypeptide chain terminates at a codon for which there is no tRNA.

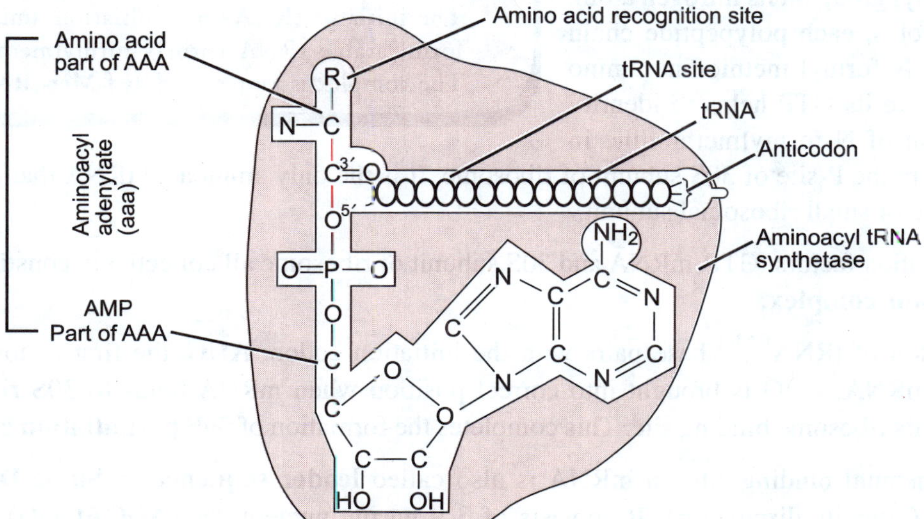

Fig. 13.7 Aminoacyl adenylate consisting of enzyme *amino acyl tRNA synthetase*, amino acid and tRNA

3. Stages during Translation

Process of translation can be separated into the following steps: (1) Initiation, (2) Elongation, and (3) Termination.

Though the process of translation includes the same three steps in both prokaryotes and eukaryotes, these are more complex in eukaryotic cells. Therefore, translation in prokaryotes and eukaryotes are discussed separately.

TRANSLATION IN PROKARYOTES

1. Initiation of Polypeptide Chain in Prokaryotes (Formation of Initiation Complex)

In prokaryotes initiation involves following four steps:

1. Three **initiation factors (IFs),** called **IF1, IF2** and **IF3** bind to the small sub-unit of ribosome with GTP attached to IF2.

2. **Initiator tRNA** with its amino acid and mRNA binds to the ribosomal sub-unit. In *E. coli* and other bacteria the initiator tRNA is called **formylatable tRNA,** and is represented as **tRNA$^{f\text{-}Met}$**. It carries **N-formyl-methionine (f-Met),** which is a modified methionine with a formyl group on its nitrogen atom. In prokaryotes, each polypeptide chain starts with N-formyl-methionine amino acid. IF_2 with its GTP helps to identify the location of N-formylmethionine in tRNA$^{f\text{-}Met}$ in the P-site of 30S subunit of ribosome. It is the only aminoacyl tRNA that can bind to the P-site of small ribosomal subunit.

Requirements for Initiation

The initiation of polypeptide chain synthesis requires:

- Associated ribosomal sub-units 30S and 50S
- mRNA to be translated
- Protein initiation factors (IFs)
- GTP
- The initiator tRNA with initiation amino acid formylatable tRNA with N-formylmethionine. The complex is represented as **f-Met-tRNA$^{f\text{-}Met}$**.

The initiation factors, GTP, mRNA and 30S subunit of ribosome all collectively constitute **30S preinitiation complex.**

3. Anticodon of tRNA$^{f\text{-}Met}$ base pairs with the initiation codon, **AUG,** the first codon on the 5′ end of mRNA. AUG is brought into correct position when mRNA binds to 30S ribosomal subunit by its ribosome binding site. This completes the formation of **30S preinitiation complex**.

The ribosomal binding site on mRNA is also called **leader sequence** or **Shine-Dalgarno Sequence** (after its discoverer). It consists of 3-9 purine nucleotides (AGGAGGU), located slightly upstream of initiation codon. This purine sequence base pairs with a pyrimidine rich complementary sequence on the 3′ end of 16S rRNA of 30S subunit of ribosome. The 3′ end of 16S rRNA was earlier called mRNA binding site.

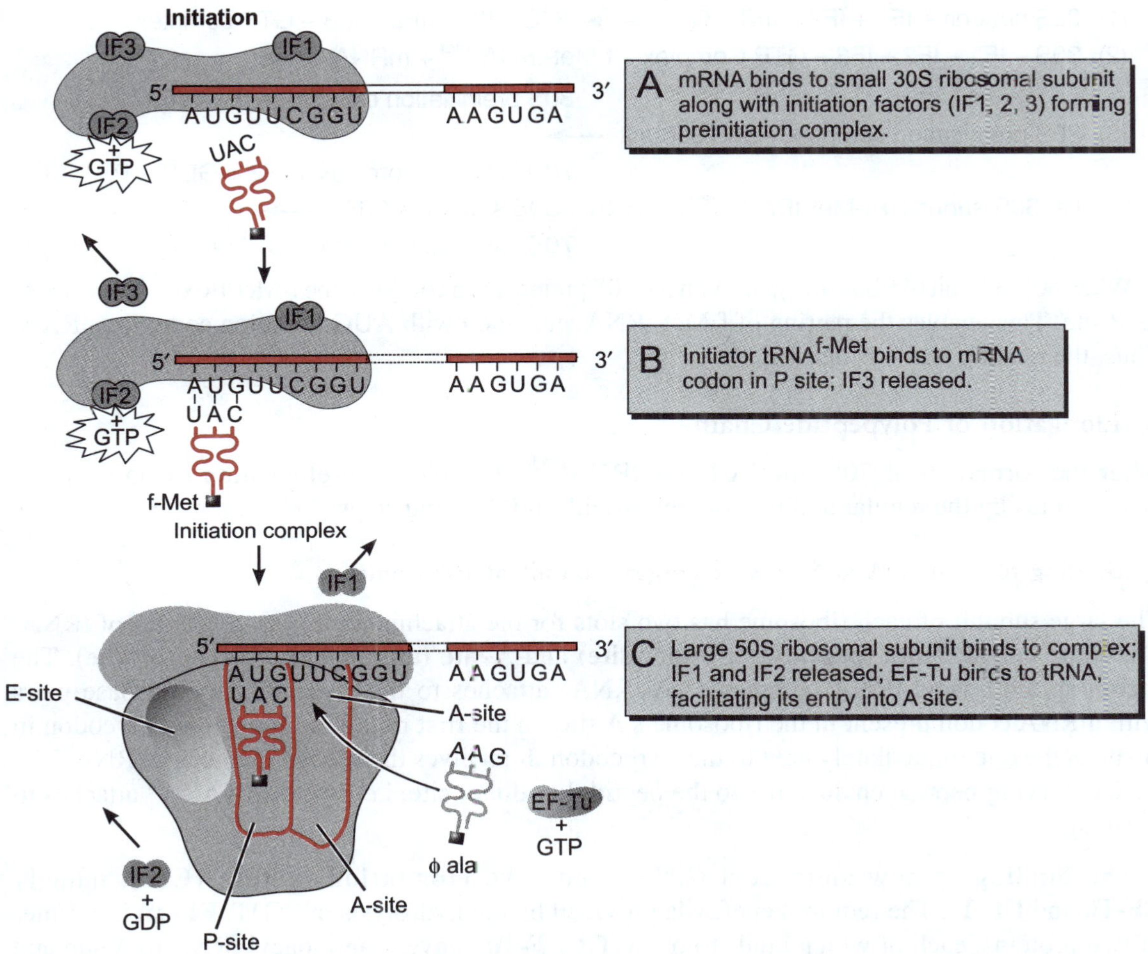

Fig. 13.8 Steps in the formation of initiation complex for the initiation of polypeptide chain in *E. coli*

Shine-Dalgarno sequence — Initiation codon

```
mRNA 5′... U G U A C U A A G G A G G U U G U A U G G A A C A A C G C...3′
                     | | | | | | | | |
16s rRNA 3′ .........A U U C C U C C A U A G C.........................5′
```

Fig. 13.9 Base pairing between Shine-Dalgarno sequence (leader sequence) in mRNA and complementary region near 3′ terminus of 16S tRNA

4. **30S preinitiation complex** formed this way joins with 50S ribosomal subunit, generating **70S initiation complex**. The energy needed for the binding of 50S subunit is provided by the hydrolysis of GTP. Mg^{2+} ions are required during this process. The entire process of polypeptide initiation in prokaryotes can be represented as follows:

(1) 30S subunit + IF1 + IF2 + IF3 + GTP $\longrightarrow$ 30S – IF1 – IF2 – IF3 – GTP complex

(2) 30S – IF1 – IF2 – IF3 – GTP Complex + f-Met-tRNA$^{f\text{-}Met}$ + mRNA $\longrightarrow$ 30S preinitiation complex + IF1 + IF3

(3) 30S preinitiation complex + 50S subunit $\longrightarrow$ 70S initiation complex + IF2 + GDP + P_i

i.e. 30S subunit + f-Met-tRNA$^{f\text{-}Met}$ + mRNA + 50S subunit + GTP $\longrightarrow$ 70S initiation complex + GDP + P_i

When 50S subunit of ribosome joins with the 30S preinitiation complex, the f-Met-tRNA$^{f\text{-}Met}$ occupies its P site. This enables the pairing of f-Met-tRNA anticodon with AUG initiation codon in mRNA. Thus, the reading frame is defined.

2. Elongation of Polypeptide Chain

After the formation of 70S mRNA-f-Met-tRNA$^{f\text{-}Met}$ complex, the elongation of polypeptide chain begins by the regular addition of amino acids in following steps:

1. Binding of AA-tRNA at Site-A of Larger Subunit of Ribosome

The large subunit of each ribosome has two slots for the attachment of two molecules of tRNA. These are called **P-site (peptidyl** or **donor site**) and **A-site (aminoacyl** or **acceptor site**). The incoming aminoacyl-tRNA complex (AA-tRNA) attaches to the acceptor site and base pairs with mRNA codon present in the ribosome's A site. In the first round of elongation, the codon in A site is the one immediately next to the start codon. It receives the second aminoacyl tRNA. The tRNA carrying peptide chain shifts to the peptidyl or donor site, i.e. f-Met-tRNA$^{f\text{-}Met}$ attaches to the P-site.

The binding of new aminoacyl tRNA needs two **elongation factors (EFs),** namely **EF-Tu** and **EF-Ts.** The required energy is provided by the hydrolysis of 2GTP. **EF-Tu** is a dimer of two proteins, each of which binds to one GTP. **EF-Tu** conveys aminoacyl tRNA to A site and is released after the transfer of aminoacyl tRNA. **EF-Ts** regenerates EF-Tu-2GTP for the next aminoacyl tRNA.

The aminoacyl tRNA binding sequence can be summarised as follows:

(1) EF-Tu + 2GTP + AA-tRNA $\longrightarrow$ EF-Tu-2GTP-AA tRNA

(2) EF-Tu-2GTP-AA-tRNA + 70S complex $\longrightarrow$ 70S complex-AA tRNA + EF-Tu-2GDP + 2Pi

(3) EF-Tu-2GDP $\xrightarrow{\text{EF – Ts}}$ EF-Tu + 2GTP

70S complex + AA-tRNA + 2GTP $\longrightarrow$ 70S complex-AA-tRNA + 2GDP + 2Pi

2. Formation of Peptide Bond

With f-Met-tRNA$^{f\text{-}Met}$ now at the **P**-site and second aminoacyl tRNA at **A**-site, a peptide bond is formed between carboxyl group of f-Met amino acid from peptidyl tRNA and amino group (NH_2^+) of amino acid from aminoacyl-tRNA of site A. As a result, two amino acids are now attached to the second tRNA, present at A-site (Fig. 13.10B).

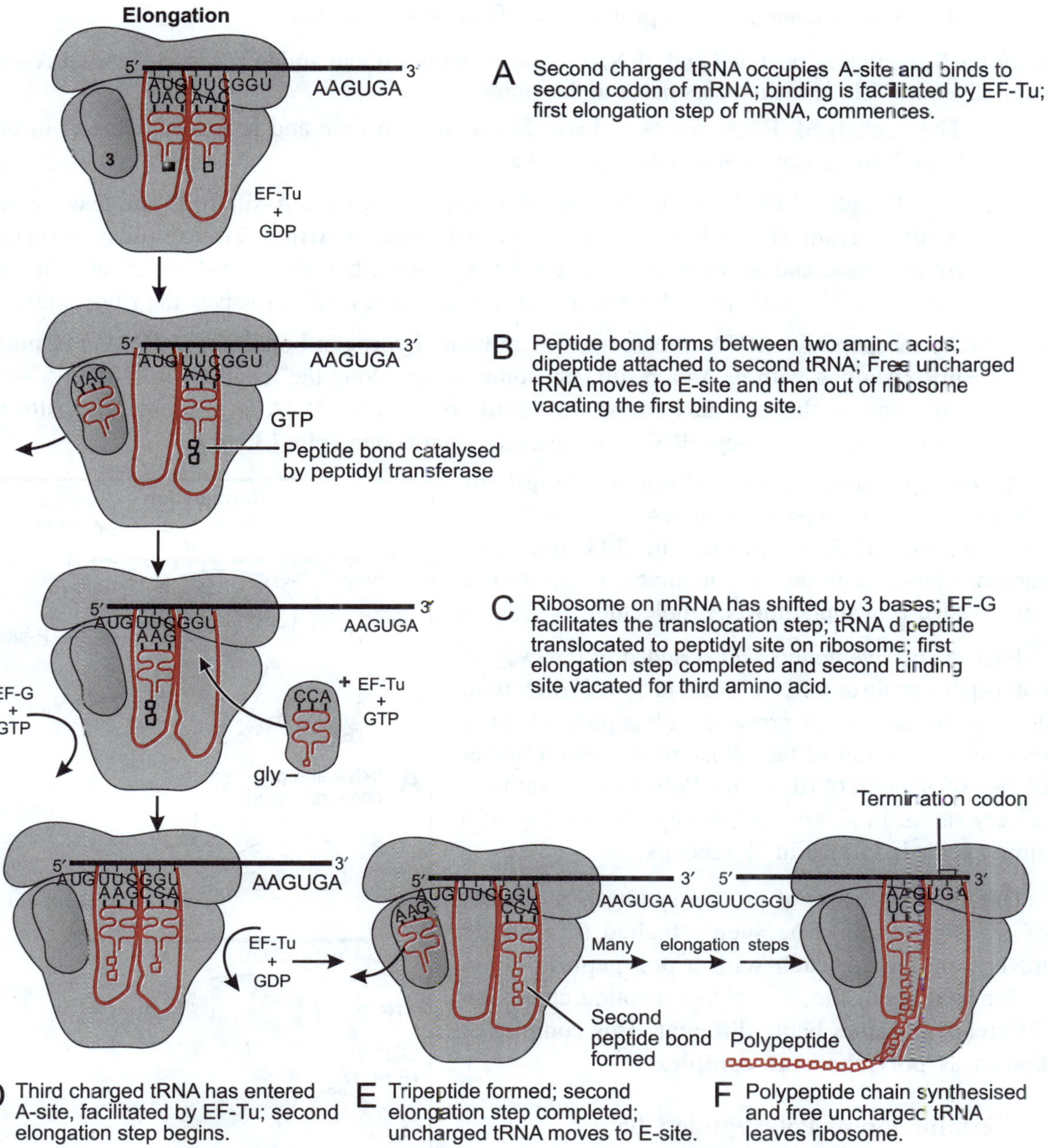

Fig. 13.10 Formation of peptide bond and elongation of growing polypeptide chain in prokaryotes

Peptide bond formation is catalysed by ***peptidyl transferase***. It is an enzymatic activity inherent in 23S rRNA of large ribosomal subunit (50S). Thus, 23S rRNA is a **ribozyme**. The energy for peptide bond formation is provided by the hydrolysis of 'high energy' ester bond by which f-Met was attached to its tRNA.

3. Translocation (Movement of peptidyl-tRNA from A-site to P-site)

After the formation of peptide bond, tRNA$^{f\text{-}Met}$ on **P**-site is without amino acid and tRNA at **A**-site has a dipeptide. Three movements occur at this point:

- The uncharged tRNA moves to leave **E**-site on ribosome and is finally released in the cytoplasm to start a new polypeptide chain.
- The **dipeptidyl tRNA** (with two amino acids) moves from **A**-site to **P**-site leaving the **A**-site vacant (Fig. 13.10C). This is called **translocation**. This requires enzyme ***translocase***, and an elongation factor **EF-G** with a bounded **GTP** which attaches to ribosome. The GTP provides energy for translocation as EF-G leaves the ribosome.
- As the peptidyl tRNA translocates, it remains hydrogen-bonded to mRNA and pulls mRNA along with it. As a result, ribosome moves along the length of mRNA in 5′→3′ direction so that the next codon (i.e. third codon) on mRNA is available at **A**-site to receive next aminoacyl tRNA having correct anticodon (Fig. 13.10D).

As one ribosome moves along the length of mRNA, the initiation point of mRNA becomes free. It can form an initiation complex with 30S subunit of another ribosome. In this way a number of ribosomes get attached to a single mRNA molecule.

Each successive amino acid is added in this way to polypeptide chain as mRNA is read in 5′ → 3′ direction. The amino end of the growing polypeptide chain is thought to pass out of the ribosome through a tunnel in the 50S subunit of ribosome. Polypeptide synthesis is very rapid. In *E. coli*, a polypeptide chain of 400 amino acids is formed in 10 seconds.

During the process of protein synthesis a number of ribosomes could be seen attached to a single mRNA molecule, each with a polypeptide chain under formation, the size of polypeptide chains on different ribosomes being different. This complex is known as **polyribosome complex**.

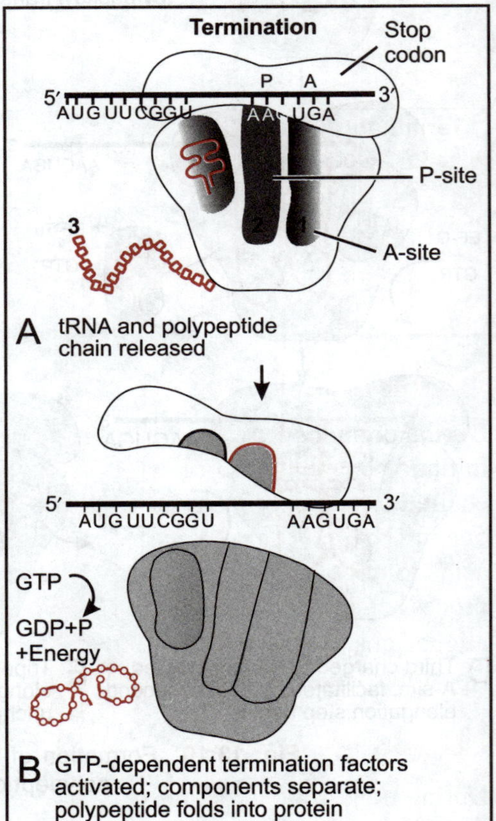

Fig. 13.11 Diagram showing termination of translation and release of polypeptide chain

3. Termination of Polypeptide Chain

Termination of polypeptide chain occurs when one of three termination codons (UAA, UAG or UGA) present on mRNA at the 3′ end of each cistron arrives at the ribosome's **A**-site. There are no tRNA molecules that recognise these termination codons. Releasing factors RF_1 and RF_2 are needed for chain termination because these recognise termination codons and terminate translation. **RF_1** is specific for UAG and **RF_2** is specific for UGA. The releasing factor with the termination codons forms a complex which induces the enzyme ***peptidyl transferase*** to catalyse the termination and release of polypeptide chain (**Lipmann**, 1973). Factor **RF_3** stimulates RF_1 and RF_2. The free ribosome now dissociates into two subunits with the help of dissociation factor IF_3.

Table 13.1 Various factors involved during translation

Process	Factors	Role
Initiation	IF_1	Stabilises 30S subunit.
	IF_2	Binds f-Met-tRNA to 30S mRNA complex; stimulates GTP hydrolysis.
	IF_3	Binds 30S subunit to mRNA, dissociates ribosomes (70S) into subunits (30S + 50S) following termination.
Elongation	EF-Tu	Brings aminoacyl-tRNA to the A-site.
	EF-Ts	Generates active EF-Tu.
	EF-G	Stimulates translocation of polypeptidyl tRNA (GTP-dependent).
Termination	RF_1	Catalyses release of the polypeptide chain from tRNA and dissociation of the translation complexes; specific for UAA and UAG termination codons.
	RF_2	Behaves like RF_1, specific for UGA and UAA codons.
	RF_3	Stimulates RF_1 and RF_2.

Polysomes or Polyribosomes

During protein synthesis, when polypeptide chain has grown to 25 amino acid residues, the **AUG** initiation codon of encoding mRNA is completely free of the ribosome to start a new polypeptide chain. Formation of a new polypeptide chain is then initiated. Now two ribosomes are attached to one mRNA molecule. When second ribosome has also moved along mRNA and has a 25 amino acids long polypeptide chain, a third ribosome also attaches to initiation codon of mRNA. This process of movement and reinitiation continues until mRNA is covered with a series of ribosomes aligned at a distance of one ribosome per 80 nucleotides. Such a large translation unit with several ribosomes attached to one mRNA is called **polyribosome** or simply a **polysome.**

Coupled Transcription-Translation

During transcription, the mRNA molecule being synthesised has a free 5′ terminus. Its ribosome binding site is transcribed first, followed by codon (AUG) that initiates synthesis of polypeptide chain, then region of mRNA associated with elongation of polypeptide chain and finally the stop codon for termination of polypeptide chain. Because, in bacteria no nuclear membrane separates DNA from cytoplasm and its ribosomes, the process of translation or polypeptide synthesis begins even before

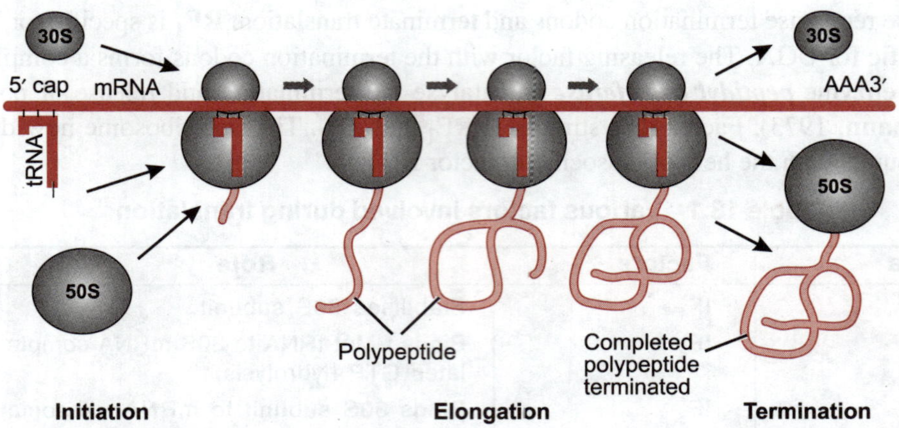

Fig. 13.12 *Polysome* showing a mRNA molecule with several ribosomes, each with individual polypeptide chain of different length. At initiation, the two subunits of ribosome are separate and at termination two subunits of ribosome are disassembled

mRNA synthesis is completed and mRNA is released from DNA. Electron micrographs have been obtained where mRNA is still attached to DNA and process of synthesis of polypeptide chain is seen. This process is called **coupled transcription-translation**. It does not occur in eukaryotes because transcription occurs inside the nucleus and translation in the cytoplasm.

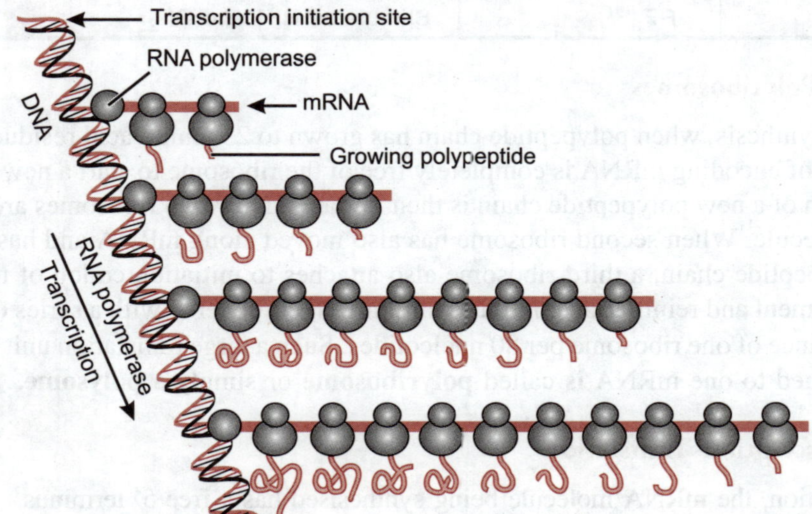

Fig. 13.13 Coupled transcription-translation in Bacteria

TRANSLATION IN EUKARYOTES

The process of polypeptide synthesis in eukaryotes follows the same general pattern as in the case of prokaryotes. It can be discussed under the following headings:

1. Initiation of Polypeptide Chain in Eukaryotes

The process of initiation of polypeptide chain in eukaryotes is more complex and needs at least twelve proteins, marked as eIFs (**eukaryotic initiation factors**). The initiation involves following steps:

- The initiation factors eIF1A and eIF3 bind to the 40S ribosomal subunit.
- GTP binds to eIF2 and this complex binds to initiator aminoacyl tRNA (methionyl $tRNA^{Met}$), forming **Met-tRNA^{Met} eIF2-GTP**. In mammals initiation factor IF2

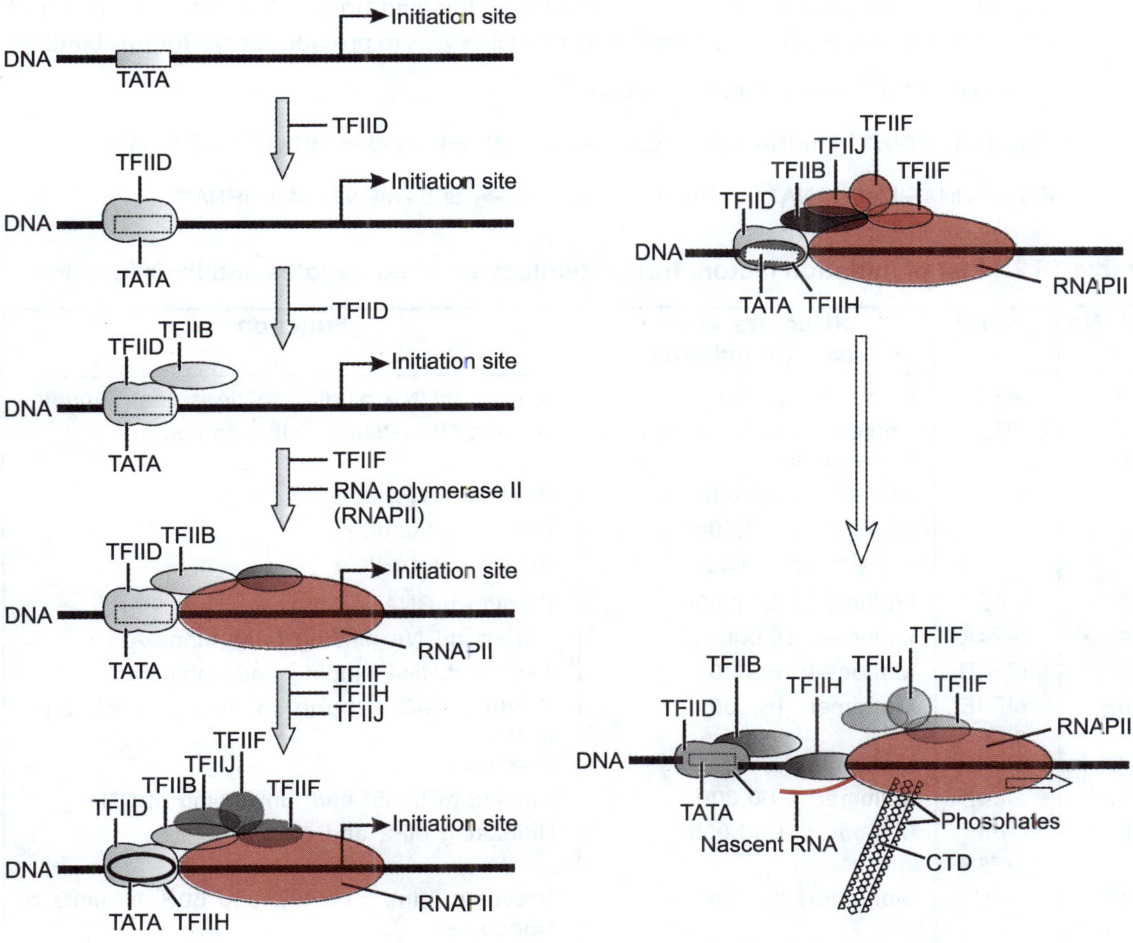

Fig. 13.14 Initiation of translation in eukaryotic cells

has three subunits α, β and γ. The α-subunit of factor eIF2 binds to eIF2. γ binds to Met-tRNA^{Met} and eIF2; β is supposed to be a recycling factor.

- Met-tRNA^{Met} -eIF2-GTP associates with 40S subunit to form **40S preinitiation complex.**
- The 40S preinitiation complex (40S-Met tRNA^{Met}-eIF2-GTP) binds to 5′ end of mRNA.
- mRNA is brought to the 40S subunit of ribosome by eIF4E.
- The eIF4E recognises and binds to the 5′ cap of mRNA.
- The eIF4G binds to both eIF4E at the 5′ cap and to PABP (the poly A binding protein) at the 3′ poly A tail and also to eIF4A and eIF4B.
- The small subunit (40S) of ribosome with initiation Met-tRNA^{Met} + eIF2α + GTP slides along mRNA to reach initiation codon AUG and gets associated with 60S subunit of ribosome. Factor eIF5 releases eIF2 and eIF3 factors and brings about the association of 60S subunit with the initiation complex. GTP hydrolyses to provide energy for this binding:

$$40S + Met\text{-}tRNA^{Met} \longrightarrow 40S - Met\text{-}tRNA^{Met}$$

$$40S + Met\text{-}tRNA^{Met} + mRNA + eIF2\text{-}GTP \longrightarrow 40S\text{-}mRNA\text{-}Met\text{-}tRNA^{Met}\text{-}eIF2\text{-}GTP$$

$$40S + mRNA - Met\text{-}tRNA^{Met} - eIF2\text{-}GTP + 60S \longrightarrow 80S\text{-}mRNA\text{-}Met\text{-}tRNA^{Met}$$

Table 13.2 List of initiation factors from reticulocytes of eukaryotes and their functions

Sl. No	Factor	Structure and Mass (in daltons)	Function
1.	eIF1	Monomer, 15,000	Assists mRNA binding to ribosomal subunit.
2.	eIF2	Trimer: Contains following three chains	Binding Met tRNA to initiation codon
		(i) α-chain; 35,000	Binds to GTP.
		(ii) β-chain, 38,000	Recycling factor.
		(iii) γ-chain; 55,000	Binds to Met-tRNA.
3.	eIF3	Multimer; 75,00,000	Binding mRNA.
4.	eIF4A	Multimer; 15,000	Assists mRNA binding (also binds ATP).
5.	eIF4B	Monomer; 15,000	Assists mRNA binding and unbinding.
6.	eIF4E eIF4G	Monomer; 15,000	Binding 60S subunit to 40S subunit and mRNA.
7.	eIF4H	Monomer	Unknown.
8.	eIF4F	Multimer; 2,00,000	Binding mRNA5′ end; unwinding of DNA.
9.	eIF5 eIF5B	Monomer; 1,50,000	Releasing eIF2 and eIF3.
10.	eIF6	Monomer; 23,000	Prevents joining of 40S and 60S subunits of ribosomes.

2. Elongation of Polypeptide Chain in Eukaryotes

The second or next codon in mRNA base pairs with the appropriate AA-tRNA at the **A**-site. This requires the presence of elongation factor I **(EFI)** and energy from GTP. A peptide bond is formed between amino group of amino acid on tRNA at A-site with carboxyl group of amino acid on tRNA at P-site. This is catalysed by ***peptidyl transferase*** enzyme present in 60S subunit of ribosome. The discharged tRNA in the **P**-site is immediately released. Enzymes ***translocase*** or ***G-factor*** also called **EF2** causes translocation of newly formed p-tRNA from **A** to **P**-site. The energy is obtained from GTP. The process is repeated again and again adding amino acids to the polypeptide chain.

3. Termination of Polypeptide Chain

When termination codon on mRNA reaches the '**A**' site of ribosome, the growth or elongation of polypeptide chain stops, because there is no tRNA for termination codons. The releasing factors **RF1** or **RF2** with the help of enzyme ***peptidyl transferase*** release polypeptide from '**P**' site and also tRNA carrying the polypeptide. Finally ribosome dissociates into 60S and 40S subunits. **RF1** recognises the termination codons UAG and UAA while $\mathbf{RF_2}$ recognises UGA and UAA.

In eukaryotes only one releasing factor **eRF1** is present. GTP is required for the binding of this factor to ribosome. On hydrolysis, GTP provides energy for the release of **eRF1** from the ribosome.

4. Modification of Released Polypeptide Chain or Polypeptide Folding

The formyl group of the first amino acid (methionine) of the released polypeptide chain is removed by the enzymes deformylases. Some other enzymes like exopeptidases remove some of the amino acids either from N-terminal end or the C-terminal end or from both ends of the polypeptide chain. Finally, this polypeptide chain alone or along with other chains gets folded to assume the tertiary or quarternary structure and is modified into a functional enzyme.

A polypeptide starts folding into its secondary and tertiary structure, while it is still being synthesised. The folding is usually spontaneous because the primary sequence of amino acids in polypeptide specifies its three-dimensional structure.

Inside the cell, protein folding is aided by a diverse class of proteins, called **molecular chaperones**. They act by inhibiting incorrect molecular interactions and incorrect folding (in fact chaperone is an English word which means a 'person' usually an elderly or married woman, who accompanies a young unmarried lady in public as a guide or protector and protects her). Two of the most widely occurring chaperone families are **Hsp70** and **Hsp60**. These are 'Heat shock proteins'. Number with them represents their molecular weight. These have ATPase activity. In addition to their role in folding of newly formed polypeptides, the chaperone proteins are also related to protein activity of the cell and to protein transport into cell organelles.

DIFFERENCES IN PROTEIN SYNTHESIS BETWEEN PROKARYOTES AND EUKARYOTES

1. In eukaryotes, most of the genes have **introns** that separate the actual message for the synthesis of one protein into small coding segment called **exons**. With the exception of a few genes in archaebacteria, genes in prokaryotes do not have introns.
2. Eukaryotes mRNA molecules are monocistronic, containing the coding sequence only for one polypeptide. In prokaryotes individual bacterial mRNA molecules are polycistronic having transcripts of several genes of a particular metabolic pathway.
3. Eukaryotic mRNA molecules are completely formed before they come out of the nucleus into the cytoplasm. In prokaryotes there is no question of mRNA molecules to move out from the place of origin.
4. In eukaryotes protein synthesis occurs in the cytoplasm. In prokaryotes protein synthesis begins even before the transcription of mRNA molecule is completed. This is called **coupled transcription-translation**.
5. The primary mRNA transcript in eukaryotes undergoes processing and splicing to change into a functional mRNA. In prokaryotes splicing of mRNA transcript does not occur.
6. In eukaryotes, mRNA molecules are modified by the addition of a 5′ G cap formed of methylated guanosine triphosphate. No such cap is formed at 5′ end of bacterial mRNA.

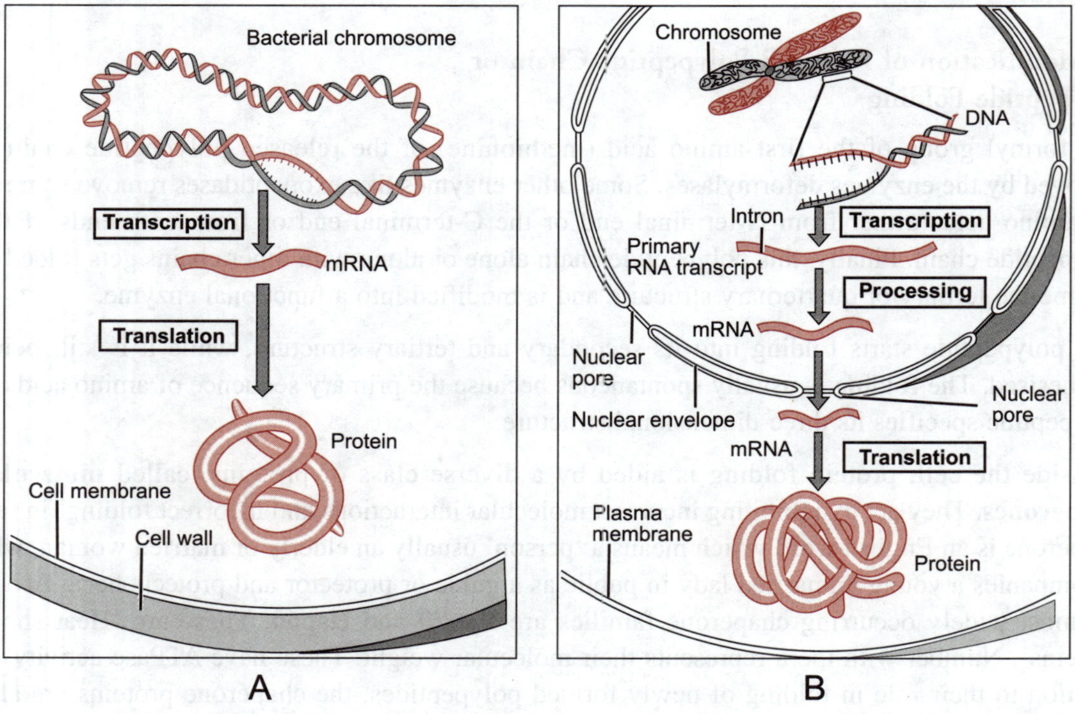

Fig. 13.15 Differences in the synthesis of protein in **A.** Prokaryotes and **B.** Eukaryotes

7. In eukaryotes 5′ G cap initiates translation by binding the mRNA to small ribosomal subunit usually at the first codon AUG. In bacteria translation begins at an AUG codon preceded by a special nucleotide sequence.

8. A **poly A tail** formed of about 200 adenine nucleotides is added at the 3′ end of mRNA in eukaryotes. No poly A tail is added to bacterial mRNA.

9. In eukaryotes initiation of polypeptide chain is brought about by special methionine-tRNA (Met-tRNA). The Met-tRNA is without formyl group whereas in prokaryotes the first or initiation amino acid is formylated methionine. (f^{Met}-$tRNA^{f\text{-}Met}$)

10. In eukaryotes, the number of initiation factors is much more than the three IFs found in prokaryotes. About ten IFs have been identified in reticulocytes and red blood cells. These are named by putting e as a prefix to denote their eukaryotic origin. These are **eIF1, eIF2, eIF3, eIF4A, eIF4B, eIF4C, eIF4D, eIF4F, eIF5 and eIF6.**

11. In eukaryotes, mRNA is without ribosome binding sequences as found in prokaryotes. The complex recognises the 5′ cap on 5′ end of mRNA.

12. In eukaryotes small subunit of ribosome (40S) gets associated with the initiator aminoacyl tRNA (**Met-tRNA$^{\mathbf{Met}}$**) without the help of mRNA. The complex joins mRNA later on. In prokaryotes 30S subunit first complexes with mRNA (30S-mRNA) which then joins with f-Met-$tRNA^{f\text{–}Met}$.

Role of Ribosomes in Protein Synthesis

Ribosomes serve as a nonspecific work-bench for translation of protein. Single ribosomes which are not involved in translation, do not occur free in the cytoplasm.

Immediately after translation, ribosomes dissociate into their large and small subunits. The two units join only during translation. The two subunits of ribosomes have the following functions:

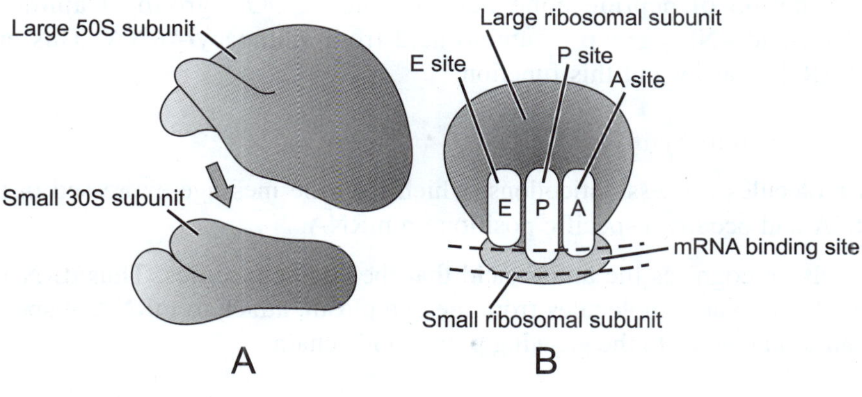

Fig. 13.16 **A.** Prokaryotic ribosome showing small and large subunits; **B.** Ribosome showing P-site and A-site for attachment of tRNA and an E-site and position of mRNA binding site

Extension: Sites on Ribosomes for Protein Synthesis

Bacterial ribosomes have four specific sites for protein synthesis. These are:

1. **mRNA binding site.**
2. **Aminoacyl site (A-site)** for the attachment of incoming amino acid carrying tRNA (AA-tRNA). This is represented by **A-site.**
3. **Peptidyl site (P-site)** for the attachment of tRNA carrying the growing polypeptide chain.
4. **Exit site (E-site)** is the site from which tRNAs leave the ribosome after they have transferred their amino acids to the growing polypeptide. E-site is confined to 50S subunit while A and P-sites are present on both 30S and 50S subunits of ribosomes.

1. Role of Small Ribosomal Subunit

The smaller subunit of ribosomes (30S in prokaryotes and 40S in eukaryotes) binds to the first codon of mRNA and forms the **initiation complex**, which initiates the linking of the first aminoacyl-tRNA (f-Methionine-tRNA$^{f\text{–}Met}$) to the first codon of mRNA (AUG). The initiation signal of 4–9 purine residues on mRNA on 5′ end located 8 to 13 base pairs prior to the initiation codon base pair with a complementary pyrimidine rich sequence near the 3′ end of 16S rRNA of 30S ribosomal subunit.

2. Role of Large Ribosomal Subunit

The large subunit of ribosome provides two slots or binding sites (**P-site** and **A-site**) for the attachment of two tRNA molecules (the peptidyl tRNA and aminoacyl tRNA). Another E-site is located exclusively on large subunit for the exit or release of tRNA after it has performed its function.

The large subunit also contains the enzyme ***peptidyl transferase*** or ***peptide synthetase*** which helps in the formation of peptide bond between the –COOH group of amino acid from polypeptide chain and $–NH_2$ group of amino acid from aminoacyl-tRNA. This enzyme is a **ribozyme**. 23S rRNA carries out this function.

Role of tRNA in Protein Synthesis

1. The tRNA molecules possess anticodons, which read the message expressed in the form of codons in mRNA and occupy a specific position on mRNA.

2. Each tRNA also recognises the amino acid that the codon specifies. Thus, tRNA molecules pick up activated amino acid molecules from the cytoplasm, attach to mRNA at specific codons and deposit their amino acids to the growing polypeptide chain.

Role of mRNA in Protein Synthesis

mRNA molecules copy the nucleotide sequence from DNA and dictate the sequence of amino acids in the growing polypeptide chain.

Inhibition of Protein Synthesis

Some antibiotics inhibit protein synthesis in bacteria. The list below shows the effect of various antibiotics:

1. **Tetracycline:** Inhibits binding of aminoacyl tRNA to ribosome.
2. **Streptomycin:** Inhibits initiation of translation.
3. **Neomycin:** Inhibits interaction between tRNA and mRNA.
4. **Chloramphenicol:** Inhibits action of enzyme ***peptidyl transferase*** and formation of peptide bonds.
5. **Erythromycin:** Inhibits translocation of ribosome along mRNA.

EFFECT OF MUTATIONS ON PROTEIN STRUCTURE AND FUNCTION

Gene mutations are caused by deletion, addition and substitution of one or more base pairs in a gene. Such mutations produce change in the structure, composition or nature of protein produced by the mutant gene. Radiations (X-rays, UV rays, gamma rays) or some chemicals cause gene mutations. **Sickle-cell anaemia, thalassemia, phenylketonuria** and **albinism,** etc. are caused by gene mutations in human beings.

Translocation of Protein

1. In case of free ribosomes, terminated protein molecules are released in the cytoplasm and are translocated to mitochondria, chloroplasts and nucleus, or to the Golgi body. Inside Golgi these may undergo **glycosylation** forming **glycoproteins**.

2. In case ribosomes are attached to ER membranes, the released protein molecules are extruded into the lumen of ER. Some of them become integral part of membranes or are transferred to Golgi complex.

In 1975, **Gunther Blobel** and his colleagues proposed **signal hypothesis** for the translocation of proteins from cytoplasm to their destination. According to this hypothesis each protein molecule has a specific sequence of 15–30 amino acids on its N-terminal which directs the protein to its specific destination. This is called **signal sequence** or **targeting sequence** and the process is called **protein targeting**. Proteins with signal sequence are transferred across or inserted into the membranes of ER. Proteins without signal sequence remain in the cell cytoplasm.

1. Translocation of Proteins through ER (Cotranslational Transport)

Proteins destined for ER are translocated into the cisternal space of ER while they are being synthesised on ribosomes attached to rough ER. They have a characteristic signal sequence on N-terminus. Their signal is rich in hydrophobic amino acids. For translocation, their signal

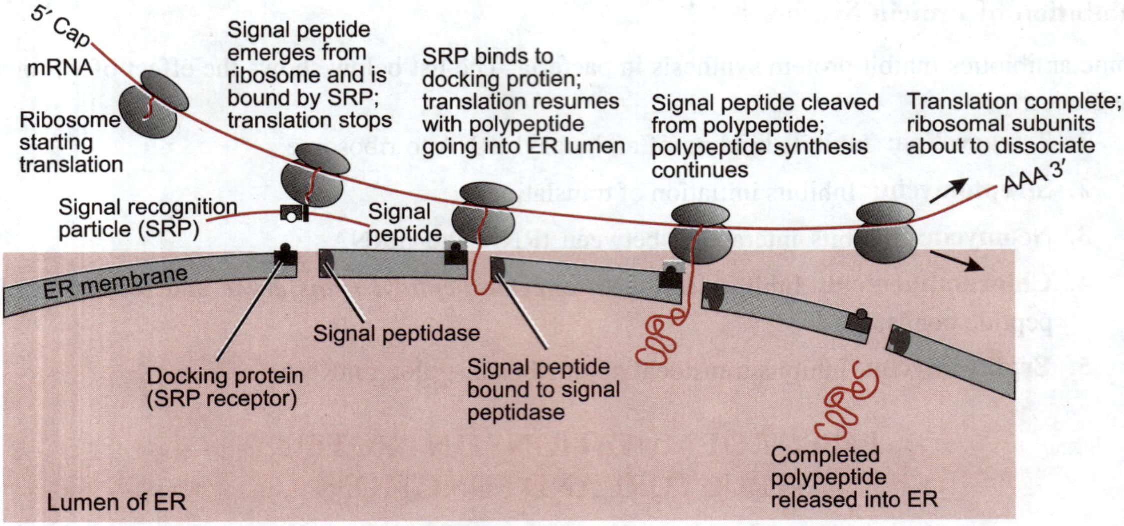

Fig. 13.17 Cotranslational transport of nascent polypeptides into central space of ER in eukaryotes

sequence is recognised by a cytoplasmic receptor particle, called **signal recognition particle (SRP).** SRP recognises an integral membrane protein of ER membranes and helps in binding of the polypeptide's signal sequence and associated ribosome to the ER membrane. The integral membrane protein is called **signal recognition protein receptor** or **docking protein**. The growing polypeptide with its signal sequence is translocated through the ER membrane into cisternal space.

> SRP stands for **signal recognition particles.** SRP is a complex of a small 7S RNA molecule and six proteins. It recognises the signal sequence present on N-terminus of polypeptide.

The transport of protein molecule into ER simultaneously with its synthesis is called **cotranslational transport**.

2. Translocation of Proteins to other Cell Organelles

Proteins destined for mitochondria, chloroplast and nucleus also have specific signal sequences to direct them to their respective organelles. Such proteins are first released from the ribosomes into the cytoplasm and are then transported to their respective organelles.

POST-TRANSLATIONAL PROCESSING OF PROTEINS

The products of translation are **polypeptides**. They are biologically inactive. Several steps of folding and biochemical modification are needed, then only they are able to perform their specific function. Also most proteins are synthesised in the cytoplasm (except those formed in mitochondria and chloroplasts). Therefore, they are to be transported to cellular regions where they have to carry out specific biological function or catalyse specific reaction.

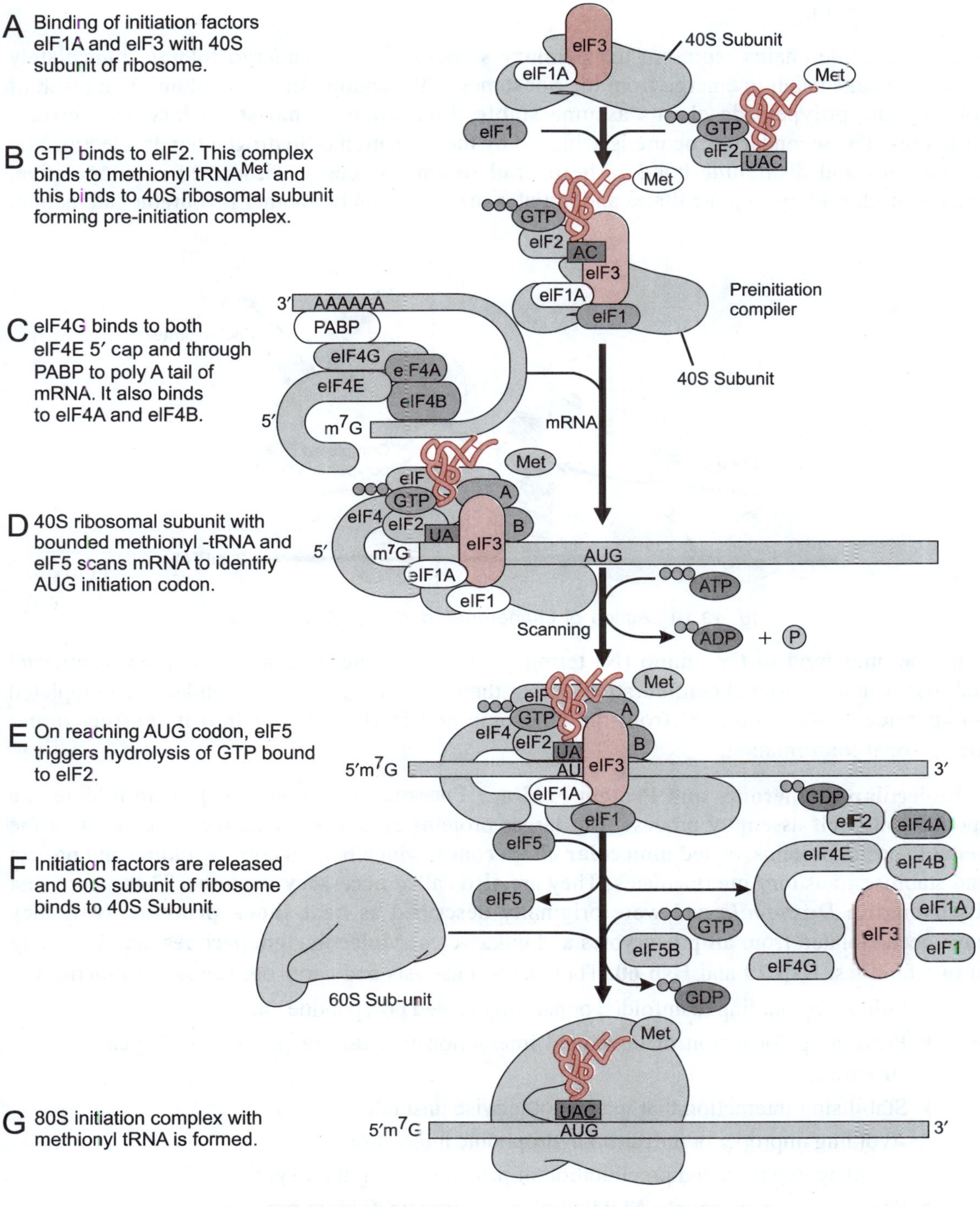

Fig. 13.18 Post-translational processing of proteins

1. Protein Folding

The polypeptide chains represent the **primary structure**. They undergo folding immediately after the chains begin to emerge from the ribosomes in the endoplasmic reticulum. As a result of folding, the polypeptide chains assume stable three-dimensional secondary and tertiary structures. The secondary structure is achieved by the formation of hydrogen bonds, electrostatic interactions and disulphide bonds. The overall folding process is described as cooperative, because each folding step facilitates and initiates the formation of another favourable interaction.

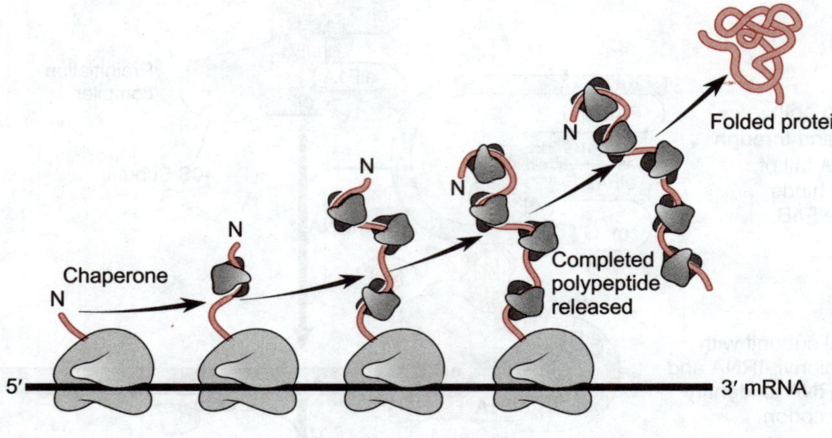

Fig. 13.19 Action of chaperones during protein folding

Chaperones bind to the amino (N) terminal portion of the nascent polypeptide chain and stabilise it in an unfolded configuration during the synthesis of the polypeptide. The completed polypeptide is then released from the ribosome and is able to fold into its correct three-dimensional conformation.

Molecular Chaperones and Protein Folding: For smaller proteins, the protein folding is a spontaneous, self assembly process. But larger proteins assume their native structure with the help of special proteins, called **molecular chaperones,** which help in correct folding and protect and stabilise transitory intermediates. They are also called **accessory proteins.** These were first discovered in *Drosophila* and were originally described as **heat-shock proteins**. Now they have been isolated from all prokaryotes and eukaryotes. **Molecular chaperones** mainly belong to two families, **Hsp 70** and **Hsp 60**. They act as catalysts and carry out following functions.

- Folding by binding to unfolded or partially folded polypeptide chains.
- Preventing formation of unwanted interaction that do not provide biologically active structure.
- Stabilising interaction that may be otherwise unstable.
- Avoiding improper or unfruitful hydrophobic interactions.
- Avoid aggregation and precipitation of proteins during their synthesis.
- Some chaperones couple ATP hydrolysis to protein-folding process.
- Bind to folding proteins to hide their exposed hydrophobic amino acid residues.

2. Biochemical Modifications

The biological changes in proteins to achieve their native functional form are:

1. Proteolytic Cleavage: The amino acid **'N-formyl-methionine'** in prokaryotes and methionine in eukaryotes attached to the N-terminus of polypeptides chain are removed by hydrolysis. This is called **proteolytic cleavage** and is an essential step. In approximately 50 percent of eukaryotic proteins an acetyl group is added to their amino end.

2. Amino Acid Modification: The activity of enzymes and some other proteins is altered by **phosphorylation, methylation** or **hydroxylation** of some amino acids. For example, hydroxyl groups (OH^-) of serine, threonine and tyrosine are modified by phosphoryl group from ATP. The hydroxylation of proline and lysine changes the protein to collagen (a structural protein).

3. Attachment of Carbohydrates: In glycoproteins, carbohydrate molecules are covalently bonded to the proteins. These provide immunological protection, cell-cell recognition and blood clotting.

4. Addition of Prosthetic Groups: Many enzymes can function in association with a covalently bound **cofactor** or **prosthetic group**. The prosthetic groups are heme, FAD, biotin and pantothenic acid.

5. Trimming of Polypeptide Chain: In some cases long polypeptide chains are trimmed by enzyme **exopeptidase.** The amino acid residues are removed one by one either from N- or C-terminus. For example, long polypeptide chain of preinsulin is cut to a small functional insulin of only 51 amino acids.

6. Signal Molecules are added or removed from the N-terminal end of some proteins. The proteins are synthesised in cytoplasm and are transported to ER, Golgi complex, mitochondria or chloroplasts or they are excreted out of the cell.

For transportation to their specific destination proteins have a sequence of up to 30 amino acids at their N-terminal. This is called **signal sequence.** Once a protein has reached its destination the signal sequence is removed before it assumes the functional status.

7. Polypeptide Chains are also complexed with metal ions or prosthetic group. The tertiary and quarternary level of protein structure is attained only when some metal atom is complexed with it. Iron in case of haemoglobin and Mg in case of chlorophyll is essential for their quarternary structure.

TEXT QUESTIONS

1. Explain the mechanism of protein synthesis.
2. What is the central dogma in molecular biology? Name the two processes represented in it. How has it been modified with the discovery of retroviruses?
3. Explain the steps involved in the synthesis of polypeptide chain. How are amino acids activated for polypeptide synthesis?

4. Describe the mechanism of initiation of polypeptide in prokaryotic cells and explain how it differs from that in eukaryotic cells.
5. Give a brief account of the different steps involved in the translation of mRNA into a polypeptide in prokaryotes. Discuss the role of differential initiation factors in this process.
6. Give the difference between transcription and translation.
7. Describe briefly reverse transcription.
8. Describe the process of amino acid activation. Why is it necessary?
9. Describe role of aminoacyl tRNA synthetase.
10. What is leader sequence or Shine-Dalgarno sequence? What is its role in initiation of polypeptide chain? Describe role of mRNA in protein synthesis.
11. Describe process of polypeptide elongation.
12. Describe the role of elongation factors in the elongation of polypeptide chain.
13. Describe how peptide bond is formed during polypeptide formation?
14. In a tabular form summarise various protein factors participating in translation.
15. Give the main differences in the initiation of translation in prokaryotic and eukaryotic cells.
16. Write short notes on:
    (a) Initiation codon
    (b) Initiation factors
    (c) Releasing factor
    (d) Role of ribosome in protein synthesis
    (e) Chaperones
    (f) Termination of polypeptide chain
    (g) Modification of polypeptide chain
    (h) Reverse transcription
    (i) Aminoacyl tRNA synthesis
    (j) Coupled transcription translation.
17. Name the three codons on mRNA that are not recognised by tRNA. What are these called?
18. Name the enzymes necessary for activation of amino acids and for attachment of activated amino acids to tRNAs.
19. Due to mistake during transcription, ATG forms UAG in mRNA. What change would occur in polypeptide chain translated from such mRNA?
20. Why tRNA are called adopter RNA?
21. Name the RNA that carries information about the sequences of amino acids in a polypeptide.
22. What is one-way flow of information? How is it described now?
23. Which enzyme controls synthesis of DNA from a single strand RNA?
24. What is central dogma reverse?

25. What is proviral DNA?
26. What is leader sequence or Shine-Dalgarno sequence?
27. Name various initiation factors that are necessary for chain initiation in prokaryotes.
28. What is the source of energy needed for the formation of initiation complex?
29. Name the first amino acid which joins the initiation codon in prokaryotes and eukaryotes.
30. How is the initiation coded on mRNA is recognised by small subunit of ribosome in prokaryotes?
31. Which ribosomal RNA functions as ribozyme?
32. What is a polysome?
33. What are Hsp 70 and Hsp 60 proteins?
34. What is the role of molecular chaperones?
35. Define the terms 'transcription' and 'translation'.
36. Name the type of bond that links the following molecules:
    (a) Nucleotide to nucleotide in nucleic acid
    (b) Amino acid to tRNA
    (c) Codon in mRNA to anticodon in aminoacyl tRNA
    (d) Amino acid to amino acid in polypeptide chain.
37. Which of the following biochemical processes are a part of post-transcriptional modification events?
    (a) Formation of disulfide bonds
    (b) Polypeptide cleavage
    (c) Addition of FAD and other prosthetic groups to proteins
    (d) Modification of amino acid residues.
38. Give functions of the following:
    (a) Aminoacyl tRNA synthesis
    (b) IF1 and IF2
    (c) AUG
    (d) N-formylmethionine
    (e) Elongation factor Ef-Tu
    (f) A-site on ribosome
    (g) RF2 and RF1
    (h) eIF4A
    (i) Enzyme translocate or G-factor.

14

Chapter

Gene: Its Expression in Prokaryotes and Eukaryotes

IN THIS CHAPTER

Gene; Types of Genes: House-keeping Genes, Luxury Genes, Structure Genes, Control Genes, Interrupted Genes or Split Genes, Alternative Genes, Overlapping Genes, Transposons, Gene Families and Pseudogenes, Smart Genes; Functions of Genes; Overview of Gene Expression: Mechanism of Gene Expression, Collinearity of Gene and Polypeptide Chain; Gene Expression in Viruses: Lytic Cycle, Lysogenic Life Cycle, Reverse Transcription (Teminism), Proto-oncogenes and Cellular Oncogenes; Gene Expression in Prokaryotes: Bacterial Genome, Transformation, Transduction and Conjugation; Lederberg and Tatum's Experiment; Gene Expression in Eukaryotes.

GENE

Genes are unit of inheritance. These are located in the chromosomes in linear fashion and are carried from parent to offspring by gametes. These control the expression of characters. **Mendel** was first to visualise a gene as a unit of inheritance and called it **factor.** The word **gene** was introduced by **Johannsen** in 1909.

Modern Definition of Gene

According to modern concept, a gene is a segment of DNA with specific nucleotide sequence which has coded information for a specific character. The specific sequence of nitrogenous bases in a gene provides information for the sequence of amino acids in the polypeptide chain during synthesis of a particular protein. A gene is now described as a unit of function **(cistron),** a unit of mutation **(muton)** and a unit of regulation **(operon).**

TYPES OF GENES

Based on the functional activity, the genes are classified as follows:

1. House-keeping Genes or Constitutive Genes

These genes are functional in all types of body cells of multicellular organisms and all the time. These are needed for basic cellular activities. Such genes are called **house-keeping genes** and their function is called as **house-keeping** or **constitutive activity.** For example, genes associated with encoding the enzymes of glycolysis are active in all cell types and all the time throughout life. Their expression is not regulated. Such enzymes are called **constitutive enzymes**.

2. Luxury Genes or Nonconstitutive Genes

These genes remain inactive for most of the time in the life span of an individual and are expressed in certain cells or at certain time only when their products are needed.

These are also called **specialist genes** or **luxury genes**. For example, some genes are functional in kidney cells, some in liver cells and some in intestine or stomach but are present in all the body cells. These are associated with the **adaptive enzyme synthesis**. It means they can be activated (induced) to increase the level of mRNA and protein product or can be deactivated (repressed) to decrease the level of mRNA and protein. These genes are regulated by the action of ***RNA polymerase*** and molecular signals such as **regulatory proteins, hormones** and **metabolites.** Therefore, these genes are also called **inducible genes** or **repressible genes.**

3. Structure Genes

These genes code for substances which are needed for the morphological or functional traits of the cell and are called **cistrons**. These occur in split form, segmented into **introns** and **exons** in eukaryotes but in prokaryotes these are continuous. The structure genes are concerned with the synthesis of a polypeptide chain or a number of polypeptide chains and also with the synthesis of different types of RNA required during protein synthesis.

The polypeptides thus synthesised may be used as components of an organelle, an enzyme, a receptor or a carrier protein of plasma membrane, as a transport protein, as a hormone or as an antigen or antibody.

4. Control Genes

Some genes produce proteins that exercise a positive or negative control on the functioning of structure genes. These are of the following types:

1. **Regulator Genes:** These code for repressor proteins which regulate transcription of **cistrons**.
2. **Operator Genes:** These act as switch and turn on or off the transcription of an structural gene as and when required by the cell.
3. **Promoter Genes:** These genes provide sites for the binding of ***RNA polymerase*** enzyme for the transcription of RNAs by structure genes.
4. **Terminator Genes:** These DNA segments lie in the end of messages and stop or terminate transcription activity of structure genes.

5. Interrupted Genes or Discontinuous Genes or Split Genes

In **prokaryotes** (eubacteria) the genes and proteins are **collinear**, i.e. the sequence of nucleotides in the gene corresponds exactly with the sequence of amino acids in the protein. Each gene is a continuous stretch of DNA whose length is related to the size of protein it synthesises. The collinearity of gene was identified in the gene that codes for ***tryptophan synthetase*** in *E. coli*.

1. **Split Genes:** In eukaryotes, the message coded in the genes is not continuous. It is found to be split into several distinct coding units separated by regions of noncoding DNA. The coding units containing biological information are called **exons** and the intervening noncoding DNA segments are called **introns**. Such genes are called **discontinuous genes** or **interrupted genes** or **split genes**. The concept of **interrupted genes** was introduced in 1977.
2. **Introns:** These are present in the genes of eukaryotes, viruses and archaebacteria. The transcription of an interrupted gene produces the primary transcript RNA. It is a faithful copy of the interrupted gene and is only the precursor. The functional RNA is formed by the removal of introns and the rejoining of its exons. This process is called **RNA splicing**.

6. Alternative Genes

These are formed when exons from different discontinuous genes get connected forming several new combinations. The alternative forms thus formed produce proteins in which one part is common while the other part is different. This concept was proposed by **Gilbert** and is known as **Gilbert hypothesis**.

7. Overlapping Genes

In overlapping genes, the same nucleotide sequence codes for more than two different proteins by employing two different reading frames. It means the same specific nucleotide sequence is shared in the transcription of two different mRNAs for the synthesis of two non-homologous proteins.

Generally, only one reading frame is used to encode a protein and genes do not overlap. But in several viruses, the same DNA sequence codes for two different proteins. For example, bacteriophage $\phi \times 174$ contains only 5386 nucleotides which code for ten different proteins. It means there are ten genes. These ten genes are called **A** to **K** genes. These are represented in concentric arcs corresponding to their reading frames.

Here **gene B** lies within the sequence of **gene A** and uses different reading frame. Similarly, **gene E** lies within **gene D** and also overlaps with **gene J**. But all the three genes use different reading frames. **Gene K** overlaps with genes A and C (Fig. 14.1).

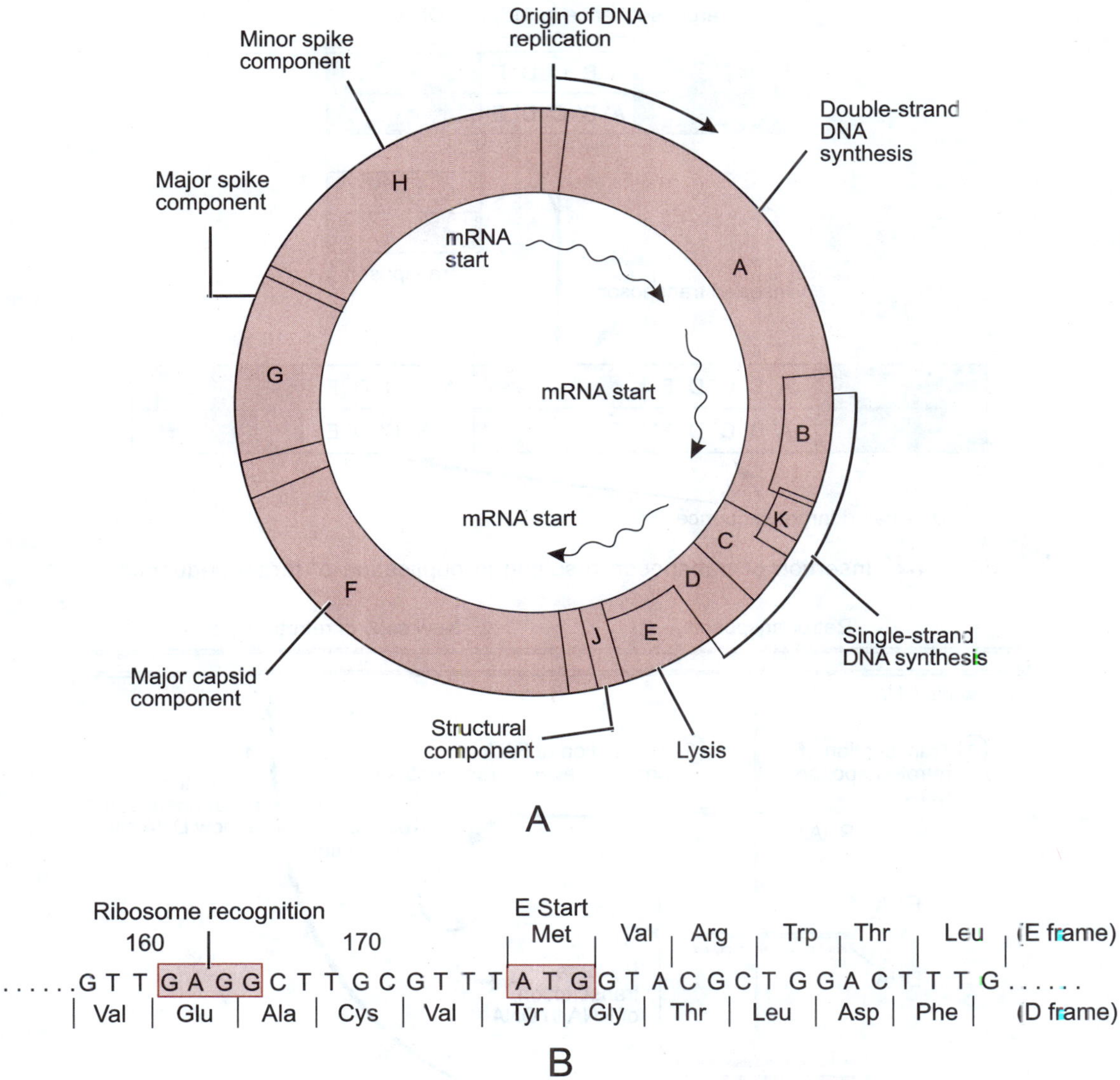

Fig. 14.1 **A.** Genetic map showing overlapping genes in circular DNA of $\phi \times 174$ bacteriophage. Their concentric arcs correspond to their reading frames; **B.** Overlap of D and E genes of $\phi \times 174$ bacteriophage

8. Transposons (Jumping Genes)

Nobel Prize winner, **Barbara McClintock**, described the occurrence of DNA segments in maize, which can move from one position to another in the genome of a cell. Recently, these have been described from snapdragon, *Drosophila*, mice and bacteria. In bacteria these are found in addition to plasmids. These can occupy different sites on plasmid DNA, on virus or on bacterial DNA.

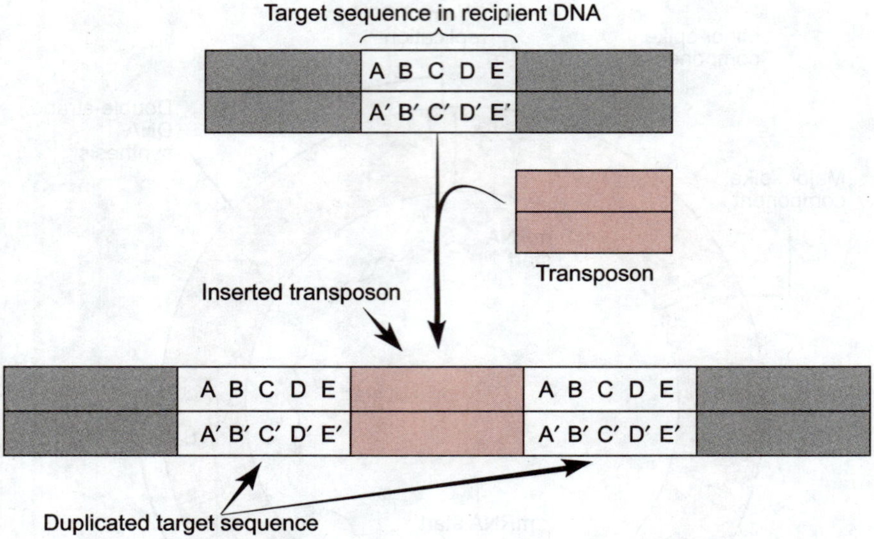

Fig. 14.2 Insertion of transposon resulting in duplication of target sequence

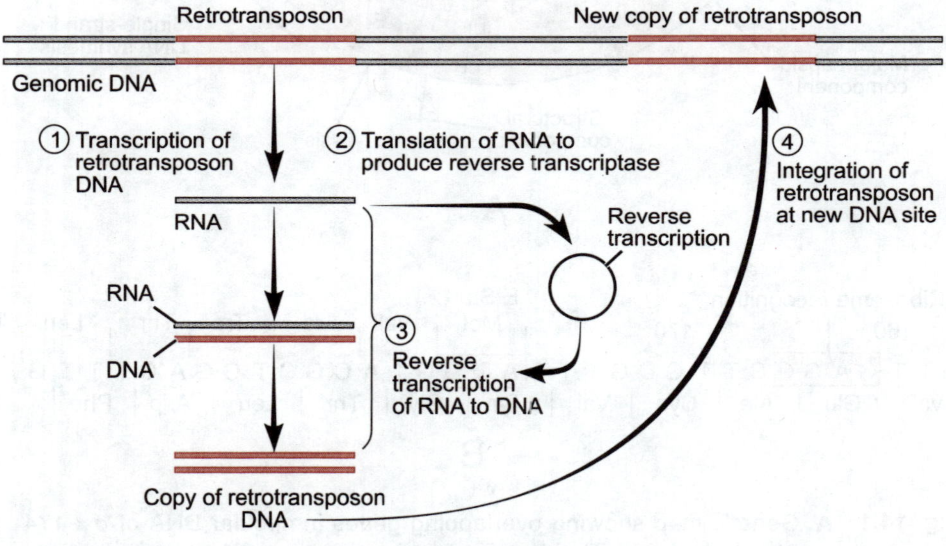

Fig. 14.3 Insertion of retrotransposon

In bacteria, these were first detected in *E. coli* and are called **insertion** or **IS elements** or **transposable elements**. The process of insertion of transposon between two base pairs in a recipient DNA molecule is called **transposition**. There are two major events during transposition:

1. There is duplication of **target sequence** in the recipient DNA molecule. The target sequence consists of 3–12 base pairs (Fig. 14.2).
2. The inserted transposon is sandwiched between the repeated target sequences.

Transposable elements mediate a number of phenomena, such as gene arrangements (inversion), deletion or fusion of DNA molecules. These can serve as switches turning genes on and off or can invert sequences containing promoters. These are also responsible for gene amplification and formation of Hfr cells in bacteria.

9. Gene Families and Pseudogenes

Prokaryotic genes are mostly represented only once in the genome. But many eukaryotic genes are present in multiple copies. These multiple copies of genes are called **gene families**. Members of many gene families are clustered in the same region of DNA, while some gene families are dispersed to different chromosomes.

Example: 1 The α and β sub-units of human haemoglobin are encoded by gene families. Different members of these families are expressed in embryonic, foetal and adult tissues (Fig. 14.4).

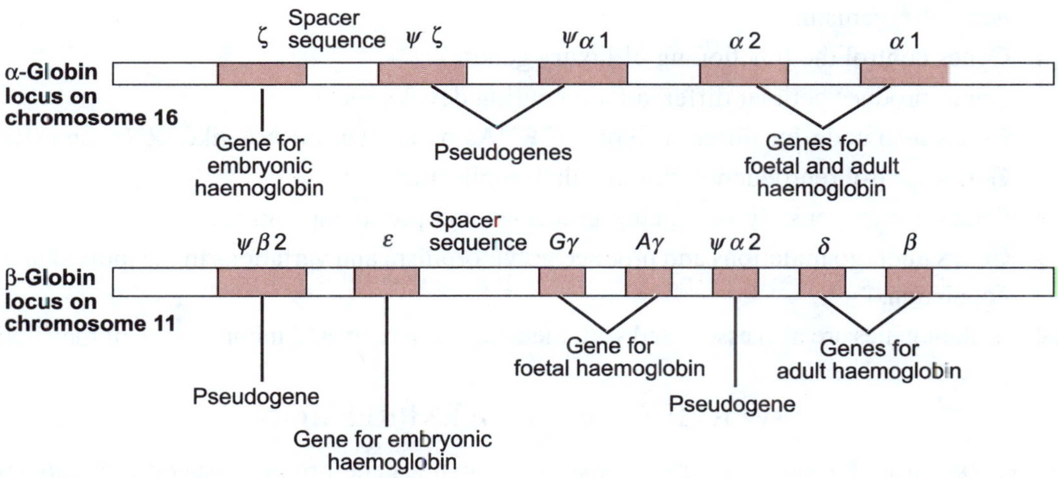

Fig. 14.4 Globin gene families showing members of human α- and β- globin genes on chromosome 16 and 11

Example: 2 Genes for ribosomal RNAs and histones occur in multiple copies, because rRNA and histones are needed in large quantities.

Origin: Gene families are regarded to have arisen by the duplication of an ancestral gene. The duplicated genes then diverged because of mutations. Such a divergence then led to the evolution of related proteins to function in different tissues or at different stages of development.

Some gene copies become non-functional as a result of mutations. Due to mutation, they fail to produce a functional product or the product formed is non-functional. Such non-functional gene copies are called **pseudogenes**. They represent evolutionary relics. They simply add to the size of eukaryotic genome. As shown in Fig. 14.4, α and β globin gene families in human beings have two pseudogenes each.

10. Smart Genes

Smart genes are artificially constructed genetic molecules composed of one or more genes that code for functional proteins or enzymes. Smart genes are expressed only under specific conditions such as presence of some particular substrate molecules. Thus a smart gene has a conditional expression. Because of this property, the constructed smart genes form ideal therapeutic vehicle for gene therapy.

Functions of Genes

Genes carry out following functions:

1. Genes are associated with the inheritance of characters. These are hereditary units.
2. Genes control phenotypes including both structural and functional characters.
3. Control genes regulate transcription of mRNA and thus control amount of proteins synthesised. They also control switching on or off of the specific genes according to the need of organism.
4. Genes control the functioning of luxury genes.
5. Genes produce cellular differentiation during development.
6. Genes also code for different types of RNAs other than mRNA like tRNA and rRNA.
7. Genes control reproduction through their replication.
8. Genes are responsible for ageing processes and producing cancer.
9. Genes undergo mutations and produce polymorphism and variations in the individuals of a population.
10. Mutant genes are also associated with metabolic disorders and inborn errors of metabolism.

OVERVIEW OF GENE EXPRESSION

Gene expression is the molecular mechanism by which a gene produces specific phenotype or produces an enzyme which controls a specific metabolic activity.

Mechanism of Gene Expression

Gene is basically a functional segment of DNA. It contains hereditary message encoded in the form of specific sequence of nitrogenous bases. It is repository and expresses its encoded information in the form of a polypeptide chain, which is synthesised according to strict instructions of a gene through its mRNA. The mRNA copies instructions from DNA.

The process is called **transcription**. Messenger RNA, then, comes out of the nucleus, attaches on to the ribosomes and directs the linking of amino acids in a specific sequence to form a polypeptide chain. Broadly speaking, gene expression involves:

1. Transcription of mRNA from the DNA segment composing the gene.
2. Transcription of rRNA and formation of ribosomal subunits.
3. Transcription of tRNA.

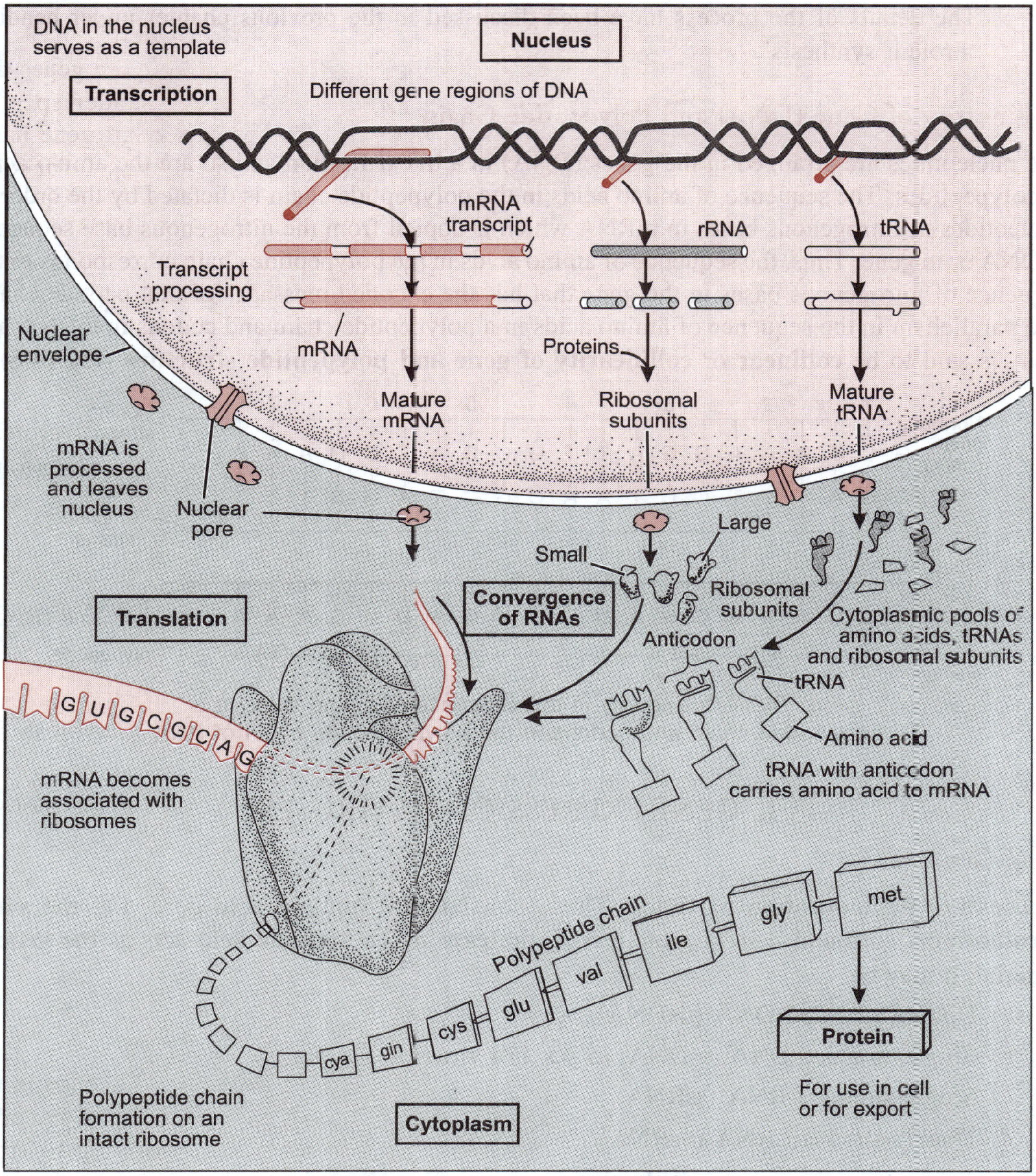

Fig. 14.5 An overview of molecular mechanism of gene expression through protein synthesis in eukaryotes

4. Translation of codons of mRNA leading to the linking of amino acids to form a polypeptide chain.
5. Termination of polypeptide chain.
6. Maturation of protein.

The details of the process have been discussed in the previous chapter under heading 'Protein synthesis'.

Collinearity of Gene (DNA) and Polypeptide Chain

The nucleotides are arranged in the genes (DNA) in a linear fashion and so are the amino acids in polypeptides. The sequence of amino acids in the polypeptide chain is dictated by the order of nucleotides or nitrogenous bases in mRNA which is copied from the nitrogenous base sequence in DNA or in gene. Thus, the sequence of amino acids in the polypeptide chain corresponds to the sequence of nitrogenous bases in the gene that has the encoded message for that peptide chain. The parallelism in the sequence of amino acids in a polypeptide chain and codons in mRNA and DNA is said to be **collinear** or **collinearity of gene and polypeptide structure** (Fig.14.6).

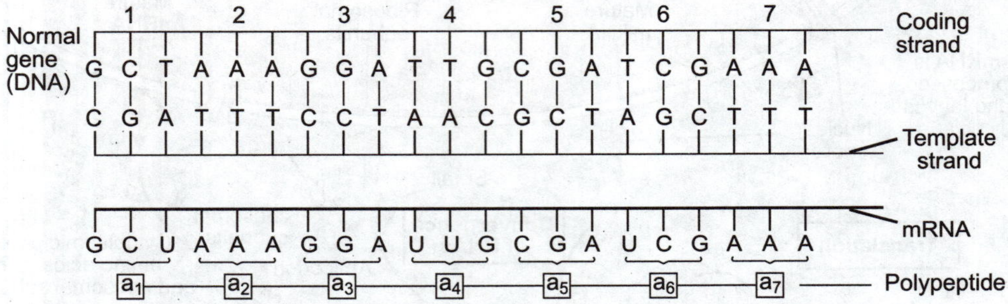

Fig. 14.6 Collinearity in the sequence of amino acids in a polypeptide chain and codons in the polynucleotide chain of DNA

I. GENE EXPRESSION IN VIRUSES

Viral Genome

Viruses are nucleoprotein particles. These consist of a nucleic acid core, i.e. the **viral chromosome**, surrounded by a protein coat, the **capsid**. The nucleic acid acts as the genetic material. It may be:

(a) Double-stranded DNA (dsDNA)
(b) Single-stranded DNA (ssDNA) in $\phi \times 174$ virus
(c) Single-stranded RNA (ssRNA)
(d) Double-stranded RNA (dsRNA).

The viral genome is very small. It contains just a few genes that code for: (*a*) synthesis of enzymes that enable them to cause lysis of host cell wall and infect them, (b) synthesis of a few proteins for the capsid, and (c) synthesis of their viral DNA.

Viruses **lack energy yielding** and **biosynthetic machinery**. These can not metabolise outside host cells and use host machinery (i.e. ribosomes, tRNA and enzymes) and raw materials (nucleotides and amino acids) and synthesise copies of their own nucleic acid and proteins. Viruses exhibit two types of life cycles:

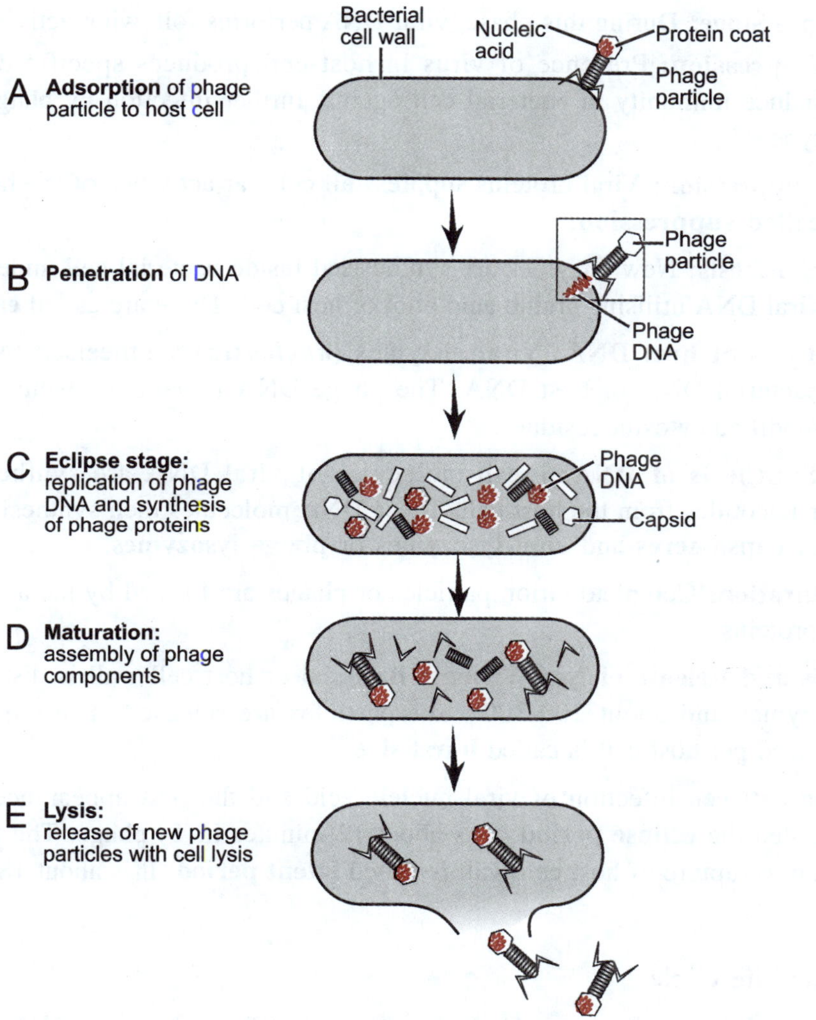

Fig. 14.7 Lytic life cycle of T_2/T_4 bacteriophage: **A.** Adsorption; **B.** Penetration; **C.** Eclipse stage; **D.** Maturation; **E.** Lysis and release of daughter phages

1. Lytic Life Cycle

Lytic life cycle is exhibited by virulent phages because at the end of the cycle, the host cell is lysed. It includes following steps:

1. **Adsorption:** The virus particles or virions get attached to the cells of specific host. Bacteriophages attach to the host bacterial cells with the help of their tail fibres.
2. **Penetration:** It involves injection of viral nucleic acid into host cell. The **lysozyme** present at the tip of tail dissolves or hydrolyses host cell wall at the point of contact making a hole. The viral nucleic acid enters host cell through this hole, and protein coat is left out. The empty protein coats of viruses are called **ghosts**.

3. **Eclipse Stage:** During this phase, viral DNA performs following activities:
    - **Repression:** Presence of virus in host cell produces specific enzymes which induce immunity in bacterial cell against further infection by phages of the same type.
    - **Suppression:** Viral proteins suppress all cellular activities of the host cell. This is called **suppression**.
    - **Synthesis:** New enzymes are synthesised inside bacterial cell under the control of viral DNA utilising amino acid pool of host cell. These are called **early proteins**.
    - **Lysis of host DNA:** Viral enzymes (***nucleases***) synthesised by phage, break bacterial DNA or host DNA. The phage DNA remains protected because it has modified cytosine residues.
    - **Synthesis of DNA:** Fresh molecules of viral DNA are synthesised utilising nucleotides from the host cell. These DNA molecules then synthesise coat proteins or **capsomeres** and viral lysozymes or **phage lysozymes**.
4. **Maturation:** Complete virion particles or phages are formed by the assembly of DNA and proteins.
5. **Lysis and Release of New Virions:** Bacterial or host cell wall is dissolved by phage lysozymes and about 200-300 virus particles are released. The number of virions produced per host cell is called **burst size**.

The period between infection of viral nucleic acid and the first appearance of new phage progeny is called the **eclipse period**. It is about 12 minutes in T_2 phage. The total time taken from infection to rupture of host cell wall is called **latent period**. It is about 18 minutes for T_2 phage.

2. Lysogenic Life Cycle

In case of some bacteriophages such as phage lambda (Phage λ), viral DNA on entering the bacterial cell may start:

- lytic cycle as discussed or
- get integrated with bacterial DNA with the help of enzyme ***integrase***

The viral DNA with bacterial DNA is called **prophage**. In this case, phage DNA becomes inactive and does not utilise host machinery. It replicates along with host DNA and passes to daughter bacterial cells. A **repressor protein** produced by prophage keeps other phage genes inactive. The existence of phage DNA as a part of host DNA is called **lysogeny**. When it breaks from host DNA, it may carry some host genes along with. These may be transferred to some other cell on infection. This phenomenon is called **transduction**.

In lysogeny, the functioning of viral DNA is associated with the functioning and multiplication of host DNA.

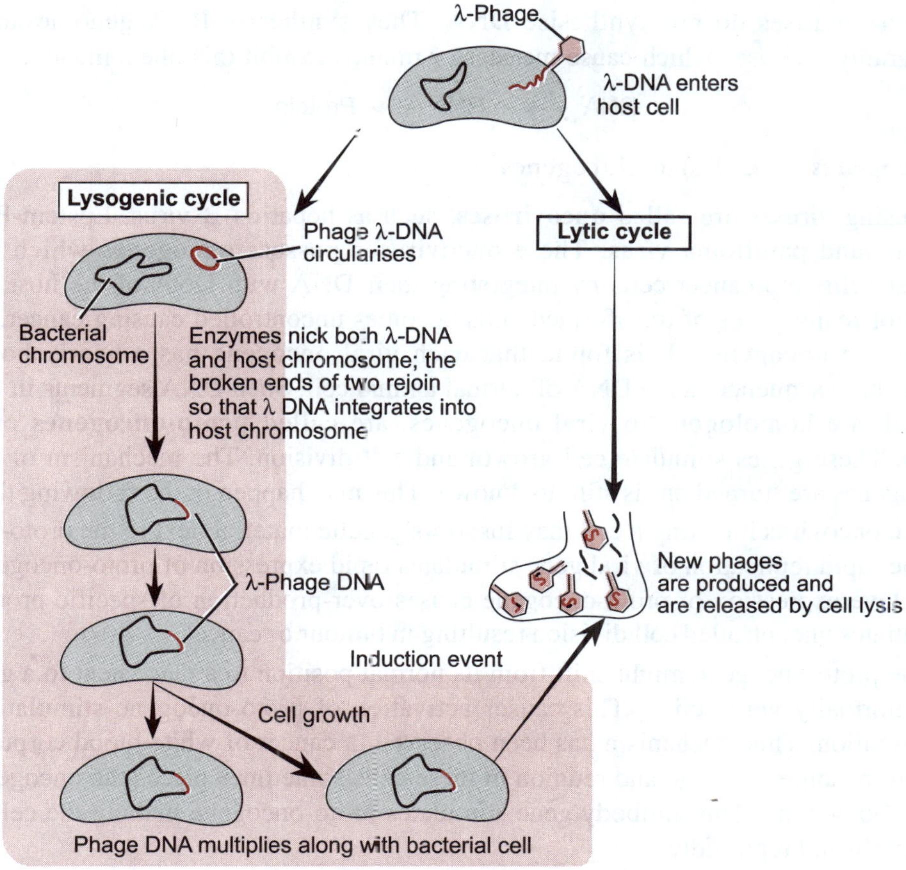

Fig. 14.8 Lysogenic life cycle of λ-phage in bacterial cell and conversion of lysogenic phage into lytic form and vice versa

In case, the repressor gene mutates, the repressor protein is not synthesised and the lysogenic phage turns lytic and enters lytic cycle. The DNA of active phage of lytic cycle is called **vege phage** or **temperate phage DNA**.

Reverse Transcription (Teminism)

All plant viruses and a few animal viruses have RNA as their genetic material. These RNA viruses have a gene which codes for enzyme ***reverse transcriptase*** or ***RNA dependent DNA polymerase***. It helps in the synthesis of single stranded DNA, which in turn forms a complementary strand. Hence, a double stranded DNA is formed. This phenomenon is called **reverse transcription**.

Viruses capable of reverse transcription are called **retroviruses**. This phenomenon was described by **Temin** and **Baltimore** in 1970. Tumour causing retroviruses and HIV-III (**human immunodeficiency virus-III**) which causes AIDS are examples of retroviruses exhibiting reverse transcription.

Some RNA viruses do not synthesise DNA. They synthesise RNA genome and mRNA directly. Paramyxoviruses which cause measles or mumps exhibit this phenomenon.

RNA ⟶ mRNA ⟶ Protein

Proto-oncogenes or Cellular Oncogenes

Cancer causing viruses are called **oncoviruses**, such as hepatitis B virus, Epstein-Barr virus, herpes virus and papilloma virus. These oncoviruses possess **oncogenes** which transform normal host cells into cancer cells by integrating their DNA with DNA of the host cells. The expression of many genes of transformed cells becomes uncontrolled causing cancer. About 20 viral genes are **oncogenes**. It is found that each viral oncogene has a nearly homologous nitrogenous base sequence in the DNA of normal animal cell. Such DNA segments in the animal cells, which are homologous to viral oncogenes, are called **proto-oncogenes** or **cellular oncogenes**. These genes stimulate cell growth and cell division. The mechanism of how these proto-oncogenes are 'turned on' is still not known. This may happen in the following three ways:

- The oncovirus infecting a cell may insert its genetic material next to the proto-oncogene. The rapid replication of viral gene stimulates rapid expression of proto-oncogene next to it. Overexpression of proto-oncogene causes over-production of specific protein which initiates uncontrolled cell division resulting in tumour or cancer.
- The proto-oncogene might shift from its normal position to a place next to a gene which is normally very active. This causes activation of proto-oncogene stimulating cancer formation. This mechanism has been observed in cancer of white blood corpuscles. The chromosome breakage and reunion in these cells sometimes places the oncogene next to antibody gene. This antibody gene stimulates proto-oncogene making the cell to divide rapidly and repeatedly.
- Proto-oncogene may mutate to oncogene. This mutation may be caused by a virus or some other gene.

II. GENE EXPRESSION IN PROKARYOTES

Bacterial Genome

Bacterial genome is represented by a circular double-stranded DNA. In *E. coli*, it comprises of about 2,000 to 3,000 genes and is approximately 1,100 μm long and 20 Å in diameter. It is folded into about 50 loops. Each loop is further coiled and supercoiled. The bacterial DNA remains highly folded and is called either **bacterial chromosome** or **nucleoid**. The DNA remains naked. The basic proteins as found associated with DNA in eukaryotic genome are absent in bacteria.

> When bacterial chromosome is treated with *protease* enzyme, its folds fail to unfold. This indicates that proteins do not participate in the folding process of bacterial chromosome. For this reason bacterial chromosome is described as naked.
>
> When treated with *RNAase*, bacterial chromosome gets unfolded. It means associated with DNA, some amount of RNA is also present.

The **pleuropneumonia-like organisms** (PPLO) or mycoplasma have a genome of a few hundred genes, which code for only a few proteins. Therefore, PPLOs usually live as parasites on other organisms and get most of their requirements from host.

Transfer of Genetic Material in Bacterial Cells

Bacteria multiply by asexual process. It includes replication of DNA, followed by division of the bacterial cell. The doubling time, also called **generation time**, may be just 20 minutes in some cases. Replication of bacterial DNA proceeds in both directions from a single **origin of replication**. This is called **bi-directional replication**. Exchange of genetic material between two different cells does occur in them. It takes place in three ways:

1. **Transformation:** The uptake of naked DNA.
2. **Transduction:** By viral infection.
3. **Conjugation:** The mating of two bacterial cells.

1. Transformation

In transformation bacterial cell takes up naked DNA segment from the environment, incorporates it into its own chromosomal DNA, and expresses the trait controlled by the incoming DNA. Transformation was described by **Griffith** and demonstrated experimentally by **Avery, MacLeod** and **McCarty** in 1944. They extracted DNA from smooth, virulent strain of *Diplococcus pneumoniae* and added it to a culture of rough, non-virulent bacteria. The culture produced some smooth, virulent bacteria. This shows that virulent DNA was taken up by some bacteria and these were transformed into virulent strain.

2. Transduction

Transfer of DNA from one bacterium to another by bacteriophage is called **transduction**. A phage particle growing in the host (bacterial) cell, may enclose a small segment of bacterial DNA (chromosome) along with its own DNA. On attacking the new bacterial cell it injects the segment of the bacterial DNA along with its own DNA. The chromosome segment of the previous host may undergo crossing over with the chromosome of the new host . This incorporates the previous host's genes into the new host's chromosome, and produces changes in the new host cell.

> 1. Conjugation in bacteria was first reported by **Lederberg** and **Tatum** (1946) in *E. coli*.
> 2. **Hayes** (1952) found a strain of *E. coli* in which recombination frequency was 100 to 1,000 times more than reported by **Lederberg**.
> 3. Conjugation is a primitive form of sexual reproduction because it lacks meiosis.
> 4. Length of chromosome or number of genes transferred depends upon the time for which the two conjugants remain joined together.

3. Conjugation

It is a process comparable to sexual mating in eukaryotes leading to the formation of new genetic combination. Two haploid bacterial cells of different strains (+ and –) come together, recognising each other by complementary macromolecules present on their surface. A copy of a part or whole of the chromosome passes from one cell, the **donor** (equivalent of male), into the

recipient (equivalent of female) cell. Crossing over then occurs between the recipient's chromosome and the donor's chromosome. This leads to genetic combination (recombination). The progeny of the recipient then expresses some of the donor's traits due to recombination.

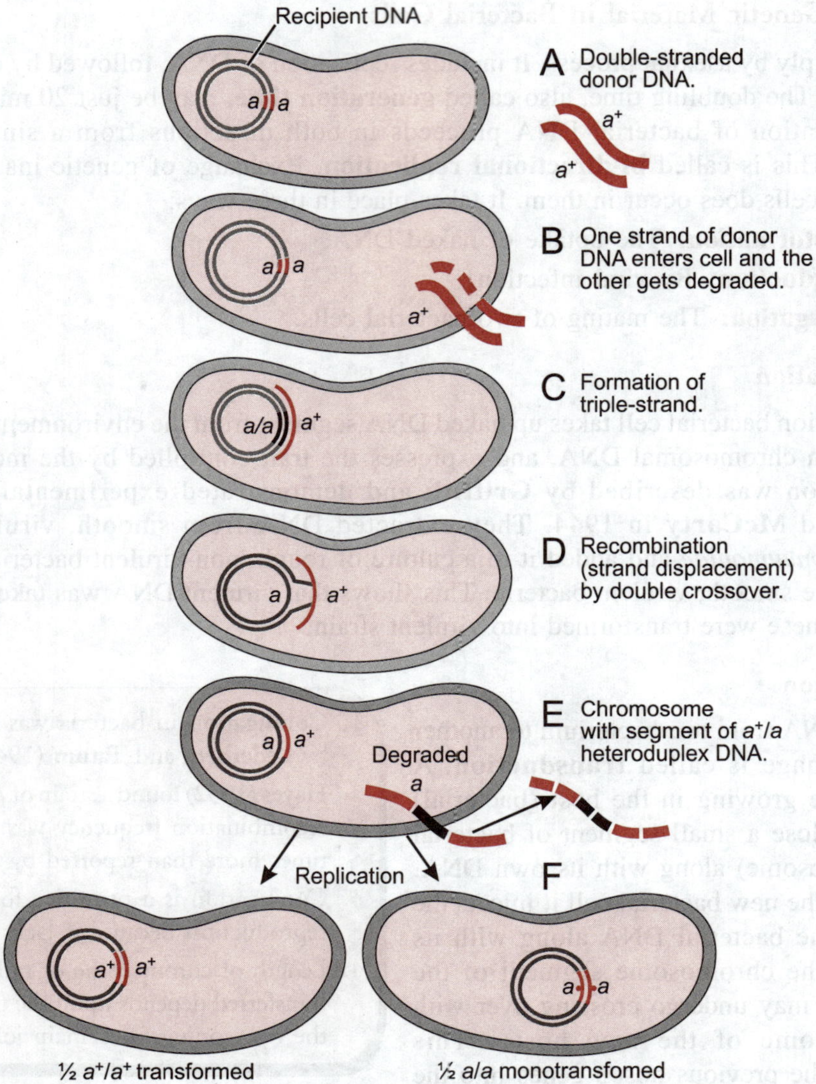

Fig. 14.9 Different steps in natural transformation of bacterial cell with DNA from the environment: **A.** Double-stranded linear DNA donor with (a^+) gene and recipient bacterial cell with DNA having 'a' allele; **B.** One of the donor DNA strand enters the recipient and other strand is degraded; **C.** Single DNA strand with (a^+) gene pairs with the homologous region of recipient chromosome; **D.** Recombination between DNA strands having (a^+) and (a) genes respectively by double crossing over; **E.** Heteroduplex chromosome formed in recipient bacterial cell having a^+/a genes. Linear DNA with gene (a) degraded; **F.** Progeny has 50% transformed and 50% nontransformed cells

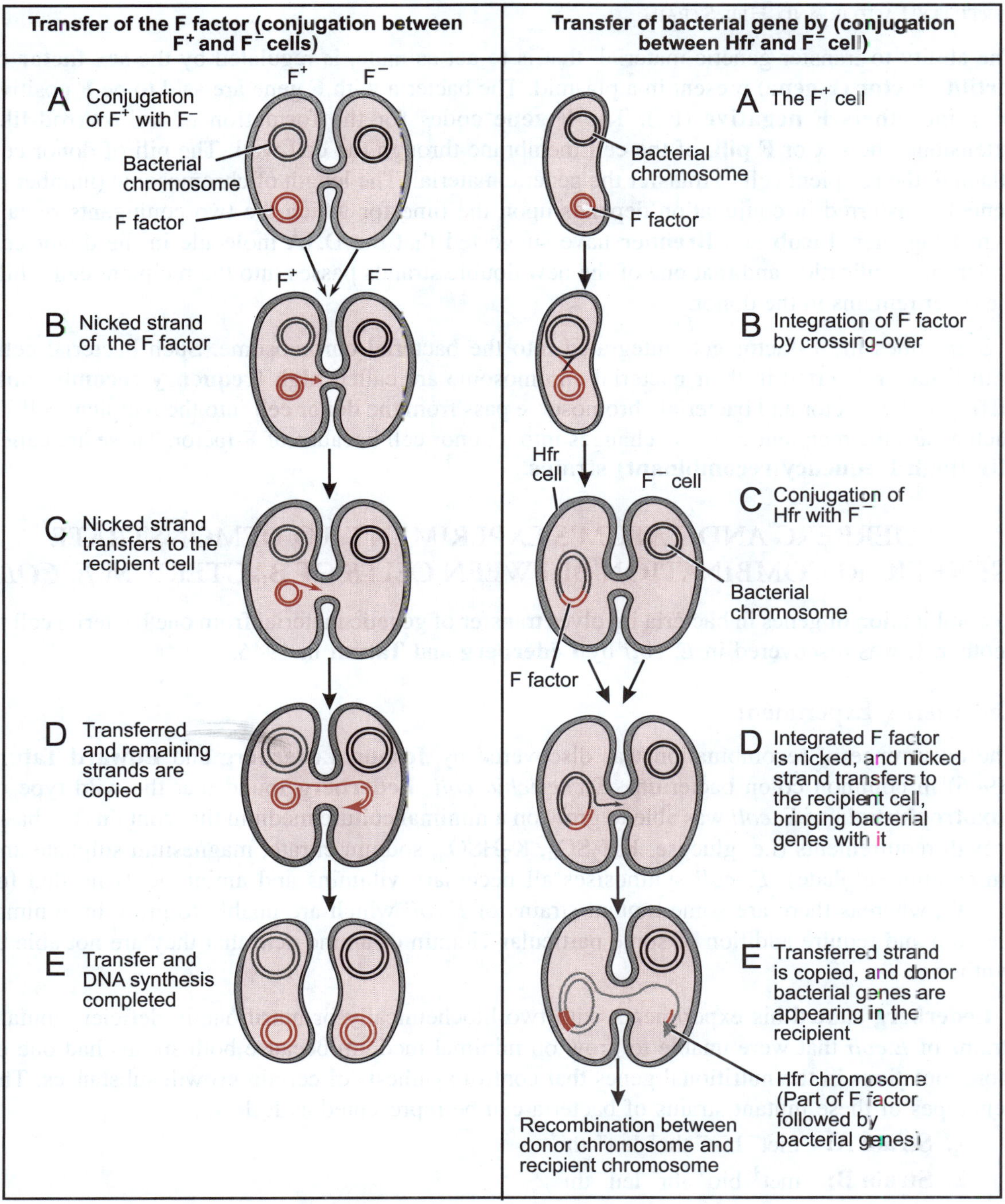

Fig. 14.10 Conjugation in bacteria: **(A)**. Conjugation between bacteria of F^+ and F^- strains of showing transfer of F factor from F^+ to F^- cells; **(B)**. Formation of Hfr strain and transfer of genes from Hfr donor to recipient F^- bacterial cell

Fertility Factor and Hfr Strain

The ability to transfer genetic material, that is to act as male, is regulated by the **sex factor** or **fertility factor (F gene)** present in a plasmid. The bacteria with **F** gene are said to be **F** positive (F^+) and others **F negative** (F^-). The **F gene** codes for the formation of hollow rod-like extensions, the **sex** or **F pili**, of the cell membrane through the cell wall. The pili of donor cell attach to the recipient cell to transfer the genetic material. The length of chromosome (number of genes) transferred in conjugation depends upon the time for which the two conjugants remain joined together. **Jacob** and **Brenner** have suggested that the DNA molecule in the donor cell undergoes replication and that one of the new double strands passes into the recipient cell while the other remains in the donor.

Sometimes the F-factor gets integrated into the bacterial chromosome. Such bacterial cells with F-factor inserted in their bacterial chromosome are called **high frequency recombinants (Hfr)**. Both F-factor and bacterial chromosome pass from the donor cell into the recipient cell. In such a case the recipient cell also changes into a donor cell because of F-factor. These are called **Hfr (high frequency recombinant) strains.**

LEDERBERG AND TATUM'S EXPERIMENT TO DEMONSTRATE GENETIC RECOMBINATION BETWEEN CELLS OF BACTERIUM *E. COLI*

Recombination of genes in bacteria involves transfer of genetic material from one bacterial cell to another. It was discovered in *E. coli* by **Lederberg** and **Tatum** in 1946.

Lederberg's Experiment

The occurrence of recombination was discovered by **Joshua Lederberg** and **Edward Tatum** (1946) in common colon bacterium, *Escherichia coli*. **Lederberg** found that the wild type or **auxotroph** strain of *E. coli* was able to grow on a minimal culture medium that contains the basic growth requirements (i.e. glucose, KH_2SO_4, K_2HSO_4, sodium citrate, magnesium sulphate and ammonium sulphate). *E. coli* synthesises all necessary vitamins and amino acids needed for growth, whereas there are some mutant strains of *E.coli* which are unable to grow in minimal medium and require addition of some particular vitamin or amino acid that they are not able to synthesise.

Lederberg started his experiments with two biochemically or nutritionally deficient mutant strains of *E.coli* that were unable to grow on minimal medium because both strains had one or more mutations in the nutritional genes that control synthesis of certain growth substances. The genotypes of these mutant strains of bacteria can be represented as follows:

1. **Strain A:** met^- bio^- thr^+ leu^+ thi^+
2. **Strain B:** met^+ bio^+ thr^- leu^- thi^-

where **met** stands for amino acid **methionine**

bio stands for vitamin **biotin**

thr stands for amino acid **threonine**

leu stands for amino acid **leucine,** and

thi stands for vitamin **thiamine**

Symbol + refers to presence of a particular gene and capability to synthesise substance.

Symbol – refers to absence of a particular gene and unability to synthesise.

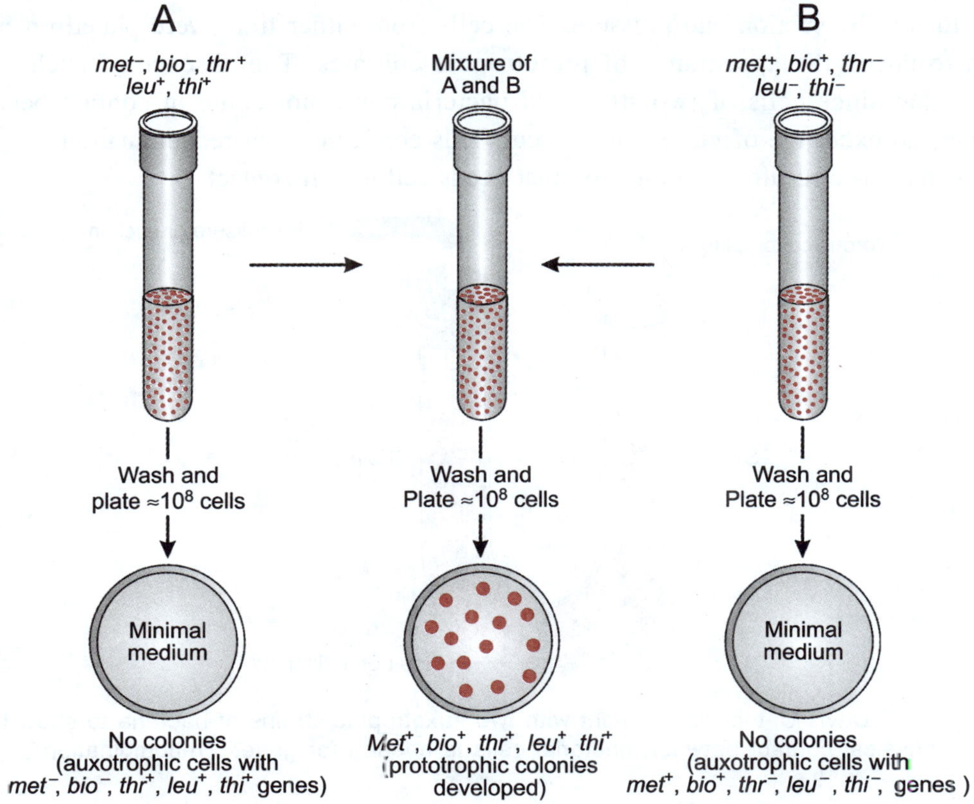

Fig. 14.11 Lederberg's experiment to show recombination in two strains of bacterium *E. coli* (Strain A and strain B)

It means the strain **A** was unable to synthesise vitamin biotin and amino acid methionine and the strain **B** was unable to synthesise amino acids threonine, leucine and vitamin B_1. When these two strains were plated separately in control experiments on minimal medium, neither strain could grow without appropriate nutritional supplement. But when both these strains were plated together in a mixed culture on complete medium containing yeast extract, peptone and glucose and after one or two generations, the suspensions of the washed cells were plated on minimal medium. It was observed that majority of cells failed to multiply and only a few of them were able to grow. **Lederberg** presumed that these cells would have been the **prototrophs** (wild type) with all the genes dominant met^+, bio^+, thr^+, leu^+, thi^+ and were formed as a result of recombination of genes from the two mutant strains.

Davis U–Tube Experiment: Bernard Davis performed an experiment using U–shaped tube to ascertain whether cell to cell contact between two strains of bacteria is necessary for above recombination. The two limbs of **U** tube were separated by a microfilter. Its pores were too small to allow bacteria to pass through. Bacteria of strain **A** and strain **B** were plalced in a liquid medium one in either limb of the tube. It was incubated for several hours, and its medium was allowed to mix by suction and pressure. The cells from either limb were plated on minimal medium to check the appearance of prototrophic colonies. The absence of such colonies indicated that since cells of two strains of bacteria could not come in contact because of microfilter, no exchange of genes could occur. This confirmed that recombination of genes in bacteria occurs as a result of conjugation that needs cell to cell contact.

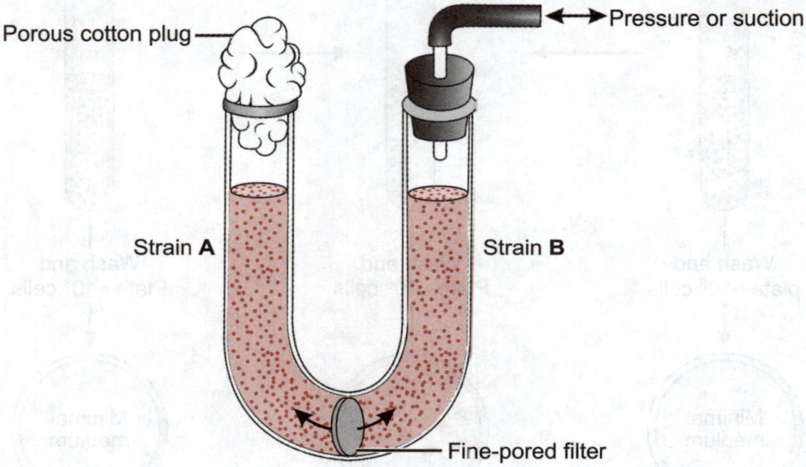

Fig. 14.12 Davis U–tube experiment with two auxotrophic strains of bacteria to show that physical contact between bacterial cells is needed for genetic recombination

III. GENE EXPRESSION IN EUKARYOTES

The genome of higher eukaryotes is very complex. This is evident from the number of genes present in the eukaryotic cells. For instance, *Drosophila* has 5,000 to 10,000 genes. Human haploid genome seems to have at least 1,00,000 genes. The eukaryotic genome has not only to control growth and division of the cells but also their differentiation into specialised tissues such as muscles, liver or heart in animals and parenchyma, chlorencyma, xylem or phloem in plants. This makes the gene expression and its regulation in eukaryotic cells a very complex but important process. The concept of collinearity of gene and protein is fully true in prokaryotes but not in eukaryotes. Eukaryotic genes are much larger than they need to be for a polypeptide. Such genes have long segments of bases which do not code for any amino acids. These are inserted between those segments of bases that code for amino acids. The coding segments of genes are called **exons** and non-coding segments as **introns**. Thus, information in eukaryotes is in pieces.

The mRNA transcribed from a eukaryotic gene is called **nuclear RNA (nRNA)** or **heterogeneous nuclear RNA(hnRNA)**. It has transcripts of both exons and introns. During processing of mRNA transcript or heterogeneous RNA, the unwanted transcripts of introns are removed by enzyme ***nucleases*** and exon transcript segments are joined together by ***ligases***. This processing of mRNA is called **splicing** and occurs inside the nucleus. The functional mRNA comes out of nucleus and codes for a polypeptide chain in the cytoplasm.

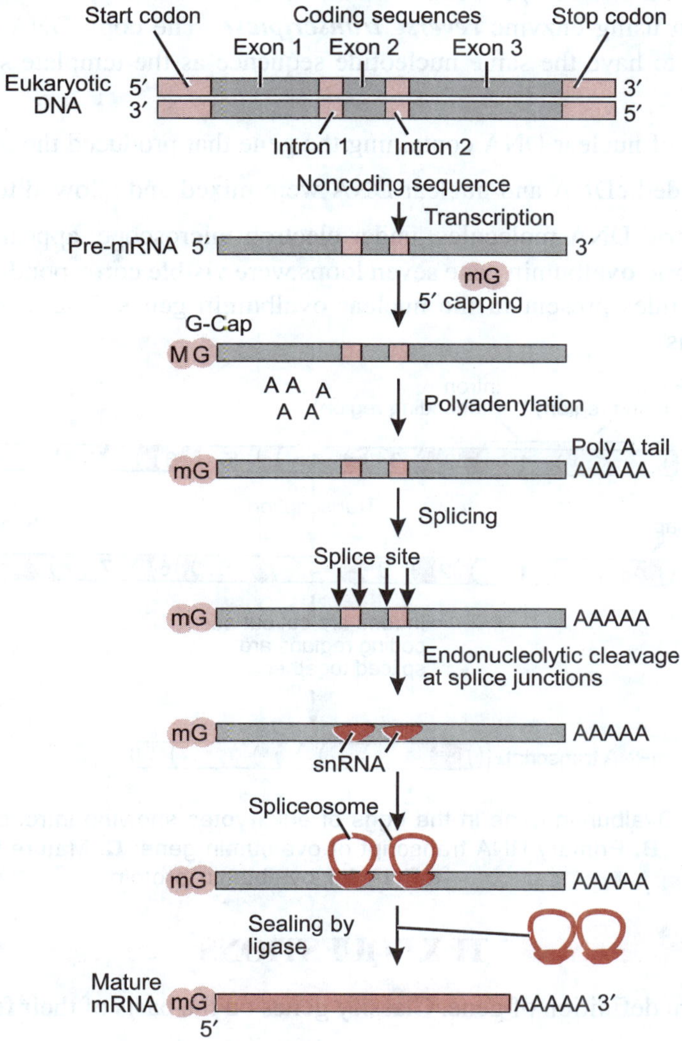

Fig. 14.13 Relationship between gene, unprocessed mRNA and final mRNA

DISCOVERY OF INTRONS

For many years, scientists believed that in all organism (prokaryotic or eukaryotic) a collinearity exists between nucleotide sequence and the sequence of amino acids in the encoded protein. But in late 1970s, it was discovered that eukaryotic proteins are encoded by RNAs that are formed by the excision of primary RNA transcript and then splicing of the excised segments. This was discovered as a result of following experiments:

1. Ovalbumin mRNA transcribed from a particular ovalbumin gene was isolated and purified.
2. Molecules of DNA complementary to the isolated mRNA were synthesised by reverse transcription using enzyme ***reverse transcriptase***. The copy DNA (cDNA) molecules were found to have the same nucleotide sequence as the template strand of ovalbumin gene.
3. The portion of nuclear DNA containing the gene that produced the mRNA was isolated.
4. Single stranded cDNA and nuclear DNA were mixed and allowed to hybridise.

The resulting hybrid DNA molecules, under electron microscope, appeared having unpaired loops. In case of hybrid ovalbumin gene seven loops were visible corresponding to the noncoding segments of nucleotides present in the nuclear ovalbumin genes. These noncoding segments were named **introns**.

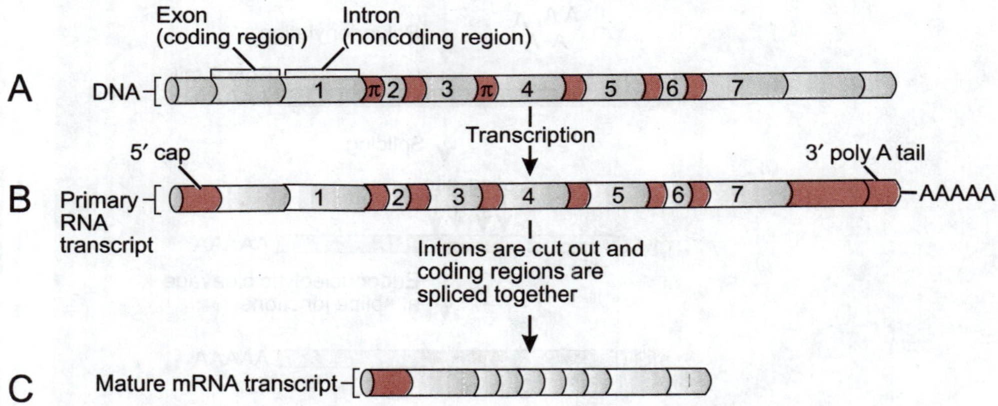

Fig. 14.14 **A.** Ovalbumin gene in the eggs of eukaryotes showing introns and exons; **B.** Primary RNA transcript of ovalbumin gene; **C.** Mature functional mRNA for ovalbumin protein

TEXT QUESTIONS

1. Give modern definition of gene. Classify genes on the basis of their function and activity.
2. Discuss the various steps in the lytic life cycle of a phage/virus.
3. Discuss in detail how viral genome expresses itself and in what forms.
4. Describe with the help of a well labelled diagram the life cycle of an oncovirus.
5. What is reverse transcription? Explain with reference to retroviruses.
6. Explain the phenomenon of transformation in bacteria.

7. Distinguish between structural gene, regulator gene and operator gene.
8. What is reverse transcription? Name the group of viruses that exhibit reverse transcription.
9. Differentiate between introns and exons. Where do they occur?
10. What are retroviruses? How has their discovery led to the modification of central dogma in molecular biology?
11. What are luxury genes?
12. What is Gilbert hypothesis?
13. Who introduced the concept of jumping genes or transposons? Explain the phenomenon of transposition.
14. Explain the term overlapping genes.
15. Summarise functions of genes.
16. Write a note on any one of the following:
    (a) Pseudogenes (b) Luxury genes
    (c) Repetitive sequences (d) Teminism
    (e) Lysogeny (f) Conjugation in bacteria.
17. What are retroviruses?
18. Define the terms:
    (a) House-keeping genes (b) Introns
    (c) Exons (d) Transcription
    (e) Reverse transcription (f) RNA splicing
    (g) Nucleoid (h) Smart genes.
19. What is RNA splicing?
20. SINEs and LINEs are used to represent what?

15
Chapter

Regulation of Gene Expression or Regulation of Protein Synthesis in Prokaryotes

IN THIS CHAPTER

Regulation of Gene Activity; Gene Regulation in Prokaryotes: Coordinate Regulation, Strategies of Gene Regulation, Gene Regulation at Transcription Level, Induction and Repression, Lactose System and *Lac* Operon System; Allosteric Regulation of *Lac* Repressor, Role of Cyclic AMP in Catabolic Repression; Tryptophan Operon System; Attenuation or Regulation after Initiation of Transcription; Autoregulation; Feedback Inhibition; Gene Regulation at Translation Level; Gene Regulation in Bacteriophages: Regulation of Lytic Pathway; Regulation of Lysogenic Pathway; Gene Regulation in Eukaryotes; At Genomic Level; at Transcription Level; Control at Post-transcription Level (Processing of RNA-Primary Transcript Level); Control on Transport of mRNA, Control at Translation of mRNA, Control at mRNA Degradation Level, Control at Protein Processing Level, Control at Protein Degradation Level, Evidence of Regulation of Gene Action in Eukaryotes.

REGULATION OF GENE ACTIVITY

Inside every cell, DNA has the coded genetic information for all the proteins which may be required at different times during cell cycle. Inspite of enormous genetic capacity, not all the genes function all the time or at any given time. As a matter of fact, out of 4,000 genes of bacterial genome and 1,00,000 genes of human genome, only a fraction is expressed at a given time. Moreover, products of certain genes are needed only in certain types of cells. It means genes are selectively activated on special occasion in response to some prevailing environmental conditions and for a limited time on demand. In all cell types, some enzymes are synthesised **constitutively**, i.e. all the time. It means that transcription of mRNA for such enzymes from constitutive genes occurs constantly. For other enzymes mRNA transcription is initiated only on demand and is switched off when need is fulfilled. The control on gene expression or on protein synthesis is called **gene regulation**.

The most efficient and economic way of controlling gene expression is the varying efficiencies of promoter recognition by ***RNA polymerase*** and initiation of translation. But this gene control can influence the efficiency of different genes of the genome. Different mechanisms will be needed to have products of those genes that are needed only occasionally and for those genes whose products are needed in significant amounts only under specific circumstances. Thus regulation of gene activity is associated with controlling:

1. Quantitative difference in the amounts of proteins/enzymes of different genes.
2. The amount of protein/enzyme produced by a gene at different times or in different tissues.

Exhaustive investigations have established that regulation of gene activity can occur at any one of the following levels:

- **Transcription level,** i.e. at the time of synthesis of primary RNA transcript from the gene.
- **Processing level** or **post-transcriptional level,** i.e. at the time of post-transcriptional processing of primary RNA transcript into mRNA.
- **Translation level,** i.e. at time of protein synthesis.
- **Post-translation level,** i.e. at the time of post-translation modification of polypeptide chain or chains into a functional protein.

Control of gene expression at the level of transcriptional initiation is the most economical way of control because this limits the number of mRNA molecules synthesised. The gene regulatory mechanisms of prokaryotes and eukaryotes are slightly different and are discussed separately.

GENE REGULATION IN PROKARYOTES

In prokaryotes, the gene regulatory systems are adapted to provide maximum growth rate in a particular environment and gene expression is controlled primarily at transcription level with some control at the translational level. Transcriptional control is accomplished by using **on-off** mechanism. It means a particular gene is switched on and synthesis of its mRNA is induced only when the gene product is needed. When the gene product is not needed, the gene is switched off and synthesis of its mRNA stops. The post-transcriptional control (i.e. control at translation level) is achieved by rapid degradation of mRNA molecules involved in this synthesis and modification of nascent polypeptide chain or chains into functional proteins.

Coordinated Gene Regulation

In bacterial systems, when several enzymes act in sequences in a single metabolic pathway, usually either all or none of these enzymes are produced. This phenomenon is called **coordinated regulation**. It results from control on the synthesis of a single polycistronic mRNA molecule that encodes all of the gene products. This type of regulation is not found in eukaryotes because eukaryotic mRNA is monocistronic.

Strategies of Gene Regulation

In bacterial cells protein synthesis or gene expression is mostly controlled at transcription level. This regulation may be **negative** or **positive**.

1. Negative Regulation: In negative regulation an **inhibitor** or **repressor** is present in the cell. The binding of inhibitor prevents or suppresses transcription of relevant gene. A signal molecule or **inducer** is needed to react with and remove the inhibitor so as to allow the initiation of transcription. The negative regulation of gene expression is also called **repression**.

2. Positive Regulation: In this system, an **effector** or **inducer** molecule activates the promoter which promotes transcription of mRNA. There is no inhibitor–inducer interaction. The positive regulation of gene action is also called **induction** and the substance that induces gene action is called **inducer**. The inducer may be a protein, a small molecule or a molecular complex.

Positive regulation can be achieved in any one of the two ways:

- Transcription will continue until a molecular signal binds to the activator molecule and causes it to dissociate from DNA.
- The activator-molecular signal complex binds to promoter. If the signal molecule dissociates, the activator falls off DNA, stopping the transcription.

Negative and positive regulations are not mutually exclusive. Some systems are both positively and negatively regulated. These utilise two different regulators to respond to different conditions of the cell. A catabolic system may be regulated positively or negatively. For catabolic pathway effectors are substrates which function as inducers. For anabolic pathways, effectors are usually the end products which function as repressors.

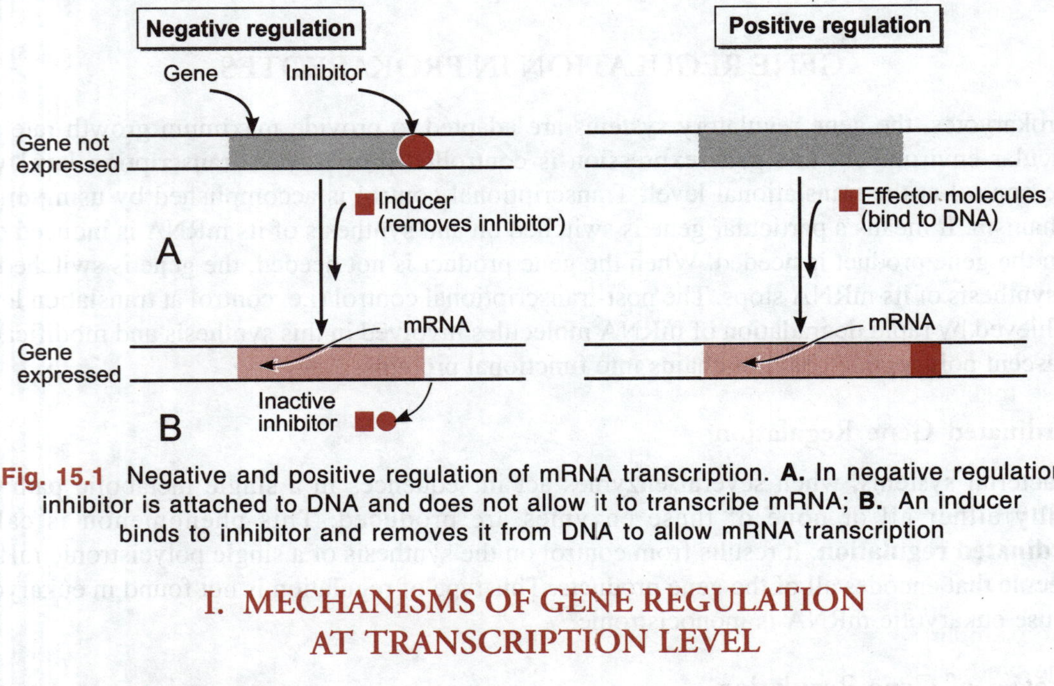

Fig. 15.1 Negative and positive regulation of mRNA transcription. **A.** In negative regulation inhibitor is attached to DNA and does not allow it to transcribe mRNA; **B.** An inducer binds to inhibitor and removes it from DNA to allow mRNA transcription

I. MECHANISMS OF GENE REGULATION AT TRANSCRIPTION LEVEL

Induction and Repression

A set of genes will be switched on when there is necessity to metabolise a new substrate. When these genes are switched on, enzymes are produced. This phenomenon is called **induction**.

Similarly, when a metabolite needed by bacterium is provided in excess in the medium, bacterium stops its synthesis and the gene associated with its metabolism is turned off. This is called **feedback repression.**

1. Induction and Inducible System (Inducible Operon)

In genetic induction the gene or genes are induced or 'switched on' to produce mRNA, needed for the synthesis of required enzyme or enzymes. The substance which induces the gene for protein synthesis or enzyme production is known as **inducer**. The phenomenon of induction can be demonstrated by growing *E. coli* in different nutritional media as follows:

- When *Escherichia coli* are grown in a culture medium containing glycerol, they produce all those enzymes, which are needed for the breakdown of glycerol. The synthesis of all other enzymes is maintained at the minimum level.
- When these *E. coli* are grown in a medium containing lactose the synthesis of enzyme ***β-galactosidase*** increases manyfolds. This enzyme hydrolyses lactose into glucose and galactose. The synthesis of other enzymes is reduced considerably.
- When *Escherichia coli*, are grown on glucose medium, they contain just traces of ***β-galactosidase*** (an enzyme that hydrolyses lactose into glucose and galactose). If these bacteria are transferred into a medium containing lactose, the concentration of this enzyme increases many-fold that enables them to metabolise lactose. Here substrate lactose has acted as inducer activating the specific gene to synthesise the enzyme in question.

In the above experiments substrates glycerol and lactose act as inducers for the synthesis of required enzymes. The enzymes whose synthesis can be induced by adding substrate are called **inducible enzymes**. The gene complex responsible for the synthesis of inducible enzymes is called **'inducible system'**.

Analysis of total mRNA present in the cells before and after addition of lactose shows that no *lac* mRNA is present before lactose is added to the medium. Addition of lactose triggers synthesis of *lac* mRNA. Here lactose is an **inducer** and genes associated with the enzymes of lactose metabolism system form ***lac* inducible system**.

2. Repression or Repressible System

In repression the activity of gene or genes is suppressed and the synthesis of specific protein is stopped or reduced. The substance which stops or suppresses protein synthesis is known as **repressor**.

For example, in minimal culture medium, *E. coli* cells can synthesise all the enzymes needed for the synthesis of different amino acids. But when a particular amino acid is added in the culture medium, the enzymes needed for its synthesis are not produced. When histidine is added in the bacterial culture medium, the production of histidine-synthesising enzymes is stopped. The enzymes, whose synthesis is checked by the addition of end product, are known as **repressible**

enzymes. The genes associated with these enzymes form **repressible system** and the substance that represses the synthesis of repressible enzymes is called a **repressor**.

As a matter of fact, some enzyme are normally present in the cell but cease to be synthesised when high concentrations of their end products are present. These end products are called **corepressors**. The regulator gene produces a substance called **aporepressor**. The aporepressor unites with corepressor to form the functional **repressor molecule**. This repressor molecule inhibits transcription of mRNA from all these structural genes which are responsible for the synthesis of enzymes of synthetic pathway. Most repressible enzyme systems occur in anabolic pathway.

Table 15.1 Differences between induction and repression

Induction	Repression
1. Induction involves switching on the specific gene to produce RNA for the synthesis of enzyme required.	1. Repression is switching off the gene or operon which was actively producing an enzyme that is no more required by the cell.
2. It is responsible for starting transcription and translation.	2. It stops transcription and consequently the translation.
3. It needs an **inducer**.	3. It needs a **suppressor** which acts as a **feedback inhibitor.**
4. When a new metabolite enters the system, it needs some new enzymes for its metabolism. Therefore, concerned genes are switched on.	4. When end products of a metabolic reaction accumulate in the system further reaction is not needed and so also the associated enzymes. Hence, concerned genes are turned off.
5. Inducer does not permit the repressor to bind with operator gene.	5. Repressor is helped by a corepressor to combine with the operator gene.
6. Induction is associated with catabolic pathway.	6. Repression acts in an anabolic pathway.

1. The Lactose System in *E. coli*

Jacob and **Monod** (1961), while studying catabolism of lactose in *E. coli*, suggested that the action of most genes is regulated at the transcription level by **induction and repression** phenomena. Three enzymes, namely, ***β-galactosidase***, ***galactose permease*** and ***thiogalactoside transacetylase*** are needed for lactose catabolism. The genes or cistrons (DNA segments) for these three enzymes are represented by ***lac* Z, *lac* Y** and ***lac* A**. These are called **structure genes** and are located in a linear sequence in the bacterial chromosome. The action of these genes is controlled in a coordinated fashion by **regulatory genes**.

Lac Operon System

Francois Jacob and **Jacques Monod** explained regulation of the activity of these *lac* genes by **operon model**. The operon model for lactose catabolism is called ***lac* operon.** It consists of two kinds of genes:

Francois Jacob in collaboration with **Jacques Monod** presented 'Operon concept' and concept of messenger RNA in 1961, and shared the 1965 **Nobel Prize** along with **Lwoff.**

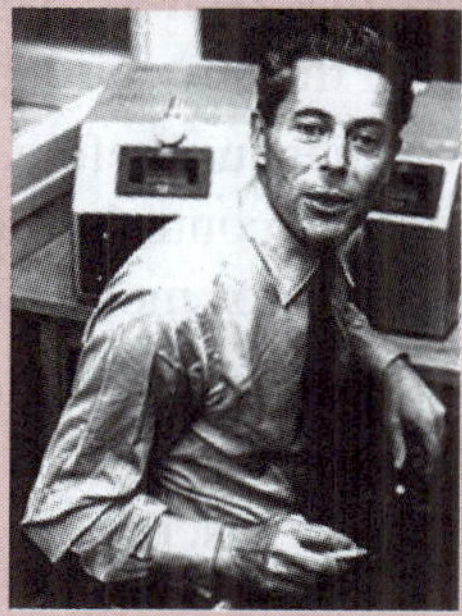

Jacques Monod

Francois Jacob

Francois Jacob and Jacques Monod served heroically during World War II, Jacob as a member of the Free French Forces, injured seriously in Normandy in August 1944 and **Monod** as a leader of the Paris Resistance Force. **Monod** had already obtained his Ph.D but **Jacob** had to wait until 1947 to gain his MD from the university of Paris. Both spent the major part of their career at the Pasteur Institute, the famous Paris Research Centre set up around Louis Pasteur in the late ninteenth century and still one of the most influential European Laboratories.

Francois Jacob contributed to some of the early works on bacteriophages with **Andre Lwoff** and **Elie Wollman**, prior to his collaboration with Monod that led to the Operon theory and the concept of messenger RNA in 1961. The two shared the 1965 Nobel Prize with Lwoff. Both authored a number of good books.

1. Structure Genes: These genes are segments of DNA that are associated with the synthesis of those enzymes that are needed for catabolism of lactose. In ***lac* operon** there are three structure genes:

- *lac Z* gene for ***β-galactosidase*** enzyme
- *lac Y* gene for ***galactose permease*** enzyme and
- *lac A* gene for ***thiogalactoside transacetylase***

These genes produce a single polycistronic mRNA molecule as shown in Fig. 15.2.

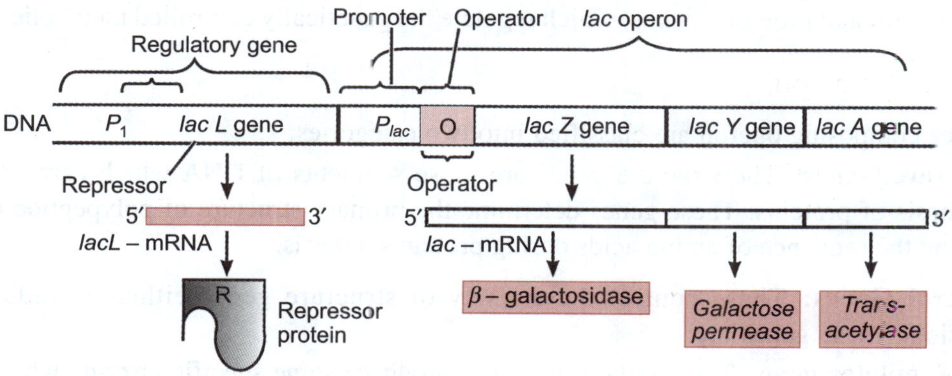

Fig. 15.2 Genetic map of *lac* operon in *E. coli*

2. Control Genes or Regulatory Genes: These genes control the activity of structure genes and lie immediately adjacent to the structure genes. There are three control genes, namely:

- **Operator gene (O):** It lies just before, i.e. upstream of first structural gene and overlaps the promoter sequence. It controls transcription of mRNA from structure genes. It is under the control of repressor molecule, produced by the regulator on repressor gene.
- **Promoter gene (P):** It lies immediately adjacent to operator gene. RNA polymerase binds at this site. Transcription of an operon begins at promoter, the site where ***RNA polymerase*** binds for the transcription of mRNA from structure genes. It controls rate of mRNA synthesis.
- **Repressor or regulator gene (L):** It lies outside the operon and produces a **repressor substance**. This repressor binds to operator gene and suppresses the transcription of structure genes.

Functioning of *Lac* Operon

Beckwith (1967), **Epstein** and **Beckwith** (1968) and **Martin** (1969) have described the operation of *lac* operon in *E. coli* in the following ways:

1. In the absence of an inducer-lactose, the regulator gene L produces a repressor protein (R) which binds strongly to the operator site and prevents its transcription. As a result, the structure genes do not synthesise mRNA and proteins are not formed.
2. When an inducer-lactose is introduced in the medium, it enters the cell and gets modified in such a way that it binds to the repressor. The repressor now fails to bind to the operator. The operator is, therefore, free and induces the ***RNA polymerase*** to bind to the initiation site on promoter and to transcribe the cistrons into a polycistronic *lac* mRNA. This codes for the three enzymes necessary for lactose catabolism.

Definition of Operon

An operon is a group of closely linked structure genes (or cistrons) and the associated control genes (operator and promoter genes) which regulate the genetically controlled metabolic acitivity.

Structure of Operon

The genes composing operon are classified into two categories:

1. Structure Genes: The structure genes are those segments of DNA which carry codes for the synthesis of proteins. These genes determine the primary structure of polypeptide chain by controlling the sequence of amino acids during protein synthesis.

2. Control Genes: These control the activity of structure genes either by induction or suppression. These genes are:

- **Regulator gene:** The regulator gene (L) produces some specific enzyme which acts as a **repressor substance**. In case of *lac* operon, the regulator gene codes for a protein which strongly binds to the operator gene and suppresses its activity.

- **Promoter gene:** The promotor gene (P) is the DNA segment at which ***RNA-polymerase*** binds and initiates the transcription of the structure genes.
- **Operator gene:** The operator gene (O) is the segment of DNA, which exercises a control over transcription. The repressor substance produced by the regulator gene binds with the operator gene. It lies close to the structural gene.

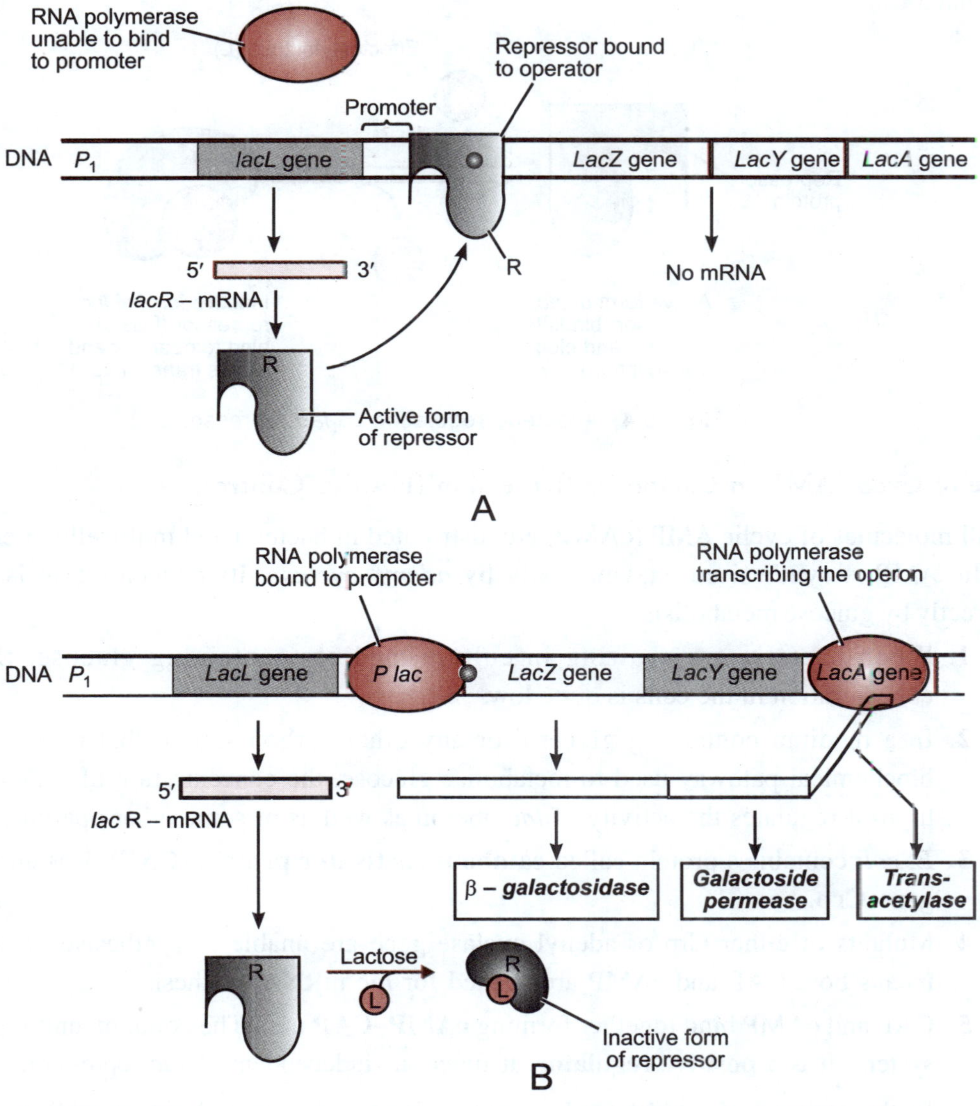

Fig. 15.3 Function of *lac* operon in *Escherichia coli*. **A.** Lactose absent, repressor active and bound to operator, operon repressed; transcription blocked and no mRNA synthesised; **B.** Lactose present, repressor bound to lactose and inactivated, repressor no longer binds the operator, operon derepressed, transcription begins and mRNA synthesised

2. Allosteric Regulation of *Lac* Repressor

Repressor proteins are **allosteric,** i.e. these exist in two conformational forms. In one form the protein is active and in the other form it is inactive. When effector molecule binds to the protein, the protein molecule undergoes conformational change and becomes inactive. But change is reversible. As soon as effector moves away, protein becomes active again by returning to the original form.

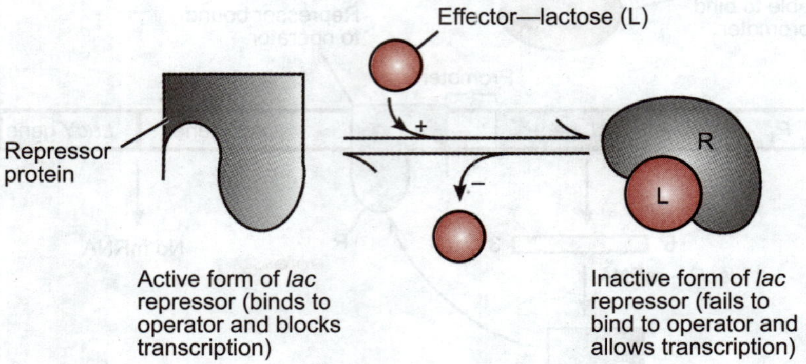

Fig. 15.4 Allosteric regulation of *lac* repressor

Role of Cyclic AMP in Catabolite Repression (Positive Control)

Small molecules of cyclic AMP (cAMP) are distributed in bacteria and multicellular eukaryotes. Cyclic AMP is synthesised enzymatically by *adenyl cyclase*. Its concentration is regulated indirectly by glucose metabolism.

1. When bacteria are growing in culture medium containing glucose, the cAMP concentration in the cells is quite low.
2. In a medium containing glycerol or any other carbon source that cannot enter the biochemical pathway used to metabolise glucose, the concentration of cAMP becomes high. It regulates the activity of *lac* operon as well as of some other operons.
3. *E. coli* contains a protein called **catabolite activator protein (CAP).** It is encoded by a gene **Crp**.
4. Mutants of either Crp or adenyl cyclase gene are unable to synthesise *lac* mRNA. It means both CAP and cAMP are needed for *lac* mRNA synthesis.
5. CAP and cAMP bind together forming **cAMP–CAP** unit. The complex unit regulates *lac* system. It is a **positive regulator**. It functions independent of *lac* repression system.
6. In the absence of cAMP–CAP complex, ***RNA polymerase*** binds weakly to promoter. The weak binding fails to initiate transcription.
7. cAMP–Crp complex binds to a base sequence in DNA in the promoter region to initiate transcription.

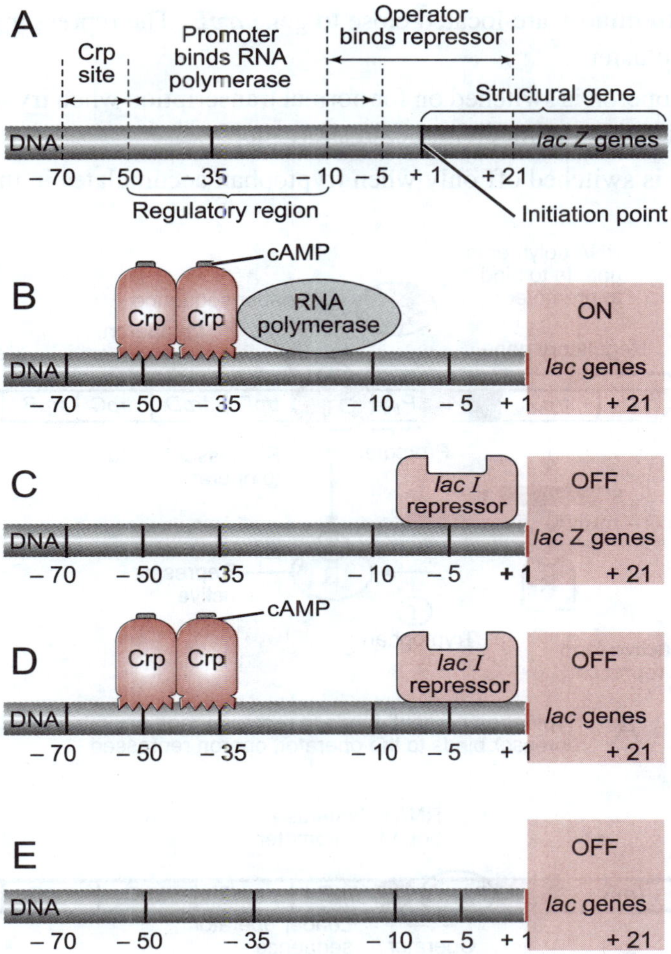

Fig. 15.5 Role of Crp (i.e. cAMP + Crps complex; Crp is cAMP receptor protein) in positive control of transcription of *lac* operon: **A.** Diagram showing Binding sites for various regulatory components; **B.** Glucose present, Crp binds to Crps site, *lac* I repressor removed, RNA polymerase binds to DNA and *lac* I gene is switched on; **C.** Glucose present, Crp absent, lactose absent, *lac* I repressor binds and *lac* I gene switched off; **D.** Both glucose and lactose are absent, *lac* I gene remains switched off, though Crp is present; **E.** Glucose absent, Crp absent, lactose present hence *lac* I repressor removed but *lac* I gene remains switched off because RNA polymerase does not bind to DNA due to the absence of Crp

Tryptophan Operon: A Repressible Operon System

Tryptophan operon (*trp* operon) in *E. coli* is responsible for controlling synthesis of amino acid **tryptophan.** Its synthesis is carried out by five enzymes. The genes encoding these enzymes are located adjacent to one another in succession and are named *trpE*, *trpD*, *trpC*, *trpB* and *trpA*. Gene *trpE* is the first one to be translated. Genes **promoter, operator** and two regions called **leader** and **attenuator** are located close to gene ***trpE***. The repressor gene ***trpR*** is present far from this gene cluster.

1. Tryptophan operon is switched on for normal transcription when tryptophan amino acid is not accumulated in the medium, i.e. in short supply.
2. This operon is switched off only when tryptophan accumulates in the medium.

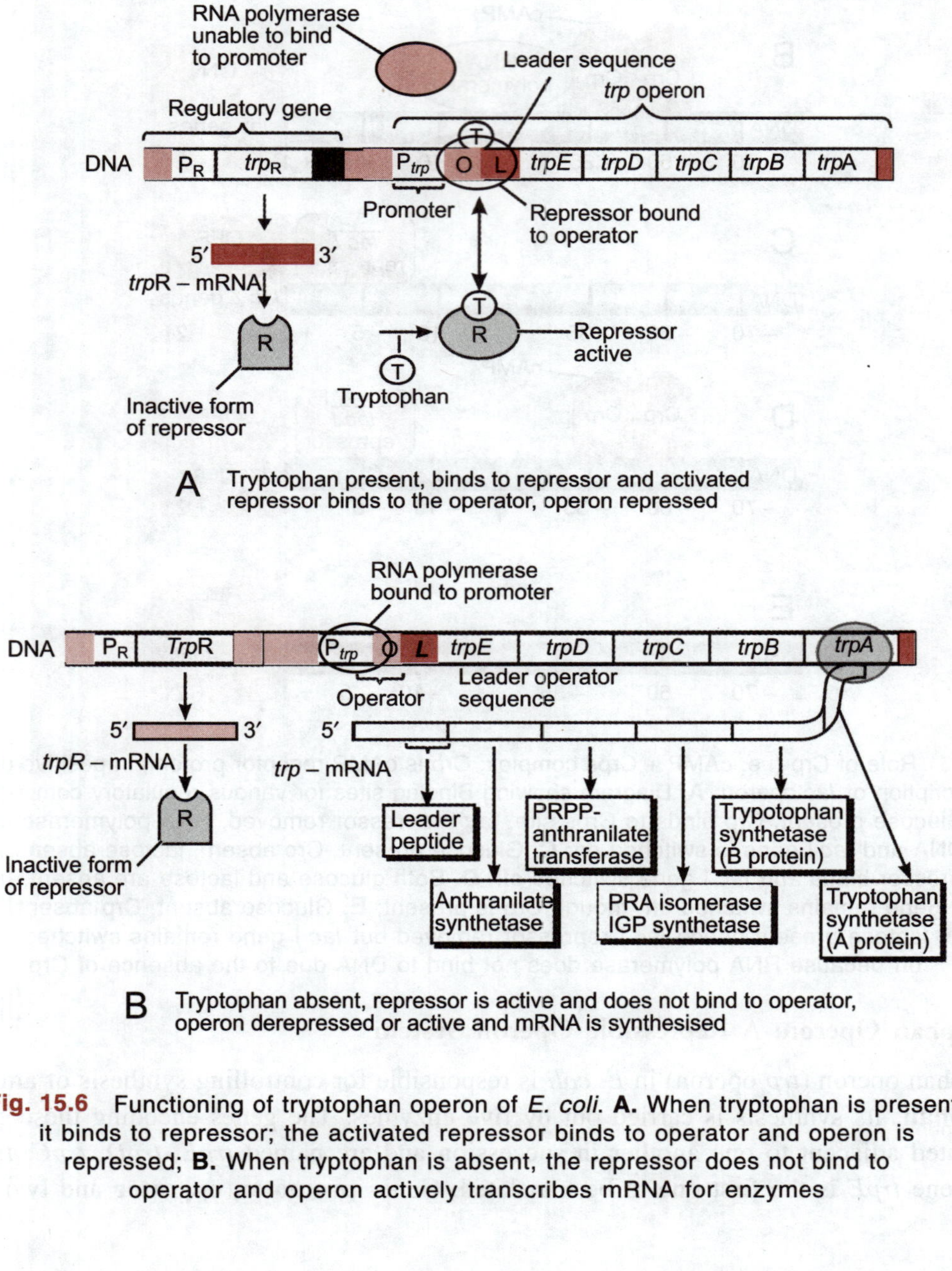

Fig. 15.6 Functioning of tryptophan operon of *E. coli*. **A.** When tryptophan is present, it binds to repressor; the activated repressor binds to operator and operon is repressed; **B.** When tryptophan is absent, the repressor does not bind to operator and operon actively transcribes mRNA for enzymes

The regulatory protein of *trp* operon is produced from the *trpR* gene. Mutations either in this gene or in the operator gene cause constitutive initiation of transcription of *trp* mRNA. This protein synthesised from mutant gene is called **aporepressor.** It does not bind to the operator unless tryptophan is present. The aporepressor and tryptophan molecule join to form active ***trp* repressor,** which binds to the operator to switch it off.

1. Aporepressor (no tryptophan) ⟶ Active operator ⟶ Transcription
2. Aporepressor + Tryptophan ⟶ Active repressor + Operator ⟶ Operator switched off ↓ No Transcription

3. Attenuation: Regulation after the Initiation of Transcription

In most cases regulatory mechanisms control the initiation of transcription. In prokaryotes, some regulatory mechanism exercises control following the initiation. **Charles Yanofsky** and his colleagues found that *trp* operon has a regulatory site between promoter/operator and the first structural gene *trp E*. This site of DNA is called **leader sequence (L).** It is non-coding DNA segment which is transcribed into leader mRNA at 5′ end of mRNA. This *trp* leader mRNA is 162 nucleotides long. It is sensitive to tryptophan level and exercises an additional control on the transcription of polycistronic *trp* mRNA. The effect of this control element, i.e. of leader sequence is called **attenuation**, because it reduces or attenuates synthesis of mRNA.

Structure of Leader mRNA

The 162 base long leader mRNA has following segments:

1. It has an initiation codon **AUG** at 5′ end and a stop codon **UGA** in the same reading frame. This region encodes a polypeptide, consisting of 14 amino acids. It is called **leader polypeptide**.
2. The leader polypeptide mRNA contains two adjacent **tryptophan codons** at position 10 to 11. It means the 10th and 11th amino acids in leader polypeptide are tryptophans.

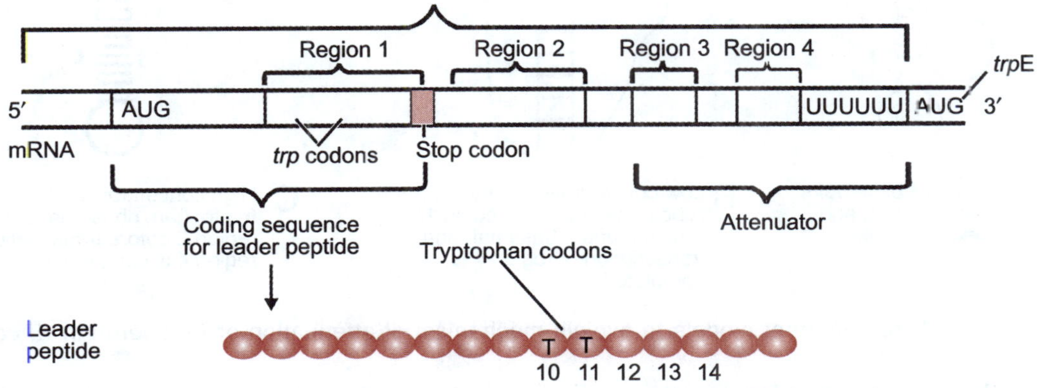

Fig. 15.7 Structure of leader segment of *trp* mRNA

3. Leader mRNA has four distinct regions denoted as 1, 2, 3 and 4. These regions can base pair with each other in different ways, forming several hairpin structures. For example:
   - In a stable form the free leader mRNA shows base pairing between 1 and 2 and between 3 and 4 regions.
   - In low concentration of tryptophan, ribosome is stalled in region 1 so that regions 2 and 3 base pair together. Transcription of region 4 and of subsequent structure genes is completed.
   - In high concentration of tryptophan in the medium, the 3 and 4 regions get base paired and synthesis of protein is interrupted on reaching the 2nd region.
4. After region 4, is present a string of eight 'U' lying just prior to the structure gene *trp E*.
5. The part of leader mRNA comprising regions 3 and 4 and a string of eight 'U' is called **attenuator**. When base pairing between 3 and 4 regions creates a hairpin, the attenuator acts as a transcription termination signal and RNA polymerase and growing RNA chain detaches from DNA.

MECHANISM OF ATTENUATION

During translation, a ribosome attaches to the first binding site (the initiation codon) on *trp* mRNA. Transcription of mRNA and its translation follow closely.

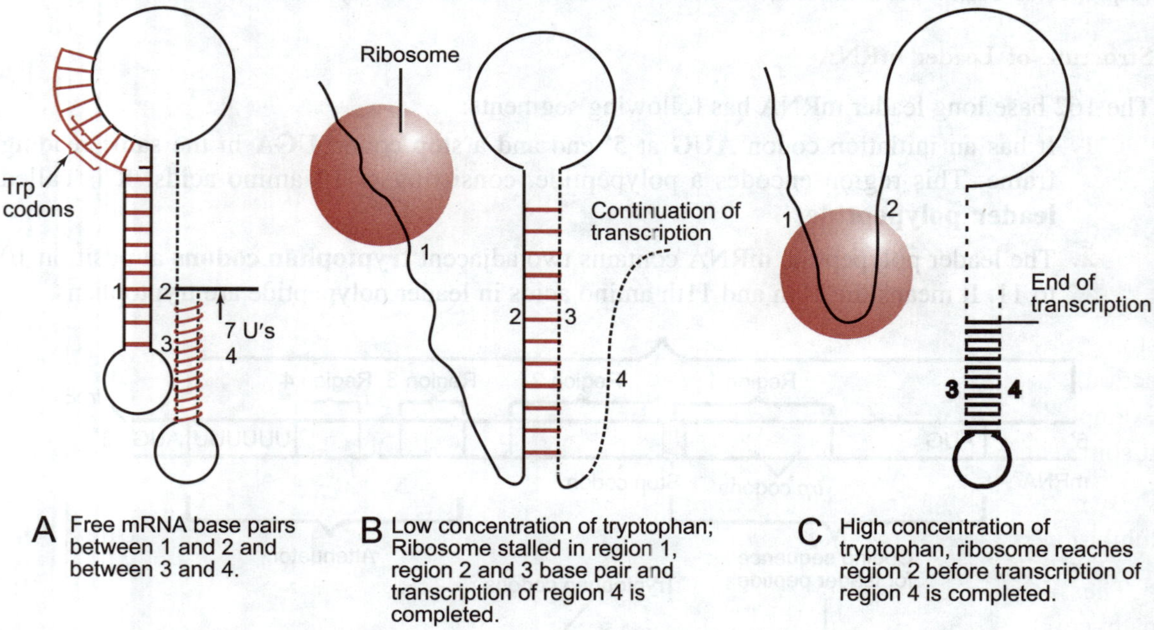

Fig. 15.8 Three different models to explain mechanism of attenuation of *trp* operon in *E. coli*

1. When tryptophan level in the medium is low, the concentration of tryptophenyl-tRNA (i.e. tRNA molecule carrying tryptophan) is also low. When ribosome arrives at the

tryptophan codons in leader sequence, it waits for the arrival of tryptophenyl—tRNA and blocks region 1. As a result, a hairpin structure is formed by pairing between regions 2 and 3.

Thus, region 3 is unable to pair and form hairpin with region 4. The termination hairpin loop is not formed and transcription of structural genes of tryptophan operon continues to produce polycistronic *trp* mRNA.

2. When tryptophan level in the medium is high, the tryptophenyl tRNA concentration is also high. The ribosome moving from the initiation codon, passes over tryptophan codons and stops at stop codon. Thus, it blocks region 2 of *trp* mRNA. This permits pairing of segments 3 and 4 and 3–4 hairpin loop is formed. This along with poly 'U' causes the termination of mRNA and structural genes are not transcribed, stopping the tryptophan synthesis.

A regulatory mechanism of this type can not exist in eukaryotes because transcription and translation can not be coupled. In eukaryotes, transcription occurs in the nucleus and translation in the cytoplasm.

4. Autoregulation

Synthesis of monocistronic mRNA can be autoregulated by gene product acting as a repressor. In an autoregulated system, when concentration of gene product exceeds than needed by the cell, the gene product molecule binds to the operator inhibiting transcription. When concentration of gene product falls below the required level either due to use or increased need, the molecule bound to the operator leaves the site and operator and promoter are free leading to the onset of transcription.

Synthesis of most repressor proteins and many enzymes needed at all times are autoregulated

5. Feedback Inhibition or End Production Inhibition

In case the end product of a biosynthetic pathway is not utilised, it accumulates inside the cell and stops further synthesis of that substance. This is known as **feedback inhibition** of the gene action. The inhibition is caused by the binding of the accumulated end product to the enzyme. For example, **Umbarger** (1961) demonstrated that **isoleucine synthesis** in *E. coli* can be inhibited if some isoleucine is added to the medium.

By the binding of isoleucine to the first enzyme, the enzyme is inactivated. This phenomenon of inhibition at the enzyme level is known as **allosteric interaction.**

Thus, in feedback or end product inhibition the accumulated end product of a biosynthetic pathway combines loosely with the first enzyme in the pathway. This complex does not bind to the operator or promoter site but the tertiary structure of enzyme is modified which makes the enzyme inactive. This **allosteric transition** of enzyme molecule blocks its enzyme activity and prevents overproduction of end product and also the intermediate metabolites.

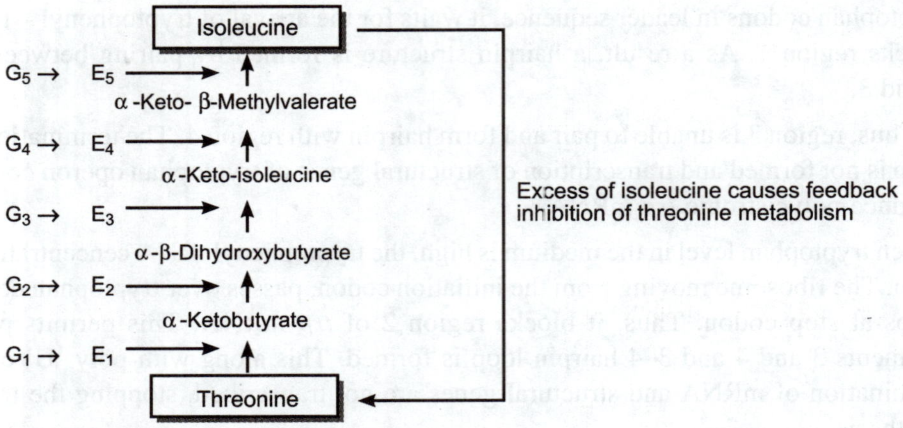

Fig. 15.9 Mechanism of feedback inhibition of synthesis of amino acid isoleucine

In end product inhibition the expression of gene is regulated after proteins have been synthesised. Thus, it is **post-translational control of gene action.**

Gene Regulation at Translation Level

Translational regulation of protein synthesis refers to the control of number of copies of a particular protein translated from a polycistronic mRNA. It varies from gene to gene and there is a gradient of translation decreasing from 5′ terminus to 3′ terminus of mRNA.

This depends on:

- Varying efficiencies of initiation of translation.
- Different spacing between chain termination codon and next initiation codon.
- Different sensitivity of various regions of mRNA to degradation.
- Inhibition of a particular gene by its gene product **(feedback inhibition)**.
- Synthesis of ribosomes.

GENE REGULATION IN BACTERIOPHAGES

Gene expression in bacteriophages is also regulated through **operons**. There are promoter and repressor genes, whose products regulate the activity of structure genes. For example, whether a phage is going to follow lytic or lysogenic pathway is determined by **genetic switch** involving repressor gene *cI* and *cro* gene. Switching on of ***cro*** gene initiates lytic pathway, while turning on of repressor gene ***cI*** leads to lysogenic pathway.

Regulation of Lytic Pathway

When repressor gene is inactive two promoter genes $\mathbf{P_L}$ (promoter left) and $\mathbf{P_R}$ (promoter right) are transcribed.

1. Gene $\mathbf{P_R}$ synthesises **cro protein** which switches on the lytic pathway.
2. Gene $\mathbf{P_L}$ transcribes protein-N. It allows RNA synthesis to proceed over certain transcription terminators. This process is called **antitermination** and gene $\mathbf{P_L}$ as **antiterminator gene.**

The products of both these genes turn on various genes of **lytic pathway**. These promote synthesis of coat protein, replication of viral DNA, assembly of new phage particles and lysis of bacterial cell.

Regulation of Lysogenic Pathway

Lysogenic pathway is initiated by proteins produced by *cII* and *cIII* genes. These activate transcription of *cI* gene which produces **λ-repressor protein**. It binds to two operators O_L and O_R and also inactivates P_L and P_R. As a result, transcription of *cro* and *N* genes is blocked.

Thus in bacteriophages, genes related to lytic or lysogenic pathways are organised into **operons**. Like bacterial operons, these operons are controlled through the interaction of regulatory proteins with operators, and these operators lie adjacent to clusters of structural genes.

TEXT QUESTIONS

1. Describe operon concept of genetic control of protein synthesis.
2. Differentiate between repressor and inducers. Give one example where metabolite acts as an inducer.
3. Describe in details the *lac* operon bacterium *E. coli*. Discuss how it controls the activity of genes which produce enzymes necessary for the catabolism of lactose.
4. Describe similarities/differences between operator and regulator genes.
5. What is the positive and negative control of gene activity? Discuss with example.
6. Define an operon. Who gave this concept? Explain this concept with regard to *lac* operon.
7. What does *lac* operon consist of? How is the operator switch turned on and off in the expression of genes in this operon? Explain.
8. Distinguish between induction and repression.
9. What are retroviruses? How has their discovery led to the modification of the central dogma in molecular biology?
10. In *E. coli*, three enzymes *p*-galactosidase, permease and transacetylase became functional as soon as lactose was added. Explain why the enzymes were not forming in the absence of lactose?

11. Name the category of virus that carries reverse transcriptase. What is the purpose of this enzyme?
12. A microbiologist found that some bacteria infected by phages, had developed the ability to make a particular amino acid that they could not make earlier. What was this new ability probably because of ? Explain this phenomenon briefly.
13. How does an excess of tryptophan cause switching off of the tryptophan operon?
14. How is the functioning of *lac* operon different from that of tryptophan operon?
15. What name is given to a virus which infects bacteria?
16. Mention any one characteristic of bacterial DNA.
17. In which condition does the phenomenon of feedback repression operate in *E. coli*?
18. Name the category of viruses whose genetic material undergoes reverse transcription.
19. Name the phenomenon by which DNA isolated from one type of cell when introduced into another is able to express some phenotypic characters in the latter.
20. Name the process in which a bacteriophage picks up a piece of DNA from one bacterial cell and transfers the same to another bacterial cell on infection.

Regulation of Gene Expression in Eukaryotes

16
Chapter

IN THIS CHAPTER

Gene Regulation in Eukaryotes; Regulation of Gene Expression at Genome Level; Regulation of Gene Expression at Transcription Level; Gene Regulation at Post-transcription Level or Gene Regulation at Processing of RNA Primary Transcript; Evidence of Regulation of Gene Action in Eukaryotes

GENE REGULATION IN EUKARYOTES

Though there are certain basic similarities between prokaryotic and eukaryotic regulation of gene expression. However, the transcriptional regulation is much more complex in eukaryotes specially multicellus larc eukaryotes. There are several dramatic differences between the two, because:

1. Eukaryotes do not have typical operons as found in bacteria.
3. Eukaryotes have many more genes than prokaryotes. Their expression is regulated differently in different tissues and at different stages of development and differentiation.
3. Transcriptional regulation in eukaryotes involves many more regulatory proteins and DNA control elements such as **promoters, enhancers, activators, insulators, mediators** and other upstream or downstream nucleotide sequences.
4. Some eukaryotic regulatory proteins interact directly with DNA but in prokaryotes, many regulatory proteins act on DNA indirectly, via interactions with other proteins.
5. Expression of eukaryotic genes requires several **activators.** These may bend to upstream regions of promoter or to enhancer sequences that may be several kilobases away from the promoter or enhancers may lie downstream of their target gene.
6. The regulatory proteins, transcription factor etc. are synthesised by ribosomes in the cytoplasm but transcription of RNA occurs in the nucleus. Therefore, the regulatory proteins are to be transported into the nucleus.
7. In eukaryotes, DNA is associated with histone proteins and form **nucleosomes.** Nucleosomes do not allow DNA transcription. Removal of histones is essential for transcription to start.
8. In eukaryotes, long sections of DNA are frequently folded tightly into heterochromatin.

In heterochromatic stage, they become inaccessible to RNA polymerase and transcription factors.

9. Introns are present in most eukaryotic genes but only few prokaryotic genes.
10. In eukaryotes, RNA is synthesised in the nucleus and is transported through nuclear pores to the cytoplasm where it is translated.
11. Eukaryotes possess mechanism for rearrangement of DNA segments in a controlled manner by increasing number of copies of genes (genes amplification).
12. Regulatory regions are much larger than those of prokaryotes and these may be away from promoters.
13. In eukaryotes, most genes are subjected to positive control. Repressors are rare and usually behave differently to those in bacteria.

Levels of Gene Regulation in Eukaryotes

In eukaryotes, gene expression is regulated at the following levels:

1. Genome level.
2. Transcription of RNA.
3. Processing of RNA transcript.
4. Transport of RNA out of nucleus.
5. Degradation of mature RNA.
6. Translation of mRNA.
7. Modification of processing of proteins.
8. Degradation of protein products.

I. REGULATION OF GENE EXPRESSION AT GENOME LEVEL

Regulation of gene expression at genome level includes:

1. Presence of Multigene Families

In eukaryotic genome, many functionally-related eukaryotic genes are grouped as a set of genes, called a **gene family**. The gene families may be **simple multigene families, complex multigene** and **developmentally controlled complex multigene families**. These provide numerous opportunities for the recombination of various genes, modification of gene expression and change in the amount of mRNA produced. The concept is also described as **gene dosage**. It means the quantity of mRNA transcribed and amount of protein synthesised are directly proportional to the number of genes involved. For example, most cells have hundreds of copies of histone genes because huge amount of histones are needed by eukaryotic cells to form chromation needed at each cell division.

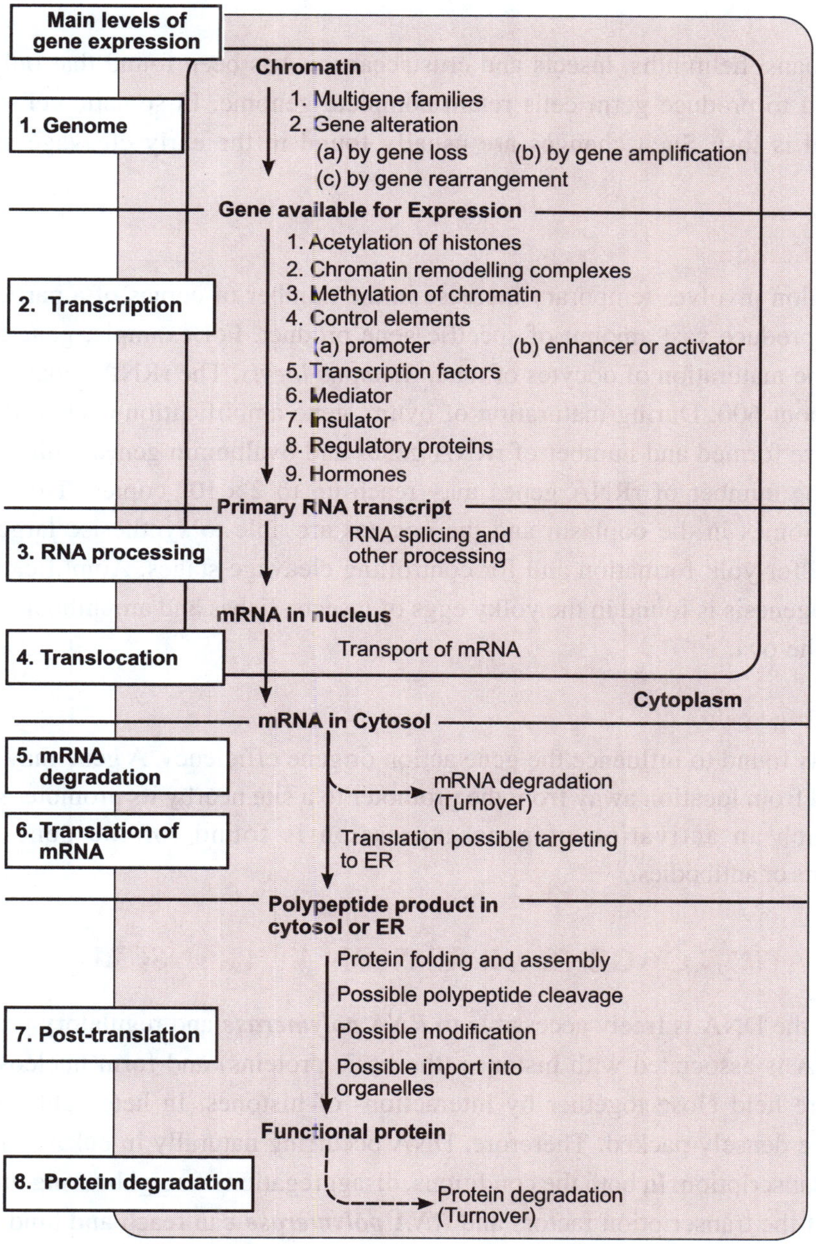

Fig. 16.1 Levels of regulation or control of gene expression in a eukaryotic cell

2. Gene Alteration

Alteration in the gene may include change in the number of copies of genes or change in the structure of gene/genes. Such changes can occur by loss, *amplification* and *rearrangement* of genes.

1. Gene Loss:

In some protozoans, helminths, insects and crustaceans, it has been found that only those cells that are destined to produce germ cells retain complete genome. In somatic cells some of the genetic material is lost. Such changes are usually found in the early cleavage stages during development.

2. Gene Amplification:

Gene amplification involves temporary increase in the number of copies of a particular gene or gene family to produce vast amount of specific gene product. For example, gene amplification occurs during the maturation of oocytes of toad, *Xenopus laevis*. The rRNA genes in the oocytes are normally about 600. During maturation of ovum, gene amplification occurs, the lampbrush chromosomes are formed and number of rRNA genes and ovalbumin genes multiply even up to 4,000 times. The number of rRNA genes may reach up to 2×10^6 copies. This increases the number of ribosomes in the ooplasm and the oocytes are able to synthesise large amounts of proteins needed for yolk formation and for controlling cleavage stages. Amplification of rRNA genes during oogenesis is found in the yolky eggs of insects, fishes and amphibians where yolk is synthesised in the ova.

3. Gene Rearrangement:

Position effect is found to influence the gene action or gene efficiency. A gene may get switched on when moved from location away from the promoter to a site nearby its promoter sequence. An example of such an activation of gene expression is found for the genes coding for immunoglobulins or antibodies.

II. REGULATION OF GENE EXPRESSION AT TRANSCRIPTION LEVEL

In prokaryotes, the DNA is freely accessible to ***RNA polymerase*** and regulatory proteins. But in eukaryotes DNA is associated with histones (the basic proteins) and form nucleosomes. These nucleosomes are held close together by interactions of histones. In heterochromatic regions, nucleosomes are densely packed. Therefore, DNA occurring naturally in eukaryotic cells is not accessible for transcription. In both the conditions, disaggregation of neighbouring nucleosomes is essential so that the transcription factors and ***RNA polymerase*** can reach and bind the promoter region to initiate transcription. Therefore, in eukaryotes transcriptional control on genes includes:

- Acetylation of histones
- Methylation of nucleotides
- Transcription factors
- Regulatory proteins
- Chromatin remodelling complexes
- Control elements
- Mediators
- Hormones

1. Transcriptional Control by Acetylation of Histones

Acetylation of histone tails promotes disaggregation of neighbouring nucleosomes. All the four types of histones (H2A, H2B, H3 and H4) can be acetylated, but H4 is acetylated the most. Acetylated histones form less condensed chromatin and provide more chances for RNA polymerase and other proteins to approach DNA helix. Acetylation is achieved by enzymes ***histone acetyl transferases*** **(HATs)** and deacetylation is carried out by ***histone deacetylases*** **(HDACs)**.

HATs activate the genes whereas HDACs suppress them. Several proteins which were previously called coactivators belong to HAT category. Similarly, several so called corepressors are HDACs proteins.

Extension: Non-acetylated and Acetylated Histones

- Non-acetylated histones form highly condensed heterochromatin which is transcriptionally inactive.
- Acetylated histones form less condensed chromatin or euchromatin which is transcriptionally active.
- Acetylation does not cause disassembling of nucleosome. It means nucleosomes as unit remain intact but become disaggregated.

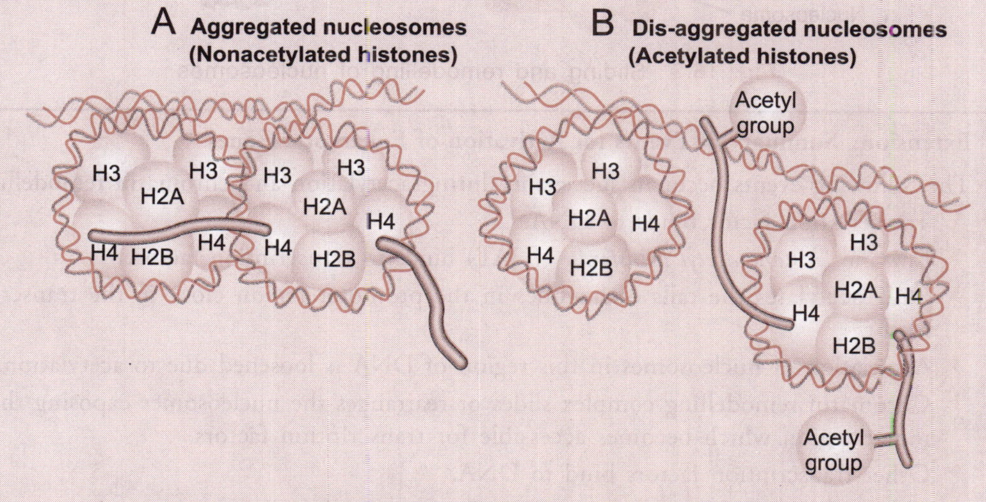

Fig. 16.2 Acetylation of histone tails leading to disaggregation of nucleosomes. **A.** Nucleosomes with nonacetylated histones are aggregated; **B.** Acetylated nucleosomes are disaggregated

2. Transcriptional Control by Chromatin Remodelling Complexes

Chromatin remodelling complexes are protein polymers that rearrange histones in the nucleosome so that promoter region is exposed and RNA polymerase can get access to it. These complexes are formed of 2 to 12 proteins. They bind strongly to nucleosome or DNA and loosen the winding of DNA. These are of two types: Small complexes formed of 2 to 6 polypeptides and are called **initiation switches** and larger complexes have 8 to 12 polypeptides and are designated as **Swi/Snf complexes**. The former bind to histones and latter to DNA. The remodelling of nucleosomes is achieved in two ways:

1. By sliding the nucleosomes along a DNA molecule and exposing nucleotide sequences for transcription,
2. By rearranging, histones loosen the ribosomes and expose nucleotide sequences.

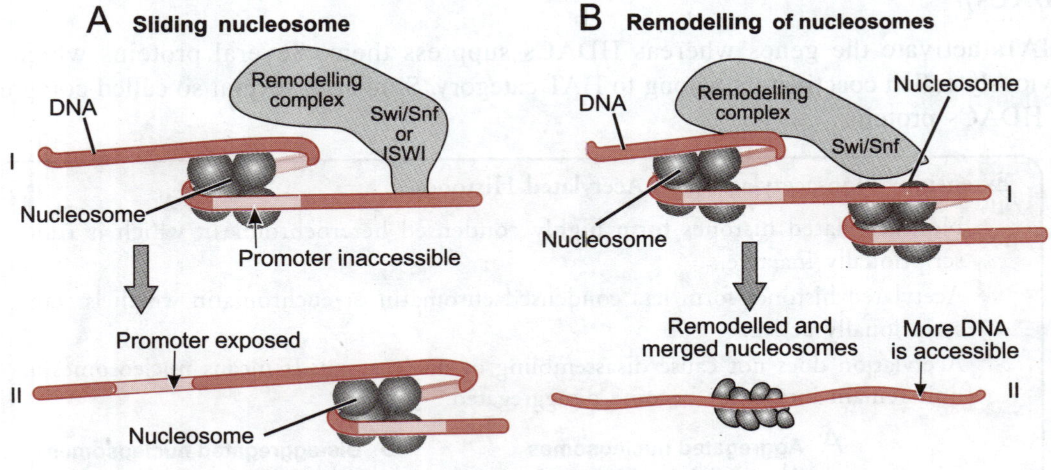

Fig. 16.3 Sliding and remodelling of nucleosomes

Extension: Summary of Events for Activation of Eukaryotic Genes

The following events occur in succession during acetylation and chromatin remodelling:

- Transcription factor binds to DNA.
- Enzyme *"histone acetyl transferase"* (HAT) binds to transcription factor.
- HAT acetylates the tails of histones in the promoter region close to the transcription factor.
- Association of nucleosomes in this region of DNA is loosened due to acetylation.
- Chromatin remodelling complex slides or rearranges the nucleosomes exposing the promoter region which becomes accessible for transcription factors.
- Other transcription factors bind to DNA.
- RNA polymerase binds to the exposed promoter region.
- Positive signal from one or more specific, transcription factors is passed on via mediator complex for the initiation of transcription.

3. Transcriptional Control by Methylation of DNA (Silencing of Gene)

Methylation of nitrogenous bases in DNA silences genes and is often used to control gene expression during development in higher organisms. Silencing involves methylation of cytosine in 5′…CG…3′ sequences.

The methyl groups project into major groove of DNA and interfere with the binding of transcription factors. Also methylated cytosine-guanine sequences are recognised by **methyl-cytosine binding proteins** (MeCPs). When bound to methylated–CG–sequences, MeCPs help in removal of acetyl groups from histones. This leads to the condensation of DNA to form heterochromatin, which is inaccessible for transcription. In animals about half of the functional genes are located close to–**CG–islands** or clusters of–CG–sequences. The housekeeping genes that are expressed in all cells and tissues, possess nonmethylated–CG–islands, and are always acetylated and active. In contrast, the–CG–islands of **tissue specific genes** get nonmethylated only in those tissues where genes are expressed.

Silencing may influence a single gene, a group of genes, a part of a chromosome or even a whole chromosome. For example, in female mammals, the whole X-chromosome becomes heterochromatic and is silenced. Methylation is prevalent in vertebrates. Frequency of methylation decreases in lower animals. Heavy methylation is found to inactivate X–chromosome (lyonisation).

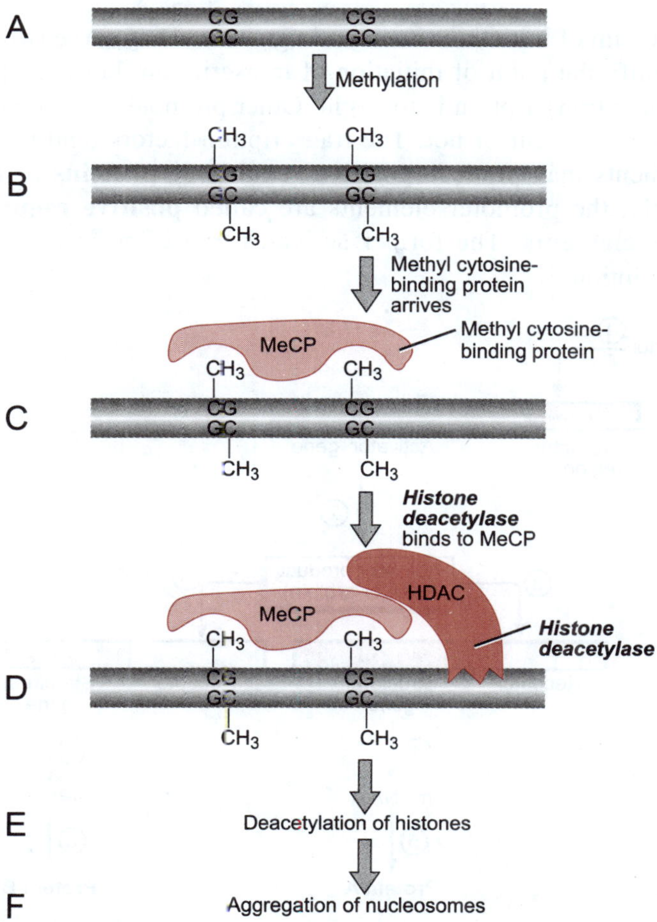

Fig. 16.4 Silencing of eukaryotic gene by methylation

Extension: Methylation causes deactivation of gene by the following methods

1. Methylation of restriction sites prevents self destruction.
2. Methylation of particular—CG—sequences prevents transcription of some genes.
3. Certain genes are highly methylated in cells in which they are not expressed but are unmethylated in cells in which these genes are expressed.
4. The **house-keeping genes** which control general cell functions and are continuously transcribed are rarely methylated.

4. Transcriptional Control by Control Elements

The protein-coding genes (structure genes) of eukaryotes have two types of controlling elements or sequences. These are **promoter elements** and **enhancer elements**.

1. Promoter Elements

They are located upstream of initiation site of coding genes. They serve two important functions. Some promoters identify the point of initiation of transcription. For example, TATA element is present at the site where transcription is to begin. Other promoters are regulatory. These control whether transcription is to occur or not. The transcription factors bind to these promoter sites. These promoter elements may produce **positive regulatory proteins** or **negative regulatory proteins**. Accordingly, the promoter elements are called **positive regulatory elements** and **negative regulatory elements**. The former activates transcription of structure genes, while latter turns off transcription.

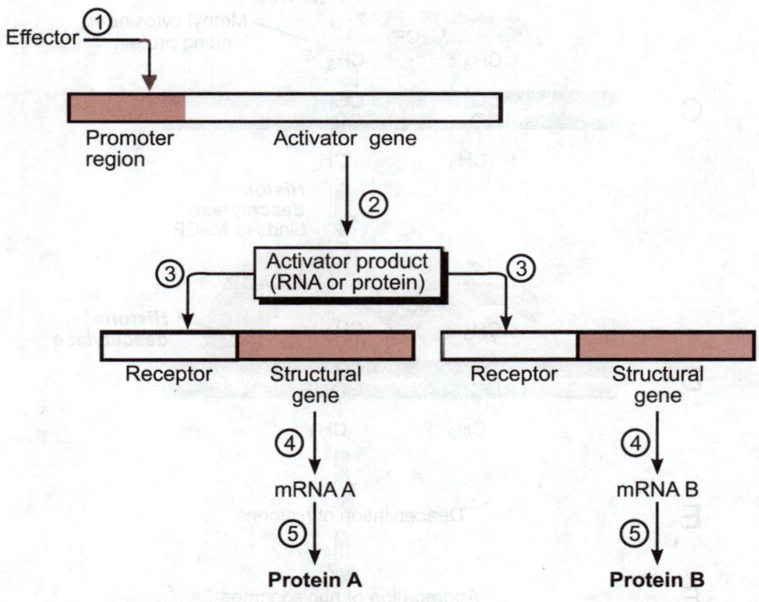

Fig. 16.5 Regulation of synthesis by promoter and activator elements

2. Enhancer or Activator Elements

The enhancers are nucleotide sequences that determine the maximal transcription of a gene by binding specific transcription factors, especially during development or in different cell types. They are located several kilobases away from the promoter either upstream or downstream. They are formed of a cluster of recognition sites and therefore bind several regulatory proteins. Binding of transcription factors to them influences polymerase activity and consequently the gene expression. Enhancers need not to lie close to its own promoter. It can increase transcription from any promoter within its neighbourhood. When an enhancer switches a gene on, the DNA between it and the promoter loops out.

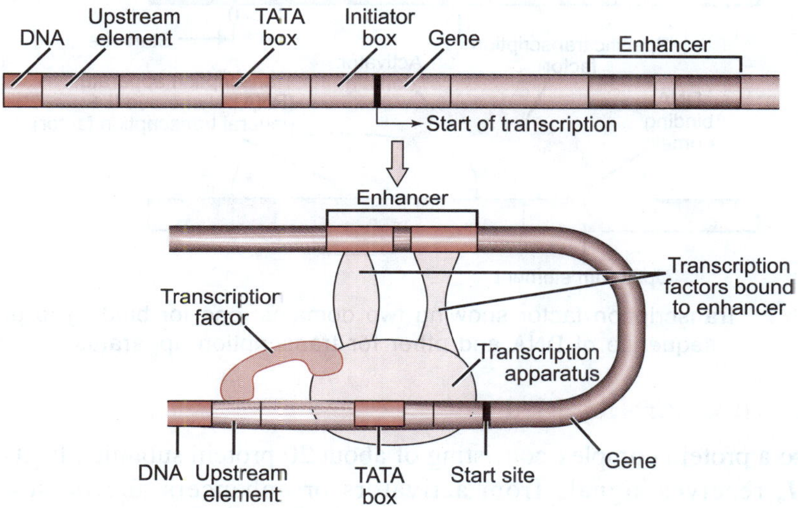

Fig. 16.6 Binding of enhancer to transcription apparatus through transcription factors

5. Transcriptional Control by Transcription Factor

In eukaryotes, initiation of transcription needs either the formation of **transcription complex** between initiation sequences of DNA, ***RNA polymerase*** and several transcription factors or regulation of binding of RNA polymerase to a pre-existing transcription complex by a transcription factor.

Transcription factors are protein complexes that respond to specific signals and in accordance regulate expression of protein coding genes by binding to the control region of DNA. They are of following two types:

1. **General Transcription Factors:** They bind initially to the TATA box and are needed for the initiation of all the transcriptions. They are TFIIB, TFIID and TFIIH. They are needed for the recognition and binding to the promoter.
2. **Specific Transcription Factors:** They bind to general transcription factors in response to specific signals to initiate transcription of specific genes. They are TFIIA, TFIIC, TFIIF. Typical specific transcription factors have following common features:
   - Transcription factors respond to some specific stimulus that signals to switch on one or more genes.

- Transcription factors can enter the nucleus through nuclear pores.
- They recognise and bind to a specific promoter sequence on DNA.
- They make contact with the transcription apparatus.
- Transcription factors have, at least, two domains. One domain binds to promoter DNA and another interacts with the transcription apparatus.

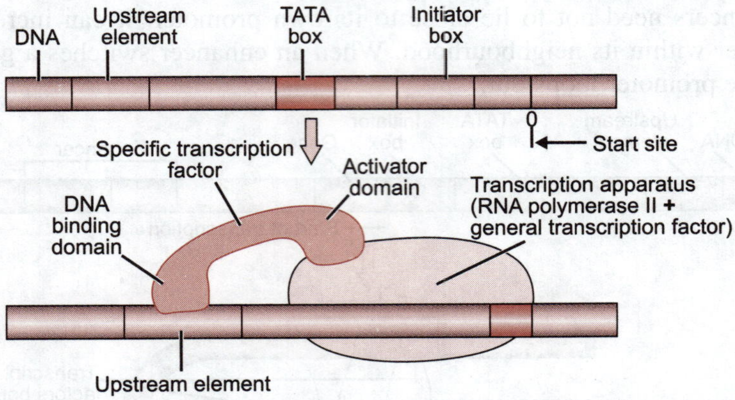

Fig. 16.7 Transcription factor showing two domains; one for binding to promoter sequence of DNA and other for transcription apparatus

6. Transcriptional Control by Mediator

Mediator is also a protein complex consisting of about 20 protein subunits. It sits on top of ***RNA polymerase II***, receives signals from activators or enhancers and passes them to RNA polymerase of the transcription apparatus. Some subunits of mediator act in a positive manner in controlling transcription of genes while others act negatively.

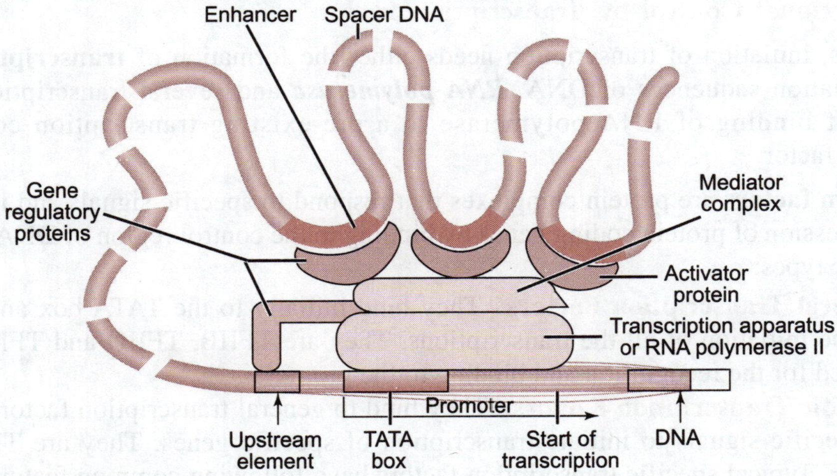

Fig. 16.8 Placement of mediator complex between enhancer and transcription apparatus to allow contact between enhancer, regulatory proteins and RNA polymerase

7. Transcriptional Control by Insulator

Insulators are regions of DNA rich in GC and CG sequences that are quickly methylated. Also insulators have clusters of nucleotide sequences that bind copies of special zinc finger proteins

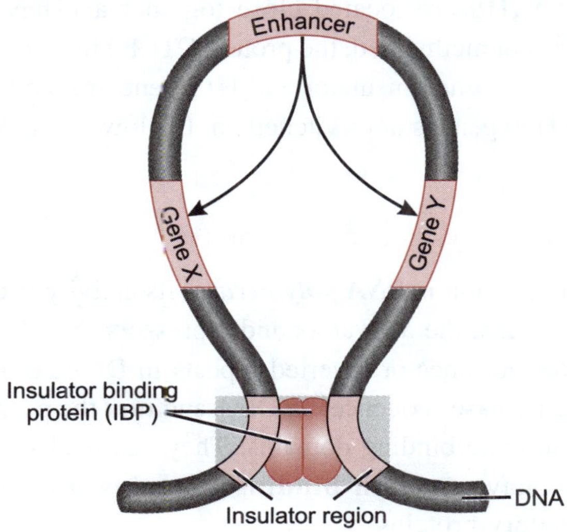

Fig. 16.9 Insulator sequences that bind to Insulator Binding Protein (IBP) and restrict the range of enhancer action

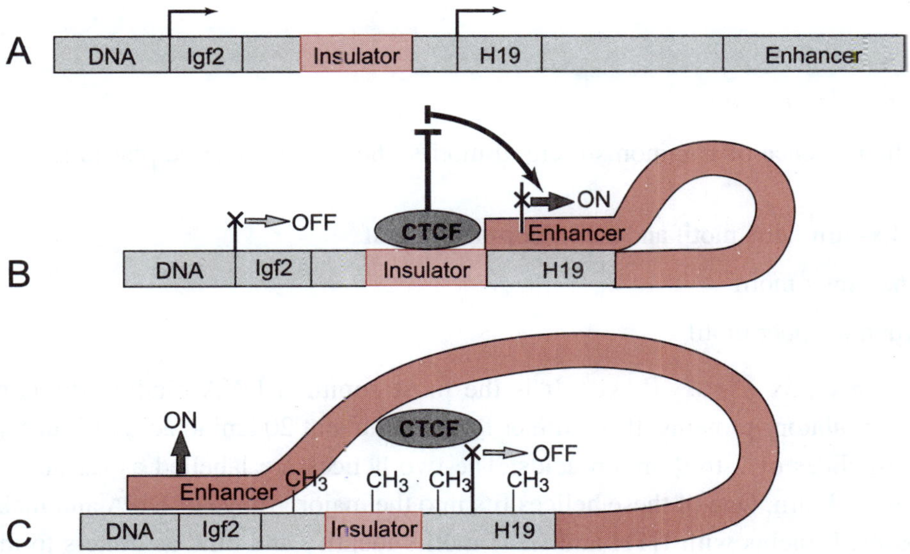

Fig. 16.10 Methylation of insulator sequences and binding of insulator binding protein CTCF: **A.** Location of gene Igf2, H19, insulator and enhancer; **B.** When insulator is not methylated, CTCF protein binds to the insulator and enhancer switches on H19 gene only; **C.** When insulator and H19 genes are methylated, CTCF protein does not bind, and enhancer activates IgF2 gene. H19 gene is switched off

which are also called **insulator binding proteins (IBPs).** On binding to insulator sequence, IBPs block enhancer but on being methylated, the insulator fails to bind to IBPs and enhancer can bind and switch on the gene. For example, in vertebrates, the Igf2 gene (encoding insulin-like growth factor-II) and gene H19 are located close together and have common insulator and enhancer. When insulator is not methylated, the protein CTCF binds to the insulator and enhancer switches on the H19 gene. But when insulator and H19 gene are methylated, the CTCF protein does not bind to them and H19 gene is not switched on. It allows the enhancer to activate the Igf2 gene.

8. Transcriptional Control by Regulatory Proteins

The key participant in transcription is ***RNA polymerase***. Its activity is controlled or mediated by two types of regulatory proteins, the activators and repressors and also the enzymes that cut or modify DNA, recognise palindromes or inverted repeats in DNA. Both types of these proteins recognise and bind to specific base sequences in DNA and turn the polymerase **'on'** or **'off'**. The regulatory proteins have descrete binding domains. They can read specific base sequences and bind to them. They also have **protein binding domains,** by which they bind to ***RNA polymerase*** or other regulatory proteins.

The action of regulatory proteins is also influenced by other molecular signals like small metabolites and hormones. Their binding can increase or decrease the affinity of protein DNA interaction.

Classes of Regulatory Proteins

Based on the presence of common structural motifs, the proteins can be placed in the following three classes:

1. Helix-turn-helix motif and helix-loop-helix motif
2. Zinc finger motif
3. Leucine zipper motif

1. Helix-Turn-Helix Motif (HTH): It is the most common DNA binding domain found in prokaryotic regulatory proteins. It is a dimer formed of about 20 amino acid residues arranged in two short α-helices of 7 to 9 amino acids. The two helices are labelled as α2 and α3 and are connected by a β turn. One of these helices fits into the major groove of DNA and makes contact with the bases. Proteins with HTH and HLH motifs identify and bind as dimers to the inverted repeats in DNA. Therefore, this helix is called **recognition helix**.

In helix-loop-helix, the turn of loop is longer. In HLH motif, the first α-helix binds to DNA, whereas in HTH motif, the second α-helix binds to DNA. HTH motif is used both in prokaryotes and eukaryotes, whereas HLH motif is used in eukaryotes only.

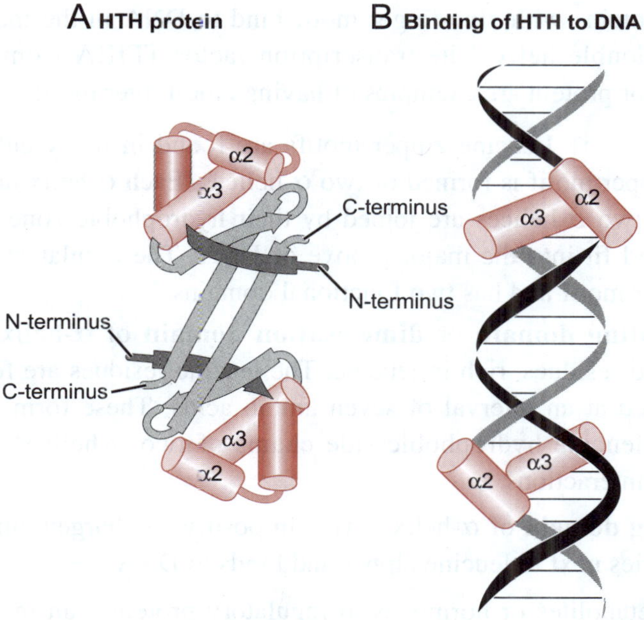

Fig. 16.11 **A.** Helix-turn-helix (HTH) molecule having two α-helices α_2 and α_3; **B.** Binding of HTH to DNA

2. Zinc Finger Motif: This motif consists of central zinc atom surrounded with 25-30 amino acid residues. Zinc is usually bound to two cystine molecules which lie in a very short piece of β-sheet having two histidines and a β-hairpin.

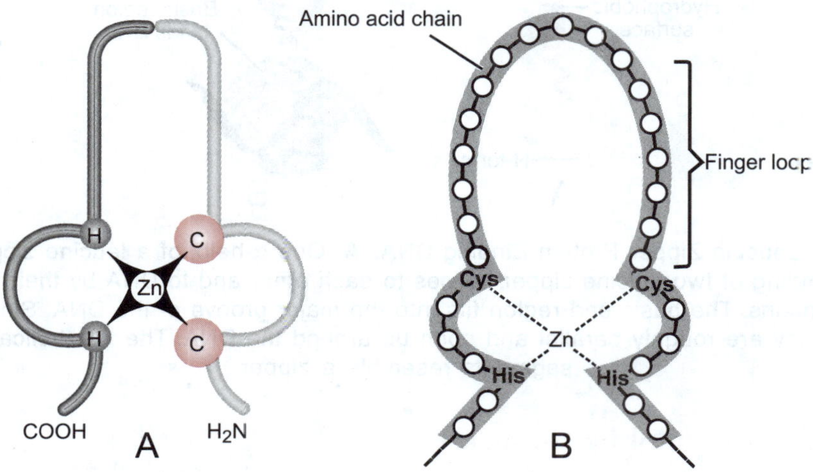

Fig. 16.12 **A.** Zinc finger DNA-binding protein; **B.** The zinc-finger motif of a loop of amino acids stabilised by the interactions between a zinc atom and the side chains of cysteine (C) and histidine (H) residues. The tip of the finger recognises and binds the edges of nucleic acid bases exposed in the DNA grooves

The regulatory poteins with zinc-finger motif bind to DNA in the major groove and wrap around the axis of double helix. The transcription factor TFIIIA from *Xenopus laevis* and glucocorticoid receptor protein are examples of having zinc-finger motif.

3. Leucine Zipper Motif: Leucine zipper motifs are found in many eukaryotic transcription factors. A leucine zipper motif is formed of two α-helices. Each α-helix has hydrophobic zones and basic ends. The two α-helices are joined by their hydrophobic zones but their basic ends remain wide apart and fit into the major groove of DNA. The regulatory protein with leucine zipper assumes zipper motif and has two functional domains.

- **Protein binding domain** or **dimerisation domain** of α-helix is formed of about 30 amino acid residues, rich in leucine. The leucine residues are four or five in number and are spaced at an interval of seven amino acids. These form the hydrophobic side chains. The leucine hydrophobic side chains of two α-helices are held together by hydrophobic interactions.
- **DNA binding domain** of α-helix is rich in positively charged amino acids (lysine and arginine). It lies next to leucine zipper and binds to DNA.

The binding of metabolites or hormones to regulatory proteins can increase or decrease the affinity of protein DNA interaction.

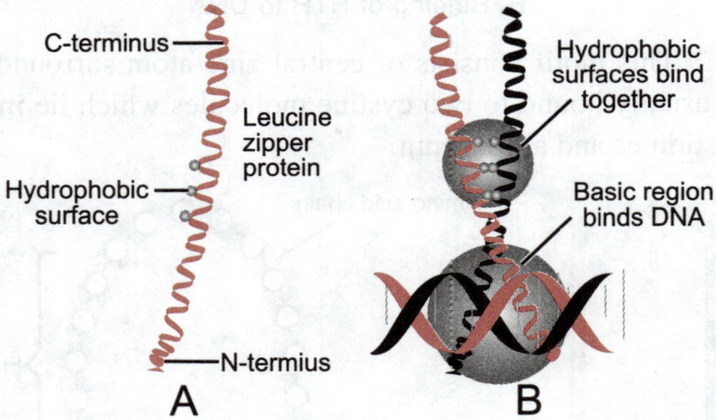

Fig. 16.13 Leucine Zipper Protein Binding DNA: **A.** One α-helix of a leucine zipper protein; **B.** Binding of two leucine zipper helices to each other and to DNA by their basic proteins. The basic end region fits into the major groove of the DNA. Since they are roughly parallel and open up around the DNA. The two helical segments resemble a zipper

9. Transcriptional Control by Hormones

In higher eukaryotes, short term gene regulation is achieved by hormones. The hormones like amino acids, amides, polypeptides and small proteins act on the cell surface and activate formation of cAMP. This acts as a **second messenger,** enters the nucleus and controls gene activity. The **steroid hormones** like estrogens, progesterone, androgens, aldosterone, etc., form a complex with specific receptor proteins. The complex then binds to the cell's genome to regulate the expression of specific genes. The specificity of hormone action is caused by the presence of hormone-receptors in only certain cell types.

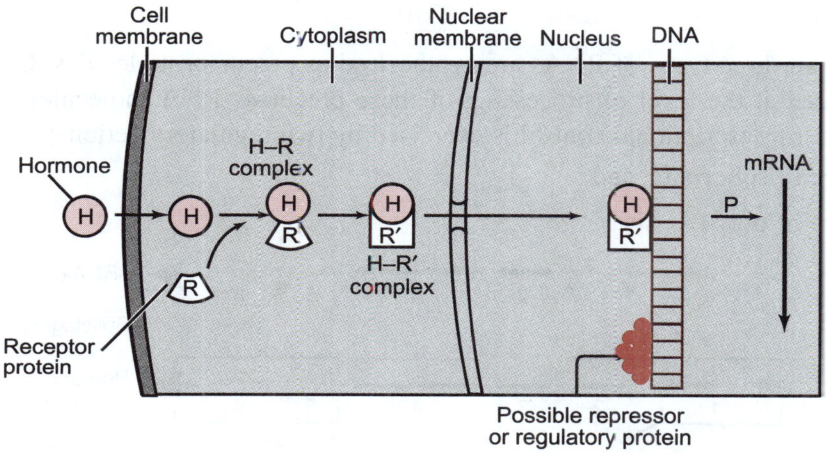

Fig. 16.14 Control of gene action by hormone: Hormone (H) reaches DNA and triggers transcription by binding to a cytoplasmic receptor (R)

Transcriptional Control at Chromosome Level

1. Interaction of Histones and Nonhistones with DNA

Eukaryotic chromosome consists of DNA complexed with **histones** to form nucleosomes. These repress transcription because TATA boxes in promoter regions are not exposed due to compacting of nucleosomes. Genes are activated when regulatory proteins bind to enhancers and disrupt nucleosomes at TATA boxes. This allows binding of appropriate regulatory proteins and transcription factors to promoters to initiate transcription.

DNA binding nonhistone proteins are acidic proteins. They interact with specific DNA sequences and exert a regulatory effect on transcription initiation. Acidic proteins inhibit separation of strands of DNA molecule and stabilise DNA.

2. DNase Sensitivity and Gene Expression

Transcriptionally active chromatin has looser DNA-protein structures than transcriptionally inactive chromatin. It is more sensitive to digestion with DNase-1. The promoter regions of transcriptionally active genes have even more loose DNA-protein structure. These are called **DNase hypersensitive sites** or **hypersensitive regions**. These are the binding sites for ***RNA polymerase*** and regulatory proteins.

III. GENE REGULATION AT POST-TRANSCRIPTION LEVEL OR GENE REGULATION AT PROCESSING OF RNA PRIMARY TRANSCRIPT

1. Post-transcriptional Control by Choice of Splice Site

In eukaryotes all three types of RNAs are synthesised as precursor molecules. Gene expression can be regulated at the level of processing of these precursor RNA molecules into functional RNAs. The post-transcriptional control is exercised by two regulatory actions:

- Choice of splice site, and
- Choice of poly (A) site.

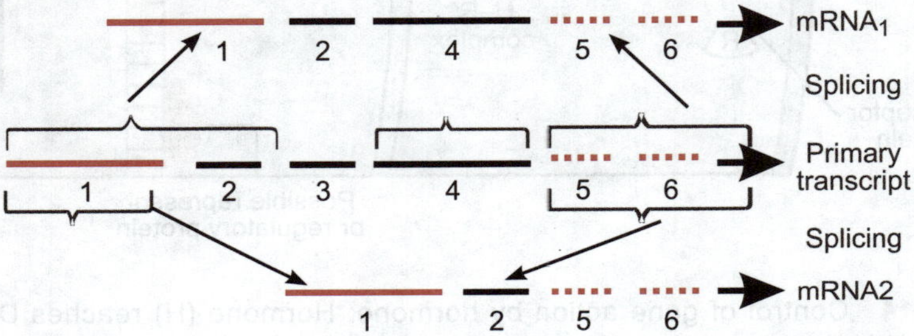

Fig. 16.15 Two alternative patterns of splicing of adenovirus EIB primary transcript

In both cases different types of mRNAs are formed, depending upon the choices made. For example, different classes of immunoglobulin molecules result from poly (A) site selection. Choice of splice site plays a key role in sex determination system in *Drosophila* and in the synthesis of α-amylase in rat. Different mRNA from the same gene are formed by use of different processing sequences.

Adenovirus has only a few promoters but makes numerous primary mRNA transcripts. From these transcripts a large number of mRNAs are generated. For example, the primary transcript for EIB region is differentially spliced to yield several different mRNAs. Formation of two of these mRNAs is shown in Fig. 16.15.

2. Polyproteins or Multiple Proteins from One mRNA

In prokaryotes coordinated regulation of the synthesis of several gene products is achieved by synthesising a single polycistronic mRNA. Similarly, in eukaryotes a polyprotein is synthesised. This polyprotein is a large polypeptide that is cleaved to yield individual proteins. Each protein segment is comparable to the product of a single gene. But these coding segments of the gene are not separated by stop and start codons. The polypeptides have some specific amino acid sequences that are recognised as cleavage of sites by specific protein cutting enzymes.

Some polyproteins are cleaved differently in different tissues. A single polyprotein in the pituitary gland is a source of several hormones:

1. In the anterior lobe of pituitary, the polyprotein **pro-opiomelanocortin** is cleaved to release β-lipoprotein from C-terminal end. A cleavage at the N-terminal leads to the formation of adrenocorticotropic hormone (ACTH).
2. In the intermediate lobe of pituitary, the β-lipoprotein is cleaved further, releasing the C-terminal polypeptide, β-endorphin. The ACTH is also cut forming α–melanotropin.

Thus, a single round of cleavage generates two hormones in the anterior lobe and a second round of cleavage generates two hormones in the intermediate lobe.

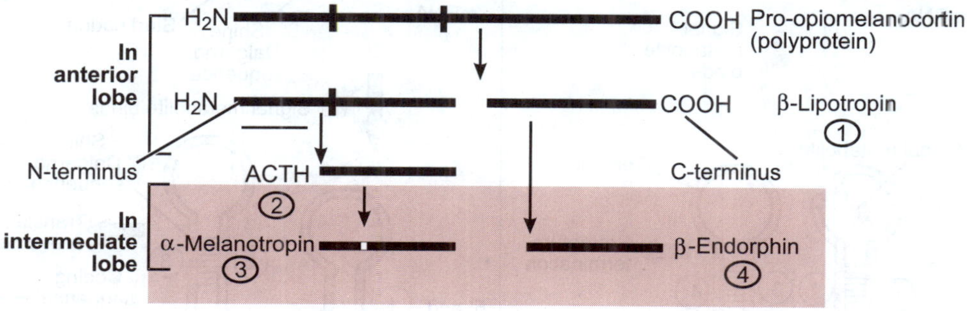

Fig.16.16 Cleavage of primary polyprotein pro-opiomelanocortin to form different hormones in anterior lobe and intermediate lobe of pituitary gland

3. Regulation of Gene Expression by RNA

Most biologists believe that only proteins regulate gene expression. The regulatory proteins sense and bind a small molecule, such as a hormone or metabolite and then bind to DNA or RNA to induce or suppress synthesis of certain proteins. It has now been discovered that some forms of RNA also control and initiate changes in gene expression. Following two types of RNAs are found to control gene expression:

1. Ribozymes or Riboswitches

Some ribozymes both in prokaryotes and eukaryotes are found to identify the presence of some metabolites or hormones in the cell and act to regulate the concentration of proteins that act on these small molecules. These unique ribozymes are described as metabolite-responsive **genetic switches** or **riboswitches**. These were discovered during vitamin synthesis in bacteria. For example, when vitamins like B_1 and B_{12} reach certain concentration inside the cell, they interact with special RNA to relay the message for slowing down the translation. The RNA segment that acts as riboswitch precedes the coding part of mRNA molecule. It makes the mRNA inactive by cleaving the biologically active mRNA or the metabolite attached to the riboswitch and switches off the translation process.

2. RNA Interference

Some RNA molecules can suppress or stop the expression of certain genes. This phenomenon is called **post-transcriptional gene silencing (PTGS)** or **RNA interference (RNAi)**.

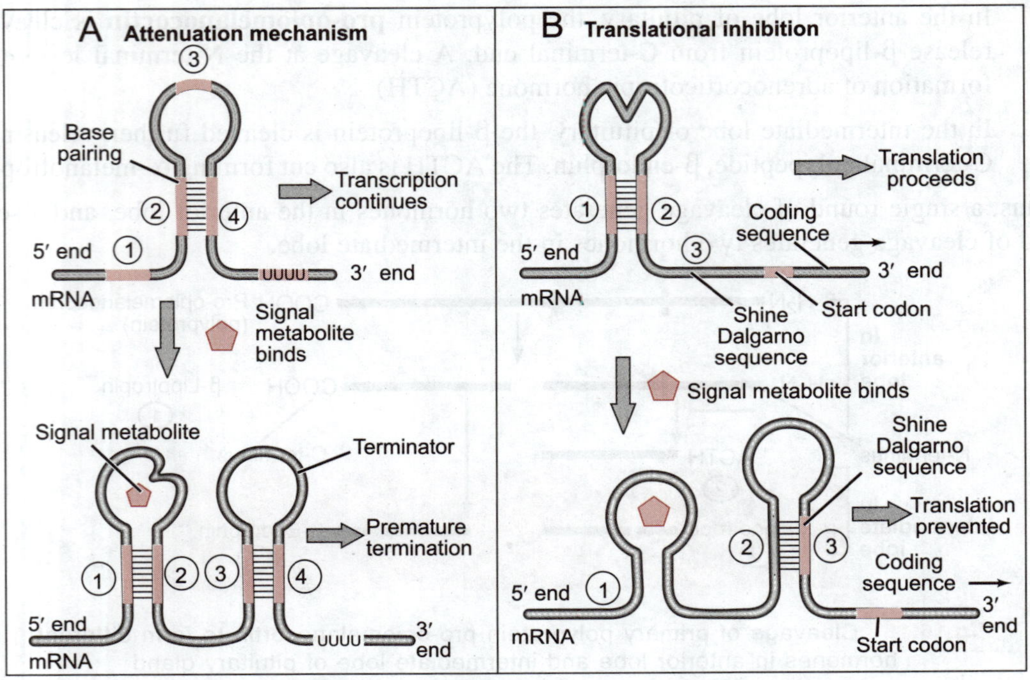

Fig. 16.17 Riboswitch mechanisms for post-transcriptional or translational control: **A.** Attenuation mechanism; **B.** Translational inhibition mechanism

This phenomenon of RNA interference was discovered by **Richard Jorgensen** and colleagues while working to darken the purple colour of petunia flowers by introducing additional copies of the concerned gene. He found that the expression of introduced genes as well as of the similar endogenous gene for pigment was suppressed instead of being enhanced.

The biochemical mechanism of RNAi has been studied in *Drosophila*. The incorporated double-stranded RNA (dsRNA) is attached by an ATP-dependent endonuclease enzyme, named **dicer**. This enzyme catalyses cleavage of the long dsRNA into small fragments of 21–23 base pairs. These dsRNA fragments are called **small interfering RNAs (siRNAs)**. They assemble into *endonuclease*-containing complexes, called **RNA-induced silencing complexes (RISCs)**. Their double-stranded RNA unwound into single strands. Finally, the complexes join with single-stranded regions of the target mRNA and cleave it to prevent translation.

4. Control on Transport of RNA

The transport of mRNAs from nucleus to the cytoplasm is another important control point in eukaryotic gene regulation. A spliceosome retention model has been proposed for this regulation. The processing of pre-mRNA involves snRNPs (small ribonucleoprotein particles). While undergoing processing pre-mRNA is retained in the nucleus by snRNAs forming spliceosomes. When all introns are removed from mRNA and all exons are joined, spliceosomes also dissociate and free mature mRNAs are released to move out of the nucleus through nuclear pores.

5. Control at Translation of mRNA

Differential translation of mRNA can also affect gene expression. For example, mRNAs are stored in unfertilised eggs. Stored mRNAs are associated with proteins that protect mRNAs and also inhibit their translation. These become active only after the eggs are fertilised. Addition of poly (A) tail is known to promote the initiation of translation. The stored inactive mRNAs have shorter poly (A) tails then active mRNAs.

6. mRNA Degradation Control

mRNA degradation is regarded to be a major control point in the regulation of gene expression. Enzyme *nucleases* are responsible for degradation of RNAs. The wide range of mRNA degradation rates is due to the structural features of individual mRNAs. For example, a group of AU-rich sequences in the 3′-untranslated regions of some short-lived mRNAs is responsible for their instability.

7. Gene Regulation at Protein Processing Level or Protein Folding Level

The regulatory mechanisms operate at post-translational level. These include processing of nascent polypeptide chain to assume secondary and tertiary structure and thus conversion of a non-functional polypeptide into folded functional protein so that the functional groups or sites are exposed on the surface and protein assumes a three-dimensional conformation. The **chaperone** proteins facilitate the folding of proteins.

Extension: Folding of Polypeptide Chain

Functional proteins are far more complex in their structure than the linear chains of amino acids. The formation of active enzymes or other proteins requires folding of polypeptide chains into precise three-dimensional conformations. Classical experiments done by **Christian Anfinsen** and his colleagues provided the answers as to how the correct protein conformation is adopted by the polypeptide and what information directs the process of protein folding?

According to **Anfinsen** and his colleagues **Michael Sela** and **Frederick H. White, Jr.,** the three-dimensional structure of a protein is determined by the interactions between the side chains of its constituent amino acids and it corresponds to its thermodynamically most stable conformation. This is called '**Thermodynamic hypothesis**' of protein folding. Since the thermodynamic stability of proteins is governed by the sequences of its constituent amino acids and the amino acid sequence in a polypeptide chain is determined by the order of nucleotides in DNA, it can be presumed that the nucleotide sequence of a gene contains all the information needed to determine the three-dimensional structure of the protein. No additional cellular factors are required for the folding of polypeptide into an active protein.

Anfinsen's conclusion was based on his observations that denatured RNase enzyme can be renatured. It can spontaneously refold *in vitro* to its active conformation. Recent studies have shown that the proper folding of proteins within the cells is mediated by the activities of other proteins, called **molecular chaperones.** The word **'chaperone'** was used by **Ron Laskey** and his colleagues. These proteins catalyse protein folding by binding to and stabilising unfolded or partially folded polypeptides.

8. Protein Degradation Control

Regulatory mechanisms operating at post translational level also determine the life span of a functional protein. Degradation of proteins is called **proteolysis** and occurs in both prokaryotes and eukaryotes. Some protein cofactor like ubiquitin is needed to prepare a protein for hydrolysis in proteolysis. Binding of ubiquitin to a protein identifies it for degradation by proteolytic enzymes.

EVIDENCE OF REGULATION OF GENE ACTION IN EUKARYOTES

Chromosome Puffs: Chromosome puffs are seen in **giant chromosomes** of salivary glands in dipteran flies and in the **lampbrush chromosomes** of amphibian oocytes. Each puff represents the lateral loop of greatly extended DNA strand which is actively busy in RNA transcription. RNA synthesised in different puffs is markedly different. Moreover, during development, puffs appear at different locations.

Injection of hormone **ecdysone** in the larvae of dipteran flies induces moulting and puffing in the giant chromosomes. These puffs are similar to the puffs which are formed in the giant chromosome prior to moulting in the normal course.

TEXT QUESTIONS

1. 'Transcriptional control is more complex in eukaryotes'. Justify the statement.
2. Explain the following terms:
   (a) Gene amplification (b) Riboswitches
   (c) Enhancers (d) Chromatin remodelling
   (e) Leucine-Zipper motif (f) Gene silencing.
3. Describe different classes of regulatory proteins which exert transcriptional control.
4. How are polyproteins or multiproteins are obtained from same mRNA?
5. What do you mean by molecular chaperones?
6. What are ribozymes?

Genetic Engineering or Recombinant DNA Technology

17
Chapter

IN THIS CHAPTER

Recombinant DNA Technology; Biological Tools for Recombinant DNA Technology; Vectors or Cloning Vehicles: Plasmid Vectors; Bacteriophage Vectors, Cosmid Vectors; Phagemid Vectors, Artificial Chromosome Vectors, Shuttle Vectors; Fosmid Vectors; Techniques used in Recombinant DNA Technology; Joining of Foreign DNA Fragment to a Cloning Vector; Introduction of Recombinant DNA into Host Cell; Selection and Screening of Transformed cells; Significance of Recombinant DNA Technology or Genetic Engineering and Human Welfare; Genetic Engineering and Biotechnology in India.

Genetic engineering or recombinant DNA technology or gene manipulation is a new born discipline of science which is associated with altering the hereditary apparatus of living cells. It involves highly sophisticated techniques for the artificial synthesis or isolation of specific genes or DNA fragments, their integration in some suitable vector and cloning these vectors in some other organism, called **host**. By this technology, we can get:

- Multiple copies of cloned DNA fragment (cDNA).
- Large quantities of proteins produced by cloned genes.
- Integration of gene in chromosome of host organism (gene transfer) so as to repair or replace the defective cistron or gene or develop new genotype.

RECOMBINANT DNA TECHNOLOGY

Discovery

Recombinant DNA is the DNA formed by the union of **foreign DNA** (from eukaryotic cell) to vector DNA. It is also called **heterologous DNA**. Recombinant DNA technology evolved because of the following discoveries:

- Denaturation and renaturation character of DNA.
- Artificial synthesis of gene *in vitro*.
- Discovery of enzyme: restriction endonucleases or restriction enzymes.
- Gene splicing and gene cloning techniques.

1. Denaturation and Renaturation of DNA

The two strands of DNA easily separate on heating by the breakdown of hydrogen bonds between their nitrogenous bases. This is called **denaturation.** On cooling, the two complementary strands reunite to form double stranded DNA (**renaturation**). Because of this property, single-stranded DNA segments from different sources can join, provided these have complementary base pairs. This is called **anealing**.

Table 17.1 Major steps in development of recombinant DNA technology

S. No.	Year	Scientist	Contribution
1.	1869	**Miescher**	Isolated DNA for the first time.
2.	1944	**Avery, MacLeod** and **McCarty**	Provided evidence that DNA, rather than protein, carries the genetic information during bacterial transformation.
3.	1953	**Watson** and **Crick**	Proposed the double helical model of DNA structure based on X-ray results of **Franklin** and **Wilkins**.
4.	1958	**Kornberg**	Discovered **DNA polymerase,** the enzyme used to produce radioactive DNA-probes.
5.	1961	**Marmur** and **Doty**	Discovered DNA renaturation and possibility of **DNA hybridisation.**
6.	1962	**Arber**	Provided first evidence for the existence of **DNA-restriction nucleases**.
7.	1967	**Gellert**	Discovered **DNA-ligase,** the enzyme used for joining DNA fragments together.
8.	1972-1973	**Boyer, Cohen** and **Berg** and **their colleagues**	Developed DNA sequencing techniques.
9.	1975	**Nathans** and **H. Smith**	Purified and used restriction endonucleases in DNA-sequencing.
10.	1975-77	**Sanger** and **Barrell, Maxam** and **Gilbert**	Developed rapid DNA sequencing technique.
11.	1981-1982	**Palmiter** and **Brinter**	Produced transgenic mice.
12.	1982	**Spradling** and **Rubin**	Produced transgenic fruit flies.

2. Artificial Synthesis of Gene

Dr. Har Gobind Khorana (1972) along with his colleagues first reported total synthesis of an artificial gene from tyrosine-tRNA *in vitro* with a potential for functioning within a living cell.

3. Restriction Endonucleases or Restriction Enzymes

> Arber, Nathans and Smith were awarded Nobel Prize in Physiology and Medicine for their discovery of restriction endonucleases in the year 1978. These enzymes are called **biological scalpel**.

The restriction endonucleases are enzymes that recognise specific nucleotide sequences in DNA and cleave the DNA double helix at or near these specific restriction sites, called the **target sites**. These enzymes are apparently elaborated in bacteria as a protection against the entry of foreign or viral DNA. These were first reported by **W. Arber** in 1962, who noticed that when DNA of a bacteriophage entered a host bacterium, it was cut into smaller pieces. However, first restriction enzyme was isolated by **Meselson** and **Yuan** from *E. coli*. Real breakthrough came when restriction enzyme **Hind II** was isolated from bacterium, *Haemophilus influenzae Rd.* by **Kelly** and **Smith** (1970) and **Smith**.

Types of Restriction Endonucleases: The restriction enzymes have been classified into three categories:

1. **Type I Restriction Endonucleases:** These identify an unmodified recognition sequence in double-stranded DNA. **These cleave only one strand of DNA and at an apparently random site away from that sequence.** These enzymes create a gap of about 75 nucleotides in length by releasing acid soluble oligonucleotides. These enzymes require Mg^{2+} ions, ATP and S-adenosyl methionine cofactors for restriction. Type I RE are not useful for the construction or analysis of recombinant DNA molecules.
2. **Type II Restriction Endonucleases:** These enzymes recognise specific DNA nucleotide sequences. They cleave both the polynucleotide chains within or near to the **palindromic sequences**. Palindromes are base pair sequences that read the same forward and backward or in $5' \rightarrow 3'$ direction in both strands of DNA. For example:

   A 5′......N N G↓A A T T C N N...... 3′
   3′......N N C T T A A↑G N N...... 5′

   B 5′......C C C↓G G G...... 3′
   3′......G G G↑C C C...... 5′

   These enzymes produce DNA fragments of defined length and sequence. These require only Mg^{2+} ions for restriction. ***Restriction enzymes type-II*** are used in constructing recombinant DNA molecules and in **gene manipulation.**
3. **Type III Restriction Endonucleases:** These cleave double-stranded DNA at well defined sites and require Mg^{2+} and ATP and S-adenosyl methionine. These enzymes are intermediate between type I and type II.

Recognition Sites or Target Sites

Restriction endonucleases cleave the DNA molecules only in the regions where the nucleotide sequence reads the same on both the strands in $5' \rightarrow 3'$ direction. These regions in DNA are called **recognition sites** or **target sites**. The sequence of nucleotides in recognition site for Type II restriction enzymes have an axis of symmetry through the midpoint. It means the base sequence from 5′ to 3′ on one strand of DNA is the same as the base sequence from 5′ to 3′ on complementary strand.

DNA Cleavage Styles

The restriction enzymes cut DNA molecules by cleavage in any one of the following two styles:

1. **Sticky End Style:** In this style of cleavage, the target sequence is asymmetrical. The staggered cuts in two strands of DNA are some nucleotides apart producing complementary

single-stranded protruding ends. These unpaired protruding ends are called **sticky** or **cohesive ends**. Enzymes *EcoRI, HindII, XmalI, etc.*, produce sticky ends as follows:

Table 17.2 Sticky end restriction sites and cleavage products

Sl. No.	Restriction Enzyme	Restriction site	Cleavage products	
1.	***EcoRI*** ***(E.coli RY03)***	5′ ...... G↓AATTC ...... 3′ 3′ ...... CTTAA↑G ...... 5′	5′ ... G 3′ ... CTTAA	AATTC ... 3′ G ... 5′
2.	***HindII*** ***(Haemophilus influenzal)***	5′ ...... A↓AGCTT ...... 3′ 3′ ...... TTCGA↑A ...... 5′	5′ ... A 3′ ... TTCGA	AGCTT ... 3′ A ... 5′
3.	***XmaII*** ***(Xanthomonas malvacearum)***	5′ ...... C↓CCGGG ...... 3′ 3′ ...... GGGCC↑C ...... 5′	5′ ... C 3′ ... GGGCC	CCGGG ... 3′ C ... 5′

2. Blunt End Style: Restriction enzymes like *AluI, HaeIII, HindII, SmaI* and *HpaI* make cuts across both strands of DNA at the same position. The DNA fragments produced have blunt ends. This enables DNA of all kinds to be broken into a wide variety of fragments depending on the restriction enzyme and also to join two defined fragments without introducing any additional material between them.

Table 17.3 Blunt end restriction sites and cleavage products

Sl. No.	Enzyme	Recognition	Cleavage products	
1.	**AluI** ***(Arthrobacter luteus)***	5′ ..AG↓CT..........3′ 3′ ..TC↑GA..........5′	5′ ..AG 3′ ..TC	CT ...3′ GA ...5′
2.	**HaeIII** ***(Haemophilus aegypticus)***	5′ ...GG↓CC..........3′ 3′ ...CC↑GG..........5′	5′ ...GG 3′ ...CC	CC ...3′ GG ...5′
3.	**SmaI** ***(Serratia marcescens)***	5′ ...CCC↓GGG...3′ 3′ ...GGG↑CCC...5′	5′ ...CCC 3′ ...GGG	GGG...3′ CCC ...5′
4.	**HpaI** ***(Haemophilus parainfluenzae)***	5′ ...GTT↓AAC...3′ 3′ ..CAA↑TTG...5′	5′ ...GTT 3′ ...CAA	AAC ...3′ TTG ...5′

BIOLOGICAL TOOLS FOR RECOMBINANT DNA TECHNOLOGY

1. Enzymes

Following types of specific enzymes are employed in recombinant DNA technology:

1. Lysing Enzymes: These are used to open the cells or dissolve the cell wall. Lysozymes are used to dissolve bacterial cell wall.

2. Cleaving Enzymes: These are used to cleave a DNA molecule into segments or to open the circular plasmid DNA. These are:

- ***Exonucleases*** which cut off nucleotides from 5′ or 3′ ends of DNA molecule.
- ***Endonucleases*** that cleave DNA duplex at any point but not at one end.
- ***Restriction endonucleases*** that cleave DNA at specific or target points and recognise **palindromic sequences** of nucleotides. They are widely used in recombinant DNA technology or genetic engineering.

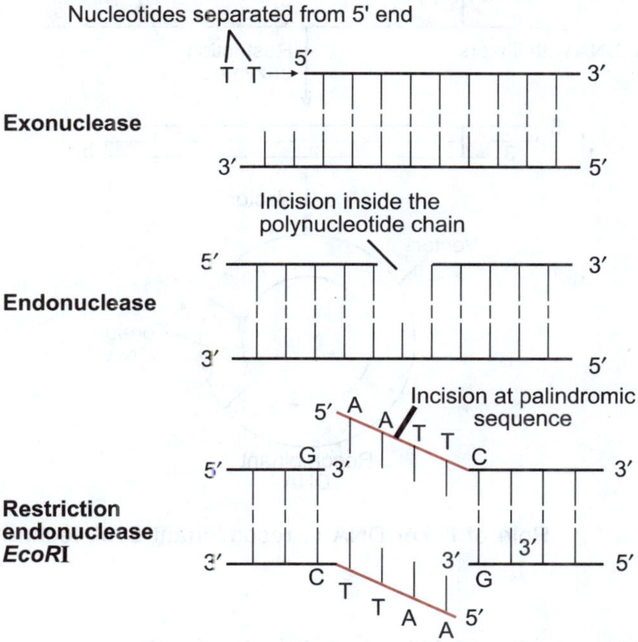

Fig. 17.1 Differences between three types of nuclease enzymes

3. Synthesising Enzymes or Synthetases: These help either in *in vitro* synthesis of DNA or in the synthesis of complementary DNA. These enzymes are:

- ***DNA polymerase*** which helps in the synthesis of complementary DNA strands on DNA templates.

- ***Reverse transcriptase*** that helps in the synthesis of complementary DNA from mRNA template.

4. **Ligase:** These seal or join together the ends of the foreign DNA and vector DNA. Ligases are present in both prokaryotes and eukaryotes.

2. Linkers

The linkers are short double-stranded DNA segments which are formed of oligonucleotides. These contain target sites for the action of one or more restriction enzymes. The linkers can be synthesised chemically, can be ligated to the blunt ends of foreign DNA or vector DNA. These are then treated with ***restriction endonuclease*** enzyme to produce cohesive ends of DNA fragments. The commonly used linkers are *EcoRI* linkers and *sal-I* linkers.

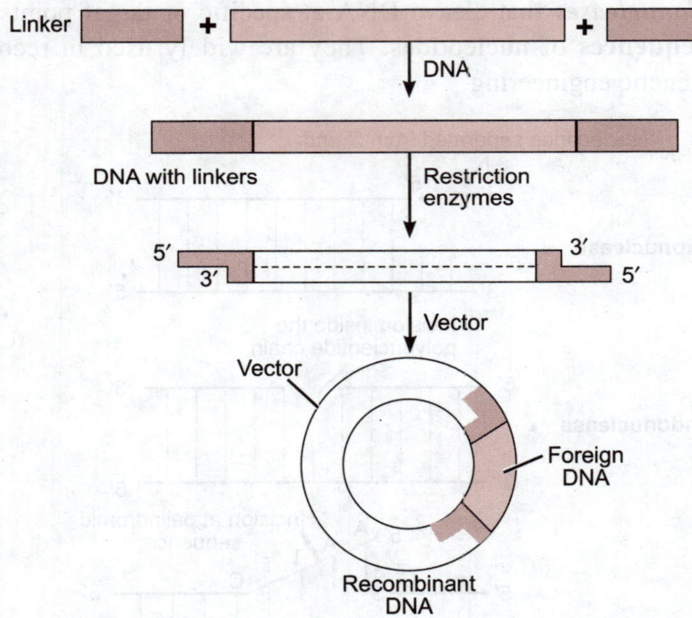

Fig. 17.2 Role of linker DNA in recombinant DNA technology

3. Foreign DNA

It involves isolation or synthesis of DNA which is to be transferred for cloning or to be used for the synthesis of recombinant DNA.

4. Cloning Vectors

For transferring and cloning a foreign DNA (gene) in a suitable host or in bacterium, some vector or carrier DNA is required. The foreign DNA is inserted in the vector DNA forming **recombinant DNA** or **recombinant vector**. Plasmids and phages are used for this purpose.

VECTORS OR CLONING VEHICLES

A vector is a small DNA molecule capable of self replication and is used as a carrier of DNA fragment inserted into it for cloning. The vector is also called **cloning vehicle** or **cloning DNA**. Vectors are of two types:

Extension: Characteristics of a Good Vector

A good vector must have the following properties:

- It should be able to replicate autonomously, so that it can generate multiple copies of itself along with DNA insert within a single host cell.
- A vector is supposed to have one **origin of replication** for DNA replication.
- A vector should be small in size and of low molecular weight. Ideally it should be less than 10 kb (kilo base) in size, because large DNA molecules are broken during purification and present difficulty during manipulation required for gene cloning.
- It should be easily isolated and purified.
- It should be easily introduced into host cell.
- Vector should afford easy transformation of host cell.
- The vector should contain a suitable marker or markers to permit its detection in the host cell and selection of transformed host cell.
- The vector should have unique target sites for as many restriction enzymes as possible, so that DNA insert can be integrated without disrupting essential functions of the vector.
- When objective is gene transfer, the vector should have the ability to integrate either itself or the DNA insert it carries into the genome of the host cell.
- When objective is expression of the DNA insert, the vector should contain suitable control elements such as promoter, operator and ribosome binding site, etc.
- The transformed cells containing DNA insert or recombinant DNA should be identifiable from those transformed by the vector molecule alone.

Cloning Vector and Expression Vector

Vector that is used only for the propagation or cloning of DNA insert inside a suitable host cell is called **cloning vector**. The cloning vectors contain relaxed replication control so that they can produce multiple copies in each transformed cell.

A vector is termed **expression vector,** when it is used to express the DNA insert producing the specified protein. Such vectors contain the regulatory sequences such as promoters, operators and ribosomal binding sites. Expression vector has prokaryotic promoter and ribosome binding site just before the eukaryotic coding sequence because the regulatory sequences of eukaryotes are not recognised in the prokaryotic cells.

The following agents are used as vectors:

1. Plasmids
2. Bacteriophages
3. Cosmids
4. Phagemids
5. Phasmids
6. Artificial chromosome vectors
7. Fosmid vectors
8. Shuttle vectors
9. Yeast vectors
10. Yeast artificial chromosome (YAC) vector

1. PLASMID VECTORS

Plasmids are extrachromosomal, self replicating, double-stranded, circular DNA molecules, found in the bacterial cells in addition to the bacterial chromosome. These range in size from 1×10^6 daltons to 200×10^6 daltons. These may exist either independently or may become integrated into bacterial chromosome. Generally, plasmids are nonessential for the bacterial cells except under specific environment. Plasmids contain genetic information for their own replication. They also contain genes for antibiotic resistance, heavy metal resistance, nitrogen fixation, and pollutant degradation, etc.

Types of Plasmid Vectors

There are several types of bacterial plasmids, but following three types are widely studied and used in genetic engineering:

1. **F plasmids** are responsible for conjugation and are also called **conjugative** or **transmissible** plasmids. Their presence in bacterial cells give them maleness.
2. **R plasmids** carry genes for resistance to antibiotics.
3. **Col plasmids** carry genes for coding colicins, the proteins that kill sensitive *E. coli* cells.

Plasmids in a bacterial cell may occur singly or in multiple copies. **Single copy plasmids** replicate and segregate with the bacterial chromosome. This is called **stringent replication**. The **multiple copy plasmids** undergo more than one replication for each replication of the bacterial chromosome. This is called **relaxed replication.**

Characteristics of a Plasmid Vector

A suitable plasmid vector must have following characteristics:

1. It must have minimum amount of DNA.
2. It must have relaxed replication control.
3. It should have at least two suitable markers for identification.
4. It should be easily isolated from the cell.

5. It should possess a single restriction site for one or more restriction enzymes.
6. Insertion of a foreign DNA molecule as one of these restriction sites does not alter its replication property.
7. The unique restriction site must be located within one of the two selectable markers.

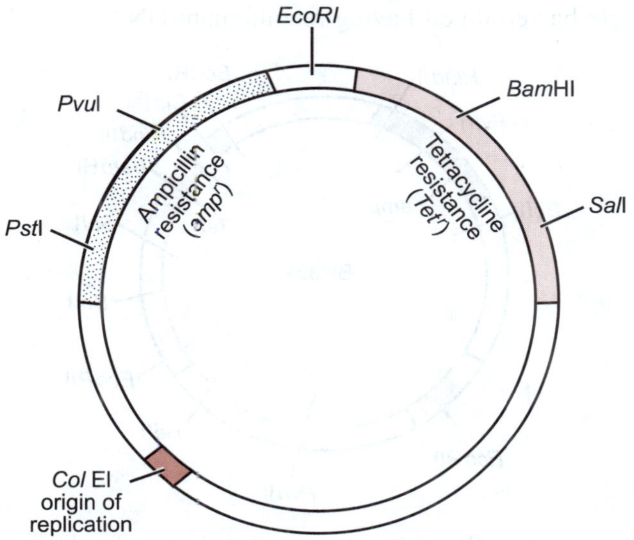

Fig. 17.3 Diagram showing essential features of a plasmid vector

Examples of Plasmid Vectors

Plasmids having all the above mentioned characters have been prepared by modifying naturally occurring plasmids by *in vitro* techniques. Some of such plasmids are:

*p*BR322 Plasmid Vector

*p*BR322 is a reconstructed plasmid. It is most widely used plasmid having 4362 base pairs. Its entire base sequence is known. It is constructed from three plasmids. It has following regions:

1. Origin of Replication: It is derived from plasmid *p*MBI. This plasmid is closely related to naturally occurring plasmid *Col* EI.

2. Gene Conferring Resistance to Antibiotics Ampicillin (amp^r) and Tetracycline (tet^r): Gene amp^r is derived from plasmid RI and gene tet^r from plasmid R6.5. Both are natural populations of antibiotic resistant plasmids. These genes are used as **markers.**

3. Unique Recognition Sites for 20 Restriction Endonucleases: Some of these sites lie within the markers amp^r and tet^r genes. The presence of restriction sites within markers tet^r and amp^r permits an easy selection of transformed cells having recombinant ***p*BR322**. Insertion of DNA fragment into the plasmid using restriction enzyme *Pst*1 or *Pvu*1 places the DNA insert (foreign DNA) with the marker gene amp^r. The DNA insert (foreign DNA) within the marker gene amp^r makes it nonfunctional. Bacterial cells with such a recombinant *p*BR322 will be unable to grow in the presence of ampicillin but will grow successfully on tetracycline. Similarly, when restriction enzyme *Bam*HI or *Sal*I is used, the DNA insert is placed within the marker gene tet^r. This makes tetracycline gene nonfunctional. Transformed bacterial cells possessing such a recombinant *p*BR322 will multiply on ampicillin but not on tetracycline. This process allows an easy selection of a single bacterial cell having recombinant DNA.

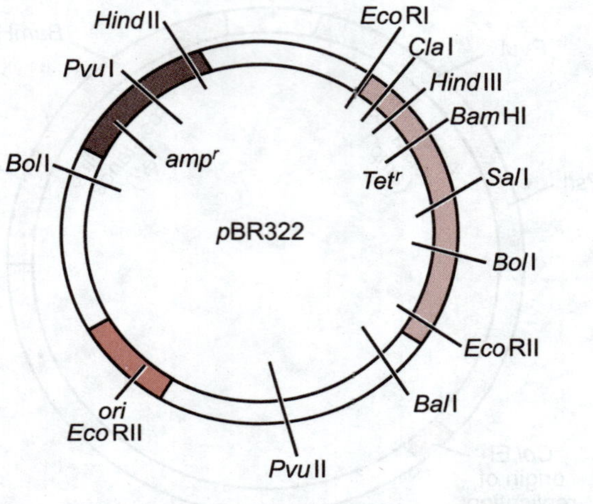

Fig. 17.4 Structure of restructured plasmid p^{BR322}

4. Nic-bom Region: *p*BR322 has a **nic-bom** region (bom basis of mobility). It is responsible for mobilisation or cell to cell transfer of the plasmid. This region interferes with the replication efficiency of extra chromosomal DNA in monkey cells.

The name *p*BR is derived from *(i)* *p* that denotes plasmid, *(ii)* **B** is from the name of scientist, **Boliver** and *(iii)* **R** from **Rodriguez**. These two scientists developed this plasmid. The numeral '322' distinguishes this plasmid from other plasmids developed in the same laboratory later on, such as *p*BR325, *p*BR327, *p*BR328, etc.

Useful Features of *p*BR322

The useful features of this plasmid are:

- Small size (4.4 kb) enables easy handling, purification and manipulation.
- Two selectable markers amp^r and tet^r permit easy selection of recombinant DNA.
- About 15 copies of them occur per cell which can be increased to 1,000 to 3,000 when protein synthesis is blocked.
- It is a cloning vector.

*p*BR327 has more advantageous features than plasmid *p*BR322. Its 30-40 copies occur in each normal cell of *E. coli*. It is an expression plasmid.

*p*UC8 Plasmid Vector

*p*UC8 is a derivative of *p*BR322. It is much smaller, only 2.7kbp long. It is named after the place of its preparation i.e., University of California. It has following two parts derived from *p*BR322:

- Ampicillin resistance gene.
- *Col* EI origin of replication.

It also has *lac*Z gene derived from *E. coli*. A polylinker sequence having unique restriction sites lies in the *lac* region. When DNA insert is cloned in this region, the *lac* gene is inactivated. The transformed bacterial cells with recombinant DNA molecule are β-galactosidase deficient and are easily selected from normal bacterial cells in a single step. *p*UC8 has become one of the most popular *E. coli* cloning vector. Other vectors in *p*UC series are *p*UC9, *p*UC12, and *p*UC13 and *p*UC18 and *p*UC19.

*p*GEM3Z Plasmid Vector

This plasmid vector is similar to a *p*UC vector. It is almost of the same size and has the same functional nodules amp^r gene, *col* EI origin and *lac* ZX sequence.

In addition to *E. coli*, some Gram negative and Gram positive bacteria are also used as cloning vector for medical and agricultural studies. PK2 plasmid is a cloning vehicle of this category. Species of *Pseudomonas* also have a variety of plasmids. Some of these plasmids are an excellent vectors for cloning small fragments of DNA and are used as tool for establishing cDNA library.

2. BACTERIOPHAGE VECTORS

Bacteriophages are viruses that attack bacteria. Several bacteriophages are used as cloning vectors. The most commonly used *E. coli* phages are **λ (lambda) phage** and **MI3 phage**. Phage vectors have following three advantages over plasmid vectors:

- Phage vectors are more efficient than plasmids for cloning large DNA fragments. Foreign DNA up to 25kb in length can be inserted into phage vector.
- DNA can be packed *in vitro* into phage particles and transduced into *E. coli* with high efficiency.
- It is easy to screen and to use in the storage of recombinant DNA.

Examples of Bacteriophage Vectors

λ (Lambda) Phage Vectors

The genome of λ-phage is 48502 bp long, i.e. about 49kb and has 50 genes. About half of these genes are essential. The genome remains linear in the phage head and has single-stranded protruding cohesive ends of 12 bases (5′GGGCGGCGACCT … 3′ and 3′…CCCGCCGCTGGA … 5′). The two cohesive ends anneal to form circular DNA molecule. The sealed cohesive ends form a ***cos* site**. It is the site of cleavage of phage DNA. For packaging in the phage head of bacteriophage, the λ-DNA must be larger than 38kb and smaller than 52kb. The genes for lysogeny are located in the middle part of λ-phage genome. For creating lambda phage vectors, this segment is removed wholly or partially. This allows the vector to accommodate large DNA inserts and avoids chances of lysogeny. Several vectors have been produced from wild type lambda genome by mutation and recombination as well as by recombinant DNA techniques.

Features of λ phage Vectors: These vectors have following two basic features:

- The vector can be propogated as phages in *E. coli* cells enabling preparation of vector DNA.
- They contain restriction sites, which allow removal of lysogenic segment and also provide insertion site for the DNA fragment to be cloned.

Types of λ Vectors: The various λ vectors are classified into two categories: **insertion vectors** and **replacement vectors**.

1. Insertion Vectors: In case of an insertion vector, a portion of nonessential region is deleted and two arms of λ genome are ligated. The foreign DNA is inserted at the unique restriction site. The DNA insert does not affect the functioning of phage. The insertion vector is of the size of about 35kb or more. It can have a DNA insert of up to 18kb. Examples of insertion vectors are **λgt10, λgt11,** and **λZAPII,** etc.

λgt10 is 43kb double-stranded DNA for cloning DNA fragments of about 7kb long. The insertion of foreign DNA inactivates *cl* (repressor) gene. **λgt11** is 43.7kb double-stranded DNA. It can clone less than 6kb long foreign DNA. It is an expression vector. Its inserted DNA is expressed as β-galactosidase fusion protein.

2. Replacement Vectors: In replacement phage vector, the nonessential part of λ phage DNA is substituted by foreign DNA. The deleted region is called the **stuffer fragment**. Replacement vectors have two recognition sites for the restriction enzyme used for cloning. These sites flank the stuffer region. The maximum size of DNA insert depends upon how much of phage DNA is non-essential. Usually 20–25kb of genome is nonessential and can be replaced with foreign DNA.

EMBL3 and **EMBL4** are two replacement vectors that are designed for replacing the central nonessential part of 44kb by foreign DNA of about 20–23kb length. These replacement vectors are used mainly for preparing genomic libraries of eukaryotic genome with cloned fragments.

Phage M13 Vectors

M13 phage is a filamentous bacteriophage of *E. coli* containing 7.2kb long single-stranded circular DNA. It has been variously modified to give rise to MP13 series of cloning vectors. Phage M13 vectors are used for obtaining single-stranded copies of cloned DNA. These are used for DNA sequencing. They are derived from 6.4kb genome of filamentous bacteriophage M13 found in *E. coli*. Phage M13 has a single-stranded linear DNA. Inside the host cells, it changes into a double-stranded circular replicative intermediate DNA. Each infected *E. coli* cell has more than 100 copies of M13 genome and about 1,000 new phage particles are formed during each generation of infected cell.

The double-stranded form of M13 genome is used for obtaining recombinant molecules. This form is obtained from host cells. A complete series of M13mp vectors (M13mp 8, M13mp 9, etc.) has been reconstructed from M13 genome.

Features of M13 Vectors

The desirable features of M13 vectors are:

- Very large DNA segments can be cloned.
- Pure single-stranded copies of double-stranded DNA inserts can be obtained.
- The single-stranded copies of both the strands are obtained because DNA inserts are accepted in either of the two orientations. It means some recombinant clones produce single-stranded copies of one strand and others form copies of its complementary strand of DNA insert.
- The bacterial cells infected with M13 vector remain viable because they do not lyse the cells.
- The recombinant DNA is obtained within stable bacteriophage particles.
- M13 infected *E. coli* cells form plaques, so they can be easily selected.

3. COSMID VECTORS

Cosmids are hybrid vectors derived from plasmids having a fragment of λ phage DNA with its ***cos*** site (the sequence yielding cohesive ends) and sequence needed for binding of and cleavage by ***terminase***. These have minimum 250bp of λ DNA.

Structure of Cosmid

A typical cosmid vector has following regions:

1. Origin of replication.
2. Unique restriction site for cleavage and insertion of foreign DNA.
3. A selectable marker from a plasmid, coding for antibiotic resistance.
4. ***Cos*** site/sites from lambda phage has 12 bases. It enables recombinant DNA to be packed in lambda particle *in vitro* due to circulation and ligation.

Characteristics of Cosmid Vectors

The typical features of a cosmid vector are:

- Cosmids have a length of about 5kb. They can be used to clone DNA inserts of up to 40–45kb.

- These can be packaged into λ-particles which infect host cells.
- Packaged cosmids infect host cell like λ-phage particles but multiply and propagate like plasmids.
- Selection for recombinant DNA is based on the procedure used for selection of plasmids.
- Cosmid vectors are amplified and maintained in the same way as the plasmids.

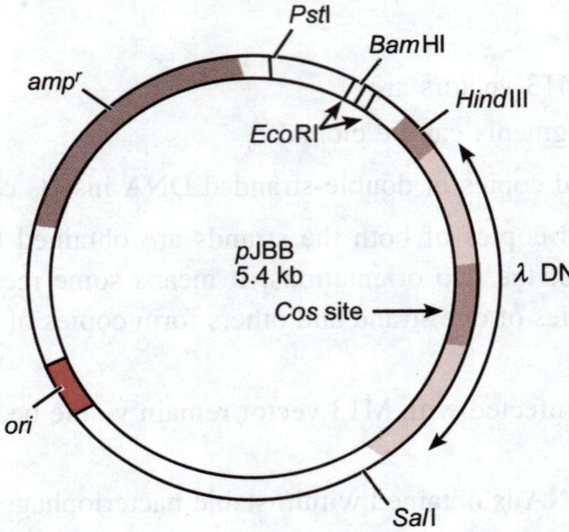

Fig. 17.5 Structure of a typical cosmid vector having λ-segment with cos site in a plasmid module

Modern cosmid vectors belonging to *p*WE and *s*Cos have (1) **multiple cloning sites** (MCS) for simple cloning using unselected DNA fragments, (2) **phage promoters** flanking MCS to generate RNA copies of DNA inserts, and (3) **unique restriction sites** (*Not*I, *Sac*II or *Sfi*III) for cutting restriction enzymes and flank MCS, and (4) vector may contain **mammalian expression modules** which may code dominant selectable markers for gene transfer to mammalian cells.

Use of Cosmids: Cosmids are used for the construction of genomic libraries of eukaryotes.

4. PHAGEMID VECTORS

Phagemids are also reconstructed plasmid vectors. Each phagemid contains **origin of replication** from a phage in addition to that of its own. For example, *p*Bluescript SK (+/–) is a phagemid vector. It has 2958bp (base pairs) derived from plasmid *p*UC19.

Structure

A typical phagemid has following structures:

1. Origin of replication from phage f1 or M13.
2. A multiple cloning site (MCS) within *lacZ* gene.

3. A portion of *lacZ* gene driven by *lac* promoters.
4. Phage T7 and T3 promoter sequences that flank MCS sequence.
5. *Col* E1 origin of replication (II).
6. amp^r gene for antibiotic resistance.

DNA insert is integrated *in vitro* into the double-stranded vector. The vector is then introduced into *E. coli* cells like any other plasmid vector. It can be used for cloning DNA insert of up to 10kb only. The recombinant DNA propagates inside *E. coli* as a plasmid using *col* EI origin of replication of plasmid and several copies of recombinant DNA are obtained. The phagemid vector can also be used as an **expression vector** because the DNA insert is transcribed along with *lacZ* gene. A fusion protein is formed which has **lac Z gene product** along with insert DNA product.

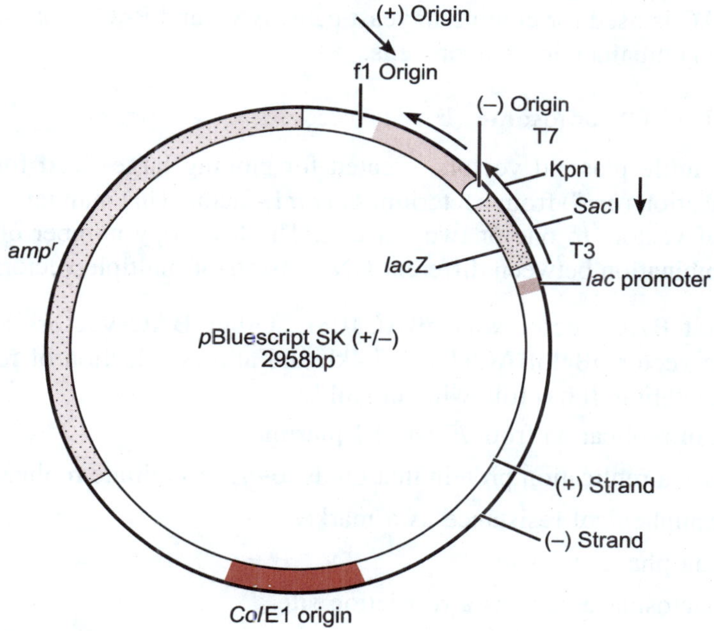

Fig. 17.6 Structure of a phagemid vector *p*Bluescript SK with T7 and T3 promoter sites on either end of *lacZ* gene respectively

*p*Bluescript vector also provides *in vitro* production of RNA copies of either strand of DNA insert. In phagemid vector phage, T7 promoter is located on one side of MCS and promotes transcription of antisense strand of *lacZ* gene and DNA insert. The phage T3 promoter is located on the other side of MCS. It promotes transcription of RNA from the sense strand of *lacZ* gene. Where these RNA transcripts are labelled with radioactivity or with nonradioactive labels they are used as RNA probes. Because of this, these phagemid vectors are also called **riboprobe vectors.**

Phagemid vectors also generate single-stranded copies of DNA insert. These single-stranded copies are used for DNA sequencing. This is facilitated by the presence of six primer sites within and around MCS. These are T7, T3, reverser primer, SK primer and M13-20 primer.

5. ARTIFICIAL CHROMOSOME VECTORS

Artificial chromosome vectors are linear or circular shuttle vectors that occur as 1 or 2 copies per cell. They allow cloning of base sequences that are several hundred kilobase pairs (up to 1,000 kbp or more). These are of following types:

1. Bacterial artificial chromosome (BAC).
2. Yeast artificial chromosome (YAC).
3. P1-derived artificial chromosome (PAC).
4. Mammalian artificial chromosome (MAC).
5. Human artificial chromosome (HAC).

Of these vectors YAC is used for cloning in yeast cells, BAC and PAC in bacteria and MAC and HAC are used for mammalian and human cells.

1. Bacterial Artificial Chromosome (BAC) Vector

BAC vectors are shuttle plasmid vectors, created for cloning large-sized foreign DNA. They have origin or replication (*oriS*) from bacterium *E. coli* F-factor. This maintains a strict control on the copy number of vectors to one or two per cell. The low copy number of BAC avoids any possibility of recombination between different DNA inserts of multiple vectors.

Structure: The first BAC vector was *p*BAC108L. Other BAC vectors are *p*BeloBAC11, *p*BACe3.6 etc. The vector *p*BeloBAC11 has 7.4kb and allows selection of recombinant clones by *lacZ*α complementation. It has following modules:

- *oriS* origin of replication from *E. coli* F1 plasmid.
- *repE* encodes a replication protein that binds to *oriS* to initiate replication.
- *CM*^{r} chloramphenicol resistance as a marker.
- *cosN* lambda phage *cos* site.
- *lacZ*, β-galactosidase gene as a restriction site.
- T7 from bacteriophage for T7 RNA polymerase driven promoter.
- SP6, bacteriophage SP6 RNA polymerase driven promoter.
- *par*A, *par*B and *par*C for the division of F plasmid during cell division.
- *lox*P site on phage P1 genome for recombination.

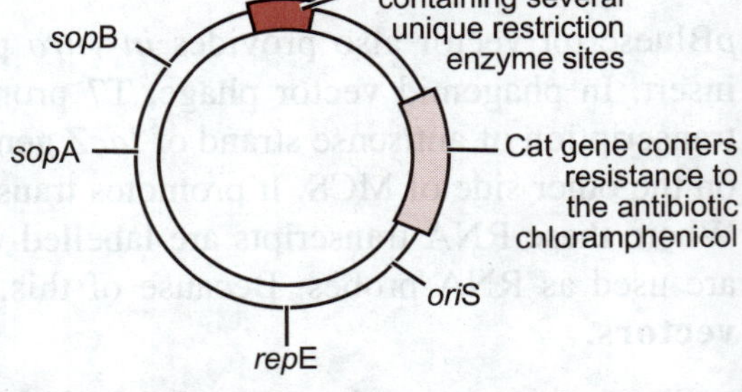

Fig. 17.7 Genetic map of a bacterial artificial chromosome (BAC) vector

Importance of BAC Vectors: BAC vectors have emerged as more popular cloning vectors because of the following features:

- BAC vectors can clone DNA inserts of up to 300kb.

- They are stable and more user friendly.
- They do not suffer from chimerism caused by variation in cloned DNA by recombination.
- The low copy number of BAC vectors per host cell maintains DNA inserts in their original form. It also avoids countersection of cloned genes.
- BAC vectors are extensively used in the analysis of genomes.

The host for BAC vectors is a mutant strain of *E. coli* in which normal restriction and modification sites are missing.

2. Yeast Artificial Chromosome (YAC) Vectors

YAC are linear plasmid vectors that behave like a yeast chromosome. Each YAC occurs in two forms. The circular form grows in bacteria and linear form multiplies in yeast cells. A typical YAC (e.g. *p*YAC3) contains following functional modules from yeast chromosome:

- *ARS* sequence for replication.
- *CEN*4 sequence for centromeric function.
- Telomeric sequences from yeast chromosome at the two ends against *exonuclease* action, consisting of 20–70 tandem repeates of 6 bases, 5′...CCCCAA...3′.
- *TRP*1 and *URA*3 as the two selectable markers.

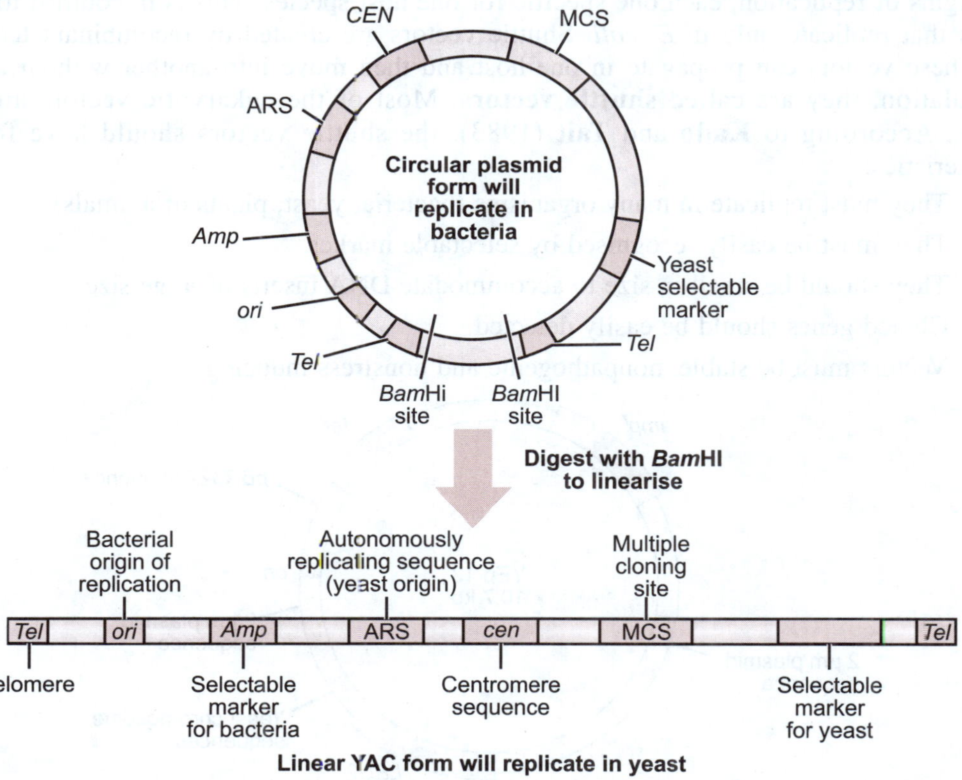

Fig. 17.8 Genetic map of a YAC vector *p*YAC3

- Sequences from *E. coli* plasmid for selection and propagation of *E. coli*.
- *SUP*4, a selectable marker where DNA insert is integrated.

The first YAC vector developed was *p*YAC3. It is essentially a *p*BR322 plasmid vector into which above mentioned yeast sequences have been integrated. It propagates in *E. coli* in circular form. The DNA insert is integrated. It propagates in *E. coli* within *SUP*4 to generate linear YAC vector. The recombinant YAC is introduced into TRPI$^-$ and URA3$^-$ yeast cells by protoplast transformation. The recombinant clones are identified by the insertional inactivation of SUP4 detected by a simple colour test.

YACs can accommodate up to 2,000kb of cloned DNA inserted into the multiple cloning site. Therefore, YACs are used for cloning very large (100 – 1,400kb) DNA segments. They help in mapping complex eukaryotic chromosomes. But they have two disadvantages:

- Cloning efficiency is very low about 1,000 clones per µg DNA. Thus, they can not be used for complete generation of genomic libraries.
- It is not possible to recover large amount of pure insert DNA from individual clones.

6. SHUTTLE VECTORS

Shuttle vectors are plasmids designed to replicate in cells of two different species. They possess two origins of replication, each one specific for one host species. This is in contrast to plasmid vectors that replicate only in *E. coli*. Shuttle vectors are created by recombinant techniques. Since these vectors can propagate in one host and then move into another without any extra manipulation, they are called **shuttle vectors**. Most of the eukaryotic vectors are shuttle vectors. According to **Kado** and **Tait** (1983), the shuttle vectors should have following characteristics:

- They must replicate in many organisms (bacteria, yeast, plants or animals).
- They must be easily recognised by selectable marker.
- They should be small in size to accommodate DNA inserts of large size.
- Cloned genes should be easily detected.
- Vectors must be stable, nonpathogenic and nonstress inducing.

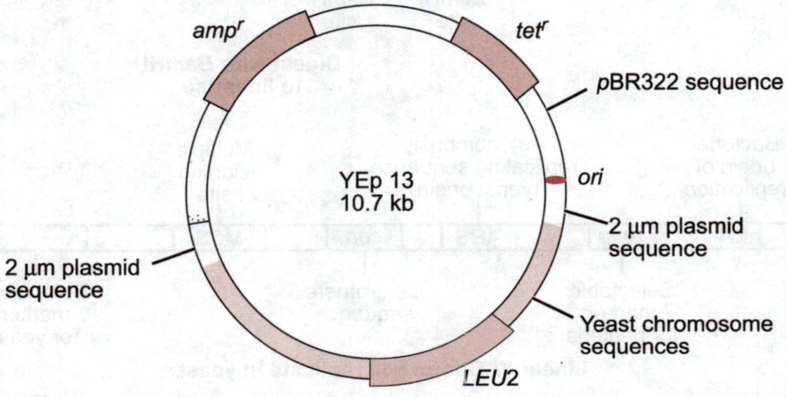

Fig. 17.9 Structure of a shuttle vector

- Vectors must effectively deliver genetic information for stable maintenance in alternative host.
- The introduced genetic information should be maintained in new host.

An example of shuttle vector is YEp (yeast episomal plasmid). Another shuttle vector is designed to replicate in *E. coli* and *Streptomyces*. It is reconstructed from *Streptomyces* plasmid and from *E. coli* plasmid. It allows initial cloning of *Streptomyces* DNA inserts in *E. coli* and their subsequent functional tests in *Streptomyces*.

7. FOSMID VECTORS

Fosmid vector is a modified BAC vector. It contains λ–phage *cos* site so that it can be packed in the head of phage lambda. An example of fosmid vector is *p*FOSI. It was formed by homologous combination between *cos* sites of *p*BAC10L and *p*UC cos. In this fosmid, the *lac*Zα region and multiple cloning sites of *p*UC vector are replaced by phage lambda *cos* sequence. The replication of this vector is driven by *p*UC origin of replication. Fosmids are mini BACs and are used to clone DNA inserts of about 40kb. They are useful in generating mini BAC libraries from small amounts of source DNA.

TECHNIQUES USED IN RECOMBINANT DNA TECHNOLOGY

Mechanism of recombinant DNA technology includes following steps:

1. Generation of DNA Fragments

Several techniques are adopted to isolate or synthesise the desired gene or the DNA fragment. These are:

- Fragmentation of DNA by cleaving with restriction enzymes or by mechanical shearing of genome.
- Artificial synthesis of gene.
- Synthesis of complementary DNA (cDNA).

Extension: Separation of Whole Chromosomes or Large DNA Molecules

DNA molecules of very large size are separated by **pulse field gel electrophoresis** (PFGE). In this technique, short pulses of electricity are used in two different directions and DNA is embedded in the form of agarose plugs to avoid fragmentation of large DNA molecules while handling. By this technique, DNA belonging to the whole chromosomes can be separated.

Counter clamped homogeneous electric field electrophoresis (CCHEFE) is another, more refined technique used for isolation of large segments of DNA.

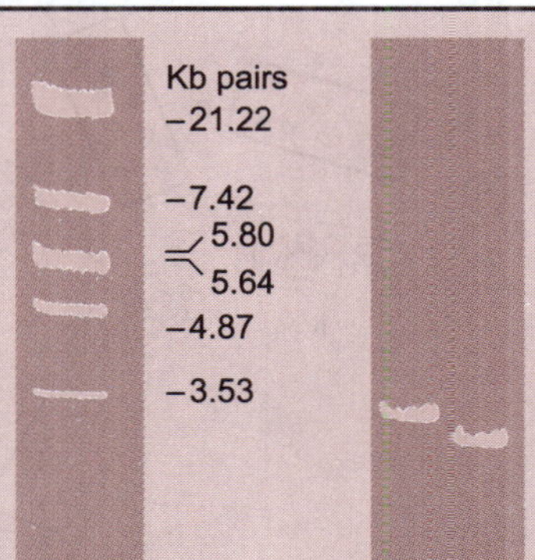

Fig. 17.10 DNA bands on agarose gel separated by electrophoresis

DNA Cleaving by Restriction Enzymes or by Mechanical Shearing of Genome

The desired segments of DNA from eukaryotic genome are cut by using ***restriction endonuclease*** or by mechanical shearing. In many cases, the desired DNA fragment is isolated from the **genomic libraries** or **gene banks**.

The process of DNA cleaving and isolation involves following steps:

1. SEPARATION OF DNA FRAGMENTS

The separation of DNA fragments is done by **gel electrophoresis**. The gel used is of two types:

- **Agarose gel** is used for separation of large segments of DNA.
- **Polyacrylamide gel** is used for separation of small DNA fragments differing in few base pairs. It is more commonly used for DNA sequencing experiments. It is also used for the separation of RNA and proteins.

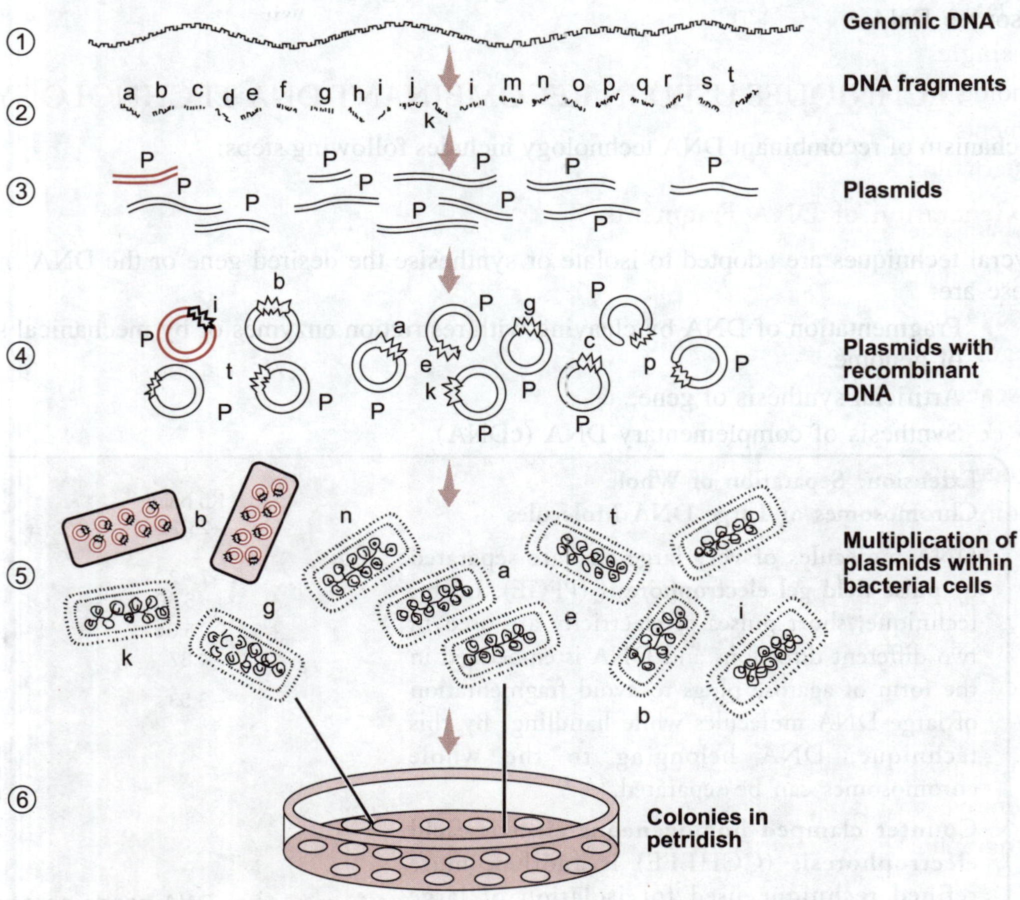

Fig. 17.11 Genomic library or gene bank

The genomic DNA is extracted from any tissue of a plant or animal species. It is digested with a restriction enzyme. It is cleaved into segments of different sizes. These segments are separated by gel electrophoresis.

The gel electrophoresis involves movement of DNA fragments or DNA molecules under a well, created on one edge of a gel slab (The gel slab is about 10 cm long and 0.5 cm thick). The rate of migration of fragments is inversely proportional to their size. The small DNA fragments migrate faster and move away, while the larger or heavier fragments remain closer to the site of loading. Fragments of different sizes separate out as bands on gel. The gel with DNA bands is soaked in fluorescent dye solution (ethidium bromide) and when it is exposed to UV radiation, DNA bands appear in fluorescent colours.

2. Identification of DNA Fragments

The DNA fragments separated as bands can be identified by using **molecular probes**. These are single-stranded radioactively labelled DNAs of known nucleotide sequence and are homologous to specific DNA segments of interest. There are three methods of identifying DNA segments of our interest. These are Southern blotting technique, Northern blotting technique and Western blotting technique.

Southern Blotting Technique

This technique was elaborated by **E. M. Southern** in 1975. For Southern blotting, the DNA bands in the gel (obtained by gel electrophoresis) are denatured into single strands by immersing the gel in sodium hydroxide solution. The alkali solution denatures the hydrogen bonds. Subsequently the gel is laid on top of a buffer-saturated filter paper placed on a glass plate. The two edges of filter paper remain immersed in the buffer. An artificially synthesised **nitrocellulose membrane** is placed on top of the gel and a stack of several sheets of dry blotting paper (paper towel) is placed on top of this membrane. A weight of 0.5 kg is placed on paper towels.

The nitrocellulose membrane with replica of DNA bands from agarose gel is used for hybridisation with labelled nucleic acid probe (DNA probe or RNA probe). The membrane is placed in a solution that contains radioactively labelled probe (the labelled single-stranded DNA with known sequence). The probe hybridises with complementary DNA (cDNA) on nitrocellulose membrane. After hybridisation is complete, the nitrocellulose membrane is washed to remove any unbounded DNA. By placing the nitrocellulose membrane in contact with photographic film, the regions of hybridisation are detected by autoradiography.

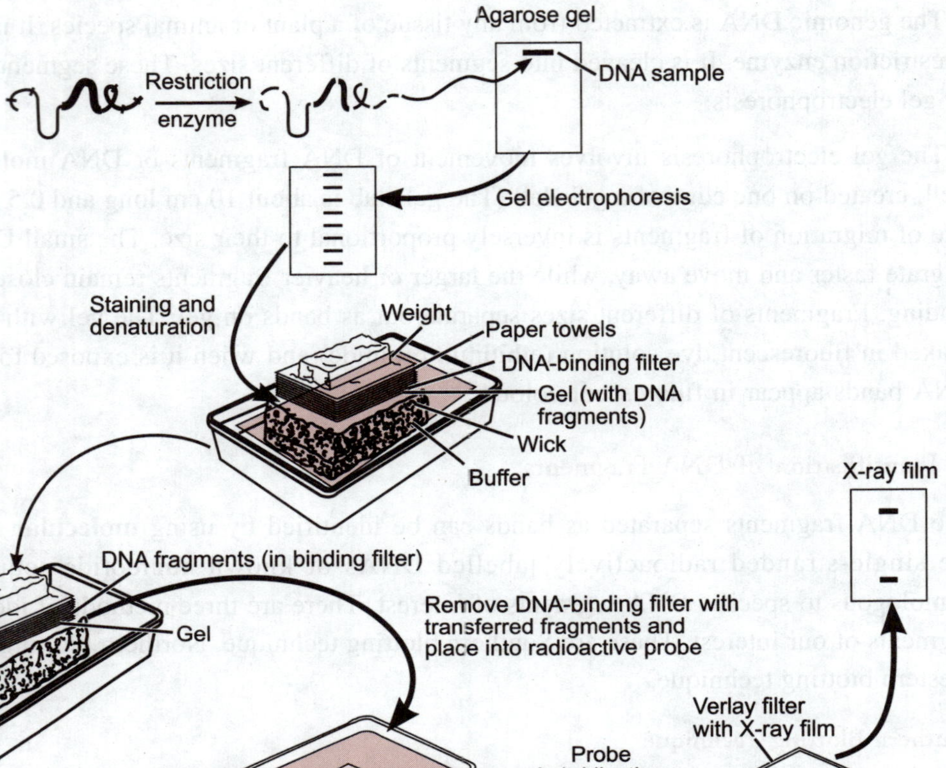

Fig. 17.12 Southern blotting method of identification of DNA fragments from the total genomic DNA by using radioactive nucleic acid probe

Northern Blotting

This method was developed by **James Alwine** *et al.* (1979). It is related to Southern blotting technique, but is used for blot transferring RNA. The mRNA separated by gel electrophoresis can not be transferred to nitrocellulose membrane because RNA does not bind to the nitrocellulose membrane.

In **Northern blotting**, the mRNA bands from agarose gel are blot transferred to chemically reactive DBM paper, prepared by diazotisation of aminobenzyloxymethyl paper (Watmans's 540 paper). The mRNA on the sheet is hybridised to a radioactive cloned DNA or RNA probe and detected by autoradiography. The single-stranded regions of the probe are removed by *nuclease* (mungbean nuclease or S_1-nuclease) and the quantitative estimation of hybridised mRNA is made.

Western Blotting

This technique is used to detect protein and to identify the translated products in the form of proteins when a transferred gene expresses inside the transformed cells. The extracted proteins are subjected to **polyacrylamide gel electrophoresis** (PAGE) and are then transferred into nitrocellulose membrane. The nitrocellulose membrane is used for probing with a specific radioactively labelled antibody, which binds to a specific protein. The antibody may be labelled with ^{125}I and detected by autoradiography or it may be detected by a second antibody tagged with an enzyme.

2. Artificial Synthesis of Gene

There are three approaches for the synthesis of gene *in vitro*.

- If the detailed structure of the gene is known, the gene can be synthesised purely by chemical method as done by **Khorana** (1970).
- If the detailed structure of gene in question is not known, ***'RNA directed DNA polymerase'*** enzyme is utilised for *in vitro* synthesis of gene. A **complementary DNA** (cDNA) is synthesised from the mRNA of the gene.
- If the amino acids sequence of a protein or polypeptide is known, the base sequence of the gene concerned with the synthesis of that protein can be determined. A polynucleotide of the same base sequence can be synthesised *in vitro*.

The process of artificial synthesis of gene involves following two steps:

- Chemical assembly of nucleotides into short stranded DNA-fragments (10–1000 nucleotides). These small fragments are called **oligonucleotides**.
- Enzymatic assembly of oligonucleotides into double-stranded clonable DNA (cDNA).

Extension: Synthesis of gene

In 1965, **R. W. Holley** and his coworkers gave a detailed structure of yeast **alanyl tRNA** *(77bp)*. **H. G. Khorana** and coworkers gave a detailed structure of the gene for this tRNA (yeast alanyl tRNA). The synthesis of such a long chain (77bp) of double-stranded DNA was rather difficult in 1965 but not now. **Michelson** (1955) for the first time chemically synthesised a simple dinucleotide in the laboratory. **Khorana** and his coworkers (1972-79) successfully synthesised two tRNA genes: **alanyl tRNA gene** and **tyrosyl tRNA gene**. It took 20 men and several years of hard work. Since then, methods of synthesis of DNA molecules of known nitrogenous base sequences have improved dramatically. With the help of **'gene-synthesis machine'**, a gene can be synthesised within few days. There are several versions of gene machines. The gene machines have a microprocessor for synthesising oligonucleotides.

Extension: Reverse transcriptase

Reverse transcriptase is a DNA polymerase enzyme. It is generally obtained from **avian myelobalstosis virus** (AMV). It is a retro virus which uses mRNA as a template for the synthesis of DNA and produces RNA: DNA hybrid molecules. RNA strand is removed from these hybrid molecules by alkaline hydrolysis or by digestion with *ribonuclease* (RNAase) leaving single-stranded cDNA. It has a short hairpin loop at its 3′ end. In the next step, this single-stranded cDNA becomes double-stranded cDNA by the synthesis of its complementary strand with the help of enzyme ***DNA polymerase***.

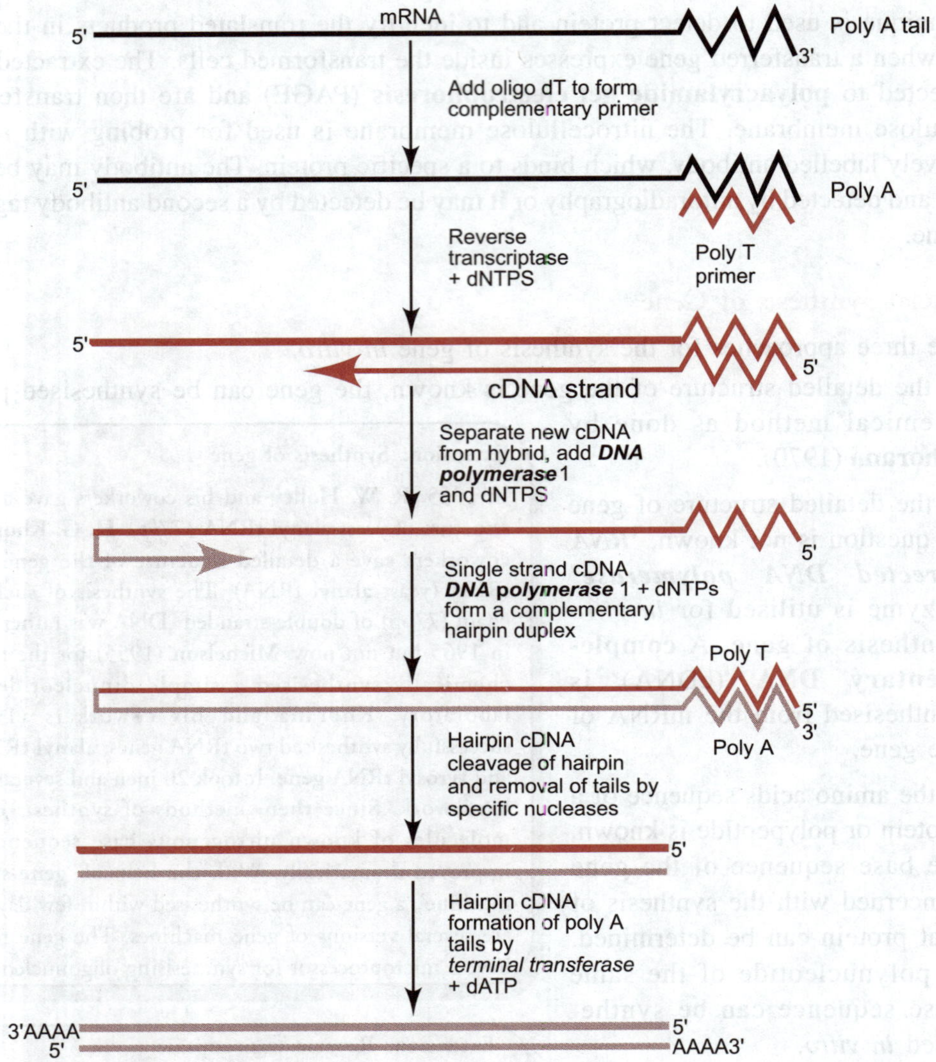

Fig. 17.13 Synthesis of cDNA from mRNA

Chemical Assembly of Oligonucleotides: The **oligonucleotides** are synthesised based on phosphodiester or phosphotriester approach. Most recently phosphite triester method has replaced the earlier two approaches. **Khorana** had followed phosphodiester approach. Presently, phosphodiester approach has been given up completely because of the following drawbacks:

- the reaction is completed in long time,
- the yield decreases with increase in the length of polynucleotide chain, and
- the purification procedures are time consuming.

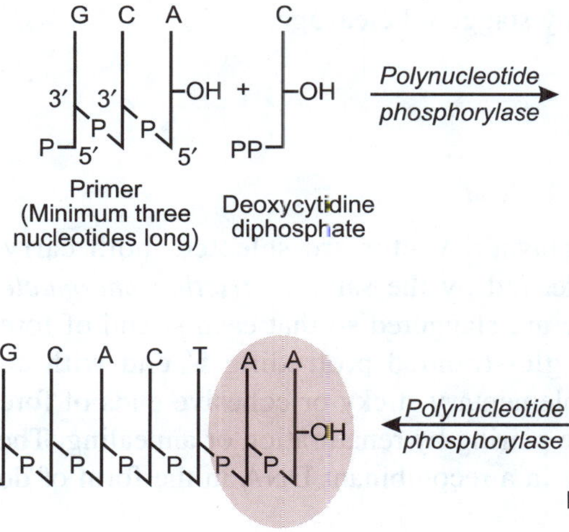

Fig. 17.14 Enzymatic assembly of dinucleotide

In **phosphotriester method,** 3′– and 5′–OH groups of deoxyribose and one OH group of the phosphate moiety are protected. In phosphate triester method compound with trivalent phosphorus is used in place of pentavalent phosphorus.

Enzymatic Assembly of Oligonucleotides or Gene Synthesis: The free hydroxyl groups present on either end of chemically synthesised oligonucleotides are 5′–phosphorylated with ATP by ***polynucleotide-kinase*** enzyme. These are then heated to high temperature (90 – 100°C). This results in annealing of oligonucleotides into double-stranded DNA with nicks at the ligation sites. These nicks are sealed by enzyme ***DNA ligase***.

3. Complementary DNA Synthesis

In this method mRNA of the desired protein is used as a template to synthesise its complementary DNA (cDNA) by using enzyme ***reverse transcriptase*** or ***RNA-directed DNA polymerase***. This was discovered by **H. Temin** and **D. Baltimore** in 1970.

For the isolation of desired mRNA, total RNA is extracted from a suitable tissue. Poly-A sequence is added as a primer to the 3′ end of desired mRNA. The cDNA, thus formed has a hairpin loop at one end. This hairpin loop is trimmed away by treatment with enzyme **S_1 nuclease** (Fig. 17.14).

JOINING OF FOREIGN DNA FRAGMENT TO A CLONING VECTOR

The isolated foreign DNA-fragment is joined to its cloning vector to create a recombinant DNA. The vector is cut open with the help of a ***restriction endonuclease***, the foreign DNA is inserted in the vector and cut ends are then sealed by ***DNA-ligase***. Because, the plasmid DNA is circular, a single cut made in the circle opens it up. DNA to be cloned is inserted, the ends are sealed or ligased and a circular recombinant DNA molecule is formed. There are three methods being used for this insertion:

- Sticky or cohesive end ligation by staggered cleavage.
- Blunt end ligation
- Homopolymer tailing.

1. Sticky or Cohesive End Ligation Method

In this method, foreign DNA and a plasmid vector are selected, both carrying the same recognition sites so that they can be cleaved by the same *restriction endonuclease*. The cuts produced by restriction by endonuclease are staggered so that each strand of foreign DNA and vector DNA has a complementary single-stranded protruding 5′ end with complementary palindromic base sequence. These complementary sticky or cohesive ends of foreign DNA and vector DNA join by complementary base pairing by renaturation or annealing. The nicks left are sealed by enzyme DNA-*ligase*, resulting in a recombinant DNA in the form of double-stranded DNA circle.

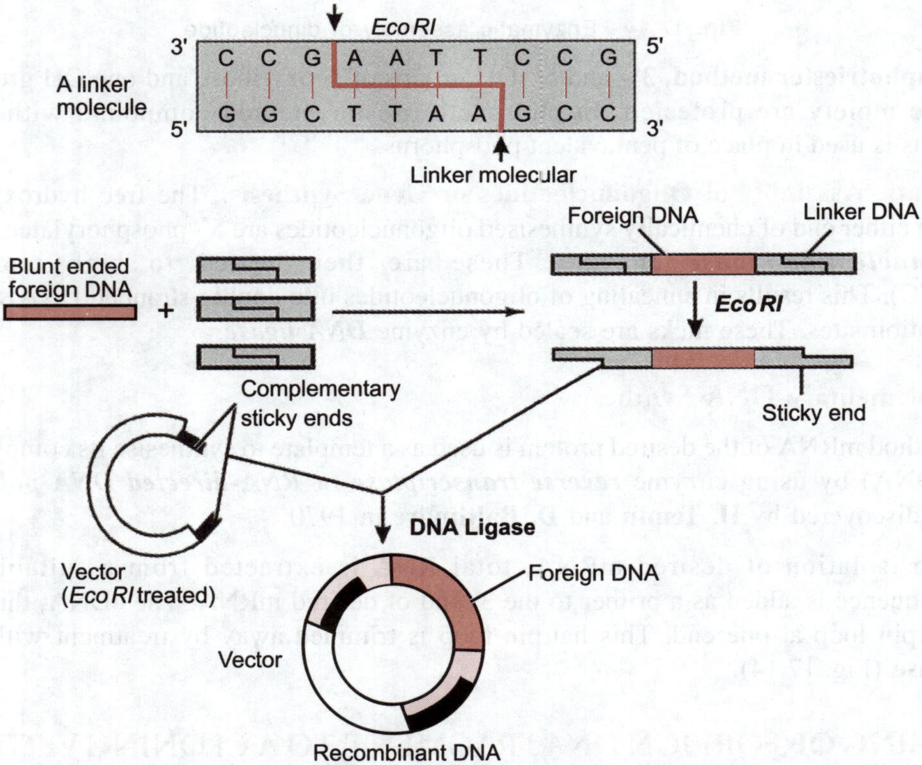

Fig. 17.15 Sticky end ligation to form recombinant DNA

2. Blunt End Ligation Method

In some cases, blunt ends of foreign DNA fragment and its cloning vector, generated by cleaving with the same restriction enzyme, join each other directly with the help of *T_4-DNA ligase* (**Sgaramella**, 1972).

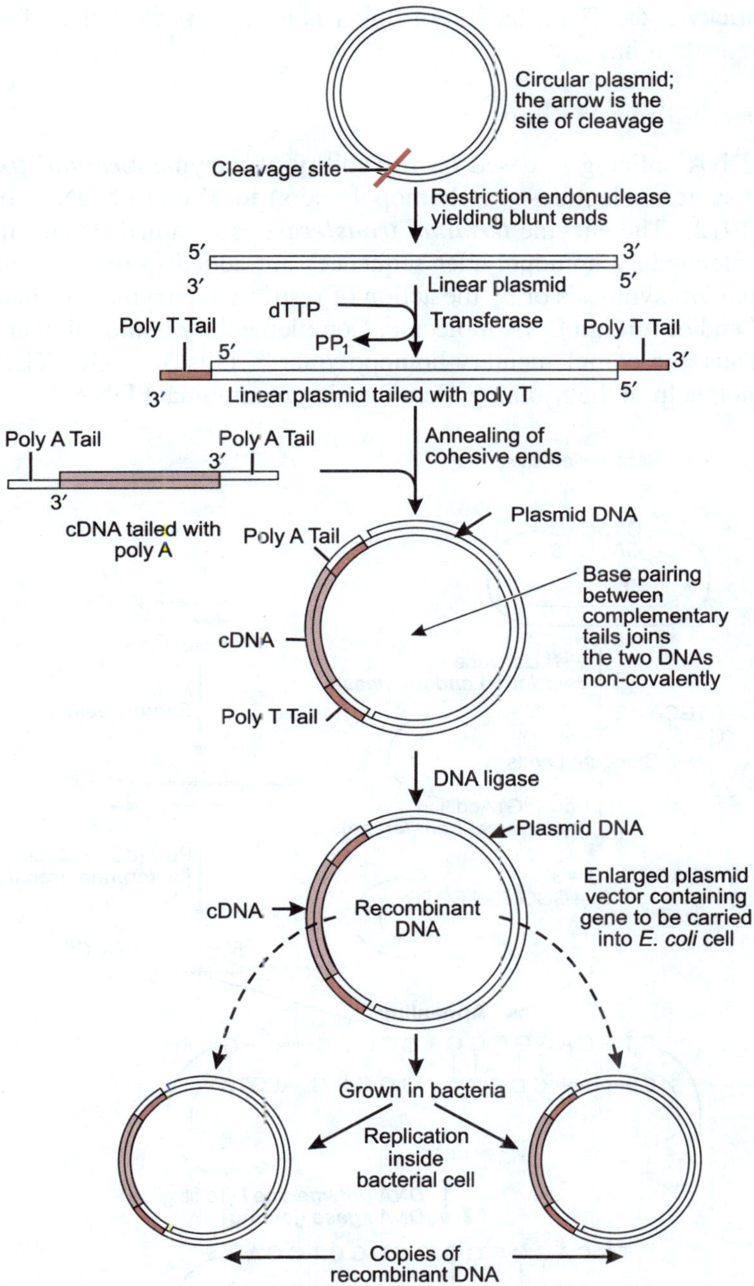

Fig. 17.16 Blunt end ligation with the help of linker molecule to form recombinant DNA

More commonly used method involves ligation of some linker DNA to both the blunt ends of the foreign DNA by ***T_4-DNA ligase***. The linker molecules have specific recognition sites for a specific restriction enzyme. By treating with ***restriction endonuclease*** the sticky ends are produced in the linkers. Thus, the DNA with sticky linkers joins the cloning vector that bears the complementary sticky ends. The selection of linker is based on the choice of vector being used and the recognition site it has.

3. Homopolymer Tailing Method

This method of DNA splicing is based on the ability of enzyme ***terminal transferase*** to add sequences of the same nucleotides (i.e. homopolymers) to 3′ end of DNA. It was applied by **Jackson** *et. al*. (1972). The enzyme ***terminal transferase*** is obtained from calf thymus. In this method, the complementary homopolymer sequences are added to the ends of DNA molecule produced either *in vitro* synthesis or by the action of restriction enzyme. For instance, polymer G is added to the 3′ end of foreign DNA molecule. Complementary homopolymer C is added to the cloning vector. Thus, the complementary homopolymer 3′ tails 5′…..GGGG...3′/3′...CCCC…5′ are generated which help in their joining, thus forming recombinant DNA.

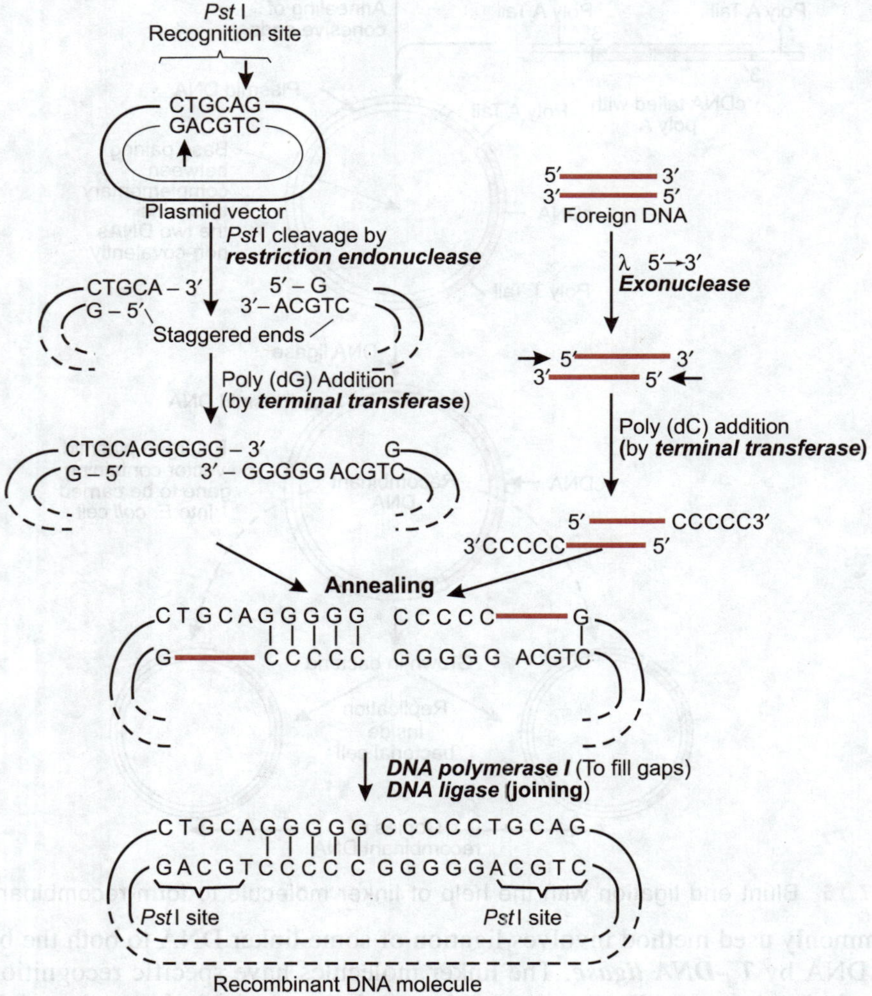

Fig. 17.17 Homopolymer tailing technique for recombinant DNA synthesis

INTRODUCTION OF RECOMBINANT DNA INTO HOST CELL

Recombinant DNA formed is to be introduced into the host cell where it can replicate and form clone. This is achieved by the following methods:

1. Transformation
2. Transduction
3. Electroporation
4. Liposomes
5. Microinjection
6. Microprojectile

Extension: Linkers and Adaptors

Linkers are short, chemically synthesised, self complementary oligonucleotides, which contain one or more recognition sites for specific *endonuclease*. For example, pCCGAATTCGG contains one **Eco RI site.** Linkers are fused with blunt ends of foreign DNA and are then cleaved with appropriate restriction enzyme to generate suitable protruding 5′ ends to pair with the ends of selected vector opened by the same restriction enzyme.

Adaptors are also chemically synthesised oligonucleotides, which are prepared from two different single-stranded oligonucleotides whose 5′ ends carry recognition sequences for different endonucleases but their 3′ ends have a self complementary base sequence. For example, oligonucleotides 5′...OHGATCCTCGAG...3′p and oligonucleotides 5′p...AATTCTCGAG...3′ are adaptors having recognition sites for *Bam HI* (GATC) and for *Eco RI* (AATT) and self-pairing complementary region 5′...CTCGAG and 3′...GAGCTC ....

Adaptors convert the protruding ends formed by one restriction enzyme into those produced by some other restriction enzyme. This enables the fusion of DNA fragments generated by one restriction endonuclease with a plasmid cut open by another endonuclease. Linkers and adaptors are used for cloning of cDNA and introduction of new cleavage sites into cloning vector.

1. Transformation

Transformation is the uptake of foreign DNA by a cell from its surroundings. This is the most usual approach used for introducing recombinant DNA into the host cell. It was developed by **Mandell** and **Higa** in 1979. Majority of cells including *E. coli,* yeast and mammalian cells, are poorly accessible to DNA molecules. But treatment with $CaCl_2$ enhances the binding of DNA to cell surface and makes the cells permeable to DNA. Growing *E. coli* cells are isolated and suspended in 50mM $CaCl_2$ solution and incubated for 12–24 hours to increase the frequency of transformation. The recombinant DNA is then added to the treated cells. It takes only few minutes for the cell transformation. The cells are plated on a suitable medium for the selection of transformed clones. The frequency of transformed cells is about one per 10,000 plasmid molecules or $10^6 - 10^7$ per µg of plasmid DNA. This frequency can further be increased by using special strains of *E. coli* and some specific conditions during transformation.

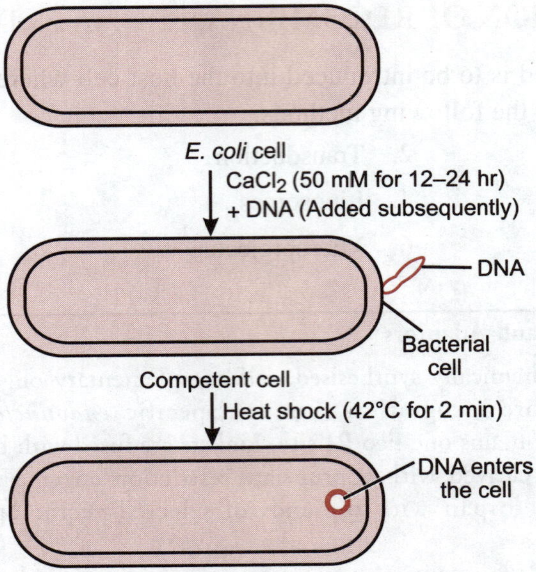

Fig. 17.18 Transformation of *E. coli* by $CaCl_2$ treatment

2. Transfection (Introduction of Recombinant DNA Packaged as Virions)

Vectors that have *cos* sequences of lambda phage (such as cosmids, plasmids and λ-phage vectors) are packaged *in vitro* into empty phage heads. The packaging produces complete λ particles with recombinant DNA. These newly synthesised phage particles are used to infect *E. coli* cells. Thus introduction of foreign DNA into bacteria by bacteriophage is called **transfection**.

The infection by phage particles containing recombinant DNA is very efficient, reaching up to 10^8 plaques per μg of DNA, while it is just 10^3 plaques per μg DNA by transformation.

The transformed bacterial cells are spread on a lawn of susceptible cells in culture, where clear area of plaques develops in the lawn. These plaques containing recombinant DNA are identified. The phage particles collected from such plaques provide **purified recombinant DNA**.

3. Electroporation

Electroporation is the introduction of foreign DNA or recombinant DNA into cell by a very brief exposure to a very high voltage electric pulses ranging from 4,000–8,000 V/cm. This induces transient pores in the cell membrane. These pores give passage through which foreign DNA can enter the cells, provided DNA is in direct contact with the membrane.

The electroporation pulse is generated by discharging a capacitor across the electrodes in a specially designed electroporation chamber. Either a high voltage pulse of short duration or a low voltage pulse of long duration is used. The protoplasts on the so called naked cells are suspended in an ionic solution containing vector DNA or recombinant DNA. They are electroporated and then plated and finally the transformed colonies are selected.

Transfer of desired genes through electroporation is carried out in the protoplasts of rice, wheat, maize, sorghum, *petunia*, tobacco, etc.

4. Liposome—Mediated Gene Transfer

Liposomes are small lipid bags in which large plasmids are enclosed. The liposomes are induced to fuse with the protoplasts and are used for gene transfer. DNA enters the protoplasts by the endocytosis of liposomes. The liposome-mediated gene transfer involves following steps:

1. Adhesion of liposomes to the surface of protoplast.
2. Fusion of liposomes at the site of adhesion.
3. Release of plasmids inside the cell.

The technique is more advantageous because it provides (1) protection of DNA/RNA from nuclease digestion, (2) stability of nucleic acids, (3) low toxicity, (4) high degree of multiplication and (5) applicability to a wide range of cell types. This technique is used in gene transfer in tobacco, carrot, *petunia*, etc.

5. Microinjection

Microinjection is the technique of delivering foreign DNA directly into the cytoplasm or nucleus of the cells or male in pronucleus of fertilised egg or into one or two celled embryo with the help of micromanipulators. The microinjection assembly consists of a low power **stereoscopic dissecting microscope** and two **micromanipulators**. The microscope is needed to view the cell or ovum during the entire process of DNA introduction. One of the micromanipulator is a glass micropipette that is used to hold the ovum by partial suction. This is called **holding pipette**. The other micromanipulator has a glass injection needle and is used to introduce DNA into the ovum or its cytoplasm or in male pronucleus. The microinjection technique can be applied to animal cells and also to the plant protoplasts.

The process of microinjection involves following steps:

- One end of a glass micropipette is heated until the glass becomes somewhat liquified. It is stretched quickly so that a very fine tip is formed at the heated end. It is about 0.5mm in diameter and resembles an injection needle.
- Cells to be microinjected are placed in a container.
- A holding pipette is placed in the field of view of microscope and manipulated to hold the target cell at its tip.
- The tip of injecting micropipette is gently pushed through the membrane of the cell.
- Contents of the injecting micropipette are delivered into the cytoplasm.

Xenopus (toad) oocytes have been widely used for the transcription by microinjection. The injected foreign DNA randomly comes in contact with the nuclear DNA. Its expression is possible only when it is attached to a suitable promoter. **Rubin** and **Spradling** (1982) successfully introduced *Drosophila* gene for ***Xanthine dehydrogenase*** into P element (parental element). Such embryos on further development produced rosy eyes instead of mosaic eyes.

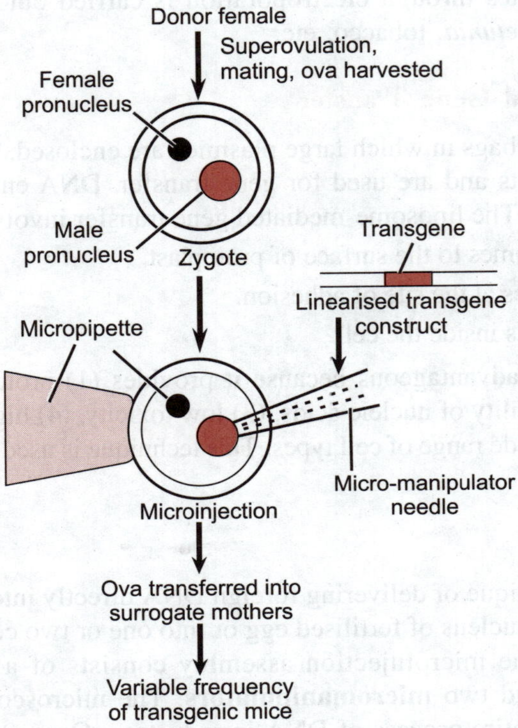

Fig. 15.19 Microinjection of cloned plasmid DNA into the male pronucleus of fertilised ovum

6. Microprojectile or Biolistic or Particle Bombardment Gun

The foreign DNA can be delivered into plant cell when heavy **microparticles** of tungsten or gold-coated with DNA are pushed into the plant cells at a very high speed (about 1,400 feet per second). The technique was developed by **Prof. Stanford** and coworkers in 1987. It is also called **particle bombardment** or **particle gun method** or **biolistic process** or **microprojectile bombardment method**.

The microprojectiles are particles of 1–3 μm diameter. They are delivered by accelerated macroprojectiles or bullet into target plant cells. They are able to penetrate plant cell walls of intact tissue. The acceleration is achieved either by an explosive charge **(cordite explosion)** or by using shock waves initiated by high voltage electric discharge.

The technique is most suitable for those plants which hardly regenerate and which do not respond to gene transfer through *Agrobacterium*. Scientists have successfully delivered foreign DNA into epidermal cells of *Allium cepa*, cell culture of rice, wheat, corn, sorghum, chickpea in bacterial cells and also in human cells and other animals cells.

SELECTION AND SCREENING OF TRANSFORMED CELLS

The clones of host cells with desired recombinant DNA need to be identified and sorted out from the mixture of various types of host cells. A number of strategies have been developed for this purpose. These are:

1. Identification of Transformed Cells by Marker Genes or Reporter Genes

A marker gene or reporter gene produces a phenotype which helps either in the easy selection or quick identification of cells in which it is present. The marker genes can be either selectable or scorable.

The selectable marker genes when present confer ability to survive under selective conditions. Genes conferring resistance to an antibiotic kanamycin are used as selectable marker. For example, when a population of bacterial cells is placed on a medium containing kanamycin, only those bacterial cells survive which have kanamycin resistance gene (kan^r) and form colony. The transformed bacterial cells with recombinant DNA having marker gene kan^r survive and are sorted out.

The **scorable marker genes** are those genes that produce distinct phenotypes. These characteristics allow an easy identification of cells with and without them. For example:

- Gene *gus* (β-glucuronidase) when present produces blue colour in presence of appropriate substrate.
- Gene *gfp* produces green fluorescent protein, and
- Gene *lux* produces protein luciferase that causes phosphorescence.

Transformed bacterial cells with any of these genes are identified from non-transformed cells.

2. Elimination of Non-transformed Cells

When vectors are introduced into the host cells, following three types of cells are formed:

1. Non-transformed host cells.
2. Transformed host cells with unaltered vector, *i.e.*, without recombinant DNA.
3. Transformed host cells with recombinant DNA having DNA inserts.

In the process of sorting, the first step is elimination of non-transformed cells. A good vector has at least two marker genes, one of them is a selectable marker. For example, vector *p*BR322 has marker genes tet^r and amp^r (for tetracycline and ampicillin resistance). The bacterial culture is plated on a medium containing tetracycline or ampicillin. All the colonies that develop on this medium will consist of antibiotic resistant transformed bacterial cells. However, these may be with unaltered vector or with recombinant DNA vector.

3. Identification of Clones having Recombinant DNA

The identification and isolation of those clones of bacterial cells that are transformed by the presence of recombinant DNA from those cells that have unaltered vector is the next step. Two methods are employed for their identification:

(i) Selection of vectors with two selectable markers: In case of vectors that have two selectable markers, the DNA insert is placed within one of these markers. For example, in case of vector *p*BR322, the two markers genes are *amp*^{r} and *tet*^{r}. The DNA insert is placed within one marker *amp*^{r}. The other marker *tet*^{r} is used for the elimination of non-transformed cells. The transformed clones are then replica-plated on ampicillin containing medium. The clones containing recombinant DNA will be sensitive to ampicillin because the DNA insert inactivates the *amp*^{r} gene. Such clones are identified and isolated from the master plate.

(ii) Selection of vectors with a gene complementary to the missing host gene: Some vectors contain a gene or a part of gene, which complements a function missing in their host cells. For example, in vectors pUC gene, *lacZα* complements certain *lacZ*$^-$ strain of *E. coli*. The same combination is used for some λ vectors and M13 phase vectors. In such cases, the DNA insert is placed in such a way that it disrupts the function of *lacZα* gene in the vector. Therefore, *E. coli* cells containing the recombinant DNA are deficient in enzyme ***β-galactosidase***. Such colonies form white plaques on a medium containing X-gal. These white colonies are selected. The *E. coli* cells without recombinant DNA (i.e. with unchanged vector) form blue colonies.

4. Selection of Clone Containing Specific DNA Insert

The population of recombinant clones obtained this way is heterogeneous having different segments of DNA. The clone with DNA insert of interest needs to be identified and separated. A number of strategies are used for this purpose. These include:

1. **Colony Hybridisation:** To identify those bacterial colonies which have desired DNA sequence or desired gene.
2. **Southern Blotting or Southern Hybridisation:** To identify and isolate the desired DNA fragment or the desired gene.
3. **Northern Hybridisation or Northern Blotting:** To separate the RNA produced by the desired DNA fragment.
4. **Western Blotting:** To identify and isolate proteins synthesised by the desired DNA segment.

5. Formation of Clones of Recombinant DNA

The transformed bacteria, containing recombinant plasmids, multiply on culture plate. The plasmids also multiply. Because a bacterial cell harbours only a single recombinant plasmid, each cell of bacterial colony or clone will contain the same cloned DNA fragment. Similarly, only one recombinant phage is likely to infect one bacterial cell and its progeny will contain the same cloned DNA fragment. Thus the multiple copies of recombinant DNA are formed. These could be isolated, purified and analysed.

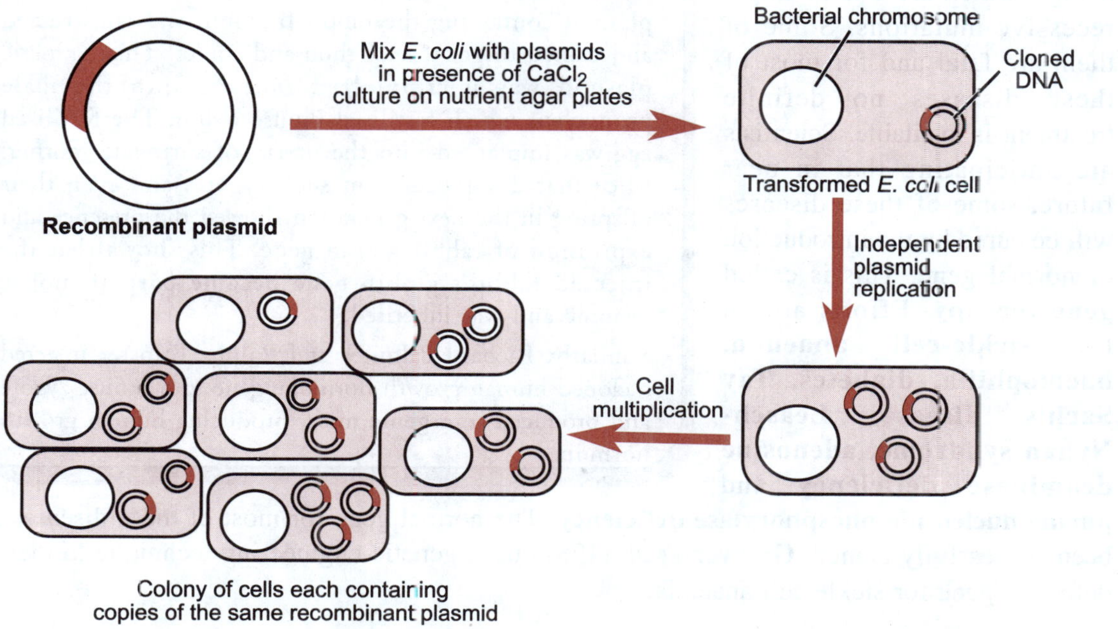

Fig. 17.20 Formation of colony of transformed bacteria having recombinant plasmid

6. Expression of Cloned Genes

To permit the expression of cloned genes in host cell, the expression vectors are constructed. Such vectors have regulatory sequences such as promoters, operators, regulators and ribosome binding sites. Since cloned eukaryotic gene lacks the sequences recognised as regulatory sites in prokaryotes, the placement of promoters and ribosomes binding sites in front of cloned eukaryotic gene ensures its translation in bacterial cells and production of associated protein. Rat insulin, human insulin, growth hormone, mammalian enzyme, ***chymosin*** (as prochymosin), have been synthesised from their respective cloned DNAs.

SIGNIFICANCE OF RECOMBINANT DNA TECHNOLOGY OR GENETIC ENGINEERING AND HUMAN WELFARE

The recombinant DNA technology has opened up wide opportunities for the scientists in the field of genetics, medicine, agriculture and industry.

1. Diagnosis and Cure of Genetic Diseases (Gene Therapy)

> **Gene Coding for β-globin**
>
> **Thomas Wagner** and a team of scientists have successfully transferred a gene coding for β-globin from rabbit into fertilised mouse eggs by microinjections. A chimeric plasmid containing the rabbit β-globin was constructed and cloned to get several thousand copies. This chimeric plasmid was microinjected *in vitro* into the male pronucleus which fertilised mouse ovum. The fertilised egg was implanted into the uterus of surrogate mother. Mice that developed from such zygotes and even their offspring in the next generation showed the presence and expression of rabbit globin gene. This shows that the injected rabbit's globin gene became part of mouse genome and was inherited.
>
> Similarly, Richard Palmiter and Ralpha Brinster injected a cloned human growth hormone gene into mouse zygote and produced **transgenic mice**, producing human growth hormone.

There are several hundred genetic diseases in human beings, which are due to single recessive mutations. Some of them are fatal and for most of these diseases no definite treatment is available. Scientists are anticipating that in near future, some of these diseases will be cured by the introduction of normal gene. This is called **gene therapy**. Efforts are on for **sickle-cell anaemia, haemophilia, diabetes, Tay Sach's disease, Leasch-Nyhan syndrome, adenosine deaminase deficiency** and **purine nucleoside phosphorylase deficiency**. The normal gene for most of these diseases has been successfully cloned. **Greever** *et al.* (1981) used genetic engineering technique to identify defective gene for sickle cell anaemia.

The plan for gene therapy involves the delivery of the cloned gene to a target cell of defective somatic tissue *in vitro*, with subsequent reinsertion of the tissue after uptake and the switch on of the transferred gene. This may also be called **'Gene surgery'**.

The objective of gene therapy is to correct a genetic disorder. The process of gene therapy involves:

- Identification of gene that plays the key role in the development of genetic disorder.
- Determination of role of its protein product in health and disease.
- Isolation and cloning of gene.
- Introduction of cloned gene in somatic tissue *in vitro*.
- Insertion of tissue and switching on of the transferred gene.

Expression of introduced gene relieves or eliminates symptoms of the disorder but the effect is not heritable because the procedure involves only somatic cells. The germ-line is not involved. The technique of *in vitro* fertilisation has contributed to the success of this technique. So far clinical trials have been carried out for treatment of cancer and blood disorders. Gene therapy can be used for the treatment of AIDS also. In case of AIDS, appropriate interleukin genes can be delivered in the patient's body to boost body's defence mechanism and in cancer patient, a toxin encoding gene can be delivered into cancer cells of the patient.

The technique of *in vitro* fertilisation has helped in gene surgery in the elaboration of transgenic animals (especially transgenic mice). Microinjection of a gene in the zygote is another method of producing transgenic animals.

2. Preparation of Molecular Maps and Chromosome Maps by Restriction Mapping, Restriction Fragment Length Polymorphisms (RFLPs) and Random Amplified Polymorphic DNA (RAPD)

1. By Restriction Mapping

At molecular level, the fine structure of a gene is studied by determining nucleotide sequence. This can be done by isolation of DNA corresponding to a gene. This can also be done by cleavage of DNA at specific sites with the help of restriction *endonucleases*. These sites of cleavage can be identified and mapped. This gives rise to **restriction maps**. A restriction map has a linear sequence of sites, each for a specific enzyme. The distance between the sites are measured as number of base pairs of DNA. In a restriction map, the nucleotide sequence and the restriction sites close to one another (300 base pair or less) can be determined.

This technique can be utilised directly on prokaryotes but not in eukaryotes.

2. By Restriction Fragment Length Polymorphisms (RFLPs)

The development of recombinant DNA technology has provided a new way to study the inheritance of allelic variants in human beings and also to map the genes. The restriction endonucleases make cut in both strands of DNA molecule at recognition sites. But single base changes in the recognition sequence can alter the pattern of cuts in DNA, producing segments of different length. These allelic variants are called **restriction fragment length polymorphs** (RFLPs) and are used as markers.

The use of RFLP markers for mapping requires:

- The selection of cloned DNA sequences that recognise variation is DNA cutting sites.
- The mapping of these sequences to specific human chromosome by using a panel of somatic cell hybrids. By using RFLPs, more than 400 human genes have been assigned to specific autosomes and about 250 genes to X-chromosome. About one thousands RFLPs are used each in tomato and maize to prepare the linkage map of their chromosomes.

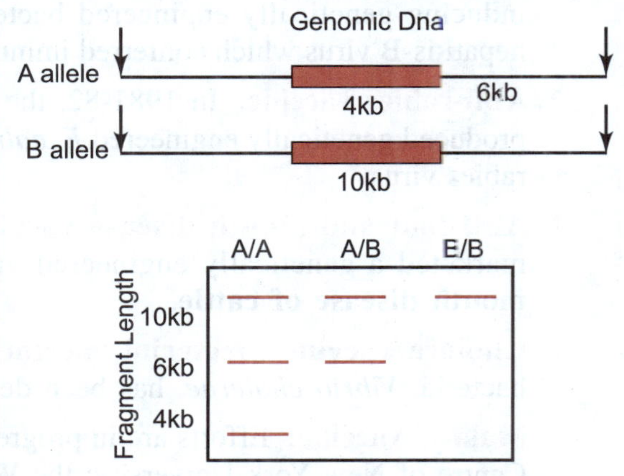

Fig. 17.22 Restriction fragment length polymorphs of a pair of alleles A and B. *Note* the restriction site in allele A in the hatched area, which is absent in allele B

3. Random Amplified Polymorphic DNA (RAPD)

Usage of polymerase chain reaction (PCR) provides another mechanism of preparing chromosome maps. Random samples of DNA markers are prepared by PCR amplification of random DNA segments. These segments are called **random amplified polymorphic DNA** (RAPD). RAPDs are used as substitute for RFLPs and have several advantages over them. RAPD technique may also be combined with RFLPs to increase the efficiency of molecular markers for genetic analysis. The main advantages of RAPD technique include:

- Same primers with arbitrary sequences can be used for different species, thereby species-specific probes are not required for different species as in RFLPs.
- Collection of data using RAPDs proceeds more quickly because fewer steps are involved in this process. The generation of RAPD data is five times quicker than RFLP.

3. Production of Rare Drugs or Pharmaceuticals

Human genes for insulin, somatostatin, thymosin, human growth hormone and blood clotting factor VIII:C have been cloned to produce pure forms of these biologically important fine chemicals or hormones. In 1984, a functional gonadotropic hormone was also obtained from transformed human cells.

4. Production of Synthetic Vaccines

A large number of viral, bacterial and protozoan vaccines are being synthesised against pathogens causing hepatitis and liver cancer, rabies, foot and mouth disease of cattle (FMDC), feline leukemia (causing cancer), cholera, malaria, etc. Gene coding for an antigen protein is isolated from the pathogen and is cloned in a bacterial host or spleen cells of mouse, where it produces the concerned antibody. This is isolated and purified to be used as a vaccine.

- **Anti-hepatitis vaccine:** A French team of Pasteur Institute in 1980 succeeded in inducing genetically engineered bacteria and mouse cells to produce the antigen of hepatitis-B virus which conferred immunity against hepatitis-B infection.
- **Anti-rabies vaccine:** In 1981-82, the scientists from a French company **'Transgene'** produced genetically engineered *E. coli* cells which could synthesise a surface protein of rabies virus.
- **Anti-foot and mouth disease vaccine for cattle:** A Dutch firm **'Akzo'** in 1982 marketed a genetically engineered vaccine against the viral disease, **anti-foot** and **mouth disease of cattle**.
- **Cholera vaccine (bacterial vaccine):** Vaccine against cholera disease caused by bacteria, *Vibrio cholerae*, has been developed.
- **Malaria vaccine:** Efforts are in progress to produce vaccine against malaria at Medical Centre of New York University; the Walter Reed Institute of Colombo, Sri Lanka and Queensland Medical Research Centre, Australia.
- **Vaccine for smallpox virus:** The cowpox virus can be used as the basis of smallpox vaccine, using recombinant DNA technology.

Extension: Gene cloning of human insulin gene

For natural gene cloning of human insulin gene in *E. coli*, the following steps are included:

- Human pancreas cells are broken, releasing many mRNAs.
- These are separated from the rest of the cell material using high speed centrifugation technique.
- Insulin mRNA is isolated by Northern Blotting.
- The enzyme, ***reverse transcriptase***, is added to the mRNA along with the primer DNA.
- Using mRNA as a template double stranded DNA is obtained. This is called complementary DNA (*c*DNA).
- *c*DNA for insulin is now introduced in plasmid (the cloning vector) of *E. coli*.
- Plasmid is now introduced in the bacterial cell. A single plasmid inside *E. coli* can multiply to yield its several copies which may be even a few dozens.
- Thus several copies of insulin gene are also produced which multiply with the bacterial multiplication. Thus a clone bacterial cells with gene for human insulin is formed.

5. Biosynthesis of Interferons

Interferons are body's first line of defence against viral infection. These are proteins, released in very minute quantities by the host cells when a virus enters. These are produced by leucocytes of blood, the fibroblasts of connective tissue and the T-lymphocytes of immune system. Interferons have antiviral and anticancer properties. In 1980, α-interferons were produced from genetically engineered *E. coli* and in 1981, these were obtained from genetically engineered yeast cells (*Saccharomyces cerevisiae*). γ-interferons were obtained from genetically engineered monkey cells in 1981.

6. Hybridoma and Monoclonal Antibodies (Mab)

Monoclonal antibodies are antibodies of only one type of homogeneous immunoglobulins or immunological reagents of defined specificity. These are used for the diagnosis and screening. Monoclonal antibodies are being synthesised by popular **hybridoma technology**.

Hybridoma technology was discovered in 1975 by **Georges Kohler** of Germany and **Cesal Milstein** of Argentina. The hybridomas are cells formed by the fusion of two different types of cells. For monoclonal antibodies, the hybridomas are formed by the fusion of:

(a) antibody producing **lymphocytes** from the spleen of mouse immunised with red blood cells or specific antigen from sheep, and

(b) **myeloma cells** (bone marrow tumour cells) from rat or rabbit which are capable of indefinite multiplication.

These hybridoma cells have the antibody-producing capability inherited from lymphocytes and ability to grow continuously from malignant cells of tumour.

Monoclonal antibodies are used in the identification of blood groups, diagnosis of diseases and cancer, in vaccine production and in immunotherapy.

Extension: Monoclonal Antibodies

The **spleen cells** produce B- and T-lymphocytes. These produce specific antibodies against specific antigens. The **myeloma cells** (tumour cells) do not produce antibodies and are mutants which can not synthesise enzyme **hypoxanthine guanine phosphoribosyl transferase** (HGPRT). The hybridoma cells of these two lines are grown in selective **hypoxanthine aminopterin thymidine** (HAT) medium containing drug aminopterin. This drug blocks one pathway for nucleotide synthesis, making the cell to follow alternative HGPRT pathway which needs enzyme HGPRT. Since the tumour cells lack this enzyme, they are unable to live and grow in this medium. The spleen cells can not survive in culture medium and die. But their hybridoma cells, which inherit HGPRT gene from B-Lymphocytes and capability to multiply in culture medium from tumour cells, survive and multiply in the HAT medium. Thus, HAT medium allows selection of hybridoma cells. By isolating single cell colonies of hybridomas and culturing them separately, monoclonal cultures are obtained. These are used in the production of large amount of monoclonal antibodies. These hybridoma cells can be freezed and stored for future use.

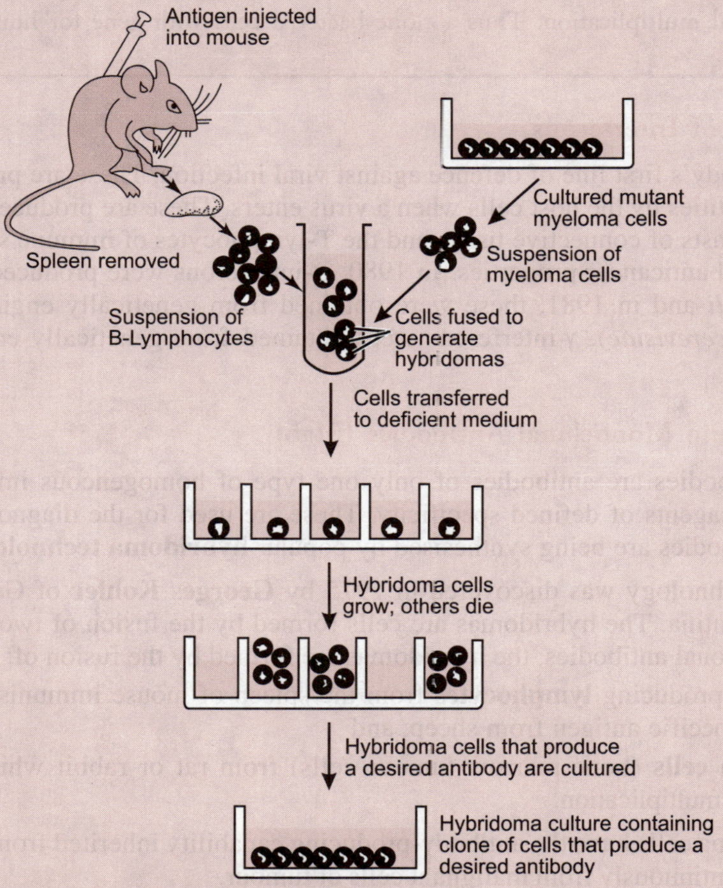

Fig. 17.21 Steps in the process of hybridoma technology for the production of monoclonal antibodies

Table 17.4 Human recombinant DNA proteins approved by FDA

S. No.	Protein	Year	Approved	Cloned in
1.	Insulin	1982	Children for insulin dependent diabetes, to replace missing insulin.	*E. coli*
2.	Growth hormone	1985	Children with pituitary dwarfism; used to replace missing growth hormone.	*E. coli.*
3.	α-interferon	1986	Cancer patients with hairy cell leukemia, AIDS- related Kaposi sarcoma, hepatitis and for genital warts.	*E. coli*
4.	Hepatitis-B-vaccine	1986	Health care workers exposed to blood, for drug users and homosexuals, and against liver infection.	Yeast
5.	Tissue plasminogen activator	1987	Heart attack patients, to dissolve blood clots blocking coronary arteries.	Hamster cells in culture
6.	Erythropoietin	1989	Patients with kidney failure, for treating severe anaemia due to deficiency of **erythropoietin** (made in kidney), for AID related anaemia.	Hamster cells in culture
7.	γ-interferon	1990	Patients with chronic granulomatous disease, a rare genetic defect of immune system, to prevent severe infections.	*E. coli*
8.	Granulocyte colony -stimulating factor	1991	For cancer patients undergoing chemo-therapy, to prevent infections by stimulating growth of WBCs.	*E. coli*
9.	Granulocyte/ macrophage colony stimulating factor	1991	Patients with leukemia or Hodgkin disease to stimulate growth of WBCs in transplanted marrow.	Yeast
10.	Interleukin-2	1992	Kidney cancer patients.	*E. coli*
11.	Clotting factor VIII	1992	Patients with haemophilia A.	Hamster cells in culture

7. Genetic Engineering and Agriculture

So far interspecific gene transfers were achieved through hybridisation. These were restricted to sexually reproducing plant species. Interspecific gene transfers are now possible through recombinant DNA technology. **Ti** plasmids of *Agrobacterium tumefaciens* and **Ri** plasmids of *A. rhizogenes* are being effectively used as vectors for gene transfer to plant cells. The plants with foreign gene are called **transgenic plants**. By the end of 1992, more than 50 species of transgenic plants were produced, many more have been added to this list since then. The transgenic plants have been produced which are resistant to insects, fungi, viruses and herbicides or which produce fruits with a longer shelf-life or have better taste or have higher yield.

Some of the successful gene transfers through Ti plasmid vector are:

- Transfer of gene **phasealine** (G_1 globin) from French bean or rajma to sunflower.

- Transfer of gene for **kanamycin resistance** from *E. coli* to tabacco and other plant species.
- Transfer of genes for **leg-haemoglobin** and **RuBP-carboxylase** from monocot wheat and maize to dicot plants.
- Transfer of genes for C_4 photosynthetic pathway to C_3 crop plants.
- Transfer of nitrogen fixation gene from the genome of bacterium *Rhizobium* into the chromosomes of nonleguminous plants. The work is also on to alter the genome of bacteria or of nonleguminous plants so that new symbiotic relationship can be developed and nonleguminous plants are also able to support these nitrogen-fixing bacteria.
- Transfer of frost-resistance genes to trees or crops so that these can be grown in snowy regions.

Extension: Transgenic Plants to Increase Shelf Life of Vegetables and Fruits

All higher plants have gene for CO_2 fixing enzyme ***Rubisco.*** This enzyme acts both as **oxygenase** (oxygen fixing enzyme used in respiration) and **carboxylase** (CO_2-fixing enzyme associated with photosynthesis). Mutant strains of purple nonsulphur bacterium, *R. rubrum* and green alga, *Chlamydomonas reinhardtii* have hightened carboxylase and lowered oxygenase activity. Microbiologists are working to isolate and introduce such genes from these bacteria and algae and incorporate them in other plant cells to reduce the rate of photorespiration which reduces the productivity of most crop plants.

8. Genetic Engineering and Industry

Recombinant DNA technology is being employed in the commercial production of industrial chemicals such as alcohols, acetone, glycerol, fructose, citric acid, gluconic acid, lactic acid, acetic acid and a variety of enzymes, etc.

Genetic engineering aims at developing microorganisms that:

(a) Produce substances in quantities beyond their natural capabilities.

(b) Can be used in food processing, conversion of inedible food to edible human or animal food.

(c) Can modify an uneconomical process to become economically competitive. Scientists have succeeded in cloning **naphthalene deoxygenase gene** from *Pseudomonas putida* into *E. coli* which is used in production of indigo dye at commercial level.

(d) Can recover valuable metals from dilute solutions.

(e) Can leach metals from low grade ores. For example, *Thiobacillus* bacteria can extract and accumulate silver from waste waters of film factories. This extracted silver can be reused.

9. Genetic Engineering and Pollution Control

Genetically engineering microorganisms are being developed to combat pollution caused by oil-spills, sewage, pesticides, herbicides, chemical effluents and heavy metals, etc. The **'superbugs'** have been developed which are capable of:

- cleaning oil spills,
- digesting and converting waste products into useful resources,
- digesting toxic substance from surroundings, and
- removing nontoxic wastes from the environment.

An Indian born American scientist, **Ananda Chakrabarty**, (1979) has created a single strain of *Pseudomonas* that can consume all the types of hydrocarbons in oil. These genetically engineered bacteria are called **'superbugs'**. These were created by introducing plasmids from different strains of *Pseudomonas* in a single cell.

Mixture of bacterial strains are being used to clean up grease deposits along the inner surface of holding tank of ships.

10. Production of Novel Proteins

Genes for completely novel proteins can be constructed and then introduced into the bacterial cell for expression. For example, a synthetic gene which produces proline-rich-protein has been expressed in *E. coli*.

11. Production of Transgenic Plants and Crop Improvements

A number of transgenic plant varieties have been developed by recombinant DNA technology. These transgenic plant varieties have novel traits like insect pest resistance, herbicide resistance, frost resistance and resistance to various biotic stresses, viruses and insects. Transgenic plants are developed to produce novel biochemicals like interferon, insulin, immunoglobulins, etc., to produce quality proteins and lipids and to produce **recombinant vaccines** or **edible vaccines,** *etc.*

12. Nitrogen Fixation

Nitrogen Fixation Gene (NIF) can be transferred from the genome of bacterium *Rhizobium* into the chromosomes of nonleguminous plants. The work is also on to alter the genome of bacteria or of nonleguminous plants so that new symbiotic relationships can be developed and nonleguminous plants are able to support these nitrogen-fixing bacteria or nonleguminous plants are able to fix nitrogen directly from the atmosphere.

13. Enzyme Technology

At present, more than 2,000 enzymes have been isolated from bacteria. Out of these about 1,000 enzymes are recommended for various uses and out of them about 50 microbial enzymes have industrial use. These enzymes have use in dairy industry, in detergent and starch industry, in brewing and wine industry and in pharmaceutical industry. They also have therapeutic use and presently the enzymes have found use as biosensors. In case a particular microorganism that produces a desired enzyme is pathogenic, the required genes for such enzymes can be transferred to a harmless and more suitable organism or bacterium.

14. Food Processing

Recombinant DNA technology or genetic engineering can be used in food industry in following ways:

- Such recombinant microorganisms can be generated that may be used to modify food and to obtain desired processed food products.
- The recombinant microorganisms may be developed that have the capability of transforming inedible biomass into food for human or animal consumption.
- Such recombinant strains of microorganisms be created that produce greater amount of protein degrading enzyme like **rennin** or **trypsin**. Rennin is obtained from calf-stomach and is used in cheese industry for curdling the milk.
- In cheese industry bacterial starter culture of *Streptococcus lactis* is used which is attacked by phages. Often cheese formation does not occur because of failure of starter. By recombinant DNA technology, phage resistant bacterial strains can be created.

15. Fermentation Technology

Using recombinant DNA technology, the microorganisms can be engineered which have higher productive capacity and can produce substances beyond their natural capability or can produce substances of desired quality.

16. Pesticide-degrading Microorganisms

Pesticide degrading plasmid mediated genetically engineered bacteria have been developed by **Chakrabarty, Gunsalus, Nagana** and others. They can degrade camphor, naphthalene, xylene, toluene, octanes and hexanes. *Opd* genes isolated from *Flavobacterium* species and *Pseudomonas diminuta* are associated with the degradation of parathion and methyl parathion. Thus, reconstruction of genetically engineered microorganisms to degrade synthetic pesticides promises freedom from such harmful chemicals.

17. Bioremediation

Use of living organisms or the genetically engineered microorganisms to degrade environmental pollutants is called **bioremediation**. It is to remove hazardous chemicals accumulated in the cells or in the environment to detoxify them into nontoxic forms. The removal of hydrocarbons, dyes, industrial wastes, heavy metals, xenobiotics, etc. by microorganisms is also bioremediation.

GENETIC ENGINEERING AND BIOTECHNOLOGY IN INDIA

Government of India has set up a **'National Biotechnology Board'** under the department of Science and Technology to coordinate and promote research in biotechnology in India.

Some of the biotechnological programmes being conducted by **Cetus**, a leading Bio-industrial Company are:

1. Automated Bioscreening.
2. Genetical improvement of pharmaceutical microorganisms.
3. Engineering of series of organisms for specific industrial uses.
4. Developing immobilised cell and enzyme systems for chemical process industries.
5. Improved production of vitamin B_{12}.
6. Manufacturing fructose from inexpensive forms of glucose.
7. Bioprocessing alkines to valuable oxides and glycols.
8. Production of ethanol by continuous fermentation.
9. Upgrading hydrocarbons microbiologically.
10. Production of xanthangun in oil fields for enhanced crude oil recovery.
11. Production of human insulin microbiologically.
12. Developing a vaccine to prevent colibacillosis, a widespread disease of newborn calves and piglets.
13. Production of monoclonal antibodies for organ transplant tissue typing.
14. Production of diagnostic kits for toxoplasmiasis identification.

All these programmes are in different stages of development.

Postgraduate and undergraduate courses in biotechnology have been introduced in many Indian universities. Biotechnology research centres have been established at **Indian Agricultural Research Institute** (IARI, New Delhi), **National Dairy Research Institute** (NDRI, Karnal), **Indian Veterinary Research Institute** (IVRI, Izatnagar), **Indian Council of Medical Research** (ICMR, New Delhi).

TEXT QUESTIONS

1. Define recombinant DNA. Discuss the role of following in recombinant DNA technology: restriction endonucleases, vectors, plasmids.
2. Explain the following terms:
   (a) Palindromic sequence
   (b) Southern blotting
   (c) Ti plasmids
   (d) Linkers
   (e) Transgenic organisms
   (f) Heterologous DNA.
3. Explain the following abbreviations:
   (a) PGE
   (b) RFLPs
   (c) cDNA
   (d) PFGE
   (e) AMV
   (f) Mab
   (g) PCR.

4. Describe the role of following:
   (a) Reverse transcriptase
   (b) Homopolymer
   (c) Nitrocellulose membrane.
5. What is a vector in recombinant DNA technology? What should be the characteristics of a good vector?
6. What do you mean by clone? In cloning the large genomes of eukaryotes, it is advantageous to make a partial digest with restriction enzyme Sau3A with a four recognition sequence (GATC). Explain.
7. In mice transfected with the rabbit-β-globin gene, the rabbit gene is active in a number of tissues including spleen, liver, brain and kidney, some mice suffer from thalassemia, caused by the imbalance in the coordinate production of α and β globulins. What problems associated with the gene therapy are illustrated by these findings?
8. What factors may be considered in deciding which vector to be used in constructing genomic library of eukaryotic DNA?
9. Discuss the factors which have been helpful in establishing recombinant DNA technology.
10. Write an essay on the application of recombinant DNA technology.
11. Enumerate the essential steps of recombinant DNA technology. Discuss various types of vectors which are used as cloning vehicles.
12. What do you understand by genomic library? How is it prepared?
13. Discuss the process of construction of genomic library. Why is it essential to prepare such libraries?
14. Describe the role of enzyme 'reverse transcriptase' in preparing genomic library.
15. Describe production of cDNA from mRNA.
16. Write short notes on:
   (a) Monoclonal antibodies
   (b) Restriction mapping
   (c) Polymerase chain reaction
   (d) Transgenic mice
   (e) Gene surgery
   (f) Superbugs.

Glossary

A (acceptor) site Binding site on the ribosome for the tRNA that brings in the next amino acid

acceptor stem Base paired stem of tRNA to which the amino acid is attached

acetylation Addition of an acetyl (CH_3CO) group

acridine orange A mutagenic agent that acts by intercalation

activator protein Protein that switches a gene on

active site Special site on a protein for binding other molecules and the chemical reaction occurs

adenine (A) A purine base that pairs with thymine, found in DNA or RNA

adenosine The nucleoside consisting of adenine plus (deoxy) ribose

A-DNA An alternative form of double stranded helical DNA with 11 base pairs per turn.

agarose A polysaccharide from seaweed that is used to form gel for separating nucleic acids by electrophoresis

agarose gel electrophoresis Technique for separation of nucleic acid (DNA and RNA) molecules by passing an electric current through agarose a gel.

alkaline phosphatase An enzyme that cleaves phosphate groups from a wide range of molecules

allele One particular version of a gene, or more broadly, a particular version of any locus on a molecules of DNA

allosteric protein Protein that changes shape when it binds a small molecule

alpha- (α-) carbon Central carbon atom of an amino acid that carries both the amino group and the carboxyl group

alpha- (α-) helix A helical secondary structure of proteins

alternative sigma factor A nonstandard sigma factor needed to recognize a specialized subset of gene

alternative splicing Alternative ways to make two or more different final mRNA molecules by using different segments from the same original gene

Ames test Test for mutagenic activity that makes use of bacteria

amino acid Monomer from which polypeptide chains are built

amino-or N-terminus The end of a polypeptide chain that has a free amino group

amino-acyl tRNA synthetase Enzyme that attaches an amino acid to tRNA

analog A chemical substance that mimics another to be mistaken by biological macromolecules, particularly enzymes, receptor proteins, or regulatory proteins.

anchor sequence Sequence added to primers or probes that may incorporate convenient restriction sites, primer binding sites for future manipulations, or primer bindings sites for subsequent PCR reactions

anticodon loop Loop of tRNA molecule that contains the anticodon

anticodon Group of three complementary bases on tRNA that recognize and bind to a codon on the mRNA

antifreeze protein Protein that prevents freezing of blood, tissue fluids or cells of organisms living at sub-zero temperatures

antiparallel Parallel, but running in opposite directions

antisense RNA An RNA molecule that is complementary to messenger RNA or another functional RNA molecule

anti-Shine-Dalgarno sequence Sequence of 16S rRNA that is complementary to the Shine-Dalgarno sequence of mRNA

anti-sigma factor Protein that binds to a sigma factor and blocks its role in the initiation of transcription

anti-terminator protein Protein that allows transcription to continue through a transcription terminator

AP-site A site in DNA where a base is missing (AP-site = apurinic site or apyrimidinic site depending on the nature of the missing base)

araBAD operon Operon that encodes proteins involved in metabolism of the sugar arabinose

arabinose A five-carbon sugar often found in plant cell wall material that can be used as a carbon source by many bacteria

asymmetric center Carbon atom with four different groups attached. This results in optical isomerism

attλ [Lambda attachment site]–site where phage lambda inserts its DNA into the bacterial chromosome.

attenuation Type of transcriptional regulation that works by premature termination and depends on alternative stem and loop structures in the leader region of the mRNA

attenuation protein Regulatory protein involved in attenuation, that binds to the leader region of mRNA

autogenous regulation Self regulation, by DNA-binding protein that regulates the expression of its own gene

autoradiography Allowing radioactive materials to take pictures of themselves by laying them flat on photographic film

azidothymidine (AZT) Nucleoside analog that acts as a DNA chain terminator during reverse transcription and is used against AIDS. Also known as zidovudine

bacterial artificial chromosome (BAC) Single copy vector based on the F-plasmid of *E. coli* that can carry very long inserts of DNA. Widely used in the human genome project

bacteriocin A toxic protein made by bacteria to kill other, closely related bacteria

bacterioferritin The bacterial analog of ferritin, an iron storage protein

bacteriophage (phage) Virus that infects bacteria

base analog Chemical mutagen that mimics a DNA base

base pair (bp) a pair of two complementary bases (A with T or G with C) held together by hydrogen bonds

base substitution Mutation in which one base is replaced by another

base Alkaline chemical substance (in molecular biology it refers to the cyclic nitrogen compounds found in DNA and RNA)

basic HLH (b/HLH) protein DNA-binding protein with a positively charged (basic) region next to a HLH-motif

beta- (β-) sheet A flat sheet-like secondary structure of proteins

beta-galactosidase or β-galactosidase (LacZ) Enzyme that splits lactose and related molecules to release galactose

B-form or B-DNA The normal form of the DNA double helix, originally described by Watson and Crick

bi-directional replication Replication that proceeds in two directions from a common origin

bioinformatics The computerized analysis of large amounts of biological sequence data

biotin Vitamin that is widely used to label or tag nucleic acids in molecular biology because it may be bound very tightly by avidin or streptavidin

blunt ends Ends of a double-stranded DNA molecule that are fully base paired and have no unpaired single-stranded overhang

branch site Site in the middle of an intron where branching occurs during splicing

CAAT box A sequence found in the upstream region of eukaryotic promoters and binds the transcription factors

calmodulin A small calcium binding protein of animal cells

cap Structure at the 5′-end of eukaryotic mRNA consisting of a methylated guanosine

capsid Shell or protective protein capsule that surrounds the DNA or RNA of a virus particle

carboxy-or C-terminus The end of a polypeptide chain that has a free carboxy-group

carrier protein Protein that transports other molecules around the body or within the cell

CAT Chloramphenicol acetyl transferase

catenane Structure in which two or more circles of DNA are interlocked

cDNA (complementary DNA) DNA copy of a gene that lacks introns and therefore consists solely of the coding sequence. Made by reverse transcription of mRNA

cDNA library Collection of genes in their cDNA form, lacking introns

cell cycle Series of stages from one cell division to the next

central dogma Basic plan of flow of genetic information in living cells which relates DNA → mRNA → proteins

centrifugation Process in which samples are spun at high speed and the centrifugal force causes the larger or heavier components to sediment to the bottom

centromere (Cen) sequence Sequence at centromere of eukaryotic chromosome that is needed for correct partition of chromosomes during cell division

centromere region of eukaryotic chromosome, usually more or less central, where the microtubules attach during mitosis and meiosis

chain termination mutation Same as nonsense mutation

chain termination sequencing Method of sequencing DNA by using dideoxynucleotides to terminate synthesis of DNA chains. Same as dideoxy sequencing

chain terminator Agent that prevents continued elongation of a strand of DNA

chaperone Sometimes "molecular chaperone"; same as chaperonin

chaperonin protein that oversees the correct folding of other proteins

charged tRNA tRNA with an amino acid attached

chiral center Same as asymmetric center

chloramphenicol Antibiotic that binds to 23S rRNA and inhibits protein synthesis

chloramphenical acetyl transferase (CAT) Enzyme that inactivates chloramphenicol by adding acetyl groups

chromatid Single double-helical DNA molecule making up whole or half of a chromosome. A chromatid also contains histones and other DNA-associated proteins.

chromatin remodelling complex A protein assembly that rearranges the histones of chromatin in order to allow transcription

chromatin Complex of DNA plus protein which constitutes eukaryotic chromosomes

Chromogenic substrate Colorless or pale substrate that is converted to a strongly colored product by an enzyme

chromosome banding technique Visualization of chromosome bands by using specific stains that emphasize regions lacking genes

chromosome walking Method for cloning neighboring regions of a chromosome by successive cycles of hybridisation using overlapping probes

chromosome Structure containing the genes of a cell and made of a single molecule of DNA

clamp-loading complex Group of proteins that loads the sliding clamp of DNA polymerase into the DNA

cloning vector Any molecule of DNA that can replicate itself inside a cell and is used for carrying cloned genes or segments of DNA. Usually a small multicopy plasmid or a modified virus

cloverleaf structure 2-D structure showing base pairing in a tRNA molecule

coding strand The strand of DNA equivalent in sequence to the messenger RNA (same as plus strand)

codon Group of three RNA or DNA bases that encodes a single amino acid

cofactor Extra chemical group bound (often temporarily) to a protein but which is not part of the polypeptide chain

ColEI plasmid Small multicopy plasmid of *Escherichia coli* that forms the basis of many cloning vectors widely used in molecular biology

colicin Toxic protein or bacteriocin made by *Escherichia coli* to kill closely related bacteria

competitive inhibitor Chemical substance that inhibits an enzyme by mimicking the true substrate well enough to be mistaken for it

complementary DNA (cDNA) Version of a gene that lacks the introns and is made from the corresponding mRNA by using reverse transcriptase

complementary sequences Two nucleic acid sequences whose bases pair with each other because A, T, G, C in one sequence correspond to T, A, C, G, respectively in the other

conditional mutation Mutation whose phenotypic effects depend on environmental conditions such as temperature or pH

conjugated protein Complex of protein plus another molecule

conjugation bridge Junction that forms between two cells and provides a channel for DNA to move from donor to recipient during conjugation

conjugation Process in which genes are transferred by cell to cell contact

conjugative transposon A transposon that is also capable of transferring itself from one bacterial cell to another by conjugation

consensus sequence Idealised base sequence consisting of the bases most often found at each position

conservative transposition same as cut-and-paste transposition

constitutive gene Gene that is expressed all the time

core enzyme The part of DNA or RNA polymerase that synthesises new DNA or RNA (i.e. lacking the recognition and/or attachment subunits)

co-repressor In prokaryotes, a small signal molecule needed for some repressor proteins to bind to DNA; in eukaryotes, an accessory protein, often a histone deacetylase, involved in gene repression

cos sequences (lambda cohesive ends) Complementary 12bp long overhangs found at each end of the linear form of the lambda genome

cosmid Small multicopy plasmid that caries lambda *cos* sites and can carry around 45kb of cloned DNA

cotransfer frequency Frequency with which two genes remain associated during transfer of DNA between cells

coupled transcription-translation when ribosomes of bacteria start translating an mRNA molecule that is still being transcribed from the DNA.

covalently closed circular DNA (cccDNA) Circular DNA with no nicks in either strand

cPABP (chloroplast polyadenylate binding protein) A translational activator protein that controls expression of chloroplast mRNA

CRP (cyclic AMP receptor protein) Bacterial protein that binds cyclic AMP and then binds to DNA

Cruciform structure Cross-shaped structure in double stranded DNA (or RNA) formed from an inverted repeat

Cryptic plasmid A plasmid that confers no identified characteristics or phenotypic properties

CTD (carboxy-terminal domain) Repetitive region at the C-terminus of RNA polymerase II that may be phosphorylated

C-terminus Carboxy-terminus. The end of a polypeptide chain that is made last and has a free carboxy-group.

cytidine The nucleoside consisting of cytosine plus (deoxy) ribose

cytokinesis Cell division

cytokinin Plant hormone that induces plant cells to divide

D- and L-forms The two isomeric forms of an optically active substance; also called D- and L-isomers

deaminase An enzyme that removes an amino group

deamination Loss of an amino group

denaturation When describing proteins or other biological polymers, refers to the loss of correct 3-D structure

denaturing gradient gel electrophoresis (DGGE) Combination of gel electrophoresis with DNA denaturation that allows separation of DNA molecules differing in sequence by only a single base

deoxynucleoside A nucleoside containing deoxyribose as the sugar

deoxynucleotide A nucleotide containing deoxyribose as the sugar

deoxyribonucleases (DNase) Enzyme that cuts or degrades DNA

doxyribonuclease I (DNase I) Non-specific nuclease that cuts DNA between any two nucleotides. Often used in footprint analysis

deoxyribonucleic acid (DNA) The nucleic acid polymer of which the genes are made

deoxyribonucleoside 5′-triphosphate (deoxyNTP) Precursor for DNA synthesis consisting of a base, deoxyribose and three phosphate groups

deoxyribose The sugar with five carbon atoms that is found in DNA

dideoxy sequencing Method of sequencing DNA by using dideoxynucleotides to terminate synthesis of DNA chains. Same as chain termination sequencing

dideoxynucleotide Nucleotide whose sugar is dideoxyribose instead of ribose or deoxyribose

dideoxyribose Derivative of ribose that lacks the oxygen of both the 2′ and the 3′ hydroxyl groups

dif site Site on bacterial chromosome used by crossover resolves to separate covalently fixed chromosomes

differential display PCR Variant of RT-PCR that specifically amplifies messenger RNA from eukaryotic cells using oligo(dT) primers

dihydrofolate (DHF) Cafactor with a variety of roles including making precursors for DNA and RNA synthesis

dihydrofolate reductase Enzyme that converts dihydrofolate back to tetrahydrofolate

dipolar ion Same as zwitter ion; a molecule with both a positive and a negative charge

directed mutagenesis deliberate alternation of the DNA sequence of a gene by any of a variety of artificial techniques

D-isomer That one of a pair of optical isomers that rotates light in a clockwise direction

disulfide bond A sulfur to sulfur bond formed between two sulfhydryl groups, in particular between those of cysteine, and which binds together two protein chains

DNA Deoxyribonucleic acid, nucleic acid polymer of which the genes are made

DNA adenine methylase (Dam) A bacterial enzyme that methylates adenine in the sequence GATC

DNA chip Chip used to simultaneously detect and identify many short DNA fragments by DNA-DNA hybridisation. Also known as DNA array or oligonucleotide array detector

DNA cytosine methylase (Dcm) A bacterial enzyme that methylates cytosine in the sequences CCAGG and CCTGG

DNA fingerprint Individually unique pattern due to multiple bands of DNA produced using restriction enzymes, separated by electrophoresis and usually visualised by Southern blotting

DNA glycosylase Enzyme that breaks the bond between a base and the deoxyribose of the DNA back-bone

DNA gyrase An enzyme that introduces negative super-coils into DNA, a member of the type II topoisomerase family

DNA helicase Enzyme that unwinds double helical DNA

DNA library Collection of cloned segments of DNA that is big enough to contain at least one copy of every gene from a particular organism Same as gene library

DNA ligase Enzyme that joins DNA fragments covalently, end to end

DNA microarray same as DNA array or DNA chip

DNA polymerase An enzyme that elongates strands of DNA, especially when chromosomes are being replicated

DNA polymerase α Enzyme that makes short segment of initiator DNA during replication of animal chromosomes

DNA polymerase δ Enzyme that makes most of the DNA when animal chromosomes are replicated

DNA polymerase eta (η) A repair polymerase in animals that can replicate past thymine dimers

DNA polymerase I (Pol I) Bacterial enzyme that makes small stretches of DNA to fill in gaps between Okazaki fragments or during repair of damaged DNA

DNA polymerase III (Pol III) Enzyme that makes most of the DNA when bacterial chromosomes are replicated

DNA polymerase Y A repaid polymerase in bacteria that can replicate past pyrimidine dimers and AP-sites

DNA virus A virus whose genome consists of DNA

DnaA protein Protein that binds to the origin of bacterial chromosomes and helps initiate replication

donor cell Cell that donates DNA to another cell

Double helix structure formed by twisting two strands of DNA spirally around each other

Ds Ac element to provide transposase

dsRNA Double-stranded RNA

effective genome the portion of the genome that consists of useful genetic information and ignores the intervening and non-coding DNA. Only applicable to eukaryotic organisms

electrophoresis Movement of charged molecules due to an electric field. Used to separate and purify nucleic acids and proteins

electroporator Device that uses a high voltage discharge to make cells competent to take up DNA

electrospray ionisation (ESI) Type of mass spectrometry in which gas-phase ions are generated from ions in solution

elongation factors Proteins that are required for the elongation of a growing polypeptide chain

enantiomers A pair of mirror-image optical isomers (i.e., D- and L-isomers)

endonuclease A nuclease that cuts a nucleic acid in the middle

error-prone repair Type of DNA repair process that introduces mutations

Escherichia coli (E. coli) A species of bacterium commonly used in genetics and molecular biology

essential amino acids those amino acids that animals are unable to synthesise for themselves (Val, Ile, Leu, Phe, Trp, Thr, Met, Lys, His)

euchromatin normal chromatin, as opposed to heterochromatin

eukaryote Higher organism with advanced cells, which have more than one chromosome within a compartment called the nucleus

eukaryotic (80S) ribosome Type of ribosome found in cytoplasm of eukaryotic cell and encoded by genes in the nucleus

excision repair system Also known as "cut and patch" repair. A DNA repair system that recognises bulges in the DNA double helix, removes the damaged strand and replaces it

excisionase Enzyme that reverses DNA integration by removing a segment of dsDNA and resealing the gap. In particular, lambda excisionase removes integrated lambda DNA

exon Segment of a gene that codes for protein and that is still present in the messenger RNA after processing is complete

exonuclease Enzyme that cleaves nucleic acid molecules at the end and usually removes just a single nucleotide at a time

3′-exonuclease An enzyme that degrades nucleic acids from the 3′-end

expression site Special location on chromosome where the chosen copy of a gene present in multiple copies may be expressed

expression vector Vector specifically designed to place a cloned gene under control of a plasmid-borne promoter

fluorescence process in which a molecule absorbs light of one wavelength and then emits light of another, longer or lower energy wavelength

fluorescence activated cell sorter (FACS) Instrument that sorts cells (or chromosomes) based on fluorescent labelling

fluorescence in situ hybridisation (FISH) Using a fluorescent probe to visualise a molecule of DNA or RNA in its natural location

fluorescence resonance energy transfer (FRET) Transfer of energy from short-wavelength fluorophore to long wavelength fluorophore so quenching the short wave emission

F-plasmid Fertility plasmid that allows *E. coli* to donate DNA by conjugation

frameshift mutation Mutation in which the reading frame of a structural gene is altered by insertion or deletion of one or a few bases

G1 phase Stage of the eukaryotic cell cycle following cell division; cell growth occurs here

G2 phase Stage of the eukaryotic cell cycle between DNA synthesis and mitosis: preparation for division

β-galactosidase Enzyme that splits lactose and related compounds

gap a break in a strand of DNA or RNA where bases are missing

GC ratio The amount of G plus C relative to all four bases in a sample of DNA. The GC ratio is usually expressed as a percentage

gel electrophoresis Electrophoresis of charged molecules through a gel meshwork in order to sort them by size

gene A unit of genetic information

gene family Group of closely related genes that arose by successive duplication and perform similar roles

gene fusion Structure in which parts of two genes are joined together, in particular when the regulatory region of one gene is joined to the coding region of a reporter gene

gene library Collection of cloned segments of DNA that is big enough to contain at least one copy of every gene from a particular organism. Same as DNA library

gene product End product of gene expression; usually a protein but includes various untranslated RNAs such as rRNA, tRNA, and snRNA

genetic code The code for converting the base sequence in nucleic acids, read in groups of three, into the sequence of a polypeptide chain

genetic element Any molecule or segment of DNA or RNA that carries genetic information and acts as a heritable unit

genome The entire genetic information of an individual organism

genomics Study of genomes as a whole rather than one gene at a time

genotype The genetic make-up of an organism

gene A unit of genetic information

gene family Group of closely related genes that arose by successive duplication and perform similar roles

gene fusion Structure in which parts of two genes are joined together, in particular when the regulatory region of one gene is joined to the coding region of a reporter gene

gene library Collection of cloned segments of DNA that is big enough to contain at least one copy of every gene from a particular organism. Same as DNA library

gene product End product of gene expression; usually a protein but includes various untranslated RNAs such as rRNA, tRNA, and snRNA

genetic code The code for converting the base sequence in nucleic acids, read in groups of three, into the sequence of a polypeptide chain

genetic element Any molecule or segment of DNA or RNA that carries genetic information and acts as a heritable unit

genome the entire genetic information of an individual organism

genomics Study of genomes as a whole rather than one gene at a time

genotype The genetic make-up of an organism

glycine The simplest amino acid

glycogen Storage carbohydrate found both in bacteria and in the livers of animals

glycoprotein Complex of protein plus carbohydrate

guanine (G) A purine base found in DNA or RNA that pairs with cytosine

guanosine The nucleoside consisting of guanine plus (deoxy)ribose

hairpin A double stranded base-paired structure formed by folding a single strand of DNA or RNA back upon itself

haploid Possessing only a single copy of each chromosome or gene

haploid genome A complete set containing a single copy of all the genes (generally used of organisms that have two or more sets of each gene)

H-DNA a form of DNA consisting of a triple helix. Its formation is promoted by acid conditions and by runs of purine bases

Helicase Enzyme that unwinds double helical DNA

Helix-loop-helix (HLH) One type of DNA-binding motif common in proteins

helix-turn-helix (HTH) One type of DNA-binding motif common in proteins

helper virus a virus that provides essential functions for defective viruses, satellite viruses and satellite RNA

hemi-methylated Methylated on only one strand

heterochromatin A highly condensed form of chromatin that cannot be transcribed because it cannot be accessed by RNA polymerase

Hfr-strain Bacterial strain that transfers chromosomal genes at high sequence due to an integrated fertility plasmid

Highly repetitive DNA DNA sequences that exist in hundreds of thousands of copies

histone Special positively charged protein that binds to DNA and helps to maintain the structure of chromosomes in eukaryotes

histone acetyle transferase (HAT) Enzyme that adds acetyl groups to histones

histone deacetylase (HDAC) Enzyme that removes acetyl groups from histones

homologous chromosomes Two chromosomes are homologous when they carry the same sequence of genes in the same linear order

Homologous recombination Recombination between two lengths of DNA that are identical, or nearly so, in sequence

Hmozygous Having two identical alleles of the same gene

horizontal gene transfer Movement of genes sideways between unrelated organism. Same as lateral gene transfer

hormone Messenger molecule that circulates inside multicellular organisms such an animals and plants

hot spots Site in DNA or RNA where mutations are unusually frequent

housekeeping genes Genes that are switched on all the time because they are needed for essential life functions

Hox genes Family of homeobox genes that control overall body layout by regulating the expression of many other regulatory genes, including those for other transcription factors

HU protein (heat-unstable nucleoid protein) A bacterial protein that binds to DNA with low specificity and is involved in bending of DNA

Human Genome Project Program to sequence all the DNA making up the human genome

human immunodeficiency virus (HIV) The retrovirus that causes AIDS

hybrid DNA Artificial double stranded DNA molecule made by pairing two single strands from two different sources

hydrogen bond Bond resulting from the attraction of a positive hydrogen atom to both or two other atoms with negative charges

hydrophilic Water-loving; readily dissolves in water

IHF (integration host factor) a bacterial protein that binds DNA so helping the initiation of transcription of certain genes; named after its role in helping the integration of bacteriophage lambda into the chromosome of *E. coli*

imprinting When the expression of a particular allele depends on whether it originally came from the father or the mother; (imprinting is a rare exception to the normal rules of genetic dominance)

incompatibility The inability of two plasmids of the same family to co-exist in the same host cell

inducer A signal molecule that turns on a gene by binding to a regulatory protein

30S initiation complex Initiation complex for translation that contains only the small subunit of the bacterial ribosome

70S initiation complex Initiation complex for translation that contains both subunits of the bacterial ribosome

initiation complex (for replication) Assemblage of proteins that binds to the origin and initiates replicating of DNA

initiation factors Proteins that are required for the initiation of a new polypeptide chain

initiator box Sequence at the start of transcription of a eukaryotic gene

initiator DNA (iDNA) Short segment of DNA made just after the RNA primer during replication of animal chromosomes

initiator tRNA The tRNA that brings the first amino acid to the ribosome when starting a new polypeptide chain

inosine A purine nucleoside, found most often in transfer RNA, that contains the unusual base hypoxanthine

insertional inactivation Inactivation of a gene by inserting a foreign segment of DNA into the middle of the conding sequence

insulator A DNA sequence that shields promoters from the action of enhancers and also prevents the spread of heterochromatin

insulator binding protein (IBP) Protein that binds to insulator sequence and is necessary for the insulator to function

integrase Enzyme that inserts a segment of dsDNA into another DNA molecule at a specific recognition sequence. In particular, lambda integrase inserts lambda DNA into the chromosome of *E. coli*

integration Insertion of a segment of dsDNA into another DNA molecule at a specific recognition sequences

Intergenic DNA Non-coding DNA that lies between genes

intergenic region DNA sequence between genes

internal eliminated segment (IES) Extra sequences in the DNA of the ciliate micronucleus that are eliminated during conversion of a micronucleus to a macronucleus

Interphase Part of the eukaryotic cell cycle between two cell divisions and consisting of Gl-, S- and G2- phases

intragenic suppression Reversion of a mutation by a second change at a different site but within the same gene

intron Segment of a gene that does not code for protein but is transcribed and forms part of the primary transcript

inverse PCR Method for using PCR to amplify unknown sequences by circularizing the template molecule

inverted repeat Sequence of DNA that is the same when read forwards as when read backwards on the complementary strand. One type of palindrome.

inverted repeats Sequences of DNA that are identical when read in opposite orientations on complementary strands.

jumping gene Popular name for a transposable element

junk DNA defective selfish DNA that is of no use to the host cell it inhabits and which can no longer move or express its genes

L- and D-forms The two isomeric forms of an optically active substance; also called L- and D-isomers

LacI protein Repressor that controls the *lac* operon

lactose permease (LacY) The transport protein for lactose

β-lactamase Enzyme that destroys antibiotics of the β-lactam class that includes penicillins and cephalosporins

lacZ gene Gene encoding β-galactosidase; widely used as a reporter gene

lagging strand The new strand of DNA which is synthesised in short pieces during replication and then joined later

lambda (or λ) Specialised transducting phage of *Escherichia coli* that may insert its DNA into the bacterial chromosome

lambda attachment site (attλ) Recognition site on DNA used during integration of lambda DNA into *E. coli* chromosome

lambda left promoter (PL) One of the promoters repressed by binding of the lambda repressor or cI protein

lambda repressor (cI protein) Repressor protein responsible for maintaining bacteriophage lambda in the lysogenic state

large subunit The larger of the two ribosomal subunits, 50S in bacteria, 60S in eukaryotes

lariat structure Branched, lariat-shaped segment of RNA generated by splicing out an intron

leader peptidase Enzyme that removes the leader sequence after protein export

leader peptide a short peptide coded by the leader region of certain genes controlled by the attenuation mechanism

leader region the region of an mRNA molecule in front of the structural genes, especially when involved in regulation by the attenuation mechanism

leading strand The new strand of DNA that is synthesised continuously during replication

leucine zipper One type of DNA-binding motif common in proteins

ligase Enzyme that joins up DNA fragments end to end

LINE Long interspersed element

LINE-1 (or L1) element A particular LINE found in many copies in the genome of humans and other mammals

long interspersed element (LINE) Long sequence found in multiple copies that makes up much of the moderately repetitive DNA of mammals

long terminal repeats (LTRs) Direct repeats of several hundred base pairs found at the ends of retroversion and some other retro-elements

lux gene Gene encoding luciferase from bacteria

lysogen A cell containing a lysogenic virus

lysogeny Type of virus infection in which the virus becomes largely quiescent, makes no new virus particles and duplicates its genome in step with the host cell. Some as latency but used of bacterial viruses

lysosome Membrane bound organelle of eukaryotic cells that contains degenerative enzymes

lysozyme An enzyme found in many bodily fluids that degrades the peptidoglycan of bacterial cells walls

lytic growth Type of infection in which a virus generates many virus particles and destroys the cell

M phase (or mitosis) Stage of the eukaryotic cell cycle in which cell division occurs

M13 Rod-shaped bacteriophage that infects *E. coli*, contains a circle of single stranded DNA, and is used to manufacture DNA for sequencing

Macromolecule Large polymeric molecule; in living cells especially DNA, RNA, protein or polysaccharide

macronucleus Large somatic nucleus of ciliates that contains multiple copies of genes that are expressed

messenger RNA (mRNA) The class of RNA molecule that carries genetic information from the genes to the rest of the cell.

metabolism The processes by which nutrient molecules are transported and transformed within the cell to release energy and to provide new cell material

mini-satellite Another term for a VNTR (variable number tandem repeats)

mirror-like palindrome Sequence of DNA that is the same when read forwards and backwards on the same strand. One type of palindrome

mismatch Wrong pairing of two bases in a double helix of DNA

mismatch repair system DNA repair system that recognises mispaired bases and cuts out part of the DNA strand containing the wrong base

missense mutation Mutation in which a single codon is altered so that one amino acid in a protein is replaced with a different amino acid

mobile DNA Segment of DNA that moves from site to site within or between other molecules of DNA

mobile genetic element A discrete segment of DNA that is able to change its location within larger DNA molecules by transposition or integration and excision

mobility shift assay Method for testing binding of a protein to DNA by measuring the change in mobility of DNA during gel electrophoresis. Same as band shift assay or gel retardation

moderately repetitive sequence DNA sequences that exist in thousands of copies (but less than a hundred thousand)

modification enzyme Enzyme that binds to the DNA at the same recognition site as the corresponding restriction enzyme but methylates the DNA

modified base Nucleic acid base that is chemically altered after the nucleic acid has been synthesised

30 nanometer fibre Chain of nucleosomes that is arranged helically, approximately 30 mm in diameter

negative feedback Form of negative regulation where the final product of a pathway inhibits the first enzyme in the pathway

negative or "minus" strand The non-coding strand of RNA or DNA

negative regulation Regulatory mode in which a repressor keeps a gene switched off until it is removed

non-coding DNA that does not code for proteins or functional RNA molecules

non-coding RNA molecule that functions without being translated into protein; includes tRNA, rRNA, snRNA, scRNA, and some regulatory RNA molecules

nonsense mutation Mutation due to changing the codon for an amino acid to a stop condon

Northern blotting Hybridisation technique in which a DNA probe binds to an RNA target molecule

novobiocin An antibiotic that inhibits type II topoisomerases, especially DNA gyrase, by binding to the B-subunit

N-terminus The end of a polypeptide chain that is made first and has a free amino group; same as amino terminus

nuclear envelope Envelope consisting of two concentric membranes that surrounds the nucleus of eukaryotic cells

nuclear matrix A mesh of filamentous proteins found on the inside of the nuclear membrane and used in anchoring DNA

nuclear membranes Pair of concentric membranes comprising the nuclear envelope that separates off the nucleus from the rest of a eukaryotic cell

nuclear pore Pore in nuclear membrane that allows proteins, RNA and other molecules into and out of the nucleus

nuclease Enzyme that cuts or degardes nucleic acid

nucleic acid Polymer made of nucleotides that carries genetic information

nucleolar organiser chromosomal region associated with the nucleolus; actually a cluster of rRNA genes

nucleoprotein Complex of protein plus nucleic acid

nucleoside The union of a purine or pyrimidine base with a pentose sugar

nucleosome Subunit of a eukaryotic chromosome consisting of DNA coiled around histone proteins

nucleotide Monomer or subunit of a nucleic acid, consisting of a pentose sugar plus a base plus a phosphate group

Okazaki fragments The short pieces of DNA that make up the lagging strand

oligo(dT) Stretch of single-stranded DNA consisting solely of dT or deoxythymidine residues

oligo(U) Stretch of single-stranded RNA consisting solely of U or uridine residues

open reading frame (ORF) Sequence of bases (either in DNA or RNA) that can be translated (at least in theory) to give a protein

operator Site on DNA to which a repressor protein binds

operon A cluster of prokaryotic genes that are transcribed together to give a single mRNA (i.e. polycistronic mRNA)

OFR See open reading frame

origin of chromosome (oriC) Origin of replication of a chromosome

origin of replication (ori) site on a chromosome or any other DNA molecule where replication begins

orthologous genes Homologous genes that are found in separate species and which diverged when the organisms containing them diverged

overlap primer PCR primer that matches small regions of two different gene segments and is used in joining segments of DNA from different sources

P (peptide) site Binding site on the ribosome for the tRNA that is holding the growing polypeptide chain

P1 artificial chromosome (PAC) Single copy vector based on the P1-phage/plasmid of *E. coli* that can carry very long inserts of DNA

PAGE Polyacrylamide gel electrophoresis. Technique for separating proteins by electrophoresis on a gel made from polyacrylamide

palindrome A sequence that reads the same backwards and forwards

palindromic Reading the same backwards as wells forwards

paralogous genes homologous genes that are located within the same organism due to gene duplication

PCNA protein The sliding clamp for the DNA polymerase of eukaryotic cells (PCNA = proliferating cell nuclear antigen)

PCR See polymerase chain reaction

PCR machine See thermonuclear

PCR primers short pieces of single-stranded DNA that match the sequences at either end of the target DNA segment and which are needed to initiate DNA synthesis in PCR

pentose A five carbon sugar, such as ribose or deoxyribose

peptide bond Type of chemical linkage holding amino acids together in a polypeptide chain

peptide nucleic acid (PNA) Artificial analog of nucleic acids with a polypeptide backbone

peptidoglycan Polymer that makes up a bacterial cell walls; consists of long chains of sugar derivatives cross-linked at intervals with short chains of amino acids

peptidyl transferase Enzyme activity on the ribosome that makes peptide bonds; actually 23S rRNA (bacterial) or 28S rRNA (eukaryotic)

permease a protein that transports nutrients or other molecules across a membrane

phosphatase an enzyme that removes phosphate groups

phosphate group Group of four oxygen atoms surrounding a central phosphorus atom found in the backbone of DNA and RNA

phosphodiester The linkage between nucleotides in a nucleic acid that consists of a central phosphate group esterified to sugar hydroxyl groups on either side

phospholipid a hydrophobic molecule found making up cell membranes and consisting of a soluble head group and two fatty acids both linked to glycerol phosphate

plasmid Self-replicating genetic elements that are some times found in both prokaryotic and eukaryotic cells. They are not chromosomes nor part of the host cell's permanent genome. Most plasmids are circular molecules of double-stranded DNA although rare linear plasmids and RNA plasmids are known

poly(A) polymerase Enzyme that adds the poly(A) tail to the end of mRNA

poly(A) tail A stretch of multiple adenosine residues found at the 3′-end of mRNA

poly(A)-binding protein (PABP) Protein that binds to mRNA via its poly(A) tail

polyacrylamide gel electrophoresis (PAGE) Technique for separating proteins by electrophoresis on a gel made from polyacrylamide

polyacrylamide Polymer used in separation of proteins or very small nucleic acid molecules by gel electrophoresis

polyadenylation complex Protein complex that adds the poly(A) tail to eukaryotic mRNA

polycistronic mRNA mRNA carrying multiple coding sequences (cistrons) that may be translated to give several different protein molecules; only found in prokaryotic (bacterial) cells

polylinker a stretch of artificially synthesised DNA that contains cut sites for seven or eight widely used restriction enzymes. Same as multiple cloning site (MCS)

polymerase Enzyme that synthesises nucleic acids

polymerase chain reaction (PCR) Amplification of a DNA sequence by repeated cycles of strand separation and replication

polymerase eta (η) A repair DNA polymerase in animals that can replicate past thymine dimers

polymerase I (Pol I) See DNA polymerase I

polymerase III (Pol III) See DNA polymerase III

polymorphism A difference in DNA sequence between two related individual organisms

polypeptide chain a polymer that consists of amino acids

polyphosphate Compound consisting of multiple phosphate groups linked by high energy phosphate bonds

polyprotein a long polypeptide that is cut up to generate several smaller proteins

positive control See positive regulation

positive or "plus" strand The coding strand of RNA or DNA

positive regulation control by an activator that promotes gene expression when it binds

post-transcriptinal gene silencing (PTGS) Plant version of the RNA interference response to double-stranded RNA that results in the degradation of mRNA or other RNA transcripts homologous to the inducing dsRNA

post-translational modification Modification of a protein or its constitutent amino acids after translation is finished

potential intrastrand triplex (PIT) Stretch of DNA that might be expected from its sequence to form H-type triplex DNA

pribnow box Another name for the –10 region of the bacterial promoter

primary structure The linear order in which the subunits of a polymer are arranged

primary transcript the original RNA molecule obtained by transcription from a DNA template, before any processing or modification has occurred

primase Enzyme that starts a new strand of DNA by making an RNA primer

primer Short segment of nucleic acid that binds to the template strand and allows synthesis of a new chain of DNA to get started. RNA primers are used in cells and DNA primers are used in PCR

primer extension Method to locate the 5′ start site of transcription by using reverse transcriptase to extend a primer bound to mRNA so locating the 5′-end of the transcript

prokaryote Lower organism, such as a bacterium, with a primitive type of cell containing a single chromosome and having no nucleus

proofreading Process that checks whether the correct nucleotide has been inserted into new DNA. Usually refers to DNA polymerase checking whether it has inserted the correct base

prophage Bacteriophage genome that is integrated into the DNA of the bacterial host cell

prosthetic group Extra chemical group, which is not part of the polypeptide chain, covalently attached to a protein

protease Same as proteinase; an enzyme that degrades proteins

protein Polymer made from amino acids; may consist of several polypeptide chains

protein A Antibody binding protein from *Staphylococcus* that is often used in making fusion proteins

protomer A single polymer chain that is itself a subunit for a higher level of assembly

provirus virus genome that is integrated into the host cell DNA

PrP (Prion protein) The prion protein found in the nervous tissue of mammals and whose misfolded form is responsible for prion diseases

Prpc (cellular PrP) The healthy, normal form of the prion protein

pseudogene Defective copy of a genuine gene

pseudouridine An isomer of uridine that is introduced into some RNA molecules by post-transcriptional modification

pulsed field gel electrophoresis (PFGE) Type of gel electrophoresis used for analysis of very large DNA molecules and which uses an electric field of "pulses" delivered from a hexagonal array of electrodes

purine Type of nitrogenous base with a double ring found in DNA and RNA

pyrimidine Type of nitrogenous base with a single ring found in DNA and RNA

radioisotope Radioactive form of an element

randomly amplified polymorphic DNA (RAPD) Method for testing genetic relatedness using PCR to amplify arbitrarily chosen sequences

rapid amplificatioin of cDNA ends (RACE) RT-PCR-based technique that gener-ates the complete 5′ or 3′ end of a cDNA starting from a partial sequence

–10 region Region of bacterial promoter 10 bases back from the start of transcription that is recognised by RNA polymerase

–35 region Region of bacterial promoter 35 bases back from the start of transcription that is recognised by RNA polymerase

regulatory protein a protein that regulates the expression of a gene or the activity of another protein

regulatory region DNA sequence in front of a gene, used for regulation rather than to encode a protein

release factor Protein that recognises a stop codon and brings about the release of a finished polypeptide chain from the ribosome

renaturation Re-annealing of single-stranded DNA or refolding of a denatured protein to give the original natural 3-D structure

repeated sequences DNA sequences that exist in multiple copies

repetitive sequences same as repeated sequences

replication Duplication of the genomic DNA (or RNA) < typically prior to cell division or virus replication

replication factor C (RFC) Eukaryotic protein that binds to initiator DNA and loads DNA polymerase δ plus its sliding clamp onto the DNA

replication fork Region where the enzymes replicating a DNA molecule are bound to untwisted, single stranded DNA

replicative form (RF) double-stranded form of the genome of a single-stranded DNA (or RNA) virus. The RF first replicates itself and is then used to generate the ssDNA (or ssRNA) to pack into the virus particles

replicative transposition Type of transposition in which two copies of the transposon are generated, one in the original site and another at a new location

replicon Molecule of DNA or RNA that contains an origin of replication and can self-replicate

replisome Assemblage of proteins (including primase, DNA polymerase, helicase, SSB protein) that replicates DNA

repressor Regulatory protein that prevents a gene from being transcribed

restriction enzyme type of endonuclease that cuts double stranded DNA at a specific sequence of bases, the recognition site

restriction fragment length polymorphism (RFLP) a difference in restriction sites between two related DNA molecules that results in production of restriction fragments of different lengths

restriction map a diagram showing the location of restriction enzyme cut sites on a segment of DNA

retrotransposon A transposable element that uses reverse transcriptase to convert the RNA form of its genome to a DNA copy

retroversions a family of animal viruses with single-stranded RNA inside two protein shells surrounded by an outer envelope. Once inside the host cell they use reverse transcriptase to convert the RNA version of the genome to a DNA copy

reverse transcriptase An enzyme that uses single-stranded RNA as a template for making double-stranded DNA

reverse transcriptase PCR (RT-PCR) Variant of PCR that allows genes to be amplified and cloned as intron-free DNA copies by starting with mRNA and using reverse transcriptase

reverse transcription The process in which single-stranded RNA is used as a template for making double-stranded DNA

Rho (ρ) protein protein factor needed for successful termination at certain transcriptional terminators

Rho-dependent terminator Transcriptional terminator that depends on Rho protein

Rho-independent terminator Transcriptional terminator that does not need Rho protein

ribonuclease a nuclease that cuts RNA

ribonuclease (RNase) Enzyme that cuts or degrades RNA

ribonuclease II A ribonuclease of bacteria whose main function is processing rRNA and tRNA precursors

ribonuclease H (RNase H) Enzyme that degrades the RNA strand of DNA : RNa hybrid double helices. In bacteria it removes the major portion of RNA primers used to initiate DNA synthesis.

ribonuclease P A ribonuclease involved in processing tRNA in bacteria that consists of an RNA ribozyme plus an accessory protein

ribonucleic acid (RNA) Nucleic acid that differs from DNA in having ribose in place of deoxyribose and having uracil in place of thymine

ribonucleoside a nucleoside whose sugar is ribose (not deoxyribose)

ribonucleotide reductase Enzyme that reduces ribonucleotides to deoxyribonucleotides

ribonucleotide A nucleotide whose sugar is ribose (not deoxyribose)

ribose the 5-carbon sugar found in RNA

ribosomal RNA (rRNA) Class of RNA molecule that makes up part of the structure of a ribosome

70S ribosome Type of ribosome found in bacterial cells

80S ribosome Type of ribosome found in cytoplasm of eukaryotic cells

ribosome The cell's machinery for making proteins

ribosome binding site (RBS) Same as Shine-Dalgarno sequence; sequence close to the front of mRNA that is recognised by the ribosome; only found in prokaryotic cells

ribosome modulation factor (RMF) protein that inactivates surplus ribosomes during slow growth or stationary phase in bacteria

ribosome recycling factor (RRE) Protein that dissociates the ribosomal subunits after a polypeptide chain has been finished and released

riboswitch Domain of messenger RNA that directly senses a signal and controls translation by alternating between two structures

ribozyme An RNA enzyme, that is an RNA molecule with catalytic activity

right-handed helix In a right-handed helix, as the observer looks down the helix axis (in either direction), each strand turns **clockwise** as it moves away from the observer

R-loop analysis Hybridisation of the DNA copy of a gene to the corresponding mRNA that results in the appearance of loops, which represent the intervening sequence in the DNA that have no partners in the mRNA

RNA editing Changing the coding sequence of a RNA molecule after transcription by altering, adding or removing bases

RNA interference Response that is triggered by the presence of double-stranded RNA and results in the degradation of mRNA or other RNA transcripts homologous to the inducing dsRNA

RNA or ribonucleic acid nucleic acid that differs from DNA in having ribose in place of deoxyribose

RNA polymerase enzyme that synthesises RNA

RNA polymerase I Eukaryotic RNA polymerase that transcribes the genes for the large ribosomal RNAs

RNA polymerase II Eukaryotic RNA polymerase that transcribes the genes encoding proteins

RNA polymerase III Eukaryotic RNA polymerase that transcribes the genes for 5S ribosomal RNA and transfer RNA

RNA primer Short segment of RNA used to initiate synthesis of a new strand of DNA during replication

RNA replicase Special RNA polymerase used by RNA viruses to replicate their RNA genomes

RNA virus a virus whose genome consists of RNA

RNA world the hypothetical stage of early life in which RNA encoded genetic information and carried out enzyme reactions without the need for either DNA or protein

RNA-dependent RNA polymerase (RdRP) RNA polymerase that uses RNA as a template and is involved in the amplification of the RNA response

rolling circle amplification technology (RCAT) Method based on rolling circle replication that uses DNA polymerase to amplify target DNA at normal temperatures

rolling circle replication Mechanisms of replicating double-stranded circular DNA that starts by nicking and unrolling one strand and using the other, still circular strand as a template for DNA synthesis. Used by some plasmids and viruses

R-plasmid or R-factor Plasmid that carries genes for antibiotic resistance

Rubisco (ribulose bisphosphate carboxylase) A critical enzyme in the fixation of carbon dioxide during photosynthesis

S1 nuclease Endonuclease from *Aspergillus oryzae* that cleaves single-stranded RNA or DNA but does not cut double-stranded nucleic acids

S1 nuclease mapping Method using S1 nuclease to locate the 5′-end or 3′-end of a transcript

satellite DNA Highly repetitive DNA of eukaryotic cells that is found as long clusters of tandem repeats and is permanently coiled tightly into heterochromatin

satellite RNA Parasitic RNA molecule that requires a helper virus for replication and capsid formation

satellite virus A defective virus that needs an unrelated helper virus to infect the same host cell in order to provide essential functions

scaffold attachment region (SAR) Site on eukaryotic DNA that binds to proteins of the chromosomal scaffold or of the chromosomal scaffold or of the nuclear matrix—same as MAR sites

scintillant Molecule that emits pulses of light when hit by a particle of radioactivity

scintillation counter Machine that detects and counts pulses of light

scintillation counting Detection and counting of individual microscopic pulses of light

selfish DNA A sequence of DNA that manages to replicate but which is of no use to the host cell it inhabits

self-splicing Splicing out of an intron by the ribozyme activity of the RNA molecule itself without the requirement for a separate protein enzyme

semi-conservative replication Mode of DNA replication in which each daughter molecule gets one of the two original strands and one new complementary strand

sense RNA RNA (such as mRNA) that has been made using the non-coding strand of DNA as a template. Equivalent to plus or coding strand RNA

sensor kinase A protein that phosphorylates itself when it senses a specific signal (often an environmental stimulus, but sometimes an internal signal)

septum Cross-wall that separates two new bacterial cells after division

Sequences Genetically modified DNA polymerase from bacteriophage T7 used for sequencing DNA

sequence tagged site (STS) a short sequence (usually 100–500bp) that is unique within the genome and can be easily detected, usually by PCR

Shine-Dalgarno (S-D) sequence Same as RBS; sequence close to the front of mRNA that is recognised by the ribosome; only found in prokaryotic cells

short interfering RNA (siRNA) Double-stranded RNA molecules of 21 – 22 nucleotides involved in triggering RNA interference in eukaryotes

short interspersed element (SINE) Short repeated sequence that makes up a significant fraction of the moderately or highly repetitive DNA of mammals

shotgun sequencing Approach in which the genome is broken into many random short fragments for sequencing. The complete genome sequence is then assembled by computerised searching for overlaps between individual sequences

shuttle vector A vector that can survive in and be moved between more than one type of host cell

sigma subunit Subunit of bacterial RNA polymerase that recognises and binds to the promoter sequence

signal molecule Molecule that exerts a regulatory effect by binding to a regulatory protein

signal sequence Short, largely hydrophobic sequence of amino acids at the front of a protein that label it for export

silencing In genetic terminology, refers to switching off genes in a relatively nonspecific manner

silent mutation An alternation in the DNA sequence that has no effect on the phenotype

SINE Short interspersed element

single nucleotide polymorphism (SNP) A difference in DNA sequence of a single base change between two individuals

single strand binding protein (SSB protein) A protein that keeps separated strands of DNA apart

site-directed mutagenesis Deliberate alternation of a specific DNA sequence by any artificial technique

site-specific recombination Recombination between two lengths of DNA that are largely unrelated. It involves specific proteins that recognise particular sequences and form crossovers between them. Same as non-homologous recombination

Slicer Ribonuclease activity of the RISC complex

sliding clamp Subunit of DNA polymerase that encircles the DNA, thereby holding the core enzyme onto the DNA

small cytoplasmic RNA (scRNA) Small RNA molecules of varied function found in the cytoplasm of eukaryotic cells

small nuclear ribonucleoprotein (snRNP) Complex of snRNA plus protein

small nuclear RNA (snRNA) Small RNA molecule that is found only in the nucleus of eukaryotes where it over sees the splicing of mRNA

small nucleolar RNA (snoRNA) Small RNA molecule that is found only in the nucleus of eukaryotes where it over sees the splicing of mRNA

small nucleolar RNA (snoRNA) Small RNA molecules that are involved in ribosomal RNA base modification in the nucleolus of eukaryotic cells

snurp snRNP or small nuclear ribonucleoprotein

sodium dodecyl sulfate (SDS) A detergent widely used to denature and solubilise proteins before separation by electrophoresis

somatic cell Cell making up the body, as opposed to the germ line

SOS system An error-prone repair system of bacteria that responds to severe DNA damage

Southern blotting A method to detect single stranded DNA that has been transferred to nylon paper by using a probe that binds DNA

South-Western blotting Detection technique in which a DNA probe binds to a protein target molecule

3′ splice site Recognition site for splicing at the down stream or 3′-end of the intron

5′ splice site Recognition site for splicing at the down stream or 5′-end of the intron

spliceosome Complex of proteins and small nuclear RNA molecules that removes introns during the processing of messenger RNA

splicing Removal of intervening sequences and rejoining the ends of a molecule; usually refers to removal of introns from RNA

start condon The special AUG condon that signals the start of a protein

stem and loop Structure made by folding an inverted repeat sequence

sticky ends Ends of a double-stranded DNA molecule that have unpaired single-stranded overhangs, generated by a staggered cut

stop condon Condon that signals the end of a protein

structural gene Sequence of DNA (or RNA) that codes for a protein or for an untranslated RNA molecule

structural protein A protein that forms part of a cellular structure

30S subunit Small subunit of a 70S ribosome

40S subunit Small subunit of an 80S ribosome

50S subunit Large subunit of a 70S ribosome

60S subunit Large subunit of an 80S ribosome

suppressor tRNA A mutant tRNA that recognises a stop condon and can insert an amino acid when it reads a stop condon on the mRNA

S-value The sedimentation coefficient is the velocity of sedimentation divided by the centrifugal field. It is dependent on mass and is measured in Svedberg units

Swi/Snf ("switch sniff") complex Larger type of chromatin remodelling complex

T4 ligase Type of DNA ligase from bacteriophage T4 and which is capable of ligating blunt ends

TA cloning vector Vector with single 3′-T overhangs (in its linearized form) that is used to clone DNA segments with single 3′-A overhangs generated by Taq polymerase

tandem mass spectrometry (MS/MS) Two successive rounds of mass spectrometry in which a parent ion is first isolated and then fragmented into daughter ions for more detailed analysis

tandem repeats Repeated sequences of DNA (or RNA) that lie next to each other

Taw polymerase Heat resistant DNA polymerase from *Thermus aquaticus* that is used for PCR

TaqMan→probe Fluorescent probe consisting of two fluorophores linked by a DNA probe sequence. Fluorescence increases only after the fluorophores are separated by degradation of the linking DNA

target DNA DNA that is the target for binding by a probe during hybridisation or the target for amplification by PCR

target sequence (a) Sequence on host DNA molecule into which a transposon inserts itself; (b) Sequence within the original DNA template that is amplified in a PCR reaction

TATA binding protein (TBP) Transcription factor that recognises the TATA box

TATA box Binding site for a transcription factor that guides RNA polymerase II to the promoter in eukaryotes

TATA box factor Another name for TATA binding protein

tautomerisation Alternation of a molecule in particular a base of a nucleic acid, between two different isomeric structures

telomerase Enzyme that adds DNA to the telomere of a eukaryotic chromosome

telomere Specific repetitive sequence of DNA found at the end of linear eukaryotic chromosomes

template strand Strand of DNA used as a guide for synthesising a new strand by complementary base pairing

Ter site Site in the terminus region that blocks movement of a replication fork

5′-terminal oligopyrimidine tract (5′-TOP) Long pyrimidine-rich tracts located between the 5′-end of mRNA and the start condon.

terminator DNA sequence at end of a gene that tells RNA polymerase to stop transcribing

terminus of chromosome (terC) The place on a chromosome where replication ends

terminus of replication (ter) The place on any DNA molecule where replication ends

theta-replication Mode of replication in which two replication forks go in opposite directions around a circular molecule of DNA

third base redundancy Situation where a set of four condons all code for the same amino acid and thus the identity of the third condon base makes no difference during translation

thymidine The nucleoside consisting of thymine plus deoxyribose

thymidylate synthetase Enzyme that adds a methyl group, so converting the uracil of dUMP to thymine

thymine (T) A pyrimidine base found in DNA that pairs with adenine

topoisomerase Enzyme that alters the level of supercoiling or catenation of DNA (i.e. changes the topological conformation)

topoisomerase IV A particular topoisomerase involved in DNA replication in bacteria

toposiomers Isomeric forms that differ in topology, i.e. their level of supercoiling or catenation

tra genes Genes needed for plasmid transfer

Tra+ Transfer positive (refers to a plasmid capable of self-transfer)

transcription Process by which information from DNA is converted into its RNA equivalent

transcription bubble Region where DNA double helix is temporarily opened up so allowing transcription to occur

transcription factor Protein that regulates gene expression by binding to DNA in the control region of the gene

transcription-coupled repair Preferential repair of the template strand of DNA that may be transcribed

transduction Process in which genes are transferred inside virus particles

transfection Process in which genes are transferred inside virus particles

transfection process in which purified viral DNA enters a cell by transformation. Often used to refer to entry of any DNA, even if not of viral origin, into an animal cell

transfer RNA (tRNA) RNA molecules that carry amino acids to the ribosome

translation making a protein using the information provided by messenger RNA

translational activator A protein that binds to mRNA and promotes its translation

translational repression Form of control in which the translation of a messenger RNA is prevented

translational repressor A protein that binds to mRNA and prevents its translation

translocase Enzyme complex that transports proteins across membranes

translocation a) Transport of a newly made protein across a membrane by means of a translocase; b) Sideways movement of the ribosome on mRNA during translation and c) Removal of a segment of DNA from its original location and its reinsertion in a different place

transport protein A protein that carries other molecules across membranes or around the body

transposable element A mobile segment of DNA that is always inserted in another host molecule of DNA. It has no origin of replication of its own and relies on the host DNA molecule for replication. Includes both DNA-based transposons and retrotransposons

transposase Enzyme responsible for moving a transposon

transposition The process by which a transposon moves from one host DNA molecule to another

transposon Some as transposable element, although the term is usually restricted to DNA-based elements that do not use reverse transcriptase

trans-splicing Splicing of a segment from one RNA molecule into another distinct RNA molecule

transversion Mutation in which a pyrimidine is replaced by a purine or vice versa

Tus protein Bacterial protein that binds to *Ter* sites and blocks movement of replication forks

twist, T The number of double helical turns in a molecule of DNA (or double-stranded RNA)

two-component regulatory system A regulatory system consisting of two proteins, a sensor kinase and a DNA-binding regulator

type I restyriction enzyme Type of restriction enzyme that cuts the DNA a thousand or more base pairs away from the recognition site

type I topoisomerse Topoisomerase that cuts a single strand of DNA and therefore changes the linking number by one

type II restriction enzyme Type of restriction enzyme that cuts the DNA in the middle of the recognition site

type II restriction enzyme Type or restriction enzyme that cuts a fixed number of bases away from its recognition site

type II topoisomerase Topoisomerase that cuts both strands of DNA and therefore changes the linking number by two

U1 Snurp (snRNP) that recognises the upstream slice site

U2 Snurp (snRNP) that recognises the branch site

U2AF (U2 accessory factor) protein involved in splicing of introns that recognises the down stream splice site

uncharged tRNA tRNA without an amino acid attached

unequal crossing over Crossing over in which the two segments that cross over are of different lengths; often due to misalignment during pairing of DNA strands

Ung protein Same as uracil-N-glycosylase

universal genetic code Version of the genetic code used by almost all organisms

3′-untranslated region (3′-UTR) Sequence at the 3′-end of mRNA, downstream of the final stop condon, that is not translated into protein

5′-untranslated region (5′-UTR) Region of an mRNA between the 5′-end and the translation start site

upstream element DNA sequence upstream of the TATA box in eukaryotic promoters that is recognised by specific proteins

upstream region Region of DNA in front (i.e. beyond the 5′-end) of a structural gene; its bases are numbered negatively counting backwards from the start of transcription

uracil (U) A pyrimidine base found in RNA that may pair with adenine

variable number tandem repeats (VNTR) Cluster of tandemly repeated sequences in the DNA, whose number of repeats differs from one individual to another

Western blotting Detection technique in which a probe, usually an antibody, binds to a protein target molecule

wobble rules Rules allowing less rigid base pairing but only for condon/anticondon pairing

yeast artificial chromosome (YAC) Single copy vector based on yeast chromosome that can carry very long inserts of DNA. Widely used in the human genome project

Z-DNA An alternative form of DNA double helix with left-handed turns and 12 base pairs per turn

zinc finger One type of DNA-binding motif common in proteins

Zoo blotting Comparative Southern blotting using DNA target molecules from several different animals to test whether the probe DNA is from a coding region

Zwitter ion Same as dipolar ion; a molecule with both a positive and a negative charge

Index

D

Q

R